THIRD-GENERATION SYSTEMS AND INTELLIGENT WIRELESS NETWORKING

THIRD-GENERATION SYSTEMS AND INTELLIGENT WIRELESS NETWORKING

smart antennas and adaptive modulation

J. S. Blogh
L. Hanzo
both of University of Southampton, UK

IEEE PRESS

IEEE Communications Society, Sponsor

JOHN WILEY & SONS, LTD

Baffins Lane, Chichester,
West Sussex, PO19 1UD, England

National 01243 779777
International (+44) 1243 779777

e-mail (for orders and customer service enquiries): cs-books@wiley.co.uk
Visit our Home Page on http://www.wiley.co.uk or http://www.wiley.com

Other Wiley Editorial Offices

John Wiley & Sons, Inc., 605 Third Avenue,
New York, NY 10158-0012, USA

WILEY-VCH Verlag GmbH
Pappelallee 3, D-69469 Weinheim, Germany

John Wiley & Sons Australia Ltd, 33 Park Road, Milton,
Queensland 4064, Australia

John Wiley & Sons (Canada) Ltd, 22 Worcester Road
Rexdale, Ontario, M9W 1L1, Canada

John Wiley & Sons (Asia) Pte Ltd, 2 Clementi Loop #02-01,
Jin Xing Distripark, Singapore 129809

IEEE Communications Society, Sponsor
COMM-S Liaison to IEEE Press, Mostafa Hashem Sherif

British Library Cataloguing in Publication Data

A catalogue record for this book is available from the British Library

ISBN 0470 84519 8

Produced from LaTeX files supplied by the author.
Printed and bound in Great Britain by T. J. International Ltd, Padstow, Cornwall.
This book is printed on acid-free paper responsibly manufactured from sustainable forestry, in which at least two trees are planted for each one used for paper production.

Contents

About the Authors

Jonathan Blogh was awarded an MEng. degree with Distinction in Information Engineering from the University of Southampton, UK in 1997. In the same year he was also awarded the IEE Lord Lloyd of Kilgerran Memorial Prize for his interest in and commitment to mobile radio and RF engineering. Between 1997 and 2000 he conducted postgraduate research and in 2001 he earned a PhD in mobile communications at the University of Southampton, UK. His current areas of research include the networking aspects of FDD and TDD mode third generation mobile cellular networks. Currently he is with Radioscape, London, UK, working as a senior software engineer.

Lajos Hanzo (http://www-mobile.ecs.soton.ac.uk) received his degree in electronics in 1976 and his doctorate in 1983. During his 25-year career in telecommunications he has held various research and academic posts in Hungary, Germany and the UK. Since 1986 he has been with the Department of Electronics and Computer Science, University of Southampton, UK, where he holds the chair in telecommunications. He has co-authored eight books on mobile radio communications, published over 300 research papers, organised and chaired conference sessions, presented overview lectures and been awarded a number of distinctions. Currently he is managing an academic research team, working on a range of research projects in the field of wireless multimedia communications sponsored by industry, the Engineering and Physical Sciences Research Council (EPSRC) UK, the European IST Programme and the Mobile Virtual Centre of Excellence (VCE), UK. He is an enthusiastic supporter of industrial and academic liaison and he offers a range of industrial courses. He is also an IEEE Distinguished Lecturer. For further information on research in progress and associated publications please refer to http://www-mobile.ecs.soton.ac.uk

Preface

Background and Overview

Wireless communications is experiencing an explosive growth rate. This high demand for wireless communications services requires increased system capacities. The simplest solution would be to allocate more bandwidth to these services, but the electromagnetic spectrum is a limited resource, which is becoming increasingly congested [1]. Furthermore, the frequency bands to be used for the Third Generation (3G) wireless services have been auctioned in various European countries, such as Germany and the UK at an extremely high price. Therefore, the efficient use of the available frequencies is paramount [1,2].

The digital transmission techniques of the Second Generation (2G) mobile radio networks have already improved upon the capacity and voice quality attained by the analogue mobile radio systems of the first generation. However, more efficient techniques allowing multiple users to share the available frequencies are necessary. Classic techniques of supporting a multiplicity of users are frequency, time, polarisation, code or spatial division multiple access [3]. In Frequency Division Multiple Access [4,5] the available frequency spectrum is divided into frequency bands, each of which is used by a different user. Time Division Multiple Access (TDMA) [4,5] allocates each user a given period of time, referred to as a time slot, over which their transmission may take place. The transmitter must be able to store the data to be transmitted and then transmit it at a proportionately increased rate during its time slot constituting a fraction of the TDMA frame duration. Alternatively, Code Division Multiple Access (CDMA) [4,5] allocates each user a unique code. This code is then used to spread the data over a wide bandwidth shared with all users. For detecting the transmitted data the same unique code, often referred to as the user signature, must be used.

The increasing demand for spectrally efficient mobile communications systems motivates our quest for more powerful techniques. With the aid of spatial processing at a cell site, optimum receive and transmit beams can be used for improving the system's performance in terms of the achievable capacity and the quality of service measures. This approach is usually referred to as Spatial Division Multiple Access (SDMA) [3,6] which enables multiple users in the same cell to be accommodated on the same frequency and time slot by exploiting the spatial selectivity properties offered by adaptive antennas [7]. By contrast, if the desired signal and interferers occupy the same frequency band and time slot, then "temporal filtering" cannot be used for separating the signal from the interference. However, the desired and interfering signals usually originate from different spatial locations and this spatial separation may be exploited, in order to separate the desired signal from the interference using a "spatially selective filter" at the receiver [8–10]. As a result, given a sufficiently large distance between two users communicating in the same frequency band, there will be negligible interference between them. The higher the number of cells in a region, due to using small cells, the more frequently the same frequency is re-used and hence higher teletraffic density per unit area can be carried.

However, the distance between co-channel cells must be sufficiently high, so that the intra-cell interference becomes lower than its maximum acceptable limit [3]. Therefore, the number of cells in a geographic area is limited by the base stations' transmission power level. A method of increasing the system's capacity is to use $120°$ sectorial beams at different carrier frequencies [11]. Each of the sectorial beams may serve the same number of users as supported in ordinary cells, while the Signal-to-Interference Ratio (SIR) can be increased due to the antenna's directionality. The ultimate solution, however, is to use independently steered high gain beams for supporting individual users [3].

Adaptive Quadrature Amplitude Modulation (AQAM) [12, 13] is another technique that is capable of increasing the spectral efficiency that may be achieved. The philosophy behind adaptive modulation is to select a specific modulation mode, from a set of modes, according to the instantaneous radio channel quality [12, 13]. Thus, if the channel quality exhibits a high instantaneous SINR, then a high-order modulation mode may be employed, enabling the exploitation of the temporarily high channel capacity. By contrast, if the channel has a low instantaneous SINR, using a high-order modulation mode would result in an unacceptable Frame Error Ratio (FER), and hence a more robust, but lower throughput modulation mode would be invoked. Hence, adaptive modulation not only combats the effects of a poor quality channel, but also attempts to maximise the throughput, whilst maintaining a given target FER. Thus, there is a trade-off between the mean FER and the data throughput, which is governed by the modem mode switching thresholds. These switching thresholds define the SINRs, at which the instantaneous channel quality requires changing the current modulation mode, i.e. where an alternative AQAM mode must be invoked.

A more explicit representation of the wideband AQAM regime is shown in Figure 1, which displays the variation of the modulation mode with respect to the pseudo SNR at channel SNRs of 10 and 20 dB. In these figures, it can be seen explicitly that the lower-order modulation modes were chosen, when the pseudo SNR was low. In contrast, when the pseudo SNR was high, the higher-order modulation modes were selected in order to increase the transmission throughput. These figures can also be used to exemplify the application of wideband AQAM in an indoor and outdoor environment. In this respect, Figure 1(a) can be used to characterise a hostile outdoor environment, where the perceived channel quality was low. This resulted in the utilisation of predominantly more robust modulation modes, such as BPSK and 4QAM. Conversely, a less hostile indoor environment is exemplified by Figure 1(b), where the perceived channel quality was high. As a result, the wideband AQAM regime can adapt suitably by invoking higher-order modulation modes, as evidenced by Figure 1(b). Again, this simple example demonstrated that wideband AQAM can be utilised, in order to provide a seamless, near-instantaneous reconfiguration between for example indoor and outdoor environments.

This book studies the network capacity gains that may be achieved with the advent of adaptive antenna arrays and adaptive modulation techniques in both FDMA / TDMA and CDMA-based mobile cellular networks. The advantages of employing adaptive antennas are multifold, as outlined below.

Reduction of Co-channel Interference: Antenna arrays employed by the base station allow the implementation of spatial filtering, as shown in Figure 2, which may be exploited in both transmitting as well as receiving modes in order to reduce co-channel inter-

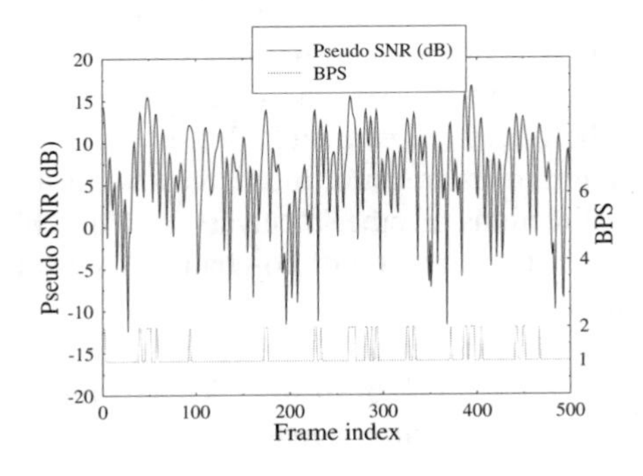

(a) Channel SNR of 10 dB

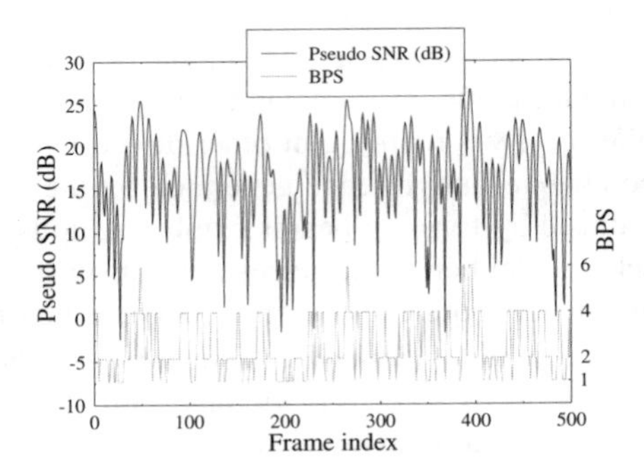

(b) Channel SNR of 20 dB

Figure 1: Modulation mode variation with respect to the pseudo SNR evaluated at the output of the channel equaliser of a wideband AQAM modem for transmission over the **TU Rayleigh fading channel**. The BPS throughputs of 1, 2, 4 and 6 represent BPSK, 4QAM, 16QAM and 64QAM, respectively.

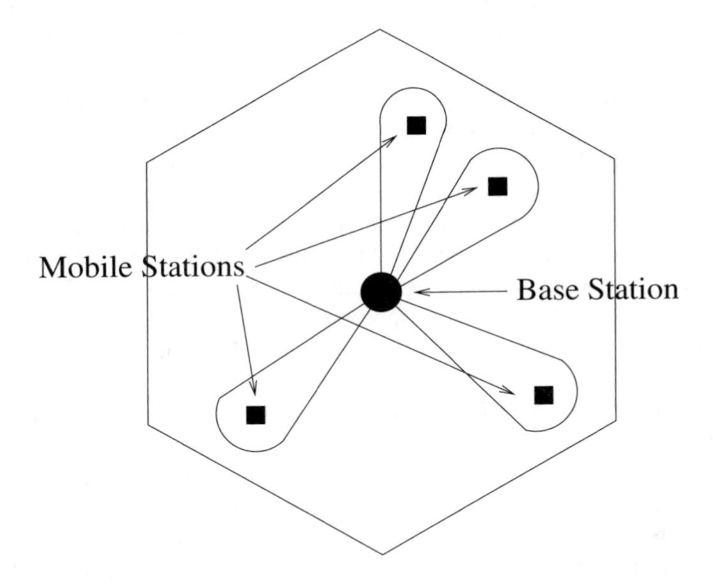

Figure 2: A cell layout showing how an antenna array can support many users on the same carrier frequency and timeslot with the advent of spatial filtering or Space Division Multiple Access (SDMA).

ferences [1, 2, 14, 15] experience in the uplink (UL) and downlink (DL) of wireless systems. When transmitting with an increased antenna gain in a certain direction of the DL, the base station's antenna is used to focus the radiated energy in order to form a high-gain directive beam in the area, where the mobile receiver is likely to be. This in turn implies that there is a reduced amount of radiated energy and hence reduced interference inflicted upon the mobile receivers roaming in other directions, where the directive beam has a lower gain. The co-channel interference generated by the base station in its transmit mode may be further reduced by forming beams exhibiting nulls in the directions of other receivers [6, 16]. This scheme deliberately reduces the transmitted energy in the direction of co-channel receivers and hence requires prior knowledge of their positions.

The employment of antenna arrays at the base station for reducing the co-channel interference in its receive mode has been also reported widely [1, 2, 6, 16–18]. This technique does not require explicit knowledge of the co-channel interference signal itself, however, it has to possess information concerning the desired signal, such as the direction of its source, a reference signal, such as a channel sounding sequence, or a signal that is highly correlated with the desired signal.

Capacity Improvement and Spectral Efficiency: The spectral efficiency of a wireless network refers to the amount of traffic a given system having a certain spectral allocation could handle. An increase in the number of users of the mobile communications system without a loss of performance increases the spectral efficiency. Channel capacity refers to the maximum data rate a channel of a given bandwidth can sustain. An improved channel capacity leads to an ability to support more users of a specified data rate, implying a better spectral efficiency. The increased quality of service that results from the reduced co-channel interference and reduced multipath fading [18, 19] upon using smart antennae may be exchanged for an increased number of users [2, 20].

Increase of Transmission Efficiency: An antenna array is directive in its nature, having a high gain in the direction where the beam is pointing. This property may be exploited in order to extend the range of the base station, resulting in a larger cell size or may be used to reduce the transmitted power of the mobiles. The employment of a directive antenna allows the base station to receive weaker signals than an omni-directional antenna. This implies that the mobile can transmit at a lower power and its battery recharge period becomes longer, or it would be able to use a smaller battery, resulting in a smaller size and weight, which is important for hand-held mobiles. A corresponding reduction in the power transmitted from the base station allows the use of electronic components having lower power ratings and therefore, lower cost.

Reduction of the Number of Handovers: When the amount of traffic in a cell exceeds the cell's capacity, cell splitting is often used in order to create new cells [2], each with its own base station and frequency assignment. The reduction in cell size leads to an increase in the number of handovers performed. By using antenna arrays for increasing the user capacity of a cell [1] the number of handovers required may actually be reduced. More explicitly, since each antenna beam tracks a mobile [2], no handover is necessary, unless different beams using the same frequency cross each other.

Avoiding Transmission Errors: When the instantaneous channel quality is low, conventional fixed-mode transceivers typically inflict a burst of transmission errors. By contrast, adaptive transceivers avoid this problem by reducing the number of transmitted bits per symbol, or even by disabling transmissions temporarily. The associated throughput loss can be compensated by transmitting a higher number of bits per symbol during the periods of relatively high channel qualities. This advantageous property manifests itself also in terms of an improved service quality, which is quantified in the book in terms of the achievable video quality.

However, realistic propagation scenarios are significantly more complex than that depicted in Figure 2. Specifically, both the desired signal and the interference sources experience **multipath propagation**, resulting in a high number of received uplink signals impinging upon the base station's receiver antenna array. A result of the increased number of received uplink signals is that the limited degrees of freedom of the base station's adaptive antenna array are exhausted, resulting in reduced nulling of the interference sources. A solution to this limitation is to increase the number of antenna elements in the base station's adaptive array, although this has the side effect of raising the cost and complexity of the array. In a macro-cellular system it may be possible to neglect multipath rays arriving at the base station from interfering sources, since the majority of the scatterers are located close to the mobile station [21]. By contrast, in a micro-cellular system the scatterers are located in both the region of the reduced-elevation base station and that of the mobile, and hence multipath propagation must be considered. Figure 3 shows a realistic propagation environment for both the up- and the downlink, with the multipath components of the desired signal and interference signals clearly illustrated, where the up- and downlink multipath components were assumed to be identical for the sake of simplicity. Naturally, this is not always the case and hence we will investigate the potential performance gains, when the up- and downlink beamforms are determined independently.

The Outline of the Book

- **Chapter 1:** Following a brief introduction to the principles of CDMA the three most important 3G wireless standards, namely UTRA, IMT 2000 and cdma 2000 are characterised. The range of various transport and physical channels, the multiplexing of various services for transmission, the aspects of channel coding are discussed. The various options available for supporting variable-rates and a range quality of service are highlighted. The uplink and downlink modulation and spreading schemes are described and UTRA and IMT 2000 are compared in terms of the various solutions standardised. The chapter is closed with a similar portrayal of the pan-American cdma 2000 system.

- **Chapter 2:** In this chapter we speculate that future standard systems as well as the extensions of the existing standard systems may become adaptive in an effort to compensate for the inevitable time-variant channel quality fluctuations of wireless channels. We commence our discourse by reviewing the state-of-the-art in near-instantaneously adaptive modulation and introduce the associated principles. We then apply the AQAM philosophy in the context of CDMA as well as OFDM and quantify the service-related benefits of adaptive transceivers in terms of the achievable video quality. The associated application examples demonstrate the potential of the proposed adaptive techniques in terms of tangible service quality improvements.

- **Chapter 3:** The principles behind beamforming and the various techniques by which it may be implemented are presented. From this the concept of adaptive beamforming is

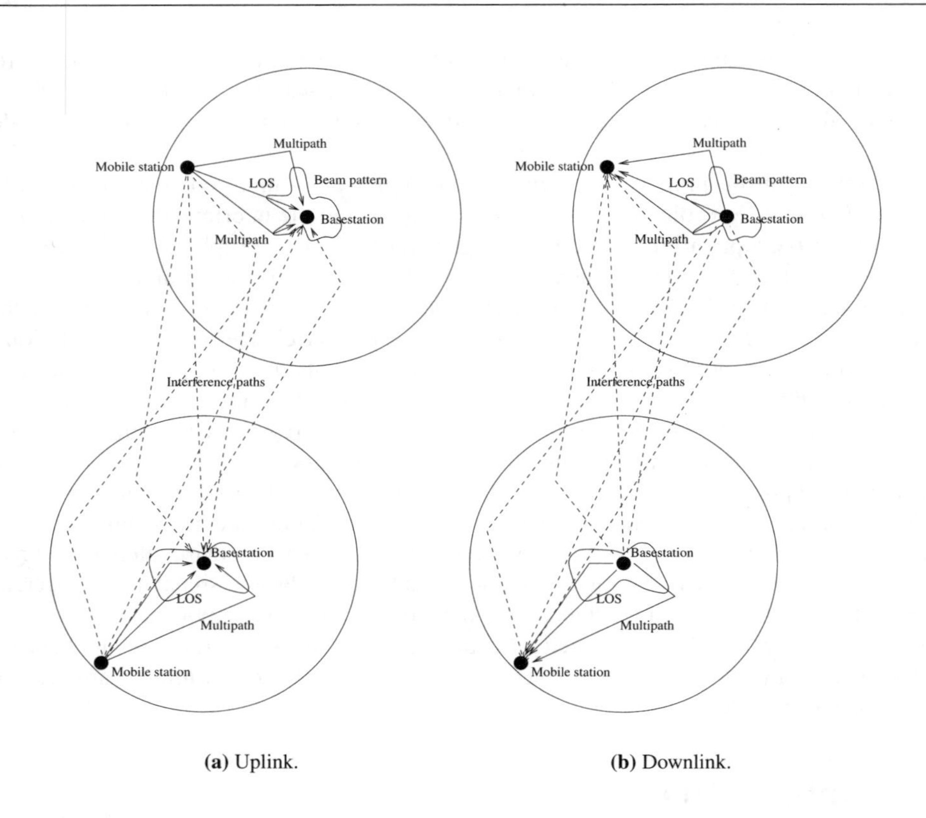

(a) Uplink. **(b)** Downlink.

Figure 3: The multipath environments of both the uplink and downlink, showing the multipath components of the desired signals, the line-of-sight interference and the associated base station antenna array beam patterns.

developed, and temporal as well as spatial reference techniques are examined. Performance results are then presented for three different temporal-reference based adaptive beamforming algorithms, namely the Sample Matrix Inversion (SMI), Unconstrained Least Mean Squares (ULMS) and the Normalised Least Mean Squares (NLMS) algorithms.

- **Chapter 4:** A brief summary of possible methods used for modelling the performance of an adaptive antenna array is provided. This is followed by an overview of fixed and dynamic channel allocation. Multipath propagation models are then considered for use in our network simulations. Metrics are then developed for characterising the performance of mobile cellular networks and our results are presented for simulations conducted under Line-Of-Sight (LOS) propagation conditions, both with and without adaptive antennas. Further results are then given for identical networks under multipath propagation conditions, which are then extended to power-controlled scenarios using both fixed and adaptive QAM techniques. These network capacity results were obtained for both "island" type simulation areas, and for an infinite plane, using wraparound techniques.

- **Chapter 5:** This chapter provides a brief description of the 3rd generation mobile cellular network, known as UTRA - the UMTS Terrestrial Radio Access - network, and then presents network capacity results obtained under various propagation conditions, in conjunction with different soft handover threshold metrics. The performance benefits of adaptive antenna arrays are then analysed, both in a non-shadowed environment, and inflicted by log-normal shadow fading having a frequencies of 0.5 Hz and 1.0 Hz. This work was then extended by the invoking of adaptive modulation techniques, which were studied when the channel conditions were impaired by shadow fading.

- **Chapter 6:** Conclusions and further work.

Contributions of the Book

- Providing an introduction to near-instantaneously adaptive modulation invoked in the context of both single- and multi-carrier modulation or OFDM, as well as CDMA.

- Quantifying the service-related benefits of adaptive transceivers in the context of wireless video telephony.

- Providing an overview of the various CDMA-based 3G wireless standards.

- Study of the network performance gains using adaptive antenna arrays at the base station in an FDMA / TDMA cellular mobile network [22, 23].

- Study of the network performance gains using adaptive antenna arrays in conjunction with power control at the base station in an FDMA/TDMA cellular mobile network [24, 25].

- Design of a combined power control and adaptive modulation assisted channel allocation algorithm, and characterisation of its performance in an FDMA / TDMA cellular mobile network [25, 26].

- Comparing the performance of various UTRA soft-handover techniques.

- Quantifying the UTRA network capacity under various channel conditions.

- Evaluating the network performance of UTRA with the aid of adaptive antenna arrays.

- Demonstrating the benefits of adaptive modulation in the context of both FDMA / TDMA and CDMA cellular mobile networks.

Our hope is that the book offers you a range of interesting topics in the era of the imminent introduction of 3G wireless networks. We attempted to provide an informative technological roadmap, allowing the reader to quantify the achievable network capacity gains with the advent of introducing more powerful enabling technologies in the physical layer. Analysing the associated system design trade-offs in terms of network complexity and network capacity is the basic aim of this book. We aimed for underlining the range of contradictory system design trade-offs in an unbiased fashion, with the motivation of providing you with sufficient information for solving your own particular wireless networking problems. Most of all however we hope that you will find this book an enjoyable and relatively effortless reading, providing you with intellectual stimulation.

Jonathan Blogh and Lajos Hanzo
Department of Electronics and Computer Science
University of Southampton

Acknowledgements

We are indebted to our many colleagues who have enhanced our understanding of the subject, in particular to Prof. Emeritus Raymond Steele. These colleagues and valued friends, too numerous to be mentioned, have influenced our views concerning various aspects of wireless multimedia communications. We thank them for the enlightenment gained from our collaborations on various projects, papers, and books. We are grateful to Jan Brecht, Jon Blogh, Marco Breiling, Marco del Buono, Sheng Chen, Peter Cherriman, Stanley Chia, Byoung Jo Choi, Joseph Cheung, Peter Fortune, Sheyam Lal Dhomeja, Lim Dongmin, Dirk Didascalou, Stephan Ernst, Eddie Green, David Greenwood, Hee Thong How, Thomas Keller, Ee Lin Kuan, W. H. Lam, Matthias Münster, C. C. Lee, M. A. Nofal, Xiao Lin, Chee Siong Lee, Tong-Hooi Liew, Jeff Reeve, Vincent Roger-Marchart, Redwan Salami, David Stewart, Clare Sommerville, Jeff Torrance, Spyros Vlahoyiannatos, William Webb, Stefan Weiss, John Williams, Jason Woodard, Choong Hin Wong, Henry Wong, James Wong, Lie-Liang Yang, Bee-Leong Yeap, Mong-Suan Yee, Kai Yen, Andy Yuen, and many others with whom we enjoyed an association.

We also acknowledge our valuable associations with the Virtual Centre of Excellence in Mobile Communications, in particular with its chief executives, Dr. Tony Warwick and Dr. Walter Tuttlebee, and other members of its Executive Committee, namely Dr. Keith Baughan, Prof. Hamid Aghvami, Prof. Ed Candy, Prof. John Dunlop, Prof. Barry Evans, Dr. Mike Barnard, Prof. Joseph McGeehan, Prof. Peter Ramsdale and many other valued colleagues. Our sincere thanks are also due to the EPSRC, UK for supporting our research. We would also like to thank Dr. Joao Da Silva, Dr Jorge Pereira, Bartholome Arroyo, Bernard Barani, Demosthenes Ikonomou, and other valued colleagues from the Commission of the European Communities, Brussels, Belgium, as well as Andy Aftelak, Mike Philips, Andy Wilton, Luis Lopes, and Paul Crichton from Motorola ECID, Swindon, UK, for sponsoring some of our recent research. Further thanks are due to Tim Wilkinson at HP in Bristol for funding some of our research efforts.

Similarly, our sincere thanks are due to our Senior Editor, Mark Hammond, Sarah Hinton, Zoe Pinnock and their colleagues at Wiley in Chichester, UK, as well as Denise Harvey, who assisted us during the production of the book. Finally, our sincere gratitude is due to the numerous authors listed in the Author Index — as well as to those, whose work was not cited due to space limitations — for their contributions to the state of the art, without whom this book would not have materialised.

Jonathan Blogh and Lajos Hanzo
Department of Electronics and Computer Science
University of Southampton

Acknowledgments

Chapter 1

Third-Generation CDMA Systems

K. Yen and L. Hanzo

1.1 Introduction

Although the number of cellular subscribers continues to grow worldwide [27], the predominantly speech-, data- and e-mail-oriented services are expected to be enriched by a whole host of new services in the near future. Thus the performance of the recently standardised Code Division Multiple Access (CDMA) third-generation (3G) mobile systems is expected to become comparable to, if not better than, that of their wired counterparts.

These ambitious objectives are beyond the capabilities of the present second-generation (2G) mobile systems such as the Global System for Mobile Communications known as GSM [28], the Interim Standard-95 (IS-95) Pan-American system, or the Personal Digital Cellular (PDC) system [29] in Japan. Thus, in recent years, a range of new system concepts and objectives were defined, and these will be incorporated in the 3G mobile systems. Both the European Telecommunications Standards Institute (ETSI) and the International Telecommunication Union (ITU) are defining a framework for these systems under the auspices of the Universal Mobile Telecommunications System (UMTS) [27,29–33] and the International Mobile Telecommunications scheme in the year 2000 (IMT-2000)[1] [30, 31, 34].

Their objectives and the system concepts will be discussed in more detail in later sections. CDMA is the predominant multiple access technique proposed for the 3G wireless communications systems worldwide. CDMA was already employed in some 2G systems, such as the IS-95 system and it has proved to be a success. Partly motivated by this successer, both the Pan-European UMTS and the IMT-2000 initiatives have opted for a CDMA-based system, although the European system also incorporates an element of TDMA. In this chapter, we provide a rudimentary introduction to a range of CDMA concepts. Then the European, American and Japanese CDMA-based 3G mobile system concepts are considered, followed by a research-oriented outlook on potential future systems.

[1] Formerly known as Future Public Land Mobile Telecommunication Systems.

The chapter is organised as follows. Section 1.2 offers a rudimentary introduction to CDMA in order to make this chapter self-contained, whereas Section 1.3 focuses on the basic objectives and system concepts of the 3G mobile systems, highlighting the European, American and Japanese CDMA-based third-generation system concepts. Finally, our conclusions are presented in Section 1.4.

1.2 Basic CDMA System

CDMA is a spread spectrum communications technique that supports simultaneous digital transmission of several users' signals in a multiple access environment. Although the development of CDMA was motivated by user capacity considerations, the system capacity provided by CDMA is similar to that of its more traditional counterparts, frequency division multiple access (FDMA), and time division multiple access (TDMA) [35]. However, CDMA has the unique property of supporting a multiplicity of users in the same radio channel with a graceful degradation in performance due to multi-user interference. Hence, any reduction in interference can lead to an increase in capacity [36]. Furthermore, the frequency reuse factor in a CDMA cellular environment can be as high as unity, and being a so-called wideband system, it can coexist with other narrowband microwave systems, which may corrupt the CDMA signal's spectrum in a narrow frequency band without inflicting significant interference [37]. This eases the problem of frequency management as well as allowing a smooth evolution from narrowband systems to wideband systems. But perhaps the most glaring advantage of CDMA is its ability to combat or in fact to benefit from multipath fading, as it will become explicit during our further discourse.

In the forthcoming sections, we introduce our nomenclature, which will be used throughout the subsequent sections. Further in-depth information on CDMA can be found in a range of excellent research papers [35, 37, 38] and textbooks [39–42].

1.2.1 Spread Spectrum Fundamentals

In spread spectrum transmission, the original information signal, which occupies a bandwidth of B Hz, is transmitted after spectral spreading to a bandwidth N times higher, where N is known as the processing gain. In practical terms the processing gain is typically in the range of $10 - 30$ dB [37]. This frequency-domain spreading concept is illustrated in Figure 1.1. The power of the transmitted spread spectrum signal is spread over N times the original bandwidth, while its spectral density is correspondingly reduced by the same amount. Hence, the processing gain is given by:

$$N = \frac{B_s}{B}, \tag{1.1}$$

where B_s is the bandwidth of the spread spectrum signal while B is the bandwidth of the original information signal. As we shall see during our further discourse, this unique technique of spreading the information spectrum is the key to improving its detection in a mobile radio environment, and it also allows narrowband signals exhibiting a significantly higher spectral density to share the same frequency band [37].

There are basically two main types of spread spectrum (SS) systems [35]:

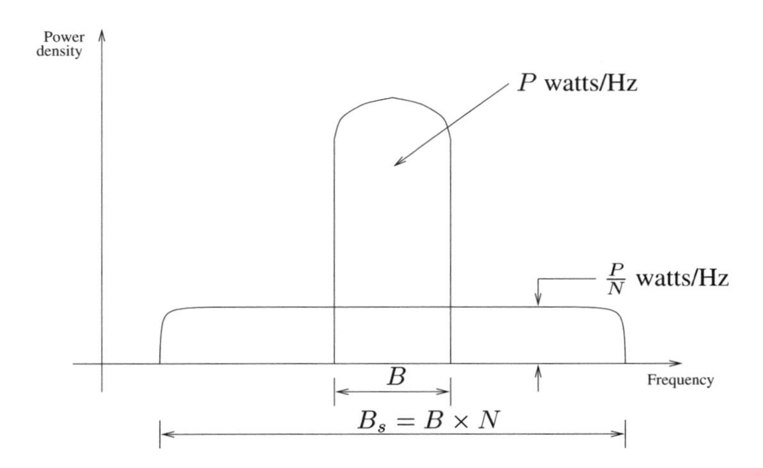

Figure 1.1: Power spectral density of signal before and after spreading.

- Direct Sequence (DS) SS systems and
- Frequency Hopping (FH) SS systems.

1.2.1.1 Frequency Hopping

In FH spreading, which was invoked in the 2G GSM system the narrowband signal is transmitted using different carrier frequencies at different times. Thus, the data signal is effectively transmitted over a wide spectrum. There are two classes of frequency hopping patterns. In fast frequency hopping, the carrier frequency changes several times per transmitted symbol, while in slow frequency hopping, the carrier frequency changes typically after a number of symbols or a transmission burst. In the GSM system each transmission burst of 114 channel-coded speech bits was transmitted on a different frequency and since the TDMA frame duration was 4.615 ms, the associated hopping frequency was its reciprocal, that is, 217 hops/sec. The exact sequence of frequency hopping will be made known only to the intended receiver so that the frequency hopped pattern may be dehopped in order to demodulate the signal [37]. Direct sequence (DS) spreading is more commonly used in CDMA. Hence, our forthcoming discussions will be in the context of direct sequence spreading.

1.2.1.2 Direct Sequence

In DS spreading, the information signal is multiplied by a high-frequency signature sequence, also known as a spreading code or spreading sequence. This user signature sequence facilitates the detection of different users' signals in order to achieve a multiple access capability in CDMA. Although in CDMA this user 'separation' is achieved using orthogonal spreading codes, in FDMA and TDMA orthogonal frequency slots or time-slots are provided, respectively.

We can see from Figure 1.2 that each information symbol of duration T_s is broken into N_c equi-spaced subintervals of duration T_c, each of which is multiplied with a different chip of

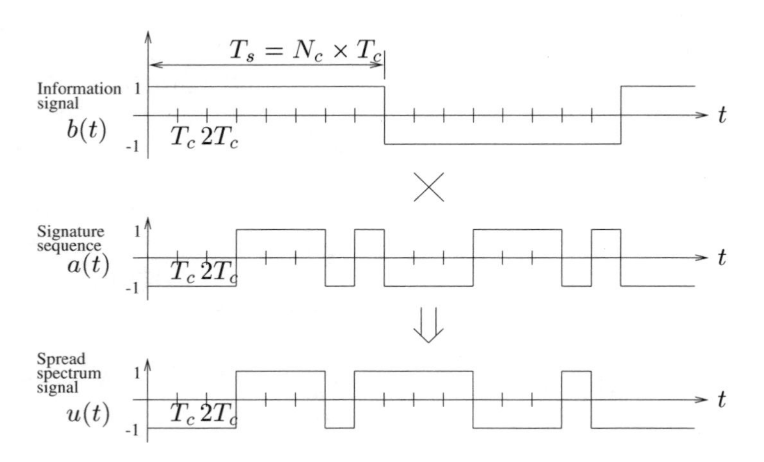

Figure 1.2: Time-domain waveforms involved in generating a direct sequence spread signal.

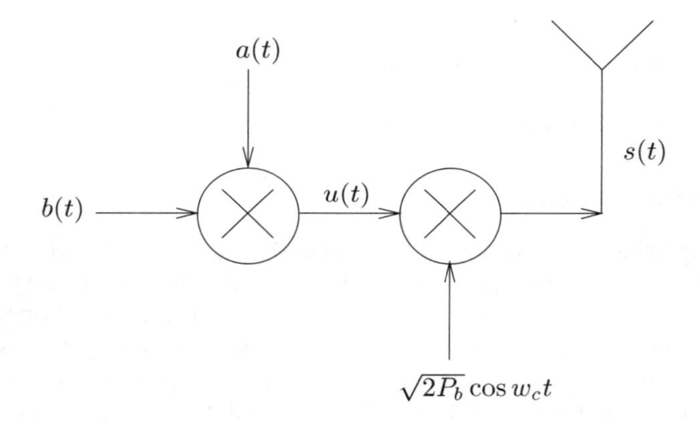

Figure 1.3: BPSK modulated DS-SS transmitter.

the spreading sequence. Hence, $N_c = \frac{T_s}{T_c}$. The resulting output is a high-frequency sequence.

For binary signalling $T_s = T_b$, where T_b is the data bit duration. Hence, N_c is equal to the processing gain N. However, for M-ary signalling, where $M > 2$, $T_s \neq T_b$ and hence $N_c \neq N$. An understanding of the distinction between N_c and N is important, since the values of N_c and N have a direct effect on the bandwidth efficiency and performance of the CDMA system.

The block diagram of a typical binary phase shift keying (BPSK) modulated DS-SS transmitter is shown in Figure 1.3. We will now express the associated signals mathematically.

The binary data signal may be written as:

$$b(t) = \sum_{j=-\infty}^{\infty} b_j \Gamma_{T_b}(t - jT_b), \tag{1.2}$$

where T_b is the bit duration, $b_j \in \{+1, -1\}$ denotes the jth data bit, and $\Gamma_{T_b}(t)$ is the pulse shape of the data bit. In practical applications, $\Gamma_\tau(t)$ has a bandlimited waveform, such as a raised cosine Nyquist pulse. However, for analysis and simulation simplicity, we will assume that $\Gamma_\tau(t)$ is a rectangular pulse throughout this chapter, which is defined as:

$$\Gamma_\tau(t) = \begin{cases} 1, & 0 \leq t < \tau, \\ 0, & \text{otherwise.} \end{cases} \tag{1.3}$$

Similarly, the spreading sequence may be written as

$$a(t) = \sum_{h=-\infty}^{\infty} a_h \Gamma_{T_c}(t - hT_c), \tag{1.4}$$

where $a_h \in \{+1, -1\}$ denotes the hth chip and $\Gamma_{T_c}(t)$ is the chip-pulse with a chip duration of T_c. The energy of the spreading sequence over a bit duration of T_b is normalised according to:

$$\int_0^{T_b} |a(t)|^2 dt = T_b. \tag{1.5}$$

As seen in Figure 1.3, the data signal and spreading sequence are multiplied, and the resultant spread signal is modulated on a carrier in order to produce the wideband signal $s(t)$ at the output:

$$s(t) = \sqrt{2P_b}\, b(t) a(t) \cos w_c t, \tag{1.6}$$

where P_b is the average transmitted power. At the intended receiver, the signal is multiplied by the conjugate of the transmitter's spreading sequence, which is known as the despreading sequence, in order to retrieve the information. Ideally, in a single-user, nonfading, noiseless environment, the original information can be decoded without errors. This is seen in Figure 1.4.

In reality, however, the conditions are never so perfect. The received signal will be corrupted by noise, interfered by both multipath fading — resulting in intersymbol interference (ISI) — and by other users, generating multi-user interference. Furthermore, this signal is delayed by the time-dispersive medium. It is possible to reduce the interference due to multipath fading and other users by innovative signal processing methods, which will be discussed in more detail in later sections.

Figure 1.5 shows the block diagram of the receiver for a noise-corrupted channel using a correlator for detecting the transmitted signal, yielding:

$$\begin{aligned} \hat{b}_i &= \operatorname{sgn}\left\{ \frac{1}{\sqrt{T_b}} \int_{iT_b}^{(i+1)T_b} a^*(t)[s(t) + n(t)] \cos w_c t \, dt \right\} \\ &= \operatorname{sgn}\left\{ \sqrt{\frac{\xi_b}{2}} b_i + \frac{1}{\sqrt{T_b}} \int_{iT_b}^{(i+1)T_b} a^*(t) n(t) \cos w_c t \, dt \right\}, \end{aligned} \tag{1.7}$$

where $\xi_b = T_b \times P_b$ is the bit energy and $\text{sgn}(x)$ is the signum function of x, which returns a value of 1, if $x > 0$ and returns a value of -1, if $x < 0$. In a single-user Additive White

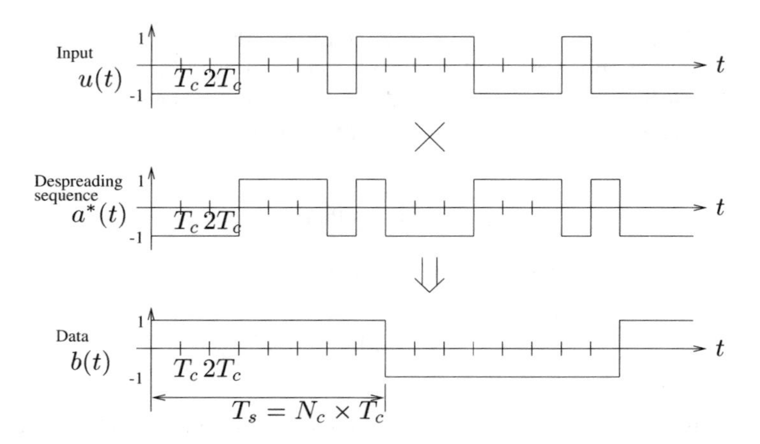

Figure 1.4: Time-domain waveforms involved in decoding a direct sequence signal.

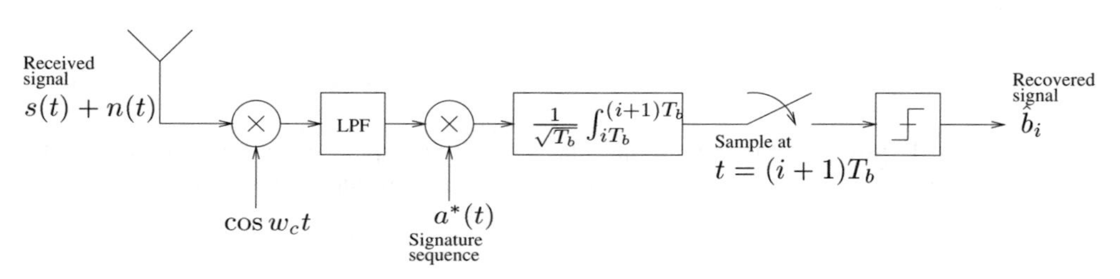

Figure 1.5: BPSK DS-SS receiver for AWGN channel.

Gaussian Noise (AWGN) channel, the receiver shown in Figure 1.5 is optimum. In fact, the performance of the DS-SS system discussed so far is the same as that of a conventional BPSK modem in an AWGN channel, whereby the probability of bit error $Pr_b(\epsilon)$ is given by:

$$Pr_b(\epsilon) = Q\left(\sqrt{\frac{2\xi_b}{N_0}}\right), \tag{1.8}$$

where

$$Q(x) = \frac{1}{\sqrt{2\pi}} \int_x^{\infty} e^{-y^2/2} dy \tag{1.9}$$

is the Gaussian Q-function. The advantages of spread spectrum communications and CDMA will only be appreciated in a multipath multiple access environment. The multipath aspects and how the so-called RAKE receiver [5, 43] can be used to overcome the multipath effects will be highlighted in the next section.

1.2.2 The Effect of Multipath Channels

In this section, we present an overview of the effects of the multipath wireless channels encountered in a digital mobile communication system, which was treated in depth for example in [11]. Interested readers may also refer to the recent articles written by Sklar in [44,45] for a brief overview on this subject.

Since the mobile station is usually close to the ground, the transmitted signal is reflected, refracted, and scattered from objects in its vicinity, such as buildings, trees, and mountains [35]. Therefore, the received signal is comprised of a succession of possibly overlapping, delayed replicas of the transmitted signal. Each replica is unique in its arrival time, power, and phase [46]. As the receiver or the reflecting objects are not stationary, such reflections will be imposed fading on the received signal, where the fading causes the signal strength to vary in an unpredictable manner. This phenomenon is referred to as multipath propagation [11].

There are typically two types of fading in the mobile radio channel [44]:

- Long-term fading
- Short-term fading

As argued in [11] long-term fading is caused by the terrain configuration between the base station and the mobile station, such as hills and clumps of buildings, which result in an average signal power attenuation as a function of distance. For our purposes the channel can be described in terms of its average pathloss, typically obeying an inverse fourth power law [35] and a log-normally distributed variation around the mean. Thus, long-term shadow fading was also referred to as log-normal fading in [11,44] .

On the other hand, short-term fading refers to the dramatic changes in signal amplitude and phase as a result of small changes in the spatial separation between the receiver and transmitter, as we noted in [11,44].

Furthermore, the motion between the transmitter and receiver results in propagation path changes, such that the channel appears to be time-variant. The time-variant frequency-selective channel was modelled as a tapped delay line in [11], where the complex low-pass impulse response can be modelled as:

$$\tilde{h}(t) = \sum_{l=1}^{L} |\alpha_l(t)| e^{j\phi_l(t)} \delta(t - \tau_l), \tag{1.10}$$

where $|\alpha_l(t)|$, $\phi_l(t)$ and τ_l are the amplitude, phase, and delay of the lth path, respectively, and L is the total number of multipath components. It was argued in [11] that the rate of signal level fluctuation is determined by the Doppler frequency, f_D, which in turn is dependent on the carrier frequency, f_c, and the speed of the mobile station v according to (see also page 16 of [47]):

$$f_D = v\frac{f_c}{c}, \tag{1.11}$$

where c is the speed of light.

Typically, the short-term fading phenomenon is modelled statistically by a Rayleigh, Rician, or Nakagami-m distribution [48]. The Rayleigh and Rician distributions were characterised for example in [11]. There have been some contrasting views in the literature as to which of these distributions best describes the fast-fading channel statistically. Although empirical results have shown that the fading statistics are best described by a Nakagami distribution [49], in most cases a Rayleigh-distributed fading is used for analysis and simulation

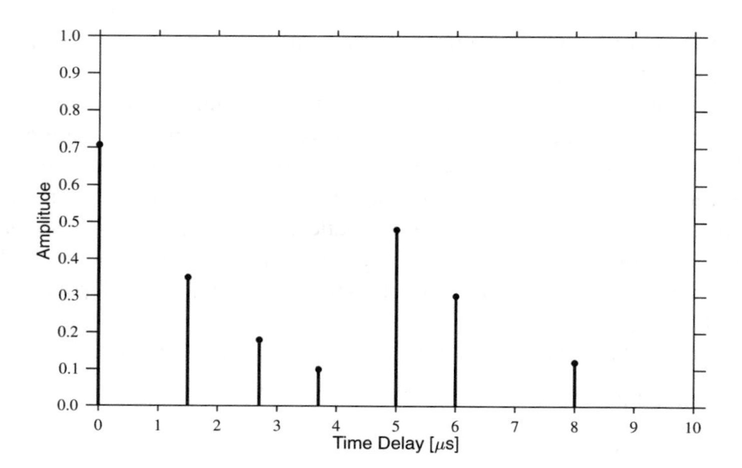

Figure 1.6: COST 207 BU impulse response.

because of simplicity, and it serves as a useful illustrative example in demonstrating the effects of fading on transmission. Moreover, the Rayleigh distribution is a special case of the Nakagami distribution, when m, known as the fading parameter, is equal to unity (see page 48 in [5]). The Rician distribution is more applicable to satellite communication, due to the presence of a dominant signal component known as the specular component [44], than to large-cell terrestrial communication, where often there is no Line-of-Sight (LOS) path between the terrestrial base station and the mobile station. However, in small microcells often the opposite is true. In our investigations in this chapter, Rayleigh-distributed frequency selective fading is assumed.

The delay is proportional to the length of the corresponding signal path between the transmitter and receiver. The delay spread due to the path-length differences between the multipath components causes Intersymbol Interference (ISI) in data transmission, which becomes particularly dominant for high data rates.

A typical radio channel impulse response is shown in Figure 1.6. This channel impulse response is known as the COST 207 bad urban (BU) impulse response [50]. It can be clearly seen that the response consists of two main groups of delayed propagation paths: a main profile and a smaller echo profile following the main profile at a delay of 5 μs. The main profile is caused by reflections of the signal from structures in the vicinity of the receiver with shorter delay times. On the other hand, the echo profile could be caused by several reflections from a larger but more distant object, such as a hill [51]. In either case, we can see that both profiles approximately follow a negative exponentially decaying function with respect to the time-delay.

Figure 1.7 shows the impairments of the spread spectrum signal travelling over a multipath channel with L independent paths, yielding the equivalent baseband received signal of:

$$r(t) = \sum_{l=1}^{L} \alpha_l(t)\tilde{s}(t - \tau_l) + n(t), \tag{1.12}$$

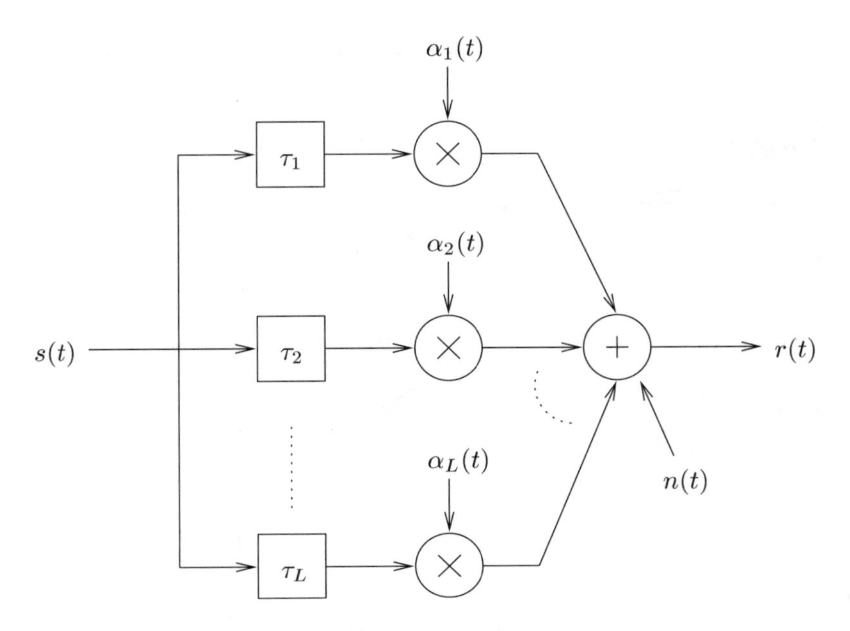

Figure 1.7: Multipath propagation model of the transmitted signal.

where $\alpha_l(t)$ is the time-variant complex channel gain, which is given by $|\alpha_l(t)|e^{j\phi_l(t)}$ in Equation 1.10 with a Rayleigh-distributed amplitude, uniformly distributed phase over the interval $[-\pi \ldots \pi]$ and $\tilde{s}(t-\tau_l)$ is the equivalent baseband transmitted spread spectrum signal from Equation 1.6 delayed by τ_l. The above equation shows that the lth path is attenuated by the channel coefficient $\alpha_l(t)$ and delayed by τ_l. Without intelligent diversity techniques [5], these paths are added together at the receiver and any phase or delay difference between the paths may result in a severely multipath interfered signal, corrupted by dispersion-induced intersymbol interference (ISI).

Figure 1.8 shows the effect of a nonfading channel and a fading channel on the bit error probability of BPSK-modulated CDMA. Without diversity, the bit error rate (BER) in a fading channel decreases approximately according to $Pr_b(\epsilon) \approx \frac{1}{4\bar{\gamma}_c}$, where $\bar{\gamma}_c$ is the average Signal-to-Noise Ratio (SNR), and hence plotted on a logarithmic scale according to $\log Pr_b(\epsilon) = -\log 4\bar{\gamma}_c$ we have a near-linear curve [5]. This is different from a nonfading, or AWGN, channel, whereby the BER decreases exponentially with increasing the SNR. Thus, in a fading channel, a high transmitted power is required to obtain a low probability of error. As we shall see in the next section, diversity techniques can be used to overcome this impediment.

1.2.3 RAKE Receiver

As mentioned previously, spread spectrum techniques can take advantage of the multipath nature of the mobile channel in order to improve reception. This is possible due to the signal's wideband nature, which has a significantly higher bandwidth than the multipath channel's coherence bandwidth [52]. In this case, the channel was termed a frequency selective

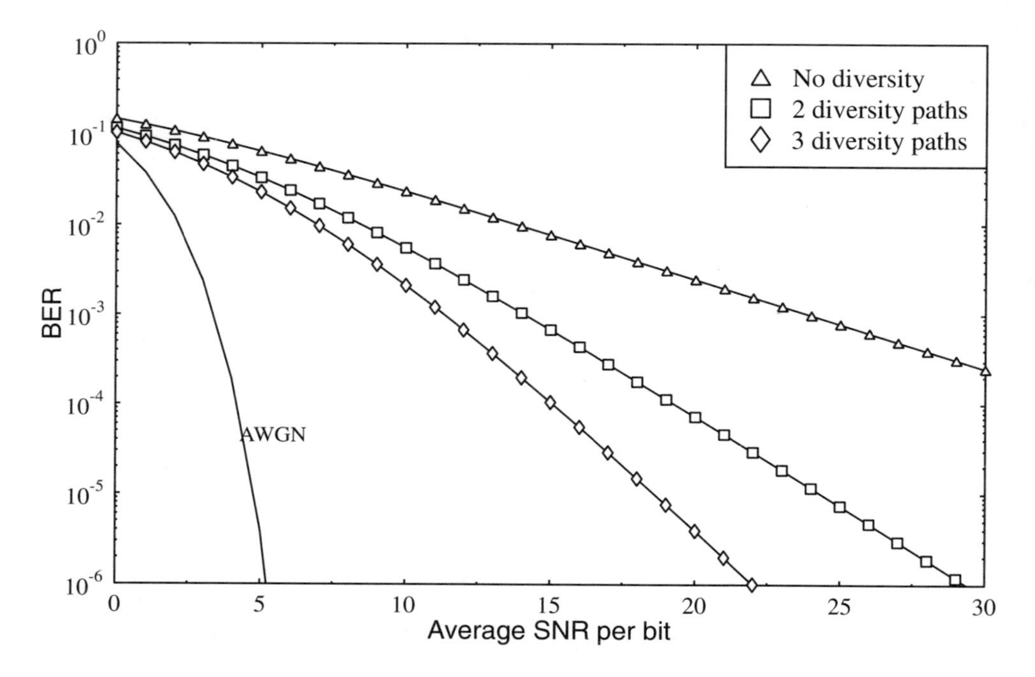

Figure 1.8: Performance of BPSK modulated CDMA over various Rayleigh-fading channels. The curves were obtained using perfect channel estimation, and there was no self-interference between diversity paths.

fading channel, since different transmitted frequencies faded differently if their separation was higher than the previously mentioned coherence bandwidth. Suppose that the spread spectrum has a bandwidth of B_s and the channel's coherence bandwidth is B_c, such that $B_s \gg B_c$. Thus, the number of resolvable independent paths — that is, the paths that fade near-independently — L_R is equal to

$$L_R = \left\lfloor \frac{B_s}{B_c} \right\rfloor + 1, \tag{1.13}$$

where $\lfloor x \rfloor$ is the largest integer that is less than or equal to x. The number of resolvable paths L_R varies according to the environment, and it is typically higher in urban than in suburban areas, since in urban areas the coherence bandwidth is typically lower due to the typically higher delay-spread of the channel [35]. More explicitly, this is a consequence of the more dispersive impulse response, since the coherence bandwidth is proportional to the reciprocal of the impulses responses delay spread, as it was argued in [52]. Similarly to frequency diversity or space diversity, these L_R resolvable paths can be employed in multipath diversity schemes, which exploit the fact that statistically speaking, the different paths cannot be in deep fades simultaneously; hence, there is always at least one propagation path, which provides an unattenuated channel. These multipath components are diversity paths.

Multipath diversity can only be exploited in conjunction with wideband signals. From Equation 1.13, for a narrowband signal, where no deliberate signal spreading takes place, the signal bandwidth B_s is significantly lower than B_c. In this case, the channel was termed frequency nonselective in [52]. Hence, no resolvable diversity paths can be observed, unlike in a wideband situation, and this renders TDMA and FDMA potentially less robust in a narrowband mobile radio channel than CDMA.

Multipath diversity is achieved, for example, by a receiver referred to as the RAKE receiver invented by Price and Green [43]. This is the optimum receiver for wideband fading multipath signals. It inherited its name from the analogy of a garden rake, whereby the fingers constitute the resolvable paths. The point where the handle and fingers meet is where diversity combining takes place. There are four basic methods of diversity combining, namely [53]:

- Selection Combining (SC).
- Maximal Ratio Combining (MRC).
- Equal Gain Combining (EGC).
- Combining of the n best signals (SCn).

The performance analysis of selection combining in CDMA can be found in [54, 55], while a general comparison of the various diversity combining techniques can be found in [53] for Rayleigh-fading channels. Maximal ratio combining gives the best performance, while selection combining is the simplest to implement. The number of resolvable paths that are combined at the receiver, represents the order of diversity of the receiver, which is denoted here as L_P. We note, however, that in practical receivers not all resolvable multipath components are combined due to complexity reasons, that is, $L_P \leq L_R$.

There are two basic demodulation techniques, namely coherent and noncoherent demodulation [5]. We will highlight the basics of coherent demodulation in this section in the context of CDMA. However, before demodulation can take place, synchronisation between the transmitter and the intended receiver has to be achieved.

Synchronisation in DS-CDMA is performed by a process known as code acquisition and tracking. Acquisition is usually carried out by invoking correlation techniques between the receiver's own copy of the signature sequence and the received signature sequence and searching for the displacement between them — associated with a specific chip epoch — that results in the high correlation [37, 56, 57]. Once acquisition has been accomplished, usually a code tracking loop [58] is employed to achieve fine alignment of the two sequences and to maintain their alignment. The details of code acquisition and tracking are beyond the scope of this chapter. Interested readers may refer to [59–62] and the references therein for an in-depth treatise on this subject. Hence, in this chapter, we will assume that the transmitter and the intended receiver are perfectly synchronised.

For optimum performance of the RAKE receiver using coherent demodulation, the path attenuation and phase must be accurately estimated. This estimation is performed by another process known as channel estimation, which will be elaborated on in Section 1.2.6. In typical low-complexity applications, known pilot symbols can be inserted in the transmitted sequence in order to estimate the channel's attenuation and phase rotation. However, for now, let us assume perfect channel estimation in order to assess the performance of the RAKE diversity combiner.

Figure 1.9 shows the block diagram of the BPSK RAKE receiver. The received signal is first multiplied by the estimated channel coefficients $\alpha_1(t), \ldots, \alpha_{L_P}(t)$ in each RAKE

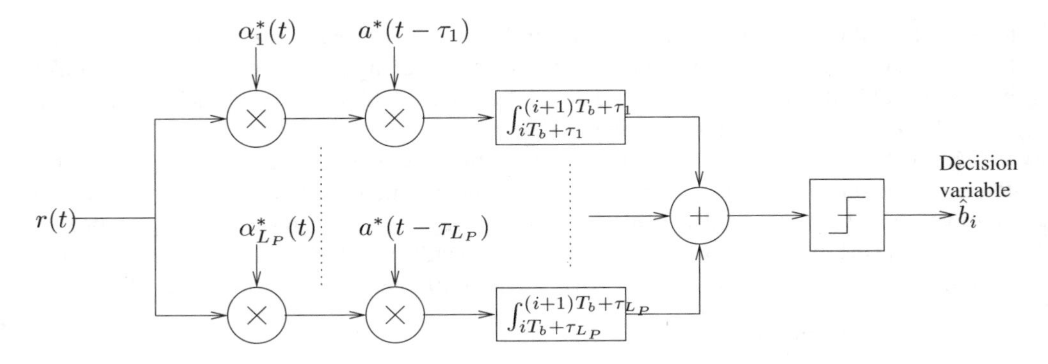

Figure 1.9: RAKE receiver.

branch tuned to each resolvable path. For optimum performance of the RAKE receiver using maximal ratio combining, these channel coefficient estimates should be the conjugates of the actual coefficients of the appropriate paths in order to invert the channel effects.[2] Note that for equal gain combining only the phase is estimated, and the various path contributions are multiplied by a unity gain before summation. The resulting signals in each RAKE branch are then multiplied by the conjugate signature sequences as we have seen in Figure 1.3, delayed accordingly by the code acquisition process. After despreading by the conjugate signature sequences $a^*(t-\tau_1), \ldots, a^*(t-\tau_{L_P})$, the outputs of the correlators in Figure 1.9 are combined in order to obtain the decoded symbol of:[3]

$$\begin{aligned}
\hat{b}_i &= \operatorname{sgn}\left\{\sum_{l=1}^{L_P}\left[\frac{1}{\sqrt{T_b}}\int_{iT_b+\tau_l}^{(i+1)T_b+\tau_l}\alpha_l^*(t)r(t)a^*(t-\tau_l)\,dt\right]\right\} \\
&= \operatorname{sgn}\left\{\sum_{l=1}^{L_P}\left[\sqrt{\frac{P_b}{T_b}}\int_{iT_b+\tau_l}^{(i+1)T_b+\tau_l}|\alpha_l(t)|^2 b(t-\tau_l)a(t-\tau_l)a^*(t-\tau_l)\,dt\right.\right. \\
&\qquad \left.\left.+\frac{1}{\sqrt{T_b}}\int_{iT_b+\tau_l}^{(i+1)T_b+\tau_l}\alpha_l^*(t)n(t)a^*(t-\tau_l)\,dt\right]\right\} \\
&= \operatorname{sgn}\left\{\sum_{l=1}^{L_P}\left[|\alpha_l(t)|^2\sqrt{\xi_b}b_i\right.\right. \\
&\qquad \left.\left.+\frac{1}{\sqrt{T_b}}\int_{iT_b+\tau_l}^{(i+1)T_b+\tau_l}\alpha_l^*(t)n(t)a^*(t-\tau_l)\,dt\right]\right\}. \qquad (1.14)
\end{aligned}$$

Normally, the first term of Equation 1.14, which contains the useful information, is much larger than the despread, noise-related second term. This is because the first term is proportional to the sum of the absolute values of the channel coefficients, whereas the second term in Equation 1.14 is proportional to the vectorial sum of the complex-valued channel

[2] $\alpha_l e^{j\phi_l} \times \alpha_l e^{-j\phi_l} = \alpha_l^2$.

[3] Here we assumed that there is no multipath interference. This interference can be considered as part of multi-user interference, which will be discussed in the next section.

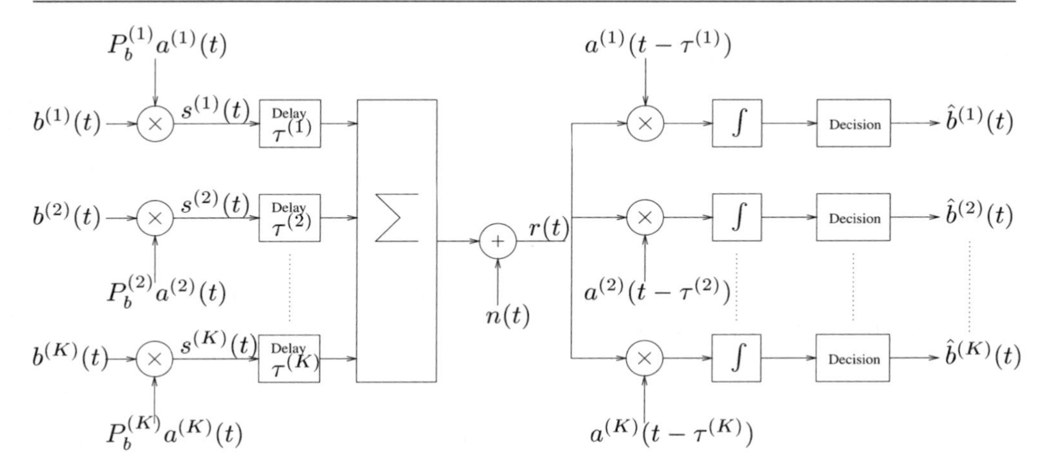

Figure 1.10: CDMA system model.

coefficients. Hence, the real part of the first term is typically larger than that of the second term. Thus, the RAKE receiver can enhance the detection of the data signal in a multipath environment.

Referring back to the BER curves of Figure 1.8, we can see that the performance of the system is improved when multipath diversity is used. Better performance is observed by increasing the number of diversity paths L_P. However, this also increases the complexity of the receiver, since the number of correlators has to be increased, which is shown in Figure 1.9.

1.2.4 Multiple Access

So far, only single-user transmission was considered. The system is simple and straightforward to implement. Let us now consider how multiple user transmission can affect the performance of the system.

Multiple access is achieved in DS-CDMA by allowing different users to share a common frequency band. Each transmitter and its intended receiver are assigned a distinct user signature sequence. Only the receivers having the explicit knowledge of this distinct sequence are capable of detecting the transmitted signal. Consider a CDMA scenario with K number of active users, transmitting simultaneously. The baseband equivalent system model is shown in Figure 1.10. For simplicity, it is assumed that there is no multipath propagation and perfect power control is maintained.

The mathematical representation of the kth user's data signal is similar to that shown in Equation 1.2, except for an additional superscript, denoting multi-user transmission. Hence, it is written as:

$$b^{(k)}(t) = \sum_{j=-\infty}^{\infty} b_j^{(k)} \Gamma_{T_b}(t - jT_b), \tag{1.15}$$

where $b_j^{(k)} \in \{+1, -1\}$. There is a distinct user signature sequence $a^{(k)}(t)$ associated with the kth user, which is similar to that of Equation 1.4, with the exception of a superscript, differentiating between users:

$$a^{(k)}(t) = \sum_{h=-\infty}^{\infty} a_h^{(k)} \Gamma_{T_c}(t - hT_c). \qquad (1.16)$$

The kth user's data signal $b^{(k)}(t)$ and signature sequence $a^{(k)}(t)$ are multiplied in order to produce an equivalent baseband wideband signal, namely,

$$s^{(k)}(t) = \sqrt{P_b^{(k)}} b^{(k)}(t) a^{(k)}(t), \qquad (1.17)$$

where $P_b^{(k)}$ is the average transmit power of the kth user's signal. The composite multi-user baseband received signal is:

$$r(t) = \sum_{k=1}^{K} \sqrt{P_b^{(k)}} b^{(k)}(t - \tau^{(k)}) a^{(k)}(t - \tau^{(k)}) + n(t), \qquad (1.18)$$

where $\tau^{(k)}$ is the propagation delay plus the relative transmission delay of the kth user with respect to other users, and $n(t)$ is the AWGN with a double-sided power spectral density of $\frac{N_0}{2}$ W/Hz.

1.2.4.1 Downlink Interference

In the downlink (base station to mobile), the base station is capable of synchronising the transmission of all the users' signals, such that the symbol durations are aligned with each other. Hence the composite signal is received at each mobile station with $\tau^{(k)} = 0$ for $k = 1, 2, \ldots, K$. This scenario is also known as symbol-synchronous transmission. Using the conventional so-called single-user detector, each symbol of the jth user is retrieved from the received signal $r(t)$ by correlating it with the jth user's spreading code in order to give:

$$\hat{b}_i^{(j)} = \mathrm{sgn}\left\{ \frac{1}{\sqrt{T_b}} \int_{iT_b}^{(i+1)T_b} r(t) a^{(j)^*}(t) dt \right\}. \qquad (1.19)$$

Substituting Equation 1.18 into Equation 1.19 yields:

$$\begin{aligned}
\hat{b}_i^{(j)} &= \mathrm{sgn}\left\{ \frac{1}{\sqrt{T_b}} \int_{iT_b}^{(i+1)T_b} \left[\sum_{k=1}^{K} \sqrt{P_b^{(k)}} b^{(k)}(t) a^{(k)}(t) + n(t) \right] a^{(j)^*}(t)\, dt \right\} \\
&= \mathrm{sgn}\left\{ \frac{1}{\sqrt{T_b}} \int_{iT_b}^{(i+1)T_b} \sqrt{P_b^{(j)}} b^{(j)}(t) a^{(j)}(t) a^{(j)^*}(t)\, dt \right. \\
&\quad + \frac{1}{\sqrt{T_b}} \int_{iT_b}^{(i+1)T_b} \sum_{\substack{k=1 \\ k \neq j}}^{K} \sqrt{P_b^{(k)}} b^{(k)}(t) a^{(k)}(t) a^{(j)^*}(t)\, dt \\
&\quad \left. + \frac{1}{\sqrt{T_b}} \int_{iT_b}^{(i+1)T_b} n(t) a^{(j)^*}(t)\, dt \right\} \\
&= \mathrm{sgn}\left\{ \underbrace{\sqrt{\xi_b^{(j)}} b_i^{(j)}}_{\text{wanted signal}} + \underbrace{\sum_{\substack{k=1 \\ k \neq j}}^{K} \sqrt{\xi_b^{(k)}} b_i^{(k)} R_{jk}}_{\text{multiple access interference}} + \underbrace{n^{(j)}}_{\text{white noise}} \right\},
\end{aligned} \tag{1.20}$$

where R_{jk} is the cross-correlation of the spreading codes of the kth and jth user for $iT_b \leq t \leq (i+1)T_b$, which is given by:

$$R_{jk} = \frac{1}{T_b} \int_0^{T_b} a^{(j)}(t) a^{(k)}(t)\, dt. \tag{1.21}$$

There will be no interference from the other users if the spreading codes are perfectly orthogonal to each other. That is, $R_{jk} = 0$ for all $k \neq j$. However, designing orthogonal codes for a large number of users is extremely complex. The so-called Walsh-Hadamard codes [63] used in the IS-95 system excel in terms of achieving orthogonality.

1.2.4.2 Uplink Interference

In contrast to the previously considered downlink scenario, in practical systems perfect orthogonality cannot be achieved in the uplink (mobile to base station), since there is no coordination in the transmission of the users' signals. In CDMA, all users transmit in the same frequency band in an uncoordinated fashion. Hence, $\tau^{(k)} \neq 0$, and the corresponding scenario is referred to as an asynchronous transmission scenario. In this case, the time-delay $\tau^{(k)}$, $k = 1, ..., K$, has to be included in the calculation. Without loss of generality, it can be assumed that $\tau^{(1)} = 0$ and that $0 < \tau^{(2)} < \tau^{(3)} < ... < \tau^{(K)} < T_b$. In contrast to the synchronous downlink scenario of Equation 1.19, the demodulation of the ith symbol of the jth user is performed by correlating the received signal $r(t)$ with $a^{(j)^*}(t)$ delayed by $\hat{\tau}^{(j)}$, yielding:

$$\hat{b}_i^{(j)} = \mathrm{sgn}\left\{ \frac{1}{\sqrt{T_b}} \int_{iT_b+\hat{\tau}^{(j)}}^{(i+1)T_b+\hat{\tau}^{(j)}} r(t) a^{(j)^*}(t-\hat{\tau}^{(j)}) dt \right\}, \tag{1.22}$$

where $\hat{\tau}^{(j)}$ is the estimated time-delay at the receiver.

Substituting Equation 1.18 into Equation 1.22 and assuming perfect code acquisition and tracking yield:[4]

$$\begin{aligned}
\hat{b}_i^{(j)} &= \mathrm{sgn}\left\{ \frac{1}{\sqrt{T_b}} \int_{iT_b+\tau^{(j)}}^{(i+1)T_b+\tau^{(j)}} \left[\sum_{k=1}^{K} \sqrt{P_b^{(k)}} b^{(k)}(t-\tau^{(k)}) a^{(k)}(t-\tau^{(k)}) \right.\right. \\
&\quad \left. + n(t) \right] \cdot a^{(j)^*}(t-\tau^{(j)}) dt \Big\} \\
&= \mathrm{sgn}\left\{ \frac{1}{\sqrt{T_b}} \left[\int_{iT_b+\tau^{(j)}}^{(i+1)T_b+\tau^{(j)}} \sqrt{P_b^{(j)}} b^{(j)}(t-\tau^{(j)}) a^{(j)}(t-\tau^{(j)}) \right.\right. \\
&\quad \times a^{(j)^*}(t-\tau^{(j)}) dt \\
&\quad + \sum_{k=1}^{j-1} \int_{(i+1)T_b+\tau^{(j)}}^{(i+1)T_b+\tau^{(k)}} \sqrt{P_b^{(k)}} b^{(k)}(t-\tau^{(k)}) a^{(k)}(t-\tau^{(k)}) \\
&\quad \times a^{(j)^*}(t-\tau^{(j)}) dt \\
&\quad + \sum_{k=1}^{j-1} \int_{iT_b+\tau^{(k)}}^{(i+1)T_b+\tau^{(j)}} \sqrt{P_b^{(k)}} b^{(k)}(t+T_b-\tau^{(k)}) a^{(k)}(t+T_b-\tau^{(k)}) \\
&\quad \times a^{(j)^*}(t-\tau^{(j)}) dt \\
&\quad + \sum_{k=j+1}^{K} \int_{iT_b+\tau^{(j)}}^{iT_b+\tau^{(k)}} \sqrt{P_b^{(k)}} b^{(k)}(t-T_b-\tau^{(k)}) a^{(k)}(t-T_b-\tau^{(k)}) \\
&\quad \times a^{(j)^*}(t-\tau^{(j)}) dt \\
&\quad + \sum_{k=j+1}^{K} \int_{iT_b+\tau^{(k)}}^{(i+1)T_b+\tau^{(j)}} \sqrt{P_b^{(k)}} b^{(k)}(t-\tau^{(k)}) a^{(k)}(t-\tau^{(k)}) \\
&\quad \left.\left. \times a^{(j)^*}(t-\tau^{(j)}) dt + \int_{iT_b+\tau^{(j)}}^{(i+1)T_b+\tau^{(j)}} n(t) a^{(j)^*}(t-\tau^{(j)}) dt \right] \right\}
\end{aligned} \tag{1.23}$$

[4]For perfect code acquisition and tracking, $\hat{\tau}^{(j)} = \tau^{(j)}$.

$$\hat{b}_i^{(j)} = \text{sgn}\left\{ \underbrace{\sqrt{\xi_b^{(j)}}b_i^{(j)}}_{\text{wanted signal}} + \underbrace{\sum_{k=1}^{j-1}\sqrt{\xi_b^{(k)}}b_i^{(k)}R_{jk}(0)}_{\text{multiple access interference}} \right.$$

$$\underbrace{+\sum_{k=1}^{j-1}\sqrt{\xi_b^{(k)}}b_{i+1}^{(k)}\hat{R}_{jk}(+1) + \sum_{k=j+1}^{K}\sqrt{\xi_b^{(k)}}b_{i-1}^{(k)}R_{jk}(-1)}_{\text{multiple access interference}}$$

$$\left. + \underbrace{\sum_{k=j+1}^{K}\sqrt{\xi_b^{(k)}}b_i^{(k)}\hat{R}_{jk}(0)}_{\text{multiple access interference}} + \underbrace{n^{(j)}}_{\text{white noise}} \right\}, \tag{1.24}$$

where $R_{jk}(i)$ and $\hat{R}_{jk}(i)$, $i \in \{+1, 0, -1\}$ represent the cross-correlation of the spreading codes due to asynchronous transmissions, which are given by [64]:

$$R_{jk}(i) = \frac{1}{T_b}\int_{\tau^{(j)}}^{\tau^{(k)}} a^{(j)}(t-\tau^{(j)})a^{(k)}(t+iT_b-\tau^{(k)})dt \tag{1.25}$$

and

$$\hat{R}_{jk}(i) = \frac{1}{T_b}\int_{\tau^{(k)}}^{T_b+\tau^{(j)}} a^{(j)}(t-\tau^{(j)})a^{(k)}(t+iT_b-\tau^{(k)})dt \tag{1.26}$$

and is limited to $+1, 0, -1$, since the maximum path delay is assumed to be limited to one symbol duration, as mentioned in Section 1.2.2.

Equations 1.24 and 1.20 represent the estimated demodulated data symbol of the jth user at the base station and mobile station, respectively. Both contain the desired symbol of the jth user. However, this is corrupted by noise and interference from the other users. This interference is known as multiple access interference (MAI). It contains the undesired interfering signals from the other $(K-1)$ users. The MAI arises due to the nonzero cross-correlation of the spreading codes. Ideally, the spreading codes should satisfy the orthogonality property such that

$$R_{jk}(\tau) = \frac{1}{T_b}\int_0^{T_b} a^{(k)}(t)a^{(j)}(t-\tau)dt = \begin{cases} 1 & \text{for } k=j, \tau=0 \\ 0 & \text{for all } k \text{ and all } \tau. \end{cases} \tag{1.27}$$

However, it is impossible to design codes that are orthogonal for all possible time offsets imposed by the asynchronous uplink transmissions. Thus there will always be MAI in the uplink. These observations are augmented by comparing the terms of Equations 1.20 and 1.24.

On the other hand, multipath interference is always present in both the forward and reverse link. Multipath interference is due to the different arrival times of the same signal via the different paths at the receiver. This is analogous to the signals transmitted from other users; hence, multipath interference is usually analysed in the same way as MAI.

As the number of users increases, the MAI increases too. Thus, the capacity of CDMA is known to be interference limited. CDMA is capable of accommodating additional users at the expense of a gradual degradation in performance in a fixed bandwidth, whereas TDMA or FDMA would require additional bandwidth to accommodate additional users. Intensive research has been carried out to find ways of mitigating the effects of MAI. Some of the methods include voice activity control, spreading code design, power control schemes, and sectored/adaptive antennas [65]. These methods reduce the MAI to a certain extent.

The most promising uplink method so far has been in the area of *multi-user detection*, which was first proposed by Verdú [66]. Multi-user detection [67–69], which will be discussed in more depth in the next chapter, invokes the knowledge of all users' signature sequences and all users' channel impulse response estimates in order to improve the detection of each individual user. The employment of this algorithm is more feasible for the uplink, because all mobiles transmit to the base station and the base station has to detect all the users' signals anyway. The topic of multi-user detection is however beyond the scope of this chapter, since it will be discussed in a little more detail in the next chapter, namely in Chapter 2. For a more indepth treatment the interested readers are referred to Verdu's excellent book [70], which provides a comprehensive discussion on the topic. A general review of the various multi-user detection schemes and further references can also be found, for example, in Moshavi's contribution [65].

Another shortcoming of CDMA systems is their susceptibility to the near-far problem to be highlighted below. If all users transmit at equal power, then signals from users near the base station are received at a higher power than those from users at a higher distance due to their different path-losses. The effects of fading highlighted in Section 1.2.2 also contribute to the power variation. Hence, according to Equation 1.24, if the jth user is transmitting from the cell border and all other users are transmitting near the base station, then the desired jth user's signal will be masked by the other users' stronger signals, which results in a high bit error rate. In order to mitigate this so-called near-far problem, power control is used to ensure that all signals from the users are received at near-equal power, regardless of their distance from the base station.

There are typically two basic types of power control [38]:

- open-loop power control
- closed-loop power control

Open-loop power control is usually used to overcome the variation in power caused by pathloss. On the other hand, closed-loop power control is used to overcome shadow fading caused by multipath. The details of the various power control techniques will not be elaborated on in this chapter. Readers may refer to [71] for more information.

1.2.4.3 Gaussian Approximation

In order to simplify any analysis involving multi-user transmission in CDMA, the MAI is usually assumed to be Gaussian distributed by virtue of the central limit theorem [72–74]. This assumption is fairly accurate even for $K < 10$ users, when the BER is 10^{-3} or higher.

We will use the standard Gaussian approximation theory presented by Pursley [72] to represent the MAI. When the desired user sequence is chip- and phase-synchronous with all the interfering sequences, where the phase-synchronous relationship is defined as in the absence of noise, the worst-case probability of error $Pr_b(\epsilon)$ performance was given by Pursley [72] as:

$$Pr_b(\epsilon) = Q\left[\sqrt{\frac{N_c}{(K-1)}}\right], \tag{1.28}$$

where $Q(\cdot)$ is the Gaussion Q-function of Equation 1.9, since the synchronous transitions do not generate pure random Gaussian-like impairments. This formula would be characteristic of the synchronous downlink scenario of Section 1.2.4.1. However, in practical uplink situations as augmented in Section 1.2.4.2, there is always some delay among the users, and each received signal will be phase-shifted independently. In this case, according to Pursley, the probability of error in the absence of noise will be [72]:

$$Pr_b(\epsilon) = Q\left[\sqrt{\frac{3N_c}{(K-1)}}\right]. \tag{1.29}$$

Equation 1.29 represents the best performance corresponding to Gaussian-like impairments. In between these two extremes are situations whereby, in the first case, the desired sequence and the interfering sequence are chip synchronous but not phase synchronous. The probability of error in the absence of noise is given by [72]:

$$Pr_b(\epsilon) = Q\left[\sqrt{\frac{2N_c}{(K-1)}}\right]. \tag{1.30}$$

In the second case, the desired sequence and interfering sequence are phase synchronous but not chip synchronous. Hence, the probability of error in the absence of noise is given by [72]:

$$Pr_b(\epsilon) = Q\left[\sqrt{\frac{3N_c}{2(K-1)}}\right]. \tag{1.31}$$

Analysing the above equations, it can be seen that by increasing the number of chips N_c per symbol, the performance of the system will be improved. However, there is a limitation to the rate of the spreading sequence based on Digital Signal Processing (DSP) technology. Figure 1.11 compares the simulated results with the numerical results given by Equations 1.28 to 1.31 for a binary system with a processing gain of 7. The figure shows that the assumption of Gaussian distributed MAI is valid, especially for a high number of users. It also demonstrates that *CDMA attains its best possible performance in an asynchronous multi-user transmission system*. This is an advantage over TDMA and FDMA because TDMA and FDMA require some coordination among the transmitting users, which increases the complexity of the system.

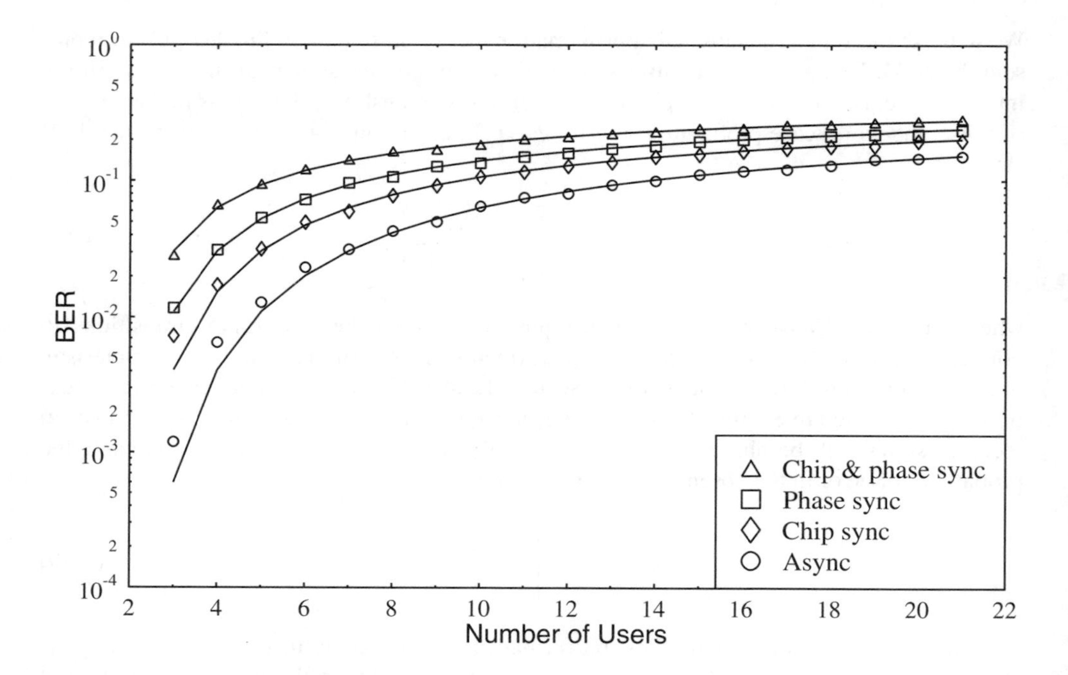

Figure 1.11: Probability of error against number of users using Equations 1.28, 1.29, 1.30, and 1.31. Markers: Simulation; solid line: Numerical computation. The processing gain is 7.

1.2.5 Spreading Codes

As seen previously, the choice of spreading codes plays an important role in DS-CDMA. The main criteria for selecting a particular set of user signature sequences in CDMA applications are that the number of possible different sequences in the set for any sequence length must be high in order to accommodate a high number of users in a cell. The spreading sequences must also exhibit low cross-correlations for the sake of reducing the multi-user interference during demodulation. A high autocorrelation main-peak to secondary-peak ratio — as indicated by Equation 1.27 — is also essential, in order to minimise the probability of so-called false alarms during code acquisition. This also reduces the self-interference among the diversity paths. Below we provide a brief overview of a few different spreading sequences.

1.2.5.1 m-sequences

Perhaps the most popular set of codes known are the m-sequences [5]. An m-sequence with a periodicity of $n = 2^m - 1$ can be readily generated by an m-stage shift register with linear feedback, as shown in Figure 1.12.

The tap coefficients $c_1, c_2, \ldots, c_m$ can be either 1 (short circuit) or 0 (open circuit). Information on the shift register feedback polynomials, describing the connections between the register stages and the modulo-2 adders can be found, for example, in [5]. Note that in spread spectrum applications, the output binary sequences of 0,1 are mapped into a bipolar sequence

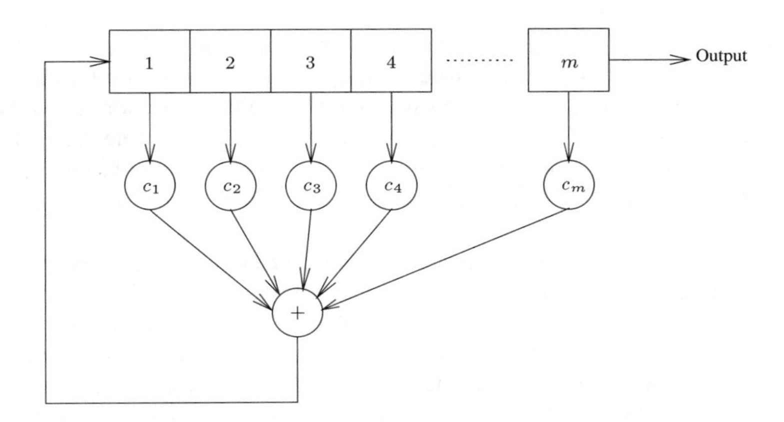

Figure 1.12: m-stage shift register with linear feedback.

m	Number of m-sequences	Peak Cross-Correlation	Number of Gold Sequences	Peak Cross-Correlation
3	2	5	$2^m + 1 = 9$	5
4	2	9	$2^m + 1 = 17$	9
5	6	11	$2^m + 1 = 33$	9
6	6	23	$2^m + 1 = 65$	17
7	18	41	$2^m + 1 = 129$	17
8	16	95	$2^m + 1 = 257$	33

Table 1.1: Properties of m- and Gold-Sequences. ©McGraw-Hill, 1995 [5]

of -1,1, respectively. Table 1.1 shows the total number of m-sequences and the associated chip-synchronous peak cross-correlation for m = 3, 4, 5, 6, 7, and 8. In this context, the peak cross-correlation quantifies the maximum number of identical chips in a pair of different spreading codes. It is desirable to have as low a number of code pairs as possible, which exhibit this peak cross-correlation. Furthermore, the peak cross-correlation has to be substantially lower than the codes' autocorrelation, which is given by the length of the code. In general, the cross-correlations of m-sequences are too high to be useful in CDMA. Another set of spreading codes, which exhibit fairly low chip-synchronous cross-correlations are the Gold sequences [5], which will be elaborated on in the next section.

1.2.5.2 Gold Sequences

Gold sequences [5] with a period of $n = 2^m - 1$ are derived from a pair of m-sequences having the same period. Out of the total number of possible m-sequences having a periodicity or length of n, there exists a pair of m-sequences, whose chip-synchronous cross-correlation equals to either $-1, -t(m)$ or $[t(m) - 2]$, where

$$t(m) = \begin{cases} 2^{(m+1)/2} + 1 & \text{odd } m \\ 2^{(m+2)/2} + 1 & \text{even } m. \end{cases} \tag{1.32}$$

This unique pair of m-sequences is commonly known as the pair of preferred codes. A set of $n = 2^m - 1$ sequences can be constructed by cyclically shifting a preferred code one chip at a time and then taking the modulo-2 summation with the other code for every chip shift. The resulting set of $n = 2^m - 1$ sequences together with the two preferred codes constitute a set of Gold sequences. Table 1.1 compares the total number of Gold sequences for m = 3, 4, 5, 6, 7, and 8, and their corresponding peak cross-correlation with the same parameters of m-sequences.

Table 1.1, shows that the Gold sequences exhibit equal or lower peak cross-correlation between different sequences of the set, in comparison to m-sequences for all m. There are also more Gold sequences than m-sequences for all values of m. Thus, Gold sequences are always preferred to m-sequences in CDMA applications, despite having a poorer asynchronous autocorrelation peak, which is a disadvantage in terms of both code acquisition and detection by correlators. Pseudo Noise (PN) sequences, such as m-sequences and Gold sequences, have periods of $N = 2^l - 1$ where l is the sequence length, which is a rather awkward number to match to the system clock requirements. Extended m-sequences having periods of 2^l solved this problem, an issue augmented below.

1.2.5.3 Extended m-sequences [75]

Extended m-sequences are derived from an m-sequence, generated by a linear feedback shift register, by adding an element into each period of the m-sequence. We will follow the notation, whereby the binary sequences of 0 and 1 are mapped to the corresponding bipolar sequences of -1 and +1, respectively. In order to arrive at zero-balanced extended m-sequences, which have a zero DC-component, the element to be inserted must be chosen so that the number of -1 s and +1 s within a period is the same. There are $2^m - 1$ positions in a period, where the additional element can be inserted. In [75], the element is inserted into the longest run of -1 s in a period. In an m-sequence of period $2^m - 1$, the longest run of -1 s is $n - 1 = 2^m - 2$. It was shown in [75] that the off-peak autocorrelation of extended sequences was similar to that of Gold sequences. However, the cross-correlation of different extended m-sequences at even-indexed chip-positions — that is, time-domain displacements — is similar to that of the m-sequences, which is much higher than that of the Gold sequences in Table 1.1. Thus, the extended m-sequences are not suitable in a multi-user environment, where the cross-correlation between the codes of different users is required to be as low as possible. Since this has a high impact on the user-capacity of cellular mobile systems, the additional hardware needed to synchronise the $N = 2^l - 1$ chip-duration m-sequences or Gold sequences with the system clock has to be tolerated and hence, extended m-sequences are not recommended in CDMA. Section 1.3.2.6 highlights the various spreading codes proposed for employment in the forthcoming 3G systems. In the next section we provide a rudimentary introduction to channel estimation for CDMA systems.

1.2.6 Channel Estimation

As mentioned earlier, accurate estimation of the channel parameters is vital in optimising the coherent demodulation. This channel parameter estimation process is an integral part of coherent demodulation, particularly in a multipath mobile radio environment. This is because the mobile radio channel changes randomly as a function of time, and thus the channel estimates have to be continuously estimated. This section describes various techniques used to estimate the channel path gains and phases, which will be referred to as channel coefficients. There are basically three practical channel coefficient estimation methods, each with their advantages and disadvantages, namely:

- Pilot-channel assisted, [76–78]
- Pilot-symbol assisted and [79]
- Pilot-symbol assisted decision-directed channel estimation [80],

which we briefly characterised in the following subsections.

1.2.6.1 Downlink Pilot-Assisted Channel Estimation

Channel estimation using a pilot channel/tone was proposed, for example, in [76–78], where a channel is dedicated solely for the purpose of estimating the multipath channel attenuations and delays. In order to prevent the pilot channel from interfering with the data channel, the pilot channel must either be allocated to a dedicated portion of the spectrum or share the spectrum with the data channel, but a spectral notch has to be created for accommodating the pilot. The former technique is known as the pilot tone-above-band (TAB) regime, while the latter is referred to as the transparent tone-in-band (TTIB) technique [76], both of which have been used in conventional single-carrier modems [12].

However, CDMA is more amenable to employing the TAB or TTIB techniques and their various derivatives, since the pilot signal can be transmitted in the same frequency band as the data signal by invoking orthogonal or quasi-orthogonal spreading codes. Hence, the pilot signal is treated as part of the MAI, and no notch filtering or additional pilot frequency band is required. In some 2G mobile systems, such as the IS-95 system this method is used on the downlink but not on the uplink. This is because it would be inefficient to have every mobile station transmitting their own pilot channel.

In 3G mobile systems, however, it was proposed [81] that a separate dedicated user control channel be transmitted simultaneously with the information channel, which could also be used as an alternative to the pilot channel – an issue to be elaborated on at a later stage. Suffice to say here that the main advantage of pilot-channel based channel estimation is that since the pilot channel is always present, the channel coefficients can be continuously estimated for every data symbol's demodulation. Hence, it is particularly useful for channels that are highly time-variant.

The block diagram of the channel estimator is shown in Figure 1.13, where $r(t)$ is the received signal and $a(t)$ is the spreading code. Assume that the known bit-stream is a continuous sequence of binary 1 s, then

$$\begin{aligned}\hat{\alpha}(k) &= \frac{1}{T_b}\int_{kT_b}^{(k+1)T_b} r(t)a^*(t)dt \\ &= \frac{1}{T_b}\int_{kT_b}^{(k+1)T_b} [\alpha(t)a(t)+n(t)]a^*(t)dt \\ &= \alpha(k)+\frac{1}{T_b}\int_{kT_b}^{(k+1)T_b} n(t)a^*(t)dt, \end{aligned} \quad (1.33)$$

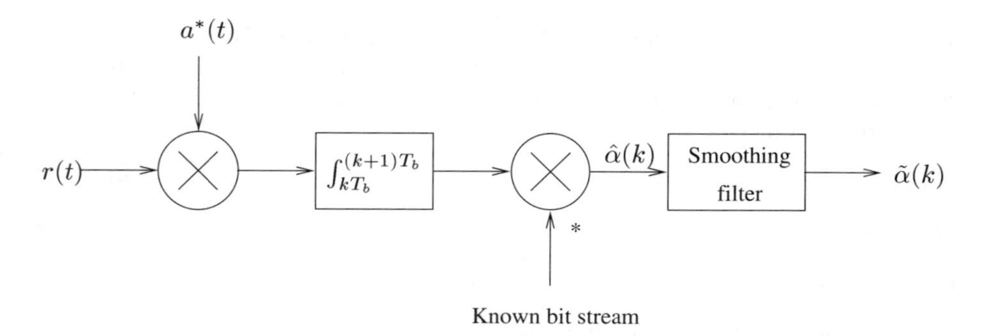

Figure 1.13: Structure of the channel estimator using known transmitted pilot symbols or bits.

where $\alpha(k)$ is the complex channel coefficient in the bit interval $kT_b \leq t < (k+1)T_b$. The variable $\hat{\alpha}(k)$ is termed the noisy channel estimate derived from the received signal contaminated by the noise element in the second term of Equation 1.33, while $\tilde{\alpha}(k)$ are estimates obtained from the output of the smoothing filter in Figure 1.13, which assists in averaging out the random effects of channel noise. Assuming that $n(t)$ is the AWGN having a zero mean (any MAI can be fairly accurately modelled also as AWGN [82]), averaging a large number of these noisy estimates will suppress the noise's influence. Several proposals have been published in the literature regarding the smoothing algorithm used in channel estimation, such as moving average [83, 84], least squares line fitting [85], low-pass filtering [79, 80, 85], and adaptive linear smoothing [86]. A more in-depth discourse on the TTIB technique was also given in Section 10.3.1 of [12] in the context of QAM. A compromise in terms of complexity and accuracy has to be made in selecting a particular algorithm. So far, only the downlink channel estimation has been elaborated on. The associated uplink issues are discussed next.

1.2.6.2 Uplink Pilot-Symbol Assisted Channel Estimation

Pilot-symbol assisted channel estimation was first proposed by Moher anand Lodge [79], and the first detailed analysis of this technique was carried out by Cavers [86]. Since then, several papers have been published, which analysed its effect on system performance [84, 85, 87]. This technique is the time-domain equivalent of the frequency-domain pilot channel-assisted TTIB method mentioned in Section 1.2.6.1, which was detailed in Section 10.3.2 of [12]. The advantage of this technique is that it dispensed with the use of a notch filter in the context of QAM modems, and so it did not result in an expanded bandwidth. However, for this technique, several parameters such as the number of pilot symbols or their periodicity has to be carefully chosen in order to trade-off the accuracy of estimation against the required pilot overhead. More explicitly, the pilot-spacing has to be sufficiently low to satisfy the Nyquist sampling theorem for the fading Doppler frequency encountered. This technique can be used for efficient coherent demodulation on the uplink, and Section 1.3.2.3 highlights how uplink channel estimation is carried out in the context of 3G systems.

The pilot symbols are multiplexed with the data stream periodically, as shown in Figure 1.14. This multiplexed stream is then transmitted to the base station from every communicating mobile station. The base station will extract the channel estimates from the known

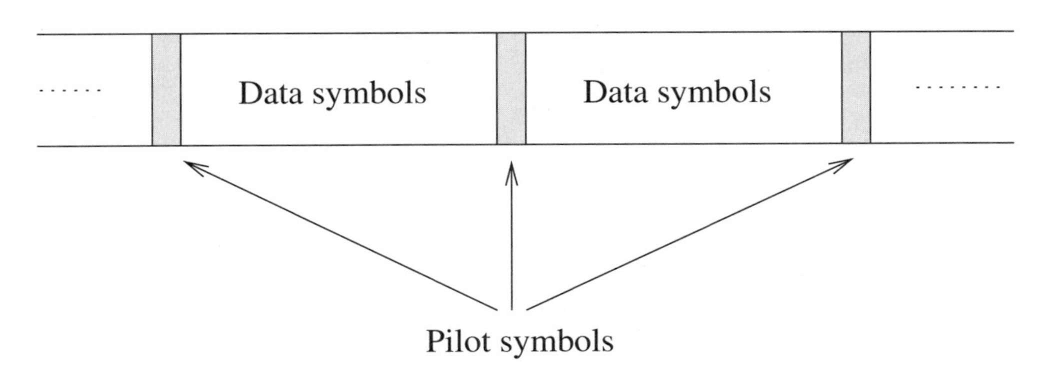

Figure 1.14: Data stream with embedded pilot symbols.

demodulated pilot symbols, and using, for example, ideal low-pass or simple linear interpolation [85], it will generate a channel magnitude and phase estimate for each uplink symbol. These channel estimates will then be used to 'de-fade', 'de-rotate', and demodulate the data symbols.

If the channel has a slow fading characteristic, such that it is more or less constant between consecutive pilot symbols, this method can be fairly accurate and of low complexity. However, the bandwidth efficiency is slightly compromised, since again, a sufficiently high number of pilots has to be incorporated in order to satisfy the Nyquist sampling criterion corresponding to the normalised Doppler frequency of the fading channel. For more information on this subject we refer to Section 10.3.2 of [12]. The above pilot-symbol assisted (PSA) concept is further developed in the next section.

1.2.6.3 Pilot-Symbol Assisted Decision-Directed Channel Estimation

Pilot-symbol assisted decision-directed channel estimation was first proposed by Irvine and McLane [80], and it was shown that it improves the accuracy of the estimation as compared with the original pilot symbol-assisted method of Section 1.2.6.2. It extends the concept of the pilot-symbol assisted channel estimation technique by using the detected data symbols in order to obtain the subsequent channel parameters, since in the absence of channel errors these demodulated data symbols can be considered to be known pilot symbols.

A decision-directed pilot-symbol assisted (PSA) scheme is illustrated in Figure 1.15, where $s(k)$ is the kth received symbol and $b(k)$ is the kth detected symbol. The signal is still transmitted in a transmission burst or frame format, similarly to that shown in Figure 1.14. At the beginning of the frame, the pilot symbols will be used to estimate the channel parameters in order to demodulate the data symbol immediately following the pilot symbol. This is performed by the pilot symbol-assisted channel estimator block of Figure 1.15. The detected data symbol $b(k)$ is then fed back and multiplied with its original but delayed received version $s(k)$, as seen in Figure 1.15. If this symbol is detected correctly, then it is analogous to a known pilot symbol and the channel coefficient corresponding to this received symbol can be estimated in the same way. This estimated channel coefficient is then passed through the smoothing filter of Figure 1.15 in order to obtain a smoothed estimate $\tilde{\alpha}(k)$ to be used in its conjugate form for de-fading and de-rotating the next symbol, as portrayed in Figure 1.15.

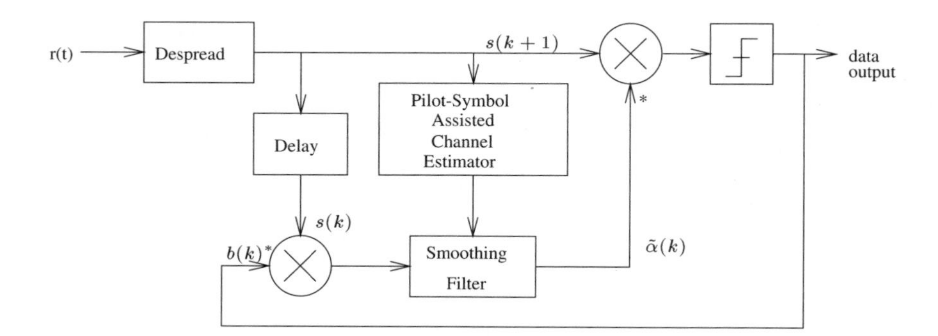

Figure 1.15: Receiver structure of PSA decision-directed channel estimation.

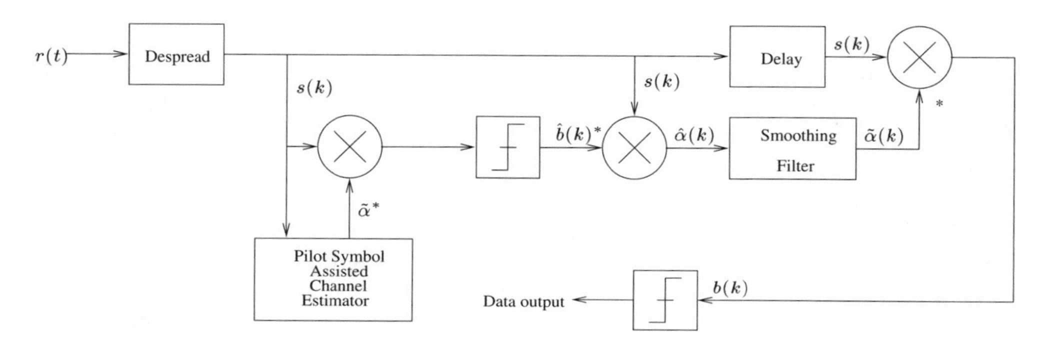

Figure 1.16: Receiver structure using decision-feedforward PSA channel estimation.

If the decision is wrong, obviously the estimated channel coefficient would be inaccurate. The effect of erroneous decisions is mitigated by the smoothing filter, which will suppress the effects of an occasional glitch due to the incorrect channel estimates. In the event that the smoothing filter is unable to average out the channel coefficient errors and its output is a complex channel coefficient, which is far from the actual value, then this error may propagate through the data stream, since the correct decoding of each data symbol is dependent on the accuracy of the previous channel coefficient estimates. In order to prevent this from happening, the smoothing process is reset when the next block of pilot symbols arrives. The averaging process will recommence with the pilot symbol-assisted channel estimates.

The schematic diagram shown in Figure 1.15 is only one of the few possibilities of implementing a decision-directed PSA channel estimation arrangement. This structure is also known as a decision-feedback PSA channel estimator because the estimated channel coefficient is used for compensating the channel's effects for the next symbol. In another version of this algorithm, shown in Figure 1.16, a tentative decision, $\hat{b}(k)$, is carried out concerning the current symbol, $s(k)$, using the pilot symbol-assisted estimate, $\tilde{\alpha}^*(k)$. Using this tentative decision concerning the received symbol $s(k)$, its corresponding channel coefficient estimate, $\hat{\alpha}(k)$, is derived from the product of $\hat{b}^*(k)$ and $s(k)$ in Figure 1.15 and averaged or smoothed with the aid of the previous estimates. The output of the smoothing filter is then multiplied with the received signal $s(k)$ again, in order to compensate the channel attenuation and phase rotation and hence to obtain the final decision, $b(k)$. Such an estimator is known as a feedforward estimator. This implementation is slightly more complicated but has the advantage of using the current estimate on the current symbol rather than tolerating a latency in the channel estimation process.

1.2.7 Summary

In this section we have briefly studied the fundamentals of a CDMA system. We have seen that several processes are vital in optimising the performance, such as spreading, channel estimation, code synchronisation, and power control. In the subsequent sections, we will make certain assumptions that will ease our analysis and simulation. These assumptions are:

- Perfect code acqusition and tracking. Hence, the transmitter and the intended receiver will always be synchronised for every path.
- Perfect channel estimation. This assumption will be used unless our focus is on the effects of imperfect estimation.
- Gaussian approximation of multi-user and multipath interference. This assumption will be used only in analysis and numerical computation, and will be validated by simulations performed in actual multi-user and multipath transmission scenarios. This also implies that random sequences will be considered instead of the deterministic sequences introduced in Section 1.2.5.
- On the uplink, the number of paths encountered by each user's signal is equal.
- Perfect power control is used. This implies that all users' signals will be received at the base station with equal power.

Following the above rudimentary considerations on PSA channel estimation, let us now review the third-generation (3G) mobile system proposals in the next section.

1.3 Third-Generation Systems

1.3.1 Introduction

The evolution of third-generation (3G) wireless systems began in the late 1980s when the International Telecommunication Union's Radiocommunication Sector (ITU-R) Task Group 8/1 defined the requirements for the 3G mobile radio systems. This initiative was then known as Future Public Land Mobile Telecommunication System (FPLMTS) [27, 34]. The frequency spectrum for FPLMTS was identified on a worldwide basis during the World Administrative Radio Conference (WARC) in 1992 [34], as the bands 1885–2025 MHz and 2110–2200 MHz.

The tongue-twisting acronym of FPLMTS was also aptly changed to IMT-2000, which refers to the International Mobile Telecommunications system in the year 2000. Besides possessing the ability to support services from rates of a few kbps to as high as 2 Mbps in a spectrally efficient way, IMT-2000 aimed to provide a seamless global radio coverage for global roaming. This implied the ambitious goal of aiming to connect virtually any two mobile terminals worldwide. The IMT-2000 system was designed to be sufficiently flexible in order to operate in any propagation environment, such as indoor, outdoor to indoor, and vehicular scenarios. It is also aiming to be sufficiently flexible to handle circuit as well as packet mode services and to handle services of variable data rates. In addition, these requirements must be fulfilled with a quality of service (QoS) comparable to that of the current wired network at an affordable cost.

Several regional standard organisations — led by the European Telecommunications Standards Institute (ETSI) in Europe, the Association of Radio Industries and Businesses (ARIB) in Japan, and the Telecommunications Industry Association (TIA) in the United States — have been dedicating their efforts to specifying the standards for IMT-2000. A total of 15 Radio Transmission Technology (RTT) IMT-2000 proposals were submitted to ITU-R in June 1998, five of which are satellite-based solutions, while the rest are terrestrial solutions. Table 1.2 shows a list of the terrestrial-based proposals submitted by the various organisations and their chosen radio access technology.

As shown in Table 1.2 most standardisation bodies have based their terrestrial oriented solutions on *Wideband-CDMA* (W-CDMA), due to its advantageous properties, which satisfy most of the requirements specified for 3G mobile radio systems. W-CDMA is aiming to provide improved coverage in most propagation environments in addition to an increased user capacity. Furthermore, it has the ability to combat — or to benefit from — multipath fading through RAKE multipath diversity combining [39–41]. W-CDMA also simplifies frequency planning due to its unity frequency reuse. A rudimentary discourse on the RTT proposals submitted by ETSI, ARIB, and TIA can be found in [11].

Recently, several of the regional standard organisations have agreed to cooperate and jointly prepare the Technical Specifications (TS) for the 3G mobile systems in order to assist as well as to accelerate the ITU process for standardisation of IMT-2000. This led to the formation of two Partnership Projects (PPs), which are known as 3GPP1 [88] and 3GPP2 [89]. 3GPP1 was officially launched in December 1998 with the aim of establishing the TS for IMT-2000 based on the evolved Global System for Mobile Telecommunications (GSM) [28] core networks and the UMTS[5] Terrestrial Radio Access (UTRA) RTT proposal. There are six organisational partners in 3GPP1: ETSI, ARIB, the China Wireless Telecommunication Standard (CWTS) group, the Standards Committee T1 Telecommunications (T1, USA), the Telecommunications Technology Association (TTA, Korea), and the Telecommunication Technology Committee (TTC, Japan). The first set of specifications for UTRA was released in December 1999, which contained detailed information on not just the physical layer aspects for UTRA, but also on the protocols and services provided by the higher layers. Here we will concentrate on the UTRA physical layer specifications, and a basic familiarity with CDMA principles is assumed.

In contrast to 3GPP1, the objective of 3GPP2 is to produce the TS for IMT-2000 based on the evolved ANSI-41 core networks, the cdma2000 RTT. 3GPP2 is spearheaded by TIA, and its members include ARIB, CWTS, TTA, and TTC. Despite evolving from completely diversified core networks, members from the two PPs have agreed to cooperate closely in order to produce a globally applicable TS for the 3G mobile systems.

This chapter serves as an overview of the UTRA specifications, which is based on the evolved GSM core network. However, information given here is by no means the final speci-

[5]UMTS, an abbreviation for Universal Mobile Telecommunications System, is a term introduced by ETSI for the 3G wireless mobile communication system in Europe.

fications for UTRA or indeed for IMT-2000. It is very likely that the parameters and technologies presented in this chapter will evolve further. Readers may also want to refer to a recent book by Ojanperä and Prasad [90], which addresses W-CDMA 3G mobile radio systems in more depth.

1.3.2 UMTS Terrestrial Radio Access (UTRA) [32, 88, 90–97]

Research activities for UMTS [27, 29, 31, 33, 91, 92, 98] within ETSI have been spearheaded by the European Union's (EU) sponsored programmes, such as the Research in Advanced Communication Equipment (RACE) [81, 99] and the Advanced Communications Technologies and Services (ACTS) [91, 98, 99] initiative. The RACE programme, which is comprised of two phases, commenced in 1988 and ended in 1995. The objective of this programme was to investigate and develop testbeds for the air interface technology candidates. The ACTS programme succeeded the RACE programme in 1995. Within the ACTS Future Radio Wideband Multiple Access System (FRAMES) project, two multiple access modes have been chosen for intensive study, as the candidates for UMTS terrestrial radio access (UTRA). They are based on Time Division Multiple Access (TDMA) with and without spreading, and on W-CDMA [30, 32, 100].

As early as January 1997, ARIB decided to adopt W-CDMA as the terrestrial radio access technology for its IMT-2000 proposal and proceeded to focus its activities on the detailed specifications of this technology [31]. Driven by a strong support behind W-CDMA worldwide and this early decision from ARIB, ETSI reached a consensus agreement in January 1998 to adopt W-CDMA as the terrestrial radio access technology for UMTS. In this section, we highlight the key features of the physical layer aspects of UTRA that have been developed since then. Most of the material in this section is based on an amalgam of [32, 88, 90–97].

1.3.2.1 Characteristics of UTRA

The proposed spectrum allocation for UTRA is shown in Figure 1.17. As can be seen, UTRA is unable to utilise the full frequency spectrum allocated for the 3G mobile radio systems during the WARC'92, since those frequency bands have also been partially allocated to the Digital Enhanced Cordless Telecommunications (DECT) systems. Also, the allocated frequency spectrum was originally based on the assumption that speech and low data rate transmission would be the dominant services offered by IMT-2000. However, this assumption has become invalid, as the trend has shifted toward services that require high-speed data transmission, such as Internet access and multimedia services. A study conducted by the UMTS Forum [101] forecasted that the current frequency bands allocated for IMT-2000 are only sufficient for the initial deployment until the year 2005. According to the current demand estimates, it was foreseen that an additional frequency spectrum of 187 MHz is required for IMT-2000 in high-traffic demand areas by the year 2010. This extension band will be identified during the World Radio Conference (WRC)-2000. Among the many candidate extension bands, the band 2520–2670 MHz has been deemed by many people to be the most likely to be chosen. Unlike other bands, which have already been allocated for use in other applications, this band was allocated to mobile services in all regions. Furthermore, the 150 MHz bandwidth available is sufficiently wide to satisfy most of the forecasted spectrum requirements.

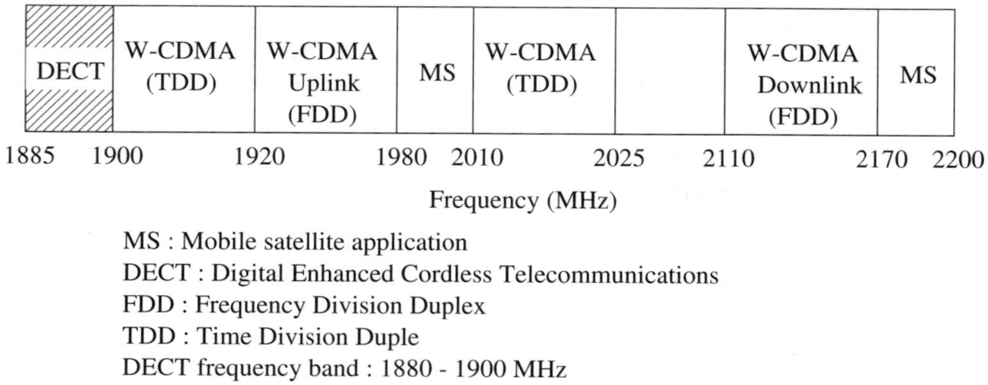

DECT	W-CDMA (TDD)	W-CDMA Uplink (FDD)	MS	W-CDMA (TDD)		W-CDMA Downlink (FDD)	MS

1885 1900 1920 1980 2010 2025 2110 2170 2200

Frequency (MHz)

MS : Mobile satellite application
DECT : Digital Enhanced Cordless Telecommunications
FDD : Frequency Division Duplex
TDD : Time Division Duple
DECT frequency band : 1880 - 1900 MHz

Figure 1.17: The proposed spectrum allocation in UTRA.

The radio access supports both *Frequency Division Duplex* (FDD) and *Time Division Duplex (TDD)* operations. The operating principles of these two schemes are augmented here in the context of Figure 1.18.

Specifically, the uplink (UL) and downlink (DL) signals are transmitted using different carrier frequencies f_1 and f_2, respectively, separated by a frequency guard band in FDD mode. On the other hand, the UL and DL messages in the TDD mode are transmitted using the same carrier frequency f_c, but in different time-slots, separated by a guard period. As seen from the spectrum allocation in Figure 1.17, the paired bands of 1920–1980 MHz and 2110–2170 MHz are allocated for FDD operation in the UL and DL, respectively, whereas the TDD mode is operated in the remaining unpaired bands [91]. The parameters designed for FDD and TDD operations are mutually compatible so as to ease the implementation of a dual-mode terminal capable of accessing the services offered by both FDD and TDD operators.

We note furthermore that recent research advocates the TDD mode quite strongly in the context of burst-by-burst adaptive CDMA modems [69], in order to adjust the modem parameters, such as the spreading factor or the number of bits per symbol on a burst-by-burst basis. This allows the system to more efficiently exploit the time-variant wireless channel capacity, hence maintaining a higher bits/s/Hz bandwidth efficiency. Furthermore, there have been proposals in the literature for allowing TDD operation in certain segments of the FDD spectrum as well, since FDD is incapable of surrendering the UL or DL frequency band of the duplex link, when the traffic demand is basically simplex. In fact, segmenting the spectrum in FDD/TDD bands inevitably results in some inefficiency in bandwidth utilisation terms, especially in case of asymmetric or simplex traffic, when only one of the FDD bands is required. Hence, the more flexible TDD link could potentially double the link's capacity by allocating all time-slots in one direction. The idea of eliminating the dedicated TDD band was investigated [102], where TDD was invoked within the FDD band by simply allowing TDD transmissions in either the UL or DL frequency band, depending on which one was less interfered. This flexibility is unique to CDMA, since as long as the amount of interference is not excessive, FDD and TDD can share the same bandwidth. This would be particularly feasible in the indoor scenario of [102], where the surrounding outdoor cell could be using FDD, while the indoor cell would reuse the same frequency band in TDD mode. The buildings'

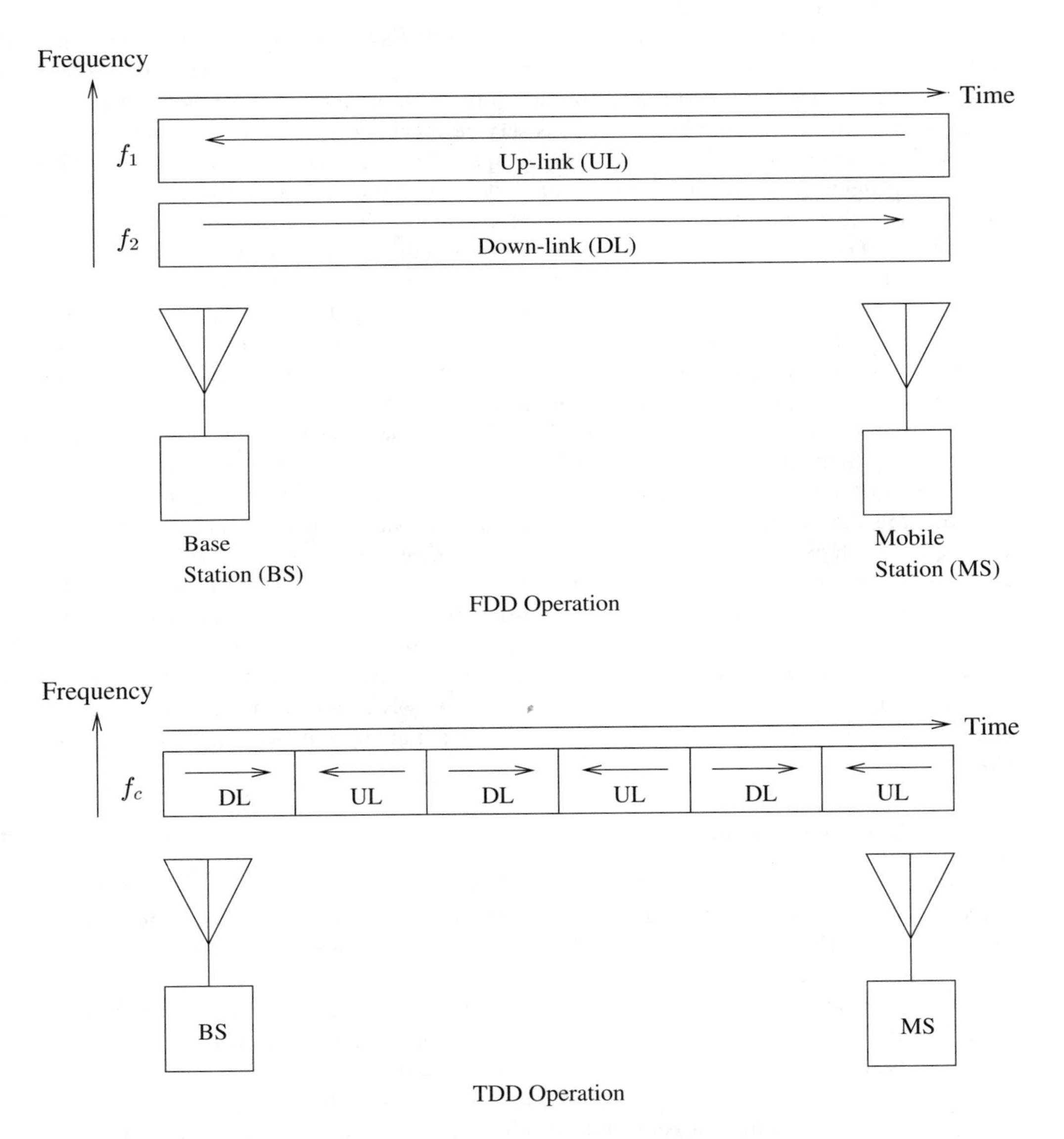

Figure 1.18: Principle of FDD and TDD operation.

walls and partitions could mitigate the interference between the FDD/TDD schemes.

Table 1.3.2.1 shows the basic parameters of the UTRA. Some of these parameters are discussed during our further discourse, but significantly more information can be gleaned concerning the UTRA system by carefully studying the table.

The UTRA system is operated at a basic chip rate of 3.84 Mcps,[6] giving a nominal bandwidth of 5 MHz, when using root-raised cosine Nyquist pulse-shaping filters with a rolloff factor of 0.22. UTRA fulfilled the requirements of 3G mobile radio systems by offering a range of user bit rates up to 2 Mbps. Various services having different bit rates and QoS can be readily supported using Orthogonal Variable Spreading Factor (OVSF) codes [103], which will be highlighted in Section 1.3.2.6.1, and service multiplexing, which will be discussed in Section 1.3.2.4. A key feature of the UTRA system, which was absent in the second-generation (2G) IS-95 system [63] was the use of a dedicated pilot sequence embedded in the users' data stream. These can be invoked in order to support the operation of adaptive antennas at the base station (BS), which was not facilitated by the common pilot channel of the IS-95 system. However, a common pilot channel was still retained in UTRA in order to provide the demodulator's phase reference for certain common physical channels, when embedding pilot symbols for each user is not feasible.

Regardless of whether a common pilot channel is used or dedicated pilots are embedded in the data, they facilitate the employment of *coherent detection*. Coherent detection is known to provide better performance than noncoherent detection [5]. Furthermore, the inclusion of short spreading codes enables the implementation of various performance enhancement techniques, such as interference cancellers and joint-detection algorithms, which results in excessive complexity in conjunction with long spreading codes. In order to support flexible system deployment in indoor and outdoor environments, *intercell-asynchronous operation* is used in the FDD mode. This implies that no external timing source, such as a reference signal or the Global Positioning System (GPS) is required. However, in the TDD mode intercell synchronisation is required in order to be able to seamlessly access the time-slots offered by adjacent BSs during handovers. This is achieved by maintaining synchronisation between the BSs.

1.3.2.2 Transport Channels

Transport channels are offered by the physical layer to the higher Open Systems Interconnection (OSI) layers, and they can be classified into two main groups, as shown in Table 1.4 [32, 91]. The Dedicated transport CHannel (DCH) is related to a specific Mobile Station (MS)-BS link, and it is used to carry user and control information between the network and an MS. Hence, the DCHs are bidirectional channels. There are six transport channels within the common transport channel group, as shown in Table 1.4. The Broadcast CHannel (BCH) is used to carry system- and cell-specific information on the DL to all MSs in the entire cell. This channel conveys information, such as the initial UL transmit power of the MS during a random access transmission and the cell-specific scrambling code, as we shall see in Section 1.3.2.7. The Forward Access CHannel (FACH) of Table 1.4 is a DL common channel used for carrying control information and short user data packets to MSs, if the system knows the serving BS of the MS. On the other hand, the Paging CHannel (PCH) of Table 1.4 is used to carry control information to an MS if the serving BS of the MS is unknown, in

[6]In the UTRA RTT proposal submitted by ETSI to ITU, the chip rate was actually set at 4.096 Mcps.

order to page the MS, when there is a call for the MS. The Random Access CHannel (RACH) of Table 1.4 is UL channel used by the MS to carry control information and short user data packets to the BS, in order to support the MS's access to the system, when it wishes to set up a call. The Common Packet CHannel (CPCH) is UL channel used for transmitting bursty data traffic in a contention-based random access manner. Lastly, as its name implies, the downlink Shared CHannel (DSCH) is a DL channel that is shared by several users.

The philosophy of these channels is fairly plausible, and it is informative as well as enlightening to explore the differences between the somewhat less flexible control regime of the 2G GSM [28] system and the more advanced 3G proposals, which we leave for the motivated reader due to lack of space. Unfortunately it is not feasible to design the control regime of a sophisticated mobile radio system by 'direct synthesis' and so some of the solutions reviewed throughout this section in the context of the 3G proposals may appear somewhat heuristic and quite ingenious. These solutions constitute an amalgam of the wireless research community's experience in the design of the existing 2G systems and of the lessons learned from their operation. Further contributing factors in the design of the 3G systems were based on solving the signalling problems specific to the favoured physical layer traffic channel solutions, namely, CDMA. In order to mention only one of them, the TDMA-based GSM system [28] was quite robust against power control inaccuracies, while the Pan-American IS-95 CDMA system [63] required an accurate power control. As we will see in Section 1.3.2.8, the power control problem was solved quite elegantly in the 3G proposals. We will also see that statistical multiplexing schemes — such as ALOHA, the original root of the recently more familiar Packet Reservation Multiple Access (PRMA) procedure — found their way into public mobile radio systems. A variety of further interesting solutions have also found applications in these 3G proposals, which are the results of the past decade of wireless system research. Let us now review the range of physical channels in the next section.

1.3.2.3 Physical Channels

The transport channels are transmitted using the physical channels. The physical channels are typically organised in terms of radio frames and time-slots, as shown in Figure 1.19. The philosophy of this hierarchical frame structure is also reminiscent to a certain degree of the GSM TDMA frame hierarchy of [28]. However, while in GSM each TDMA user had an exclusive slot allocation, in W-CDMA the number of simultaneous users supported is dependent on the users' required bit rate and their associated spreading factors. The MSs can transmit continuously in all slots or discontinuously, for example, when invoking a voice activity detector (VAD). Some of these issues will be addressed in Section 1.3.2.4.

As seen in Figure 1.19, there are 15 time-slots within each radio frame. The duration of each time-slot is $\frac{2}{3}$ ms, which gives a duration of 10 ms for the radio frame. As we shall see later in this section, the configuration of the information in the time-slots of the physical channels differs from one another in the UL and DL, as well as in the FDD and TDD modes. The 10 ms frame duration also conveniently coincides, for example, with the frame length of the ITU's G729 speech codec for speech communications, while it is a 'submultiple' of the GSM system's various full- and half-rate speech codecs' frame durations [28]. We also note that a convenient mapping of the video stream of the H.263 videophone codec can be arranged on the 10 ms-duration radio frames for supporting interactive video services, while on the move. Furthermore, the spreading factor (SF) can be varied on a 10 ms burst-by-burst (BbB)

Proposal	Description	Multiple Access	Source
DECT	Digital Enhanced Cordless Telecommunications	Multicarrier TDMA (TDD)	ETSI Project (EP) DECT
UWC-136	Universal Wireless Communications	TDMA (FDD and TDD)	USA TIA TR45.3
WIMS W-CDMA	Wireless Multimedia and Messaging Services Wideband CDMA	Wideband CDMA (FDD)	USA TIA TR46.1
TD-CDMA	Time Division Synchronous CDMA	Hybrid with TDMA/CDMA/ SDMA (TDD)	Chinese Academy of Telecommunication Technology (CATT)
W-CDMA	Wideband CDMA	Wideband DS-CDMA (FDD and TDD)	Japan ARIB
CDMA II	Asynchronous DS-CDMA	DS-CDMA (FDD)	South Korean TTA
UTRA	UMTS Terrestrial Radio Access	Wideband DS-CDMA (FDD and TDD)	ETSI SMG2
NA: W-CDMA	North America Wideband CDMA	Wideband DS-CDMA (FDD and TDD)	USA T1P1-ATIS
cdma2000	Wideband CDMA (IS-95)	DS-CDMA (FDD and TDD)	USA TIA TR45.5
CDMA I	Multiband synchronous DS-CDMA	Multiband DS-CDMA	South Korean TTA

Table 1.2: Proposals for the Radio Transmission Technology of Terrestrial IMT-2000 (obtained from ITU's web site: http://www.itu.int/imt).

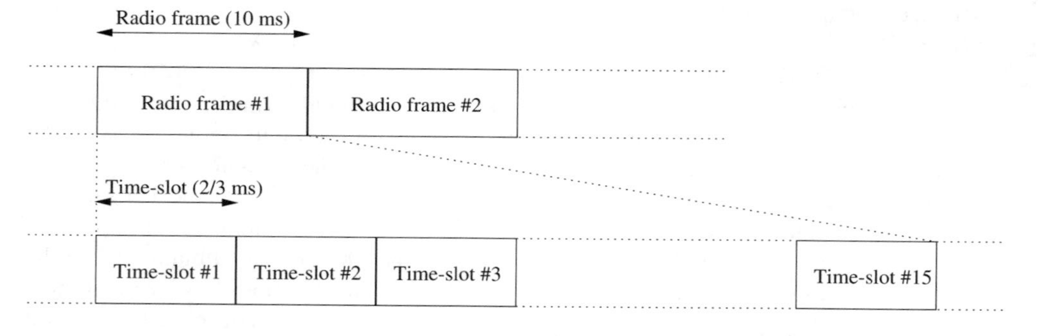

Figure 1.19: UTRA physical channel structure.

Radio Access Technology	FDD : DS-CDMA TDD : TDMA/CDMA
Operating environments	Indoor/Outdoor to indoor/Vehicular
Chip rate (Mcps)	3.84
Channel bandwidth (MHz)	5
Nyquist rolloff factor	0.22
Duplex modes	FDD and TDD
Channel bit rates (kbps)	FDD (UL) : 15/30/60/120/240/480/960 FDD (DL) : 15/30/60/120/240/480/960/1920 TDD (UL)[†] : variable, from 366 to 6624 TDD (DL)[†] : 366/414/5856/6624
Frame length	10 ms
Spreading factor	FDD (UL) : variable, 4 to 256 FDD (DL) : variable, 4 to 512 TDD (UL) : variable, 1 to 16 TDD (DL) : 1, 16
Detection scheme	Coherent with time-multiplexed pilot symbols Coherent with common pilot channel
Intercell operation	FDD : Asynchronous TDD : Synchronous
Power control	Inner-loop Open loop (TDD UL)
Transmit power dynamic range	80 dB (UL), 30 dB (DL)
Handover	Soft handover Inter-frequency handover

[†] Channel bit rate per time-slot.

Table 1.3: UTRA Basic Parameters

Dedicated Transport Channel	**Common Transport Channel**
Dedicated CHannel (DCH) (UL/DL)	Broadcast CHannel (BCH) (DL) Forward Access CHannel (FACH) (DL) Paging CHannel (PCH) (DL) Random Access CHannel (RACH) (UL) Common Packet CHannel (CPCH) (UL) downlink Shared CHannel (DSCH) (DL)

Table 1.4: UTRA transport channels.

basis, in order to adapt the transmission mode in harmony with channel quality fluctuations, while maintaining a given target bit error rate. Although it is not part of the standard proposal, we found that it was more beneficial to adapt the number of bits per symbol on a BbB basis than varying the SF [69].

In the FDD mode, a DL physical channel is defined by its spreading code and frequency. Furthermore, in the UL, the modem's orthogonal in-phase (I) and quadrature-phase (Q) branches are used to deliver the data and control information simultaneously in parallel

Dedicated Physical Channels	Transport Channels
Dedicated Physical Data CHannel (DPDCH) (UL/DL) †	DCH
Dedicated Physical Control CHannel (DPCCH) (UL/DL)	

Common Physical Channels	Transport Channels
Physical Random Access CHannel (PRACH) (UL)	RACH
Physical Common Packet CHannel (PCPCH) (UL)	CPCH
Common PIlot CHannel (CPICH) (DL)	
Primary Common Control Physical CHannel (P-CCPCH) (DL)	BCH
Secondary Common Control Physical CHannel (S-CCPCH) (DL)	FACH, PCH
Synchronisation CHannel (SCH) (DL)	
Physical Downlink Shared CHannel (PDSCH) (DL)	DSCH
Acquisition Indication CHannel (AICH) (DL)	
Page Indication CHannel (PICH) (DL)	

† On the DL, the DPDCH and DPCCH are time-multiplexed in each time slot to form a single Dedicated Physical CHannel (DPCH).

Table 1.5: Mapping the transport channels of Table 1.4 to the UTRA physical channels.

(as will be augmented in Figure 1.37). Thus, knowledge of the relative carrier phase, namely whether the I or Q branch is involved, constitutes part of the physical channel's identifier. On the other hand, in the TDD mode, a physical channel is defined by its spreading code, frequency, and time-slot.

Similarly to the transport channels of Table 1.4, the physical channels in UTRA can also be classified as dedicated and common channels. Table 1.5 shows the type of physical channels and the corresponding mapping of transport channels on the physical channels in UTRA.

1.3.2.3.1 Dedicated Physical Channels The dedicated physical channels of UTRA shown in Table 1.5 consist of the Dedicated Physical Data CHannel (DPDCH) and Dedicated Physical Control CHannel (DPCCH), both of which are bidirectional. The time-slot structures of the UL and DL dedicated physical channels are shown in Figures 1.20 and 1.21, respectively. Notice that on the DL, as illustrated by Figure 1.21, the DPDCH and DPCCH are interspersed by time-multiplexing to form a single Dedicated Physical CHannel (DPCH), as will be augmented in the context of Figure 1.38. On the other hand, the DPDCH and DPCCH on the UL are transmitted in parallel on the I and Q branches of the modem, as will become more explicit in the context of Figure 1.37 [32]. The reason for the parallel transmission on the UL is to avoid Electromagnetic Compatibility (EMC) problems due to Discontinuous Transmission (DTX) of the DPDCH of Table 1.5 [31]. DTX occurs when temporarily there are no data to transmit, but the link is still maintained by the DPCCH. If the UL DPCCH is time-multiplexed with the DPDCH, as in the DL of Figure 1.21, this can create short, sharp energy spikes. Since the MS may be located near sensitive electrical equipment, these spikes may affect this equipment.

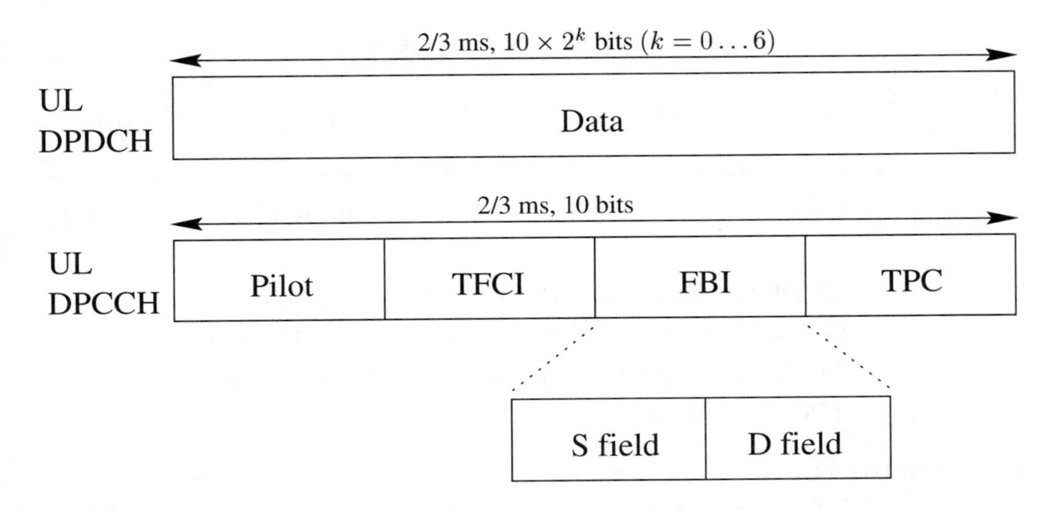

Figure 1.20: UTRA UL FDD dedicated physical channels time-slot configuration, which is mapped to the time-slots of Figure 1.19. The UL DPDCH and DPCCH messages are transmitted in parallel on the I and Q branches of the modem of Figure 1.37. By contrast, the DPDCH and DPCCH bursts are time-multiplexed on the DL as shown in Figure 1.21.

The DPDCH is used to transmit the DCH information between the BS and MS, while the DPCCH is used to transmit the Layer 1 information, which includes the pilot bits, Transmit Power Control (TPC) commands, and an optional Transport-Format Combination Indicator (TFCI), as seen in Figures 1.20 and 1.21. In addition, on the UL the Feedback Information (FBI) is also mapped to the DPDCH in Figure 1.20. The pilot bits are used to facilitate coherent detection on both the UL and DL as well as to enable the implementation of performance enhancement techniques, such as adaptive antennas and interference cancellation. Since the pilot sequences are known, they can also be used as frame synchronisation words in order to maintain transmission frame synchronisation between the BS and MS. The TPC commands support an agile and efficient power control scheme, which is essential in DS-CDMA using the techniques to be highlighted in Section 1.3.2.8. The TFCI carries information concerning the instantaneous parameters of each transport channel multiplexed on the physical channel in the associated radio frame. The FBI is used to provide the capability to support certain transmit diversity techniques. The FBI field is further divided into two smaller fields as shown in Figure 1.20, which are referred to as the S field and D field. The S field is used to support the *Site Selection DiversiTy* (SSDT), which can reduce the amount of interference caused by multiple transmissions during a soft handover operation, while assisting in fast cell selection.

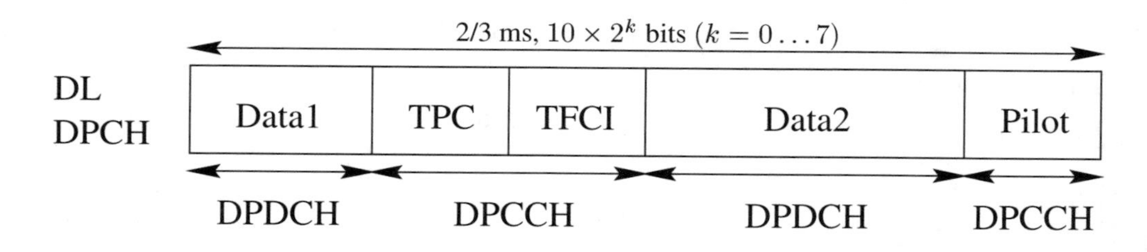

DPDCH : Dedicated Physical Data CHannel
DPCCH : Dedicated Physical Control CHannel
TFCI : Transport-Format Combination Indicator
TPC : Transmit Power Control

Figure 1.21: UTRA DL FDD dedicated physical channels time-slot configuration, which is mapped to the time-slots of Figure 1.19. The DPDCH and DPCCH messages are time-multiplexed on the DL, as it will be augmented in Figure 1.38. By contrast, the UL DPDCH and DPCCH bursts are transmitted in parallel on the I and Q branches of the modem as shown in Figure 1.20.

On the other hand, the D field is used to provide attenuation and phase information in order to facilitate *closed-loop transmit diversity*, a technique highlighted in Section 1.3.4.1.3. Given that the TPC and TFCI segments render the transmission packets 'self-descriptive', the system becomes very flexible, supporting burst-by-burst adaptivity, which substantially improves the system's performance [69], although this side-information is vulnerable to transmission errors.

The parameter k in Figures 1.20 and 1.21 determines the number of bits in each time-slot, which in turn corresponds to the bit rate of the physical channel. Therefore, the channel bit rates available for the UL DPDCH are 15/30/60/120/240/480/960 kbps, due to the associated 'payload' of 10×2^k bits per $\frac{2}{3}$ ms burst in Figure 1.20, where $k = 0 \ldots 6$. Note that the UL DPCCH has a constant channel bit rate of 15 kbps. Similarly, the channel bit rates available for the DL DPCH are 15/30/60/120/240/480/960 and 1920 kbps. However, since the user data are time-multiplexed with the Layer 1 control information, the actual user data rates on the DL will be slightly lower than those mentioned above. Even higher channel bit rates can be achieved using a technique known as multicode transmission [104], which will be highlighted in more detail in the context of Figure 1.35 in Section 1.3.2.5. Let us now consider the common physical channels summarised in Table 1.5.

1.3.2.3.2 Common Physical Channels

1.3.2.3.2.1 Common Physical Channels of the FDD Mode The Physical Random Access CHannel (PRACH) of Table 1.5 is used to carry the RACH message on the UL. A random access transmission is activated whenever the MS has data to transmit and wishes to establish a connection with the local BS. Although the procedure of this transmission will be elaborated on in Section 1.3.2.7, here we will briefly highlight the structure of a random

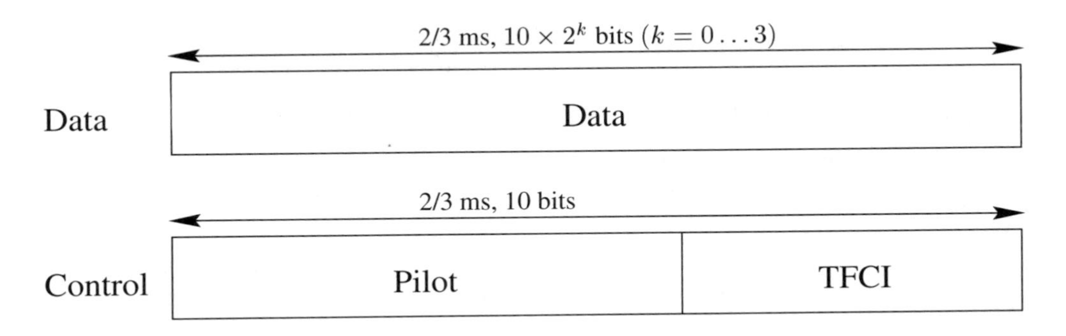

TFCI : Transport-Format Combination Indicator

Figure 1.22: The time-slot configuration of the message part during a random access transmission in UTRA, which are mapped to the frame structure of Figure 1.19. The data and control information are multiplexed on the I/Q channels of the modulator and the frame is transmitted at the beginning of an access slot, as it will be augmented in Section 1.3.2.7.1.

access transmission burst. Typically, a random access burst consists of one or several so-called preambles and a message. Each preamble contains a signature that is constructed of 256 repetitions of a 16-chip Hadamard code, which yields a $256 \times 16 = 4096$-chip-long signature. Similarly to the UL dedicated physical channels of Figure 1.20, the message part of the random access transmission consists of data information and control information that are transmitted in parallel on the I/Q channels of the modulator, as shown in Figure 1.22. The channel bit rates available for the data part of the message are 15/30/60/120 kbps. By contrast, the control information, which contains an 8-bit pilot and a 2-bit TFCI, is transmitted at a fixed rate of 15 kbps. Obviously in this case, no FBI and TPC commands are required, since transmission is initiated by the MS.

The Physical Common Packet CHannel (PCPCH) of Table 1.5 is used to carry the CPCH message on the UL, based on a Digital Sense Multiple Access-Collision Detection (DSMA-CD) random access technique. A CPCH random access burst consists of one or several Access Preambles (A-P), one Collision Detection Preamble (CD-P), a DPCCH Power Control Preamble (PC-P), and a message. The length of both the A-P and CD-P spans a total of 4096 chips, while the duration of the PC-P can be equivalent to either 0 or 8 time-slots. Each time-slot of the PC-P contains the pilot, the FBI, and the TPC bits. The message part of the CPCH burst consists of a data part and a control part, which is identical to the UL dedicated physical channel shown in Figure 1.20 in terms of its structure and available channel bit rates. A 15 kbps DL DPCH is always associated with an UL PCPCH. Hence, both the FBI and TPC information are included in the message of a CPCH burst in order to facilitate a DL transmit diversity and power control, unlike a RACH burst. The procedure of a CPCH transmission will be further elaborated in Section 1.3.2.7.

The DL Primary Common Control Physical CHannel (P-CCPCH) of Table 1.5 is used by the BS in order to broadcast the BCH information at a fixed rate of 30 kbps to all MSs in the cell. The P-CCPCH is transmitted only after the first 256 chips of each slot, as shown in Figure 1.23. During the first 256 chips of each slot, the Synchronisation CHannel (SCH)

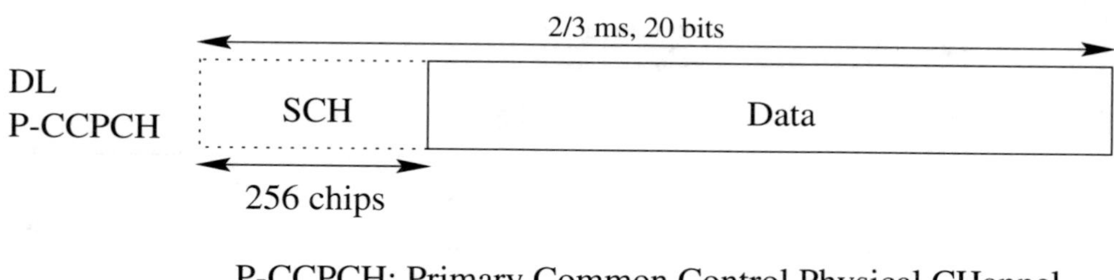

Figure 1.23: UTRA DL FDD Primary Common Control Physical CHannel (P-CCPCH) time-slot configuration, which is mapped to the time-slots of Figure 1.19.

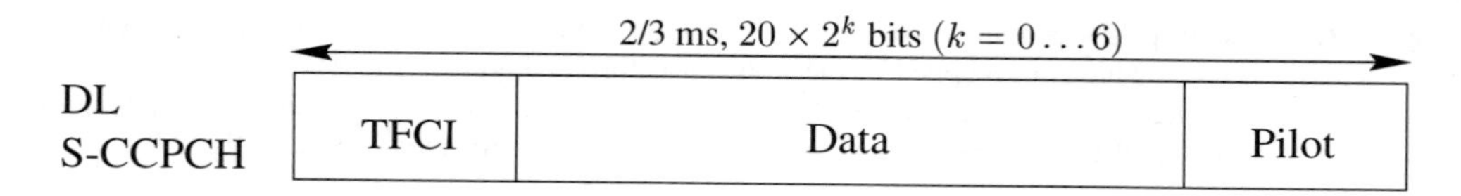

Figure 1.24: UTRA DL FDD Secondary Common Control Physical CHannel (S-CCPCH) time-slot configuration, which is mapped to the time-slots of Figure 1.19.

message is transmitted instead, as will be discussed in Section 1.3.2.9. The P-CCPCH is used as a timing reference directly for all the DL physical channels and indirectly for all the UL physical channels. Hence, as long as the MS is synchronised to the DL P-CCPCH of a specific cell, it is capable of detecting any DL messages transmitted from that BS by listening at the predefined times. For example, the DL DPCH will commence transmission at an offset, which is a multiple of 256 chips from the start of the P-CCPCH radio frame seen in Figure 1.23. Upon synchronisation with the P-CCPCH, the MS will know precisely when to begin receiving the DL DPCH. The UL DPDCH/DPCCH is transmitted 1024 chips after the reception of the corresponding DL DPCH.

The Secondary Common Control Physical CHannel (S-CCPCH) of Table 1.5 carries the FACH and PCH information of Table 1.4 on the DL, and they are transmitted only when data are available for transmission. The S-CCPCH will be transmitted at an offset, which is a multiple of 256 chips from the start of the P-CCPCH message seen in Figure 1.23. This will allow the MS to know exactly when to detect the S-CCPCH, as long as the MS is synchronised to the P-CCPCH. The time-slot configuration of the S-CCPCH is shown in Figure 1.24. Notice that the S-CCPCH message can be transmitted at a variable bit rate, namely, at 30/60/120/240/480/960/1920 kbps.

At this stage it is worth mentioning that the available control channel rates are significantly higher in the 3G systems than in their 2G counterparts. For example, the maximum BCH signalling rate in GSM [28] is more than an order of magnitude lower than the above-

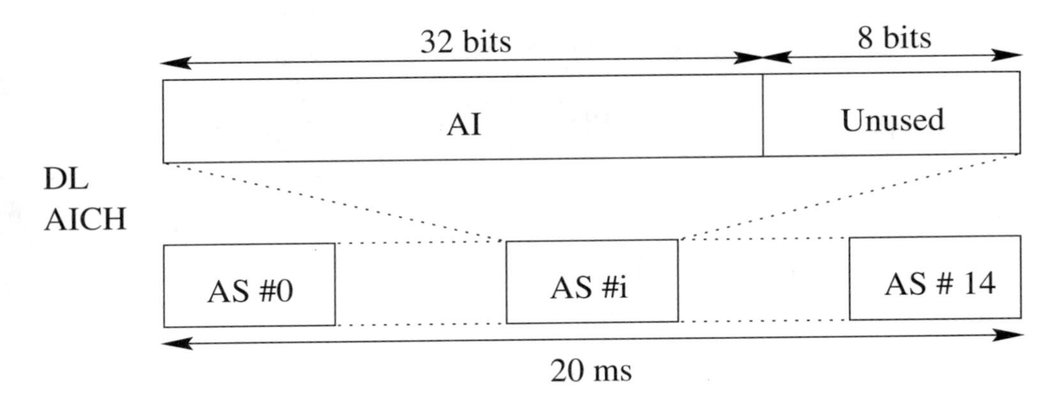

Figure 1.25: UTRA DL Acquisition Indicator CHannel (AICH) Access Slot (AS) configuration, which is mapped to the corresponding AS of the AICH. Due to its duration of 20 ms, it is mapped to every other 10 ms frame in Figure 1.19.

mentioned 30 kbps UTRA BCH rate. In general, this increased control channel rate will support a significantly more flexible system control than the 2G systems.

The Physical downlink Shared CHannel (PDSCH) of Table 1.5 is used to carry the DSCH message at rates of 30/60/120/240/480/960/1920 kbps. The PDSCH is shared among several users based on code multiplexing. The Layer 1 control information is transmitted using the associated DL DPCH.

The Acquisition Indicator CHannel (AICH) of Table 1.5 and the Page Indicator CHannel (PICH) are used to carry Acquisition Indicator (AI) and Page Indicator (PI) messages, respectively. More specifically, the AI is a response to a PRACH or PCPCH transmission, and it corresponds to the signature used by the associated PRACH preamble, a PCPCH A-P or a PCPCH CD-P, which were defined above. The AICH consists of a repeated sequence of 15 consecutive Access Slots (AS). Each AS consists of a 32-symbol AI part and an eight-symbol unused part, as shown in Figure 1.25. The AS#0 will commence at the start of every other 10 ms P-CCPCH radio frame seen in Figure 1.19, since its duration is 20 ms.

A PI message is used to signal to the MS on the associated S-CCPCH that there are data addressed to it, in order to facilitate a power-efficient sleep-mode operation. A PICH, illustrated in Figure 1.26, is a 10 ms frame consisting of 300 bits, out of which 288 bits are used to carry PIs, while the remaining 12 bits are unused. Each PICH frame can carry a total of N PIs, where $N = 18, 36, 72,$ and 144. The PICH is also transmitted at an offset with respect to the start of the P-CCPCH, which is a multiple of 256 chips. The associated S-CCPCH will be transmitted 7680 chips later.

Finally, the Common PIlot CHannel (CPICH) of Table 1.5 is a 30 kbps DL physical channel that carries a predefined bit sequence. It provides a phase reference for the SCH, P-CCPCH, AICH, and PICH, since these channels do not carry pilot bits, as shown in Fig-

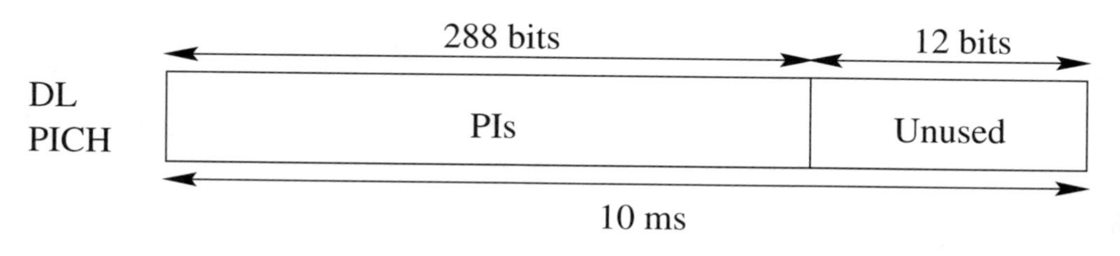

Figure 1.26: UTRA DL Page Indicator CHannel (PICH) configuration. Each PICH frame can carry a total of N PIs, where $N = 18, 36, 72$, and 144.

ures 1.23, 1.25, and 1.26, respectively. The PICH is transmitted synchronously with the P-CCPCH.

1.3.2.3.2.2 Common Physical Channels of the TDD Mode In contrast to the previous FDD structures of Figures 1.20–1.26, in TDD operation the burst structure of Figure 1.27 is used for all the physical channels, where each time-slot's transmitted information can be arbitrarily allocated to the DL or UL, as shown in the three possible TDD allocations in Figure 1.28. Hence, this flexible allocation of the UL and DL burst in the TDD mode enables the use of an adaptive modem [69, 105] whereby the modem parameters, such as the spreading factor or the number of bits per symbol can be adjusted on a burst-by-burst basis to optimise the link quality. A symmetric UL/DL allocation refers to a scenario in which an approx-

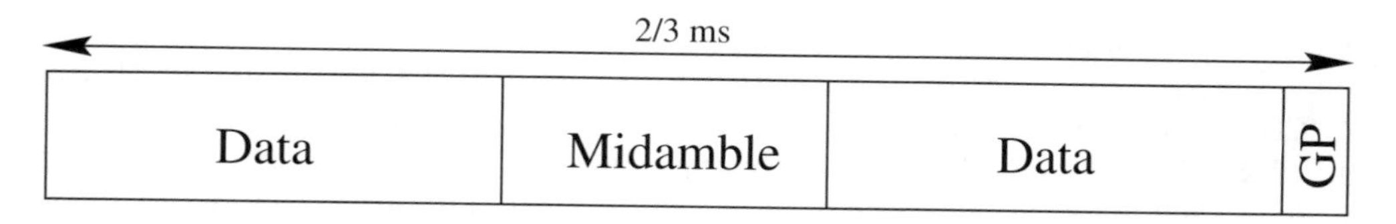

Figure 1.27: Burst configuration mapped on the TDD burst structure of Figure 1.28 in the UTRA TDD mode. Two different types of TDD bursts are defined in UTRA, namely, Burst Type 1 and Burst Type 2.

imately equal number[7] of DL and UL bursts are allocated within a TDD frame, while in

[7] Since there are 15 time-slots per frame, there will always be one more additional DL or UL burst per frame in a symmetric allocation.

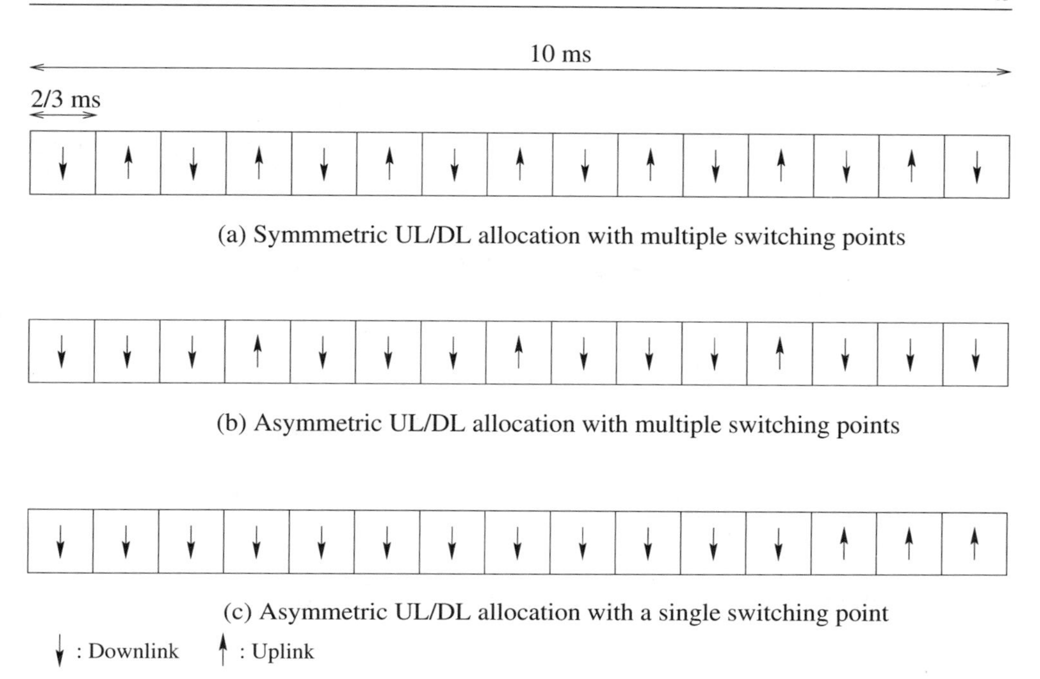

Figure 1.28: uplink/downlink allocation examples for the 15 slots in UTRA TDD operation using the time-slot configurations of Figure 1.27.

asymmetric UL/DL allocation, there is an unequal number of UL and DL bursts, such as, for example, in 'near-simplex' file download from the Internet or video-on-demand.

In UTRA, two different TDD burst structures, known as *Burst Type 1* and *Burst Type 2*, are defined, as shown in Figure 1.27. The Type 1 burst has a longer midamble (512 chips) than the Type 2 burst (256 chips). However, both types of bursts have an identical *Guard Period* (GP) of 96 chips. The midamble sequences that are allocated to the different TDD bursts in each time-slot belong to a so-called *midamble code set*. The codes in each midamble code set are derived from a unique *Basic Midamble Code*. Adjacent cells are allocated different midamble code sets, that is, different basic midamble code. This can be exploited to assist in cell identification.

Unlike in the FDD mode, there is only one type of Dedicated Physical CHannel (DPCH) in the TDD mode. Hence, the Layer 1 control information — such as the TPC command and the TFCI information — will be transmitted in the data field of Figure 1.27, if required. The TDD burst structures that incorporate the TFCI information as well as the TFCI+TPC information are shown in Figure 1.29. This should be contrasted with their corresponding FDD allocations in Figures 1.20 and 1.21. The TFCI field is divided into two parts, which reside immediately before and after the midamble (or after the TPC command, if power control is invoked) in the data field. The TPC command is always transmitted immediately after the midamble, as portrayed in Figure 1.29. As a result of these control information segments, the amount of user data is reduced in each time-slot. Note that the TPC command is only transmitted on the UL and only once per 10 ms frame for each MS.

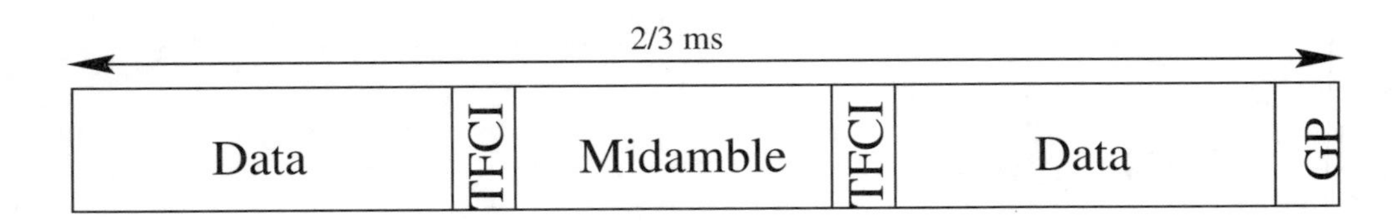

Burst structure with TFCI information only

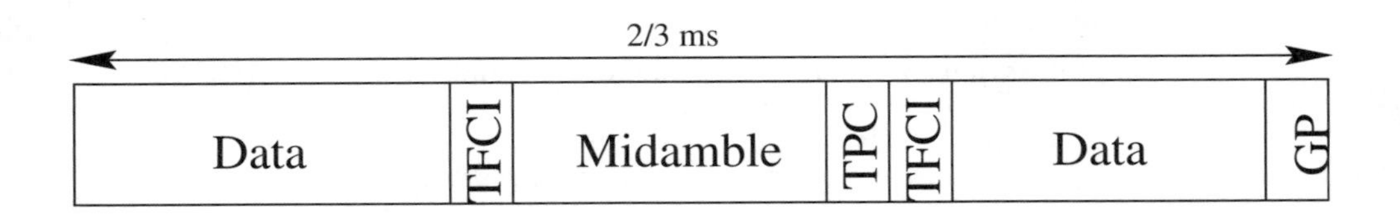

Burst structure with TFCI and TPC information

GP : Guard Period
TFCI : Transport Format Combination Indicator
TPC : Transmit Power Control

Figure 1.29: Burst configuration mapped on the TDD burst configuration of Figure 1.28 in the UTRA TDD mode incorporating TFCI and/or TPC information.

In contrast to the FDD mode, the SCH in the TDD mode is not time-multiplexed onto the P-CCPCH of Table 1.5. Instead, the SCH messages are transmitted on one or two time-slots per frame.[8] The P-CCPCH will be code-multiplexed with the first SCH time-slot in each frame.

Having highlighted the basic features of the various UTRA channels, let us now consider how the various services are error protected, interleaved, and multiplexed on to the physical channels. This issue is discussed with reference to Figures 1.30 and 1.31 in the context of UTRA.

1.3.2.4 Service Multiplexing and Channel Coding in UTRA

Service multiplexing is employed when multiple services of identical or different bit rates requiring different QoS belonging to the same user's connection are transmitted. An example would be the simultaneous transmission of a voice and video service for a multimedia application. Each service is represented by its corresponding transport channels, as described in

[8]If two time-slots are allocated to the SCH per frame, they will be spaced seven slots apart.

Transport Channels	Channel-Coding Schemes	Coding Rate
BCH, PCH, RACH	Convolutional code	1/2
CPCH,DCH,DSCH,FACH	Convolutional code Turbo code No coding	1/3, 1/2 1/3

Table 1.6: UTRA Channel-Coding Parameters for the Channels of Table 1.4

Section 1.3.2.2. The coding and multiplexing of the transport channels are performed in sets of transport blocks that arrived from the higher layers at fixed intervals of 10, 20, 40 or 80 ms. These intervals are known as the *Transmission Time Interval* (TTI). Note that the number of bits on each transport channel can vary between different TTIs, as well as between different transport channels. A possible method of transmitting multiple services is by using code-multiplexing with the aid of orthogonal codes. Every service could have its own DPDCH and DPCCH, each assigned to a different orthogonal code. This method is not very efficient, however, since a number of orthogonal codes would be reserved by a single user, while on the UL it would also inflict self-interference when the multiple DPDCH and DPCCH codes' orthogonality is impaired by the fading channel. Alternatively, these services can be time-multiplexed into one or several DPDCHs, as shown in Figures 1.30 and 1.31 for the UL and DL, respectively. The function of the individual processing steps is detailed below.

1.3.2.4.1 CRC Attachment A Cyclic Redundancy Checksum (CRC) is first calculated for each incoming transport block within a TTI. The CRC consists of either 24, 16, 12, 8, or 0 parity bits, which is decided by the higher layers. The CRC is then attached to the end of the corresponding transport block in order to facilitate reliable error detection at the receiver. This facility is very important, for example, for generating the video packet acknowledgement flag in wireless video telephony using standard video codecs, such as H.263 [106].

1.3.2.4.2 Transport Block Concatenation Following the CRC attachment, the incoming transport blocks within a TTI are serially concatenated in order to form a code block. If the number of bits exceeds the maximum code block length, denoted as Z, then the code block is segmented into shorter ones and filler bits (zeros) are added to the last code block, if neccessary, in order to generate code blocks of the same length. The maximum code block length Z is dependent on the type of channel-coding invoked. For convolutional coding $Z = 504$, while for turbo coding $Z = 5114$, since turbo codes require a long coded block length [107]. If no channel-coding is invoked, then the code block can be of unlimited length.

1.3.2.4.3 Channel-Coding Each of the code blocks is then delivered to the channel-coding unit. Several Forward Error Correction (FEC) techniques are proposed for channel-coding. The FEC technique used is dependent on the QoS requirement of that specific transport channel. Table 1.6 shows the various types of channel-coding techniques invoked for different transport channels. Typically, *convolutional coding* is used for services with a bit error rate requirement on the order of 10^{-3}, for example, for voice services. For services requiring a lower BER, namely, on the order of 10^{-6}, *turbo coding* is applied. Turbo coding is known to guarantee a high performance [108] over AWGN channels at the cost of increased

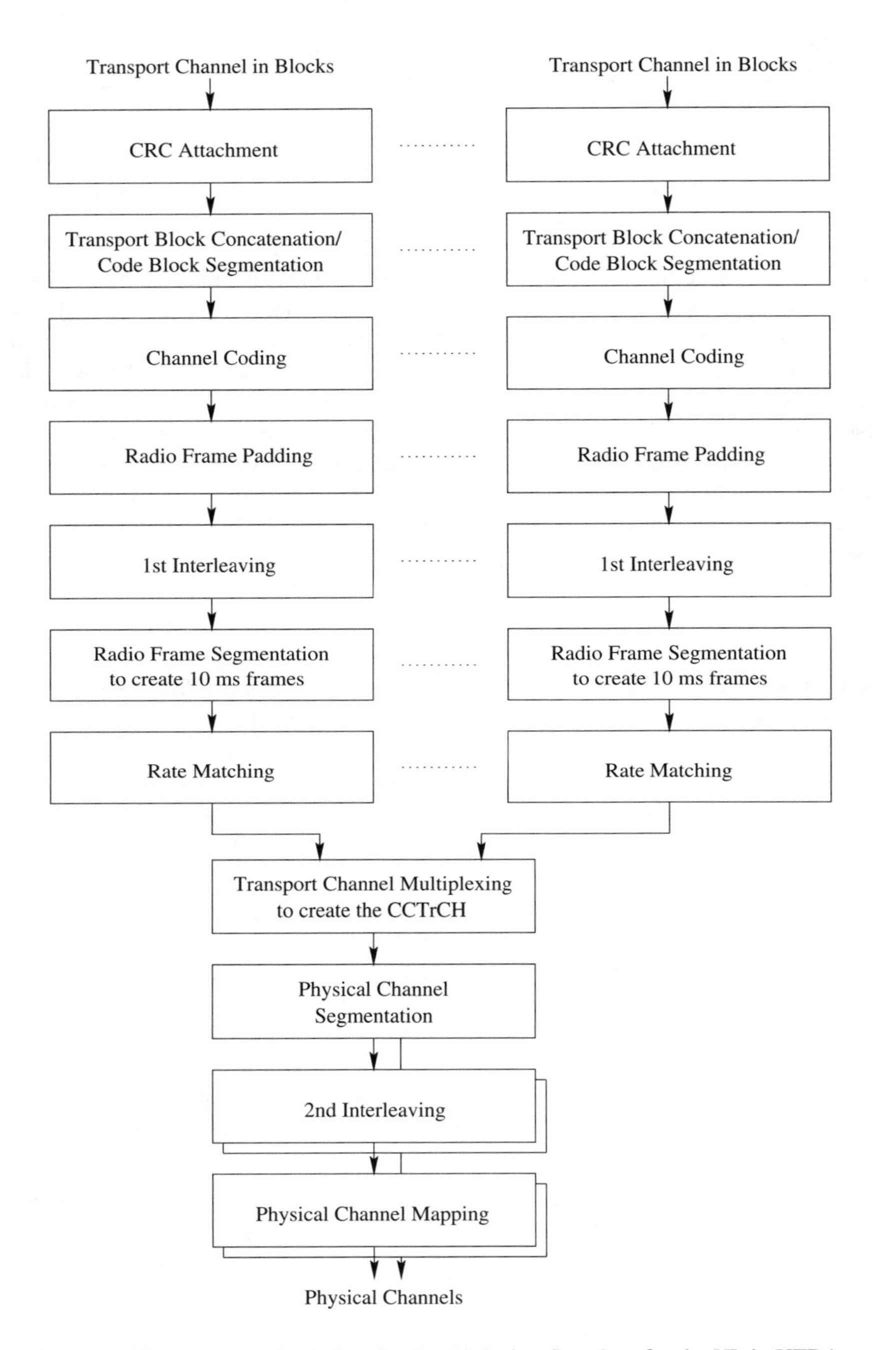

Figure 1.30: Transport channel-coding/multiplexing flowchart for the UL in UTRA.

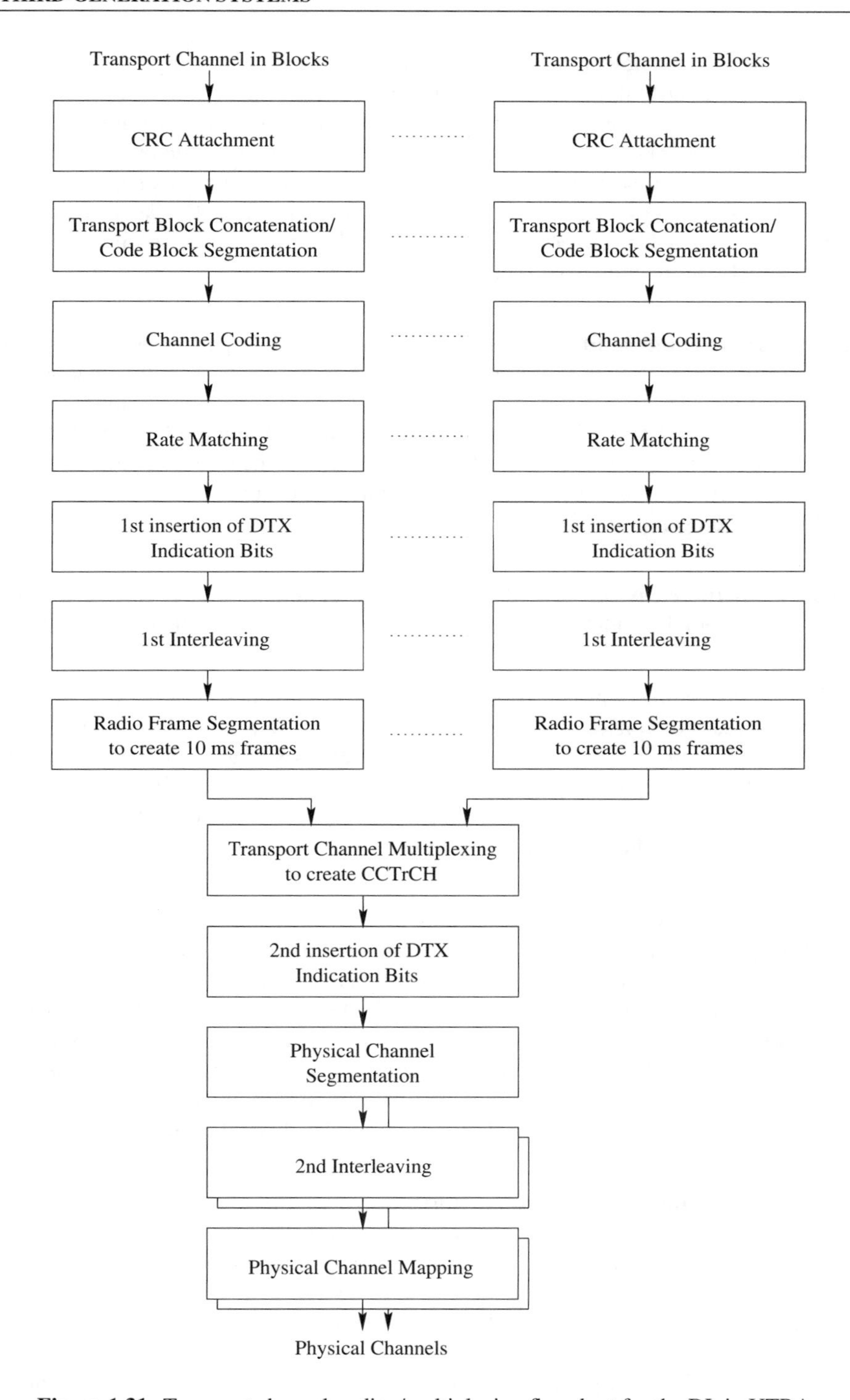

Figure 1.31: Transport channel-coding/multiplexing flowchart for the DL in UTRA.

interleaving-induced latency or delay. The implementational complexity of the turbo codec (TC) does not necessarily have to be higher than that of the convolutional codes (CC), since a constraint-length $K = 7$ or $K = 9$ CC is often invoked, while the constraint-length of the turbo codes employed may be as low as $K = 3$. In somewhat simplistic but plausible terms, one could argue that a $K = 3$ TC using two decoding steps per iteration and employing four iterations has a similar complexity to a $K = 6$ CC, since they are associated with the same number of trellis states. The encoded code blocks within a TTI are then serially concatenated after the channel-coding unit, as seen in Figures 1.30 and 1.31.

1.3.2.4.4 Radio Frame Padding Radio frame padding is only performed on the UL whereby the input bit sequence (the concatenated encoded code blocks from the channel-coding unit) is padded in order to ensure that the output can be segmented equally into (TTI/10 ms) number of 10 ms radio frames. Note that radio frame padding is not required on the DL, since DTX is invoked, as seen in Figure 1.31. This process was termed Radio Frame Equalisation in the standard. However, in order to avoid confusion with channel equalisation, we used the terminology 'padding'.

1.3.2.4.5 First Interleaving The depth of this first interleaver seen in Figures 1.30 and 1.31 may range from one radio frame (10 ms) to as high as 80 ms, depending on the TTI.

1.3.2.4.6 Radio Frame Segmentation The input bit sequence after the first interleaving is then segmented into consecutive radio frames of 10 ms duration, as highlighted in Section 1.3.2.3. The number of radio frames required is equivalent to (TTI/10). Because of the Radio Frame Padding step performed prior to the segmentation on the UL in Figure 1.30 and also because of the Rate Matching step on the DL in Figure 1.31, the input bit sequence can be conveniently divided into the required number of radio frames.

1.3.2.4.7 Rate Matching The rate matching process of Figures 1.30 and 1.31 implies that bits on a transport channel are either repeated or punctured in order to ensure that the total bit rate after multiplexing all the associated transport channels will be identical to the channel bit rate of the corresponding physical channel, as highlighted in Section 1.3.2.3. Thus, rate matching must be coordinated among the different coded transport channels, so that the bit rate of each channel is adjusted to a level that fulfills its minimum QoS requirements [91]. On the DL, the bit rate is also adjusted so that the total instantanenous transport channel bit rate approximately matches the defined bit rate of the physical channel, as listed in Table 1.3.2.1.

1.3.2.4.8 Discontinuous Transmission Indication On the DL, the transmission is interrupted if the bit rate is less than the allocated channel bit rate. This is known as discontinuous transmission (DTX). DTX indication bits are inserted into the bit sequence in order to indicate when the transmission should be turned off. The first insertion of the DTX indication bits shown in Figure 1.31 is performed only if the position of the transport channel in the radio frame is fixed. In this case, a fixed number of bits is reserved for each transport channel in the radio frame. For the second insertion step shown in Figure 1.31, the DTX indication bits are inserted at the end of the radio frame.

1.3.2.4.9 Transport Channel Multiplexing One radio frame from each transport channel that can be mapped to the same type of physical channel is delivered to the transport channel multiplexing unit of Figures 1.30 and 1.31, where they are serially multiplexed to form a *Coded Composite Transport CHannel* (CCTrCH). At this point, it should be noted that the bit rate of the multiplexed radio frames may be different for the various transport channels. In order to successfully de-multiplex each transport channel at the receiver, the TFCI — which contained information about the bit rate of each multiplexed transport channel — can be transmitted together with the CCTrCH information (which will be mapped to a physical channel), as highlighted in Section 1.3.2.3. Alternatively, **blind transport format detection** can be performed at the receiver without the explicit knowledge of the TFCI, where the receiver acquires the transport format combination through some other means, such as, for example, the received power ratio of the DPDCH to the DPCCH.

1.3.2.4.10 Physical Channel Segmentation If more than one physical channel is required in order to accommodate the bits of a CCTrCH, then the bit sequence is segmented equally into different physical channels, as seen in Figures 1.30 and 1.31. A typical example of this scenario would be where the bit rate of the CCTrCH exceeds the maximum allocated bit rate for the particular physical channel. Thus, multiple physical channels are required for its transmission. Furthermore, restrictions are imposed on the number of transport channels that can be multiplexed onto a CCTrCH. Hence, several physical channels are required to carry any additional CCTrCHs.

1.3.2.4.11 Second Interleaving The depth of the second interleaving stage shown in Figures 1.30 and 1.31 is equivalent to one radio frame. Hence, this process does not increase the system's delay.

1.3.2.4.12 Physical Channel Mapping Finally, the bits are mapped to their respective physical channels summarised in Table 1.5, as portrayed in Figures 1.30 and 1.31.

Having highlighted the various channel-coding and multiplexing techniques as well as the structures of the physical channels illustrated by Figures 1.20–1.27, let us now consider how the services of different bit rates are mapped on the UL and DL dedicated physical data channels (DPDCH) of Figures 1.20 and 1.21, respectively. In order to augment the process, we will present three examples. Specifically, we consider the mapping of two multirate services on a UL DPDCH and an example of the mapping of a 4.1 kbps data service on a DL DPDCH in the FDD mode. We will then use the same parameters as employed in the first example and show how the multirate services can be mapped to the corresponding UL DPCH in TDD mode.

1.3.2.4.13 Mapping Several Multirate Services to the UL Physical Channels in FDD Mode [88] In this example, we assume that a 4.1 kbps speech service and a 64 kbps video service are to be transmitted simultaneously on the UL. The parameters used for this example are shown in Table 1.7. As illustrated in Figure 1.32, a 16-bit CRC checksum is first attached to each transport block of DCH#1, that is, #1a,...,#1d, as well as the transport block of DCH#2 for the purpose of error detection. As a result, the number of bits in the transport block of Service 1 and Service 2 is increased to 640 + 16 = 656 bits and 164 + 16 = 180 bits,

	Service 1, DCH#1	Service 2, DCH#2
Transport Block Size	640 bits	164 bits
Transport Block Set Size	4 * 640 bits	1 * 164 bits
TTI	40 ms	40 ms
Bit Rate	64 kbps	4.1 kbps
CRC	16 bits	16 bits
Coding	Turbo Rate: 1/3	Convolutional Rate: 1/3

Table 1.7: Parameters for the Multimedia Communication Example of Section 1.3.2.4.13.

respectively. The four transport blocks of Service 1 are then concatenated, as illustrated in Figure 1.32. Notice that no code block segmentation is invoked, since the total number of bits in the concatenated transport block is less than $Z = 5114$ for turbo coding, as highlighted in Section 1.3.2.4.2. Since the video service typically requires a low BER — unless specific measures are invoked for mitigating the video effects of transmission errors [105] — turbo coding is invoked, using a coding rate of $\frac{1}{3}$. Hence, after turbo coding and the attachment of tailing bits, the resulting 40 ms segment would contain $(656 \times 4) \times 3 + 12 = 7884$ bits, as shown in Figure 1.32. By contrast, the speech service can tolerate a higher BER. Hence, convolutional coding is invoked. First, a block of $4 + 4 = 8$ tail bits is concatenated to the transport block in order to flush the assumed constraint-length $K = 5$ shift registers of the convolutional encoder. Thus, a total of 180 + 8 = 188 bits are conveyed to the convolutional encoder of DCH#1, as shown in Figure 1.32. Again, no code block segmentation is invoked, since the total number of bits in the transport channel is less than $Z = 504$ for convolutional coding, as highlighted in Section 1.3.2.4.2. A coding rate of $\frac{1}{3}$ is used for the convolutional encoding of DCH#1, as exemplified in Table 1.7. The output of the convolutional encoder of DCH#1 will have a total of $188 \times 3 = 564$ bits per 40 ms segment. Since the TTI of these transport channels is 40 ms, four radio frames are required to transmit the associated data. At this stage, notice that there are a total of 7884 bits and 564 bits for DCH#1 and DCH#2, respectively. Since these numbers are divisible by four, they can be divided equally into four radio frames. Thus, no padding is required as illustrated in the Radio Frame Padding step of Figure 1.32. Interleaving is then performed across the 40 ms segment for each transport channel before being segmented into four 10 ms radio frames.

At this point, we note that these two transport channels can be mapped to the same DPDCH, since they belong to the same MS. Hence, the 10 ms radio frames, marked 'A' in Figure 1.32 will be multiplexed, in order to form a CCTrCH. Similarly, the frames marked 'B', 'C' (not shown in Figure 1.7 due to lack of space), and 'D' will be multiplexed, in order to form another three CCTrCHs. The rate of these CCTrCHs must be matched to the allocated channel bit rate of the physical channel. Without rate matching, the bit rate of these CCTrCHs is (1971 + 141)/10 ms = 211.2 kbps, which does not fit any of the available channel bit rates of the UL DPDCH, as listed in Table 1.3.2.1. Hence, the Rate Matching step of Figures 1.30, 1.31, and 1.32 must be invoked in order to adapt the multiplexed bit rate to one of the available UL DPDCH bit rates of Table 1.3.2.1. Let us assume that the allocated channel bit rate is 240 kbps. Thus, a number of bits must be punctured or repeated for each service, in order to increase the total number of bits per 10 ms segment after multiplexing

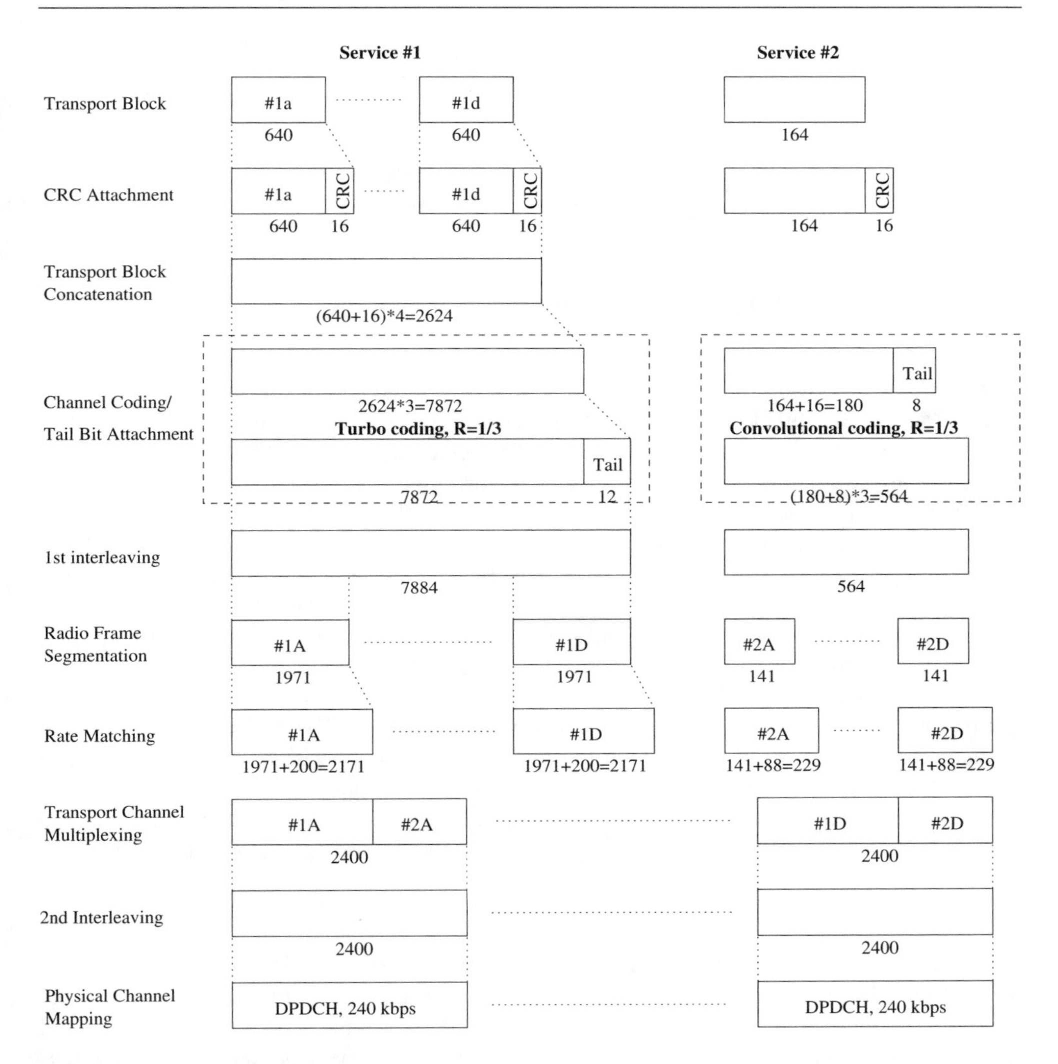

Figure 1.32: Mapping of several multimedia services to the UL dedicated physical data channel of Figure 1.20 in FDD mode. The corresponding schematic diagram is seen in Figure 1.30.

from 2171 to 2400. This would require coordination among the different services, as it was highlighted in Section 1.3.2.4.7. After multiplexing the transport channels, a second interleaving is performed across the 10 ms radio frame before finally mapping the bits to the UL DPDCH.

1.3.2.4.14 Mapping of a 4.1 Kbps Data Service to the DL DPDCH in FDD Mode The parameters for this example are shown in Table 1.8. In this context, we assume that a single DCH consisting of one transport block within a TTI duration of 40 ms is to be transmitted on the DL. As illustrated in Figure 1.33, a 16-bit CRC sum segment is appended to the transport

	Service 1, DCH#1
Transport Block Size	164 bits
TTI	40 ms
Bit Rate	4.1 kbps
CRC	16 bits
Coding	Convolutional Rate: 1/3

Table 1.8: Parameters for the example of Section 1.3.2.4.14.

block. A $4+4=8$-bit tailing block is then attached to the end of the segment in order to form a 188-bit code block. Similarly to the previous example, the length of the code block is less than $Z=504$, since CC is used. Hence, no segmentation is invoked. The 188-bit data block is convolutional coded at a rate of $\frac{1}{3}$, which results in a $3\times 188=564$-bit segment. According to Figure 1.31, rate matching is invoked for the encoded block. Since the TTI duration is 40 ms, four radio frames are required to transmit the data. Without rate matching, the bit rate per radio frame is 564/40 ms = 14.1 kbps, which does not fit any of the available bit rates listed in Table 1.4 for the DL. Note that for the case of the DL dedicated physical channels, the channel bit rate will include the additional bits required for the pilot and TPC, as shown explicitly in Figure 1.21. Since there is only one transport channel in this case, no TFCI bits are required. We assume that an 8-bit pilot and a 2-bit TPC per slot are assigned to this tranmission, which yields a total rate of 15 kbps for the DPCCH. Hence all the bits in the encoded block will be repeated in order to increase its bit rate of 15 kbps to 30 kbps for the DL DPCH. In this case the number of padding bits appended becomes $N=36$. After the second interleaving stage of Figure 1.31, the segmented radio frames are mapped to the corresponding DPDCH, which are then multiplexed with the DPCCH, as shown in Figure 1.33.

1.3.2.4.15 Mapping Several Multirate Services to the UL Physical Channels in TDD Mode [88] In this example, we will demonstrate how the multirate multimedia services, considered previously in the example of Section 1.3.2.4.13 in an FDD context, are mapped to the corresponding dedicated physical channels (DPCH) in the TDD mode. The channel-coding/multiplexing process is identical in the FDD and TDD mode, and so both are based on Figures 1.30 and 1.31. The only difference is in the mapping of the transport channels to the corresponding physical channels seen at the bottom of Figures 1.30 and 1.31, since the FDD and TDD modes have a different frame structure, as shown previously in Figures 1.20–1.26 and Figure 1.27, respectively. In this example, we are only interested in the process of service mapping to the physical channel, which follows the second interleaving stage of Figure 1.34. Here we assumed that for the TDD UL scenario of Table 1.7 the total number of bits per segment after DCH multiplexing is 2186 as a result of rate matching. In the FDD example of Section 1.3.2.4.13, this was 2400. Each segment is divided into two bursts, which can be transmitted either by orthogonal code multiplexing onto a single time-slot, or using two time-slots within a 10 ms radio frame. Note that only one burst in each segment is required to carry the TFCI and the TPC information.

Following these brief discussions on service multiplexing, channel coding, and interleav-

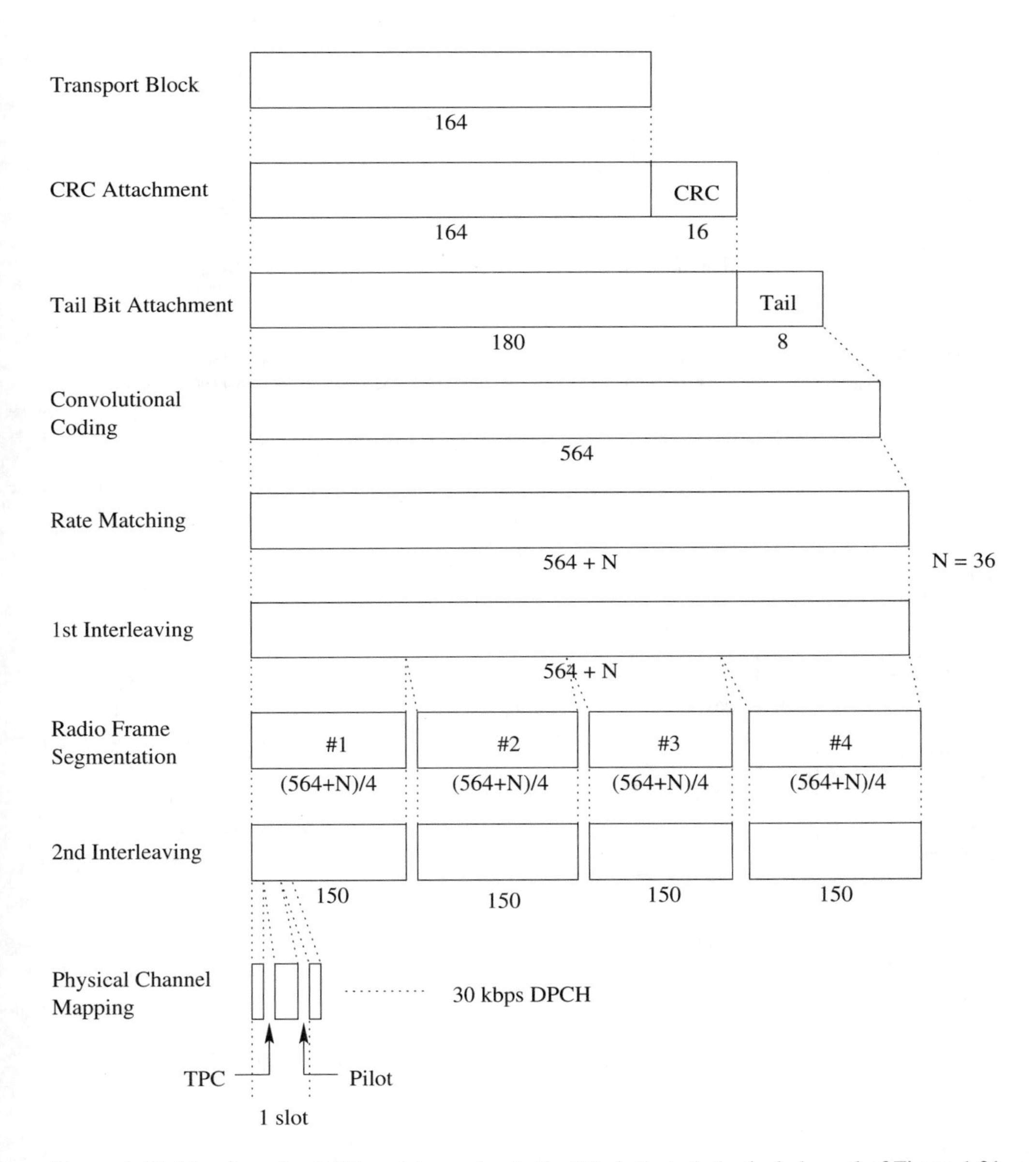

Figure 1.33: Mapping of a 4.1 kbps data service to the DL dedicated physical channel of Figure 1.21 in FDD mode. The corresponding schematic diagram is seen in Figure 1.31.

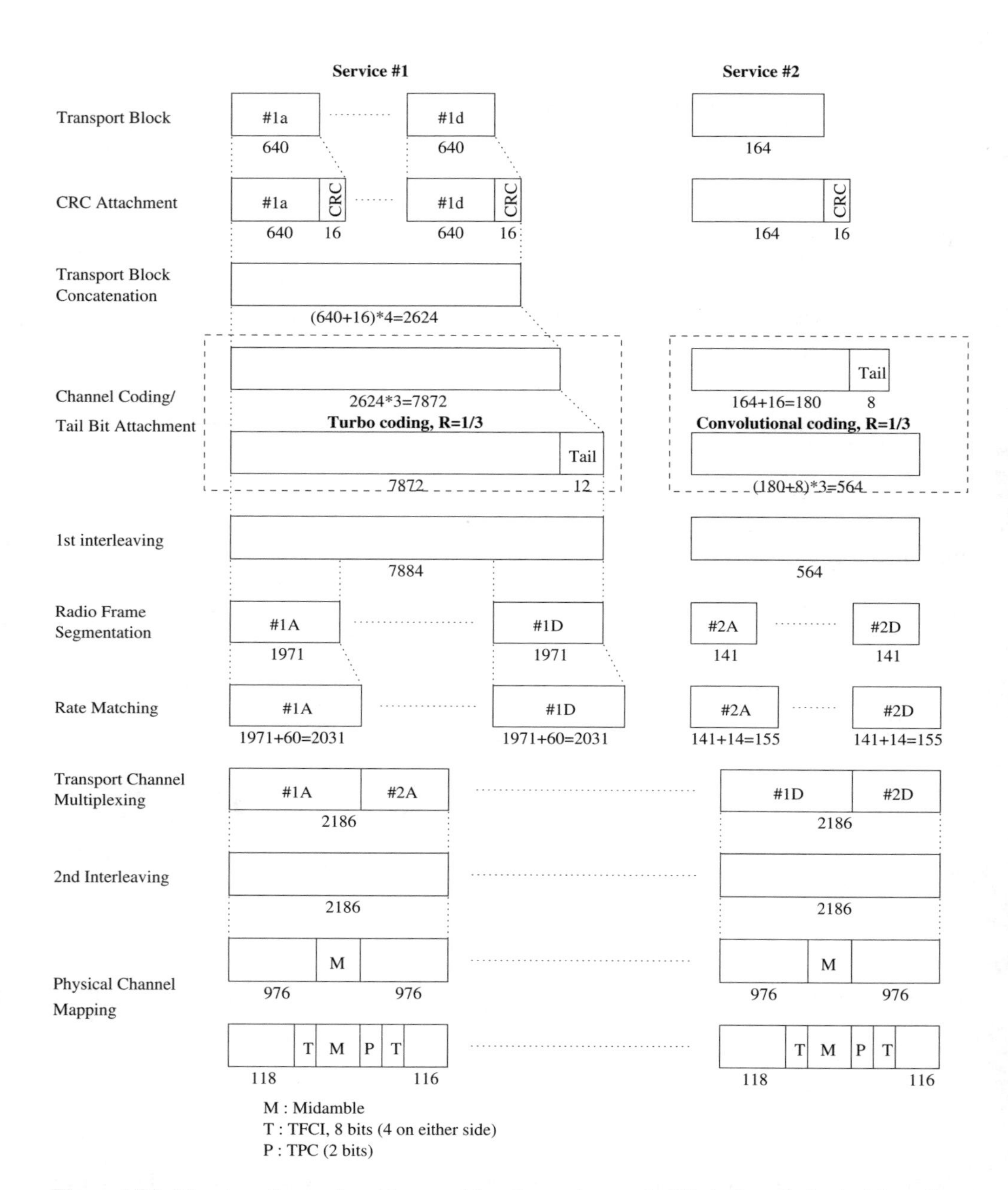

Figure 1.34: Mapping of several multirate multimedia services to the UL dedicated physical data channel of Figure 1.20 in TDD mode. The corresponding schematic diagram is seen in Figure 1.30.

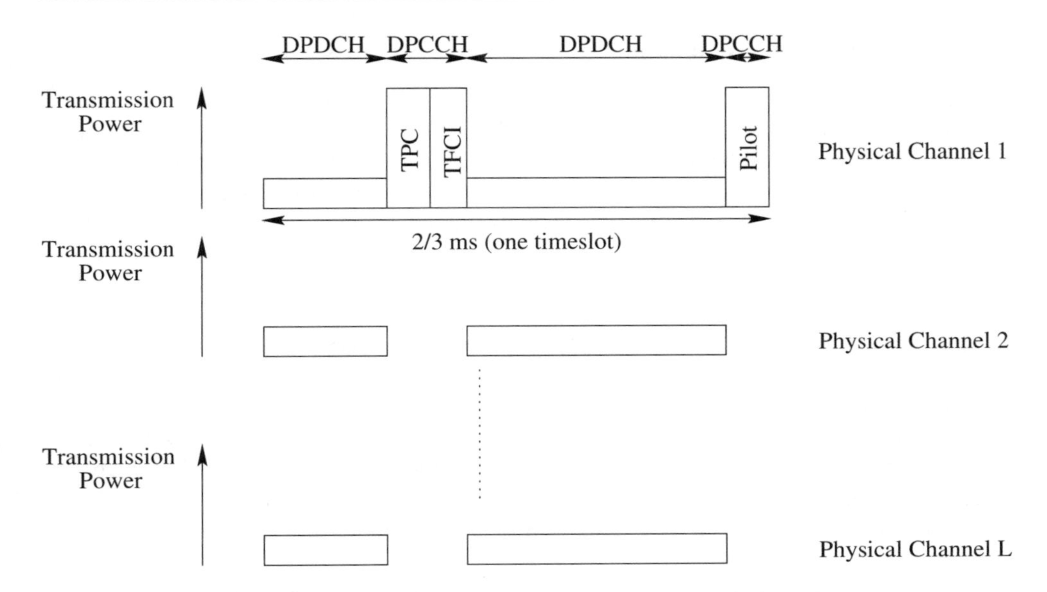

Figure 1.35: DL FDD slot format for multicode transmission in UTRA, based on Figure 1.21, but dispensing with transmitting DPCCH over all multicode physical channels.

ing, let us now concentrate on the aspects of variable-rate and multicode transmission in UTRA in the next section.

1.3.2.5 Variable-Rate and Multicode Transmission in UTRA

Three different techniques have been proposed in the literature for supporting variable-rate transmission, namely, multicode-, modulation-division multiplexing- (MDM), and multiple processing gain (MPG)-based techniques [109]. UTRA employs a number of different processing gains, or variable spreading factors, in order to transmit at different channel bit rates, as highlighted previously in Section 1.3.2.3. The spreading factor (SF) has a direct effect on the performance and capacity of a DS-CDMA system. Since the chip rate is constant, the SF — which is defined as the ratio of the spread bandwidth to the original information bandwidth — becomes lower, as the bit rate increases. Hence, there is a limit to the value of the SF used, which is SF = 4 in FDD mode in the proposed UTRA standards. Multicode transmission [104, 109, 110] is used if the total bit rate to be transmitted exceeds the maximum bit rate supported by a single DPDCH, which was stipulated as 960 kbps for the UL and 1920 kbps for the DL. When this happens, the bit rate is split among a number of spreading codes and the information is transmitted using two or more codes. However, only one DPCCH is transmitted during this time. Thus, on the UL one DPCCH and several DPDCH are code-multiplexed and transmitted in parallel, as it will be augmented in the context of Figure 1.37. On the DL, the DPDCH and DPCCH are time-multiplexed on the first physical channel associated with the first spreading code as seen in Figure 1.35. If more physical channels are required, the DPCCH part in the slot will be left blank again, as shown in Figure 1.35. The transmit power of the DPDCH is also reduced.

1.3.2.6 Spreading and Modulation

It is well known that the performance of DS-CDMA is interference limited [72]. The majority of the interference originates from the transmitted signals of other users within the same cell, as well as from neighbouring cells. This interference is commonly known as *Multiple Access Interference* (MAI). Another source of interference, albeit less dramatic, is a result of the wideband nature of CDMA, yielding several delayed replicas of the transmitted signal, which reach the receiver at different time instants, thereby inflicting what is known as *interpath interference*. However, the advantages gained from wideband transmissions, such as multipath diversity and the noise-like properties of the interference, outweigh the drawbacks.

The choice of the spreading codes [111, 112] used in DS-CDMA will have serious implications for the amount of interference generated. Suffice to say that the traditional measures used in comparing different codes are their *cross-correlations* (CCL) and *autocorrelation* (ACL). If the CCL of the spreading codes of different users is nonzero, this will increase their interference, as perceived by the receiver. Thus a low CCL reduces the MAI. The so-called out-of-phase ACL of the codes, on the other hand, plays an important role during the initial synchronisation between the BS and MS, which has to be sufficiently low to minimise the probability of synchronising to the wrong ACL peak.

In order to reduce the MAI and thereby improve the system's performance and capacity, the UTRA physical channels are spread using two different codes, namely, the *channelisation code* and a typically longer so-called *scrambling code*. In general, the channelisation codes are used to maintain orthogonality between the different physical channels originating from the same source. On the other hand, the scrambling codes are used to distinguish between different cells, as well as between different MSs. All the scrambling codes in UTRA are in complex format. Complex-valued scrambling balances the power on the I and Q branches. This can be shown by letting c_s^I and c_s^Q be the I and Q branch scrambling codes, respectively. Let $d(t)$ be the complex-valued data of the transmitter, which can be written as:

$$d(t) = d_I + jd_Q, \tag{1.34}$$

where d_I and d_Q represent the data on the I and Q branches, respectively. Let us assume for the sake of argument that the power level in the I and Q branches is unbalanced due to, for instance, their different bit rates or different QoS requirements. If only real-valued scrambling is used, then the output becomes:

$$s(t) = c_s^I \left(d_I + jd_Q\right), \tag{1.35}$$

which is also associated with an unbalanced power level on the I and Q branches. By contrast, if complex-valued scrambling is used, then the output would become:

$$\begin{aligned} s(t) &= (d_I + jd_Q) \cdot (c_s^I + jc_s^Q) && (1.36) \\ &= c_s^I \cdot d_I - c_s^Q \cdot d_Q + j\left(c_s^Q \cdot d_I + c_s^I \cdot d_Q\right). && (1.37) \end{aligned}$$

As can be seen, the power on the I and Q branches after complex scrambling is the same, regardless of the power level of the unscrambled data on the I and Q branches. Hence, complex scrambling potentially improves the power amplifier's efficiency by reducing the peak-to-average power fluctuation. This also relaxes the linearity requirements of the UL power amplifier used.

	Channelisation Codes	Scrambling Codes
Type of codes	OVSF (Section 1.3.2.6.1)	UL : Gold codes, S(2) codes (Section 1.3.2.6.2) DL : Gold codes (Section 1.3.2.6.3)
Code length	Variable	UL : 10 ms of $(2^{25}-1)$-chip Gold code DL : 10 ms of $(2^{18}-1)$-chip Gold code
Type of spreading	BPSK (UL/DL)	QPSK (UL/DL)

Table 1.9: UL/DL Spreading and Modulation Parameters in UTRA.

Table 1.9 shows the parameters and techniques used for spreading and modulation in UTRA, which will be discussed in depth in the following sections.

1.3.2.6.1 Orthogonal Variable Spreading Factor Codes The channelisation codes used in the UTRA systems are derived from a set of orthogonal codes known as *Orthogonal Variable Spreading Factor* (OVSF) codes [103]. OVSF codes are generated from a tree-structured set of orthogonal codes, such as the Walsh-Hadamard codes, using the procedure shown in Figure 1.36. Specifically, each channelisation code is denoted by $c_{N,n}$, where $n = 1, 2, \ldots, N$ and $N = 2^x, x = 2, 3, \ldots 8$. Each code $c_{N,n}$ is derived from the previous code $c_{(N/2),n}$ as follows [103]:

$$\begin{bmatrix} c_{N,1} \\ c_{N,2} \\ c_{N,3} \\ \vdots \\ c_{N,N} \end{bmatrix} = \begin{bmatrix} c_{(N/2),1} | c_{(N/2),1} \\ c_{(N/2),1} | \bar{c}_{(N/2),1} \\ c_{(N/2),2} | c_{(N/2),2} \\ \vdots \\ c_{(N/2),(N/2)} | \bar{c}_{(N/2),(N/2)} \end{bmatrix}, \quad (1.38)$$

where [|] at the right-hand side of Equation 1.38 denotes an augmented matrix and $\bar{c}_{(N/2),n}$ is the binary complement of $c_{(N/2),n}$. For example, according to Equation 1.38 and Figure 1.36, $c_{N,1} = c_{8,1}$ is created by simply concatenating $c_{(N/2),1}$ and $c_{(N/2),1}$, which doubles the number of chips per bit. By contrast, $c_{N,2} = c_{8,2}$ is generated by attaching $\bar{c}_{(N/2),1}$ to $c_{(N/2),1}$. From Equation 1.38, we see that, for example, $c_{N,1}$ and $c_{N,2}$ at the left-hand side of Equation 1.38 are not orthogonal to $c_{(N/2),1}$, since the first half of both was derived from $c_{(N/2),1}$ in Figure 1.36, but they are orthogonal to $c_{(N/2),n}, n = 2, 3, \ldots, (N/2)$. The code $c_{(N/2),1}$ in Figure 1.36 is known as the mother code of the codes $c_{N,1}$ and $c_{N,2}$, since these two codes are derived from $c_{(N/2),1}$. The codes on the 'highest'-order branches ($k = 6$) of the tree at the left of Figure 1.36 have a spreading factor of 4, and they are used for transmission at the highest possible bit rate for a single channel, which is 960 kbps. On the other hand, the codes on the 'lowest'-order branches ($k = 0$) of the tree at the right of Figure 1.36 result in a spreading factor of 256, and these are used for transmission at the lowest bit rate, which is 15 kbps. It is worth noting here that an intelligent BbB adaptive scheme may vary its SF on a 10 ms frame basis in an attempt to adjust the SF on a near-instantaneous channel-quality motivated basis [69, 105]. Orthogonality between parallel transmitted channels of the same bit rate is preserved by assigning each channel a different orthogonal code accordingly. For channels with different bit rates transmitting in parallel, orthogonal codes are assigned, ensuring that no code is the mother-code of the other. Thus, OVSF channelisation codes provide total isolation between different users' physical channels on the DL that are transmitted synchronously and hence eliminate MAI among them. OVSF channelisation codes also provide

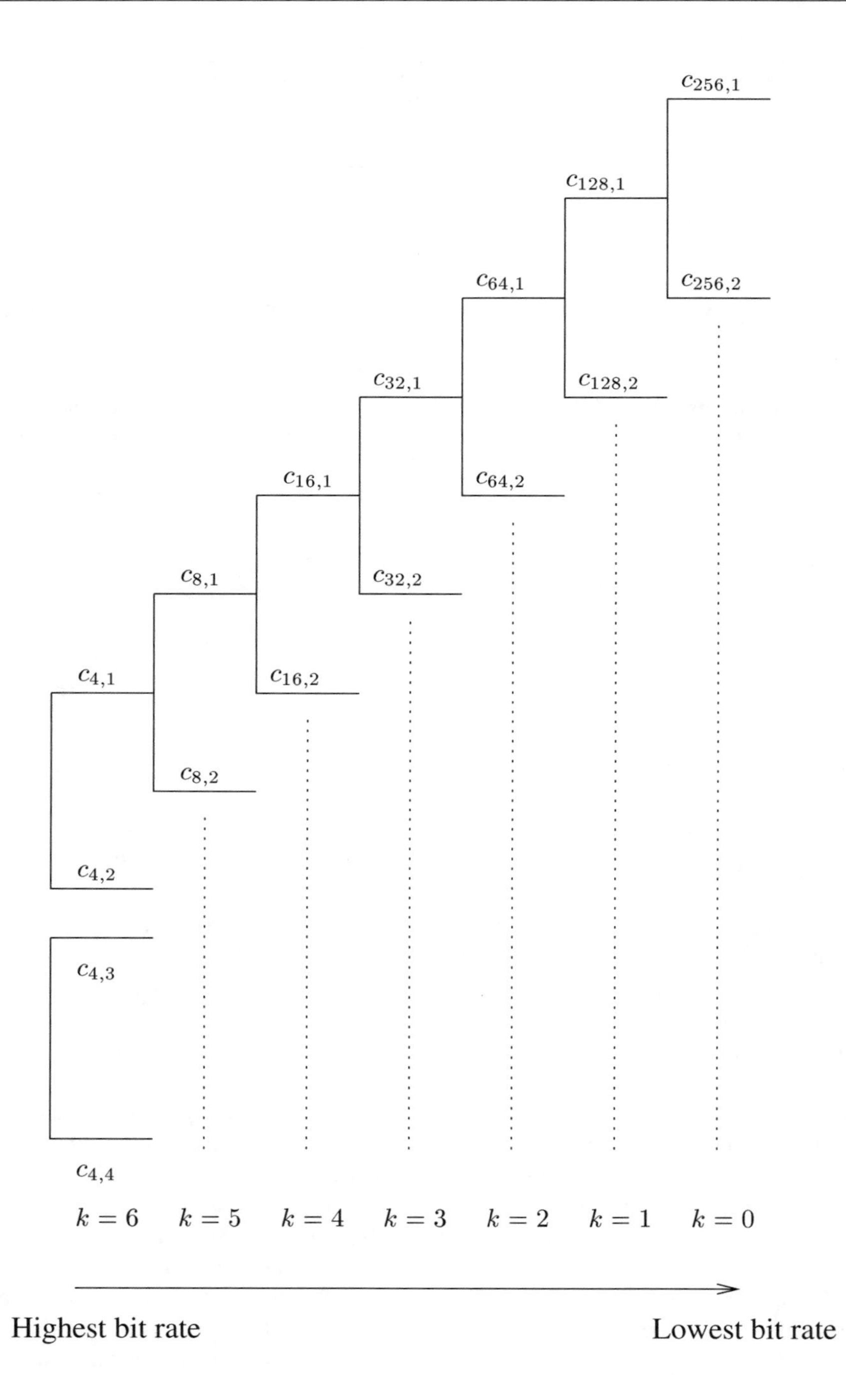

Figure 1.36: Orthogonal variable-spreading factor code tree in UTRA according to Equation 1.38. The parameter k in the figure is directly related to that found in Figures 1.20–1.24.

orthogonality between the different physical channels seen in Figure 1.35 during multicode transmission.

Since there is only a limited set of OVSF codes, which is likely to be insufficent to support a large user-population, while also allowing identification of the BSs by the MSs on the DL, *each cell will reuse the same set of OVSF codes.* Statistical multiplexing schemes such as packet reservation multiple access (PRMA) can be used to allocate and de-allocate the OVSF codes on a near-instantaneous basis, for example, depending on the users' voice activity in the case of DTX-based communications [113]. However, orthogonal codes, such as the orthogonal OVSF codes, in general exhibit poor out-of-phase ACL and CCL properties [114]. Therefore, the correlations of the OVSFs of adjacent asynchronous BSs will become unacceptably high, degrading the correlation receiver's performance at the MS. On the other hand, certain long codes such as Gold codes exhibit low CCL, which is advantageous in CDMA applications [39]. Hence in UTRA, cell-specific long codes are used in order to reduce the intercell interference on the DL. On the UL, MAI is reduced by assigning different scrambling codes to different users.

1.3.2.6.2 Uplink Scrambling Codes The UL scrambling codes in UTRA can be classified into long scrambling codes and short scrambling codes. A total of 2^{24} UL scrambling codes can be generated for both the long and short codes.

Long scrambling codes are constructed from two m-sequences using the polynomials of $1+X^3+X^{25}$ and $1+X+X^2+X^3+X^{25}$, following the procedure highlighted by Proakis [5] in order to produce a set of **Gold codes** for the I branch. The Q-branch Gold code is a shifted version of the I-branch Gold code, where a shift of 16,777,232 chips was recommended. Gold codes are rendered different from each other by assigning a unique initial state to one of the shift registers of the m-sequence. The initial state of the other shift register is a sequence of logical 1. Although the Gold codes generated have a length of $2^{25} - 1$ chips, only 38,400 chips (10 ms at 3.84 Mcps) are required in order to scramble a radio frame.

Short scrambling codes are defined from a family of periodically extended S(2) codes. This 256-chip S(2) code was introduced to ease the implementation of multi-user detection at the BS [31]. The multi-user detector has to invert the so-called system matrix [68], the dimension of which is proportional to the sum of the channel impulse response duration and the spreading code duration. Thus, using a relatively short scrambling code is an important practical consideration in reducing the size of the system-matrix to be inverted.

1.3.2.6.3 Downlink Scrambling Codes Unlike the case for the UL, only Gold codes are used on the DL. The DL Gold codes on the I branch are constructed from two m-sequences using the polynomials of $1 + X^7 + X^{18}$ and $1 + X^5 + X^7 + X^{10} + X^{18}$. These Gold codes are shifted by 131,072 chips in order to produce a set of Gold codes for the Q branch.

Although a total of $2^{18} - 1 = 262,143$ Gold codes can be generated, only 8192 of them will be used as the DL scrambling code. These codes are divided into 512 groups, each of which contains a *primary scrambling code* and 15 *secondary scrambling codes.* Altogether there are 512 primary scrambling codes and $8192 - 512 = 7680$ secondary scrambling codes. Each cell is allocated one primary scrambling code, which is used on the CPICH and P-CCPCH channels of Table 1.5. This primary scrambling code will be used to identify the BS for the MS. All the other physical channels belonging to this cell can use either the primary scrambling code or any of the 15 secondary scrambling codes that belong to the same group, as the primary scrambling code. In order to facilitate fast cell or BS identification, the set of 512 primary scrambling codes is further divided into 64 subsets, each consisting of eight primary scrambling codes, as will be shown in Section 1.3.2.9.

1.3.2.6.4 Uplink Spreading and Modulation A model of the UL transmitter for a single DPDCH is shown in Figure 1.37 [32]. We have seen in Figure 1.20 that the DPDCH and DPCCH are transmitted in parallel on the I and Q branches of the UL, respectively. Hence, to avoid I/Q channel interference in case of I/Q inbalance of the quadrature carriers, different orthogonal spreading codes are assigned to the DPDCH and DPCCH on the I and Q branch, respectively. These two channelisation codes for DPDCH and DPCCH, denoted by $c_{D,1}$ and c_C in Figure 1.37, respectively, are allocated in a predefined order. From Figure 1.20, we know that the SF of the DPCCH is 256. Hence, $c_C = c_{256,1}$ in the context of Figure 1.36. This indicates that the high SF of the DPCCH protects the vulnerable control channel message against channel impairments. On the other hand, we have $c_{D,1} = c_{SF,2}$, depending on the SF of the DPDCH. In the event of multicode transmission portrayed by the dashed lines in Figure 1.37, different additional orthogonal channelisation codes, namely, $c_{D,2}$ and $c_{D,3}$, are assigned to each DPDCH for the sake of maintaining orthogonality, and they can be transmitted on either the I or Q branch. In this case, the BS and MS have to agree on the number of channelisation codes to be used. After spreading, the BPSK modulated I and Q branch signals are summed in order to produce a complex Quadrature Phase Shift Keying (QPSK) signal. The signal is then scrambled by the complex scrambling code, c_{scramb}. The pulse-shaping filters, $p(t)$, are root-raised cosine Nyquist filters using a roll-off factor of 0.22.

The transmitter of the UL PRACH and PCPCH message part is also identical to that shown in Figure 1.37. As we have mentioned in Section 1.3.2.3.2 in the context of Figure 1.22, the PRACH and the CPCH message consist of a data part and a control part. In this case, the data part will be transmitted on the I branch, and the control part on the Q branch. The choice of the channelisation codes for the data and control part depends on the signature of the preambles transmitted beforehand. As highlighted in Section 1.3.2.3.2, the preamble signature is a 256-chip sequence generated by the repetition of a 16-chip Hadamard code. This 16-chip code actually corresponds to one of the OVSF codes, namely, to $c_{16,n}$, where $n = 1, \ldots, 16$. The codes in the subtree of Figure 1.36 below this specific 16-chip code n will be used as the channelisation codes for the data part and control part.

1.3.2.6.5 Downlink Spreading and Modulation The schematic diagram of the DL transmitter is shown in Figure 1.38. All the DL physical channel bursts (except for the SCH) are first QPSK modulated in order to form the I and Q branches, before spreading to the chip rate. In contrast to the UL of Figure 1.37, the same OVSF channelisation code c_{ch} is used on the I and Q branches. Different physical channels are assigned different channelisation codes in order to maintain their orthogonality. For instance, the channelisation codes for the CPICH and P-CCPCH of Table 1.5 are fixed to the codes $c_{256,1}$ and $c_{256,2}$ of Figure 1.36, respectively. The channelisation codes for all the other physical channels are assigned by the network.

The resulting signal in Figure 1.38 is then scrambled by a cell-specific scrambling code c_{scramb}. Similarly to the DL, the pulse-shaping filters are root-raised cosine Nyquist filters using a rolloff factor of 0.22.

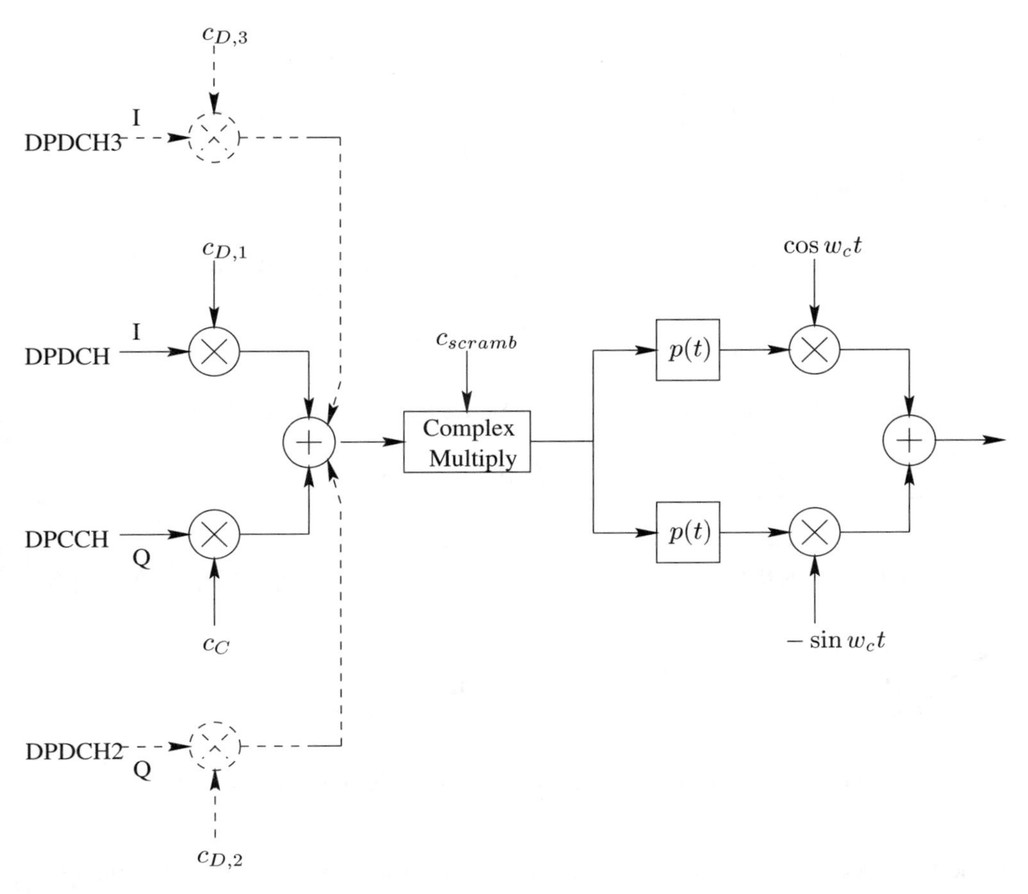

Figure 1.37: uplink transmitter in UTRA using the frame structure of Figure 1.20. Multicode transmissions are indicated by the dashed lines.

In TDD mode, the transmitter structure for both the UL and DL are similar to that of a FDD DL transmitter of Figure 1.38. Since each time-slot can be used for transmitting several TDD bursts from the same source or from different sources, the OVSF codes are invoked in order to maintain orthogonality between the burst of different TDD/CDMA users/messages. An advantage of the TDD/CDMA mode is that the user population is separated in both the time and the code domain. In other words, only a small number of CDMA users/services will be supported within a TDD time-slot, which dramatically reduces the complexity of the multi-user detector that can be used in both the UL and DL for mitigating the MAI or multi-code interference.

1.3.2.7 Random Access

1.3.2.7.1 Mobile-Initiated Physical Random Access Procedures If data transmission is initiated by an MS, it is required to send a random access request to the BS. Since such

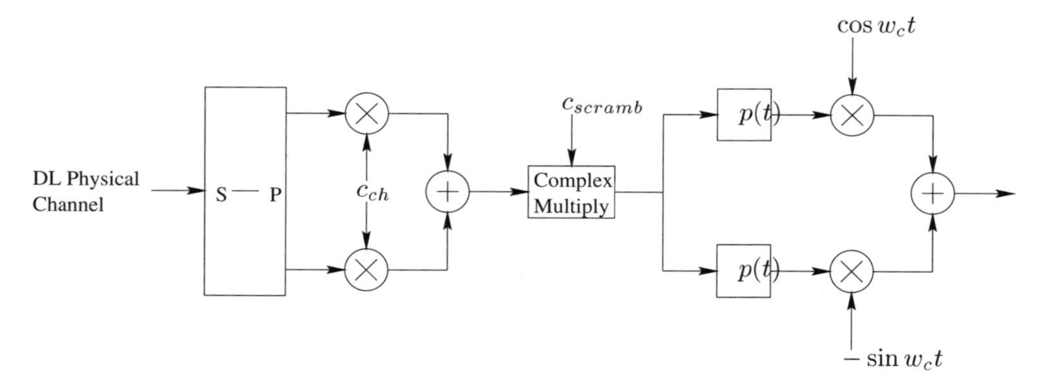

Figure 1.38: downlink transmitter in UTRA using the frame structure of Figure 1.21.

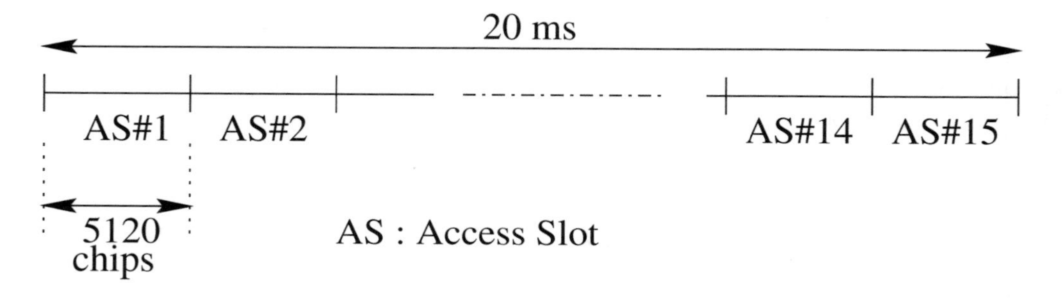

Figure 1.39: ALOHA-based physical UL random access slots in UTRA.

requests can occur at any time, collisions may result when two or more MSs attempt to access the network simultaneously. Hence, in order to reduce the probability of a collision, the random access procedure in UTRA is based on the slotted ALOHA technique [91].

Random access requests are transmitted to the BS via the PRACH of Table 1.5. Each random access transmission request may consist of one or several preambles and a message part, whose time-slot configuration was shown in Figure 1.22. According to the regime of Figure 1.39, the preambles and the message part can only be transmitted at the beginning of one of those 15 so-called *access slots*, which span two radio frames (i.e., 20 ms). Thus, each access slot has a length equivalent to 5120 chips or $\frac{4}{3}$ ms.

Before any random access request can be transmitted, the MS has to obtain certain information via the DL BCH transmitted on the P-CCPCH of Table 1.5 according to the format of Figure 1.23. This DL BCH/PCCPCH information includes the identifier of the cell-specific scrambling code for the preamble and message part of Figure 1.22, the available preamble signatures, the available access slots of Figure 1.39, which can be contended for in ALOHA mode, the initial preamble transmit power, the preamble power ramping factor, and the maximum number of preamble retransmissions necessitated by their decoding failure due to collisions at the BS. All this information may become available once synchronisation is achieved, as will be discussed in Section 1.3.2.9. After acquiring all the necessary information, the MS will randomly select a preamble signature from the available signatures and transmit a preamble at the specific power level specified by the BS on a randomly selected access slot chosen from the set of available access slots seen in Figure 1.39. Note that the preamble is formed by multiplying the selected signature with the preamble scrambling code.

After the preamble is transmitted, the MS will listen for the acknowledgement of reception transmitted from the BS on the AICH of Table 1.5. Note that the AICH is also transmitted at the beginning of an access slot and the phase reference for coherent detection is obtained from the DL CPICH of Table 1.5. The acknowledgement is represented by an AI in the AICH of Table 1.5 that corresponds to the selected preamble signature. If a negative acknowledgement is received, the random access transmission will recommence in a later access slot. If a positive acknowledgement is received, the MS will proceed to transmit the message part at the beginning of a predefined access slot. However, if the MS fails to receive any acknowledgement after a predefined time-out, it will retransmit the preamble in another randomly selected access slot of Figure 1.39 with a newly selected signature, provided that the maximum number of preamble retransmissions was not exceeded. The transmit power of the preamble is also increased, as specified by the above-mentioned preamble power ramping factor. This procedure is repeated until either an acknowledgement is received from the BS or the maximum number of preamble retransmissions is reached.

1.3.2.7.2 Common Packet Channel Access Procedures The transmission of the CPCH of Table 1.5 is somewhat similar to that of the RACH transmission regime highlighted in Figure 1.39. Before commencing any CPCH transmission, the MS must acquire vital information from the BCH message transmitted on the P-CCPCH. This information includes the scrambling codes, the available signatures and the access slots for both the A-P and CD-P messages introduced in Section 1.3.2.3.2.1, the scrambling code of the message part, the DL AICH and the associated DL DPCCH channelisation code, the initial transmit power of the preambles, the preamble power ramping factor, and the maximum allowable number of retransmissions.

The procedure of the A-P transmission is identical to that of the random access transmission highlighted in Section 1.3.2.7.1. We will accordingly omit the details here.

Once a positive acknowledgement is received from the BS on the DL AICH, the MS will transmit the CD-P on a randomly selected access slot of Figure 1.39 using a randomly selected signature. Upon receiving a positive acknowledgment from the BS on the AICH, the MS will begin transmitting the PC-P followed immediately by the message part shown in Figure 1.20 at a predefined access slot of Figure 1.39.

1.3.2.8 Power Control

Accurate power control is essential in CDMA in order to mitigate the so-called *near-far problem* [115, 116]. Furthermore, power control has a dramatic effect on the coverage and capacity of the system: we will therefore consider the UTRA power control issues in detail.

1.3.2.8.1 Closed-Loop Power Control in UTRA Closed-loop power control is employed on both the UL and DL of the FDD mode through the TPC commands that are conveyed in the UL and DL according to the format of Figures 1.20 and 1.21, respectively. Since the power control procedure is the same on both links, we will only elaborate further on the UL procedure.

UL closed-loop power control is invoked in order to adjust the MS's transmit power such that the received Signal-to-Interference Ratio (SIR) at the BS is maintained at a given target SIR. The value of the target SIR depends on the required quality of the connection. The BS measures the received power of the desired UL transmitted signal for both the DPDCH and the DPCCH messages shown in Figure 1.20 after RAKE combining, and it also estimates the total received interference power in order to obtain the estimated received SIR. This SIR estimation process is performed every $\frac{2}{3}$ ms, or a time-slot duration, in which the SIR estimate is compared to the target SIR. According to the values of the estimated and required SIRs, the BS will generate a TPC command, which is conveyed to the MS using the burst of Figure 1.21. If the estimated SIR is higher than the target SIR, the TPC command will instruct the MS to lower the transmit power of the DPDCH and DPCCH of Figure 1.20 by a step size of Δ_{TPC} dB. Otherwise, the TPC command will instruct the MS to increase the transmit power by the same step size. The step size Δ_{TPC} is typically 1 dB or 2 dB. Transmitting at an unnecessarily high power reduces the battery life, while degrading other users' reception quality, who — as a consequence — may request a power increment, ultimately resulting in an unstable overall system operation.

In some cases, BS-diversity combining may take place, whereby two or more BSs transmit the same information to the MS in order to enhance its reception quality. These BSs are known as the active BS set of the MS. The received SIR at each BS will be different and so the MS may receive different TPC commands from its active set of BSs. In this case, the MS will adjust its transmit power according to a simple algorithm, increasing the transmit power only if the TPC commands from all the BSs indicate an 'increase power' instruction. Similarly, the MS will decrease its transmit power if all the BSs issue a 'decrease power' TPC command. Otherwise, the transmit power remains the same. In this way, the multi-user interference will be kept to a minimum without significant deterioration of the performance, since at least one BS has a good reception. Again, the UL and DL procedures are identical, obeying the TPC transmission formats of Figures 1.20 and 1.21, respectively.

1.3.2.8.2 Open-Loop Power Control in TDD Mode As mentioned previously in Section 1.3.2.3, in contrast to the closed-loop power control regime of the FDD mode, no TPC commands are transmitted on the DL in TDD mode. Instead, open-loop power control is used to adjust the transmit power of the MS. Prior to any data burst transmission, the MS would have acquired information about the interference level measured at the BS and also about the BS's P-CCPCH transmitted signal level, which are conveyed to the MS via the BCH according to the format of Figure 1.27. At the same time, the MS would also measure the power of the received P-CCPCH. Hence, with knowledge of the transmitted and received power of the P-CCPCH, the DL path-loss can be found. Since the interference level and the estimated path-loss are now known, the required transmitted power of the TDD burst can be readily calculated based on the required SIR level. Let us now consider how the MS identifies the different cells or BSs with which it is communicating.

1.3.2.9 Cell Identification

1.3.2.9.1 Cell Identification in the FDD Mode System- and cell-specific information is conveyed via the BCH transmitted by the P-CCPCH of Table 1.5 in the context of Figure 1.23 in UTRA. This information has to be obtained before the MS can access the network. The

P-CCPCH information broadcast from each cell is spread by the system-specific OVSF channelisation code $c_{256,2}$ of Figure 1.36. However, each P-CCPCH message is scrambled by a cell-specific primary scrambling code as highlighted in Section 1.3.2.6.3 in order to minimise the intercell interference as well as to assist in identifying the corresponding cell. Hence, the first step for the MS is to recognise this primary scrambling code and to synchronise with the corresponding BS.

As specified in Section 1.3.2.6.3, there are a total of 512 DL primary scrambling codes available in the network. Theoretically, it is possible to achieve scrambling code identification by cross-correlating the P-CCPCH broadcast signal with all the possible 512 primary scrambling codes. However, this would be an extremely tedious and slow process, unduly delaying the MS's access to the network. In order to achieve a fast cell identification by the MS, UTRA adopted a three-step approach [117], which invoked the SCH broadcast from all the BSs in the network. The SCH message is transmitted during the first 256 chips of the P-CCPCH, as illustrated in Figure 1.23. The concept behind this three-step approach is to divide the set of 512 possible primary scrambling codes into 64 subsets, each containing a smaller set of primary scrambling codes, namely, eight codes. Once knowledge of which subset the primary scrambling code of the selected BS belongs to is acquired, the MS can proceed to search for the correct primary scrambling code from a smaller subset of the possible codes.

The frame structure of the DL SCH message seen in Figure 1.23 is shown in more detail in Figure 1.40. It consists of two subchannels, the *Primary SCH* and *Secondary SCH*, transmitted in parallel using code multiplexing. As seen in Figure 1.40, in the Primary SCH a so-called *Primary Synchronisation Code* (PSC), based on a generalised hierarchical Golay sequence [118] of length 256 chips, is transmitted periodically at the beginning of each slot, which is denoted by c_p in Figure 1.40. The same PSC is used by all the BSs in the network. This allows the MS to establish slot-synchronisation and to proceed to the frame-synchronisation phase with the aid of the secondary SCH. On the secondary SCH, a sequence of 15 *Secondary Synchronisation Codes* (SSCs), each of length 256 chips, is transmitted with a period of one 10 ms radio frame duration, that is, 10 ms, as seen in Figure 1.40. An example of this 15-SSC sequence would be:

$$c_1^1\, c_1^2\, c_2^3\, c_8^4\, c_9^5\, c_{10}^6\, c_{15}^7\, c_8^8\, c_{10}^9\, c_{16}^{10}\, c_2^{11}\, c_7^{12}\, c_{15}^{13}\, c_7^{14}\, c_{16}^{15}, \tag{1.39}$$

where each of these 15 SSCs is selected from a set of 16 legitimate SSCs. The specific sequence of 15 SSCs denoted by $c_i^1, \ldots, c_i^{15}$ — where $i = 1, \ldots, 16$ in Figure 1.40 — is used as a code in order to identify and signal to the MS which of the 64 subsets the primary scrambling code used by the particular BS concerned belongs to. The parameter a in Figure 1.40 is a binary flag used to indicate the presence ($a = +1$) or absence ($a = -1$) of a Space Time Block Coding Transmit Diversity (STTD) encoding scheme [119] in the P-CCPCH, as will be discussed in Section 1.3.4.1.1. Specifically, when each of the 16 legitimate 256-chip SSCs can be picked for any of the 15 positions in Figure 1.40 and assuming no other further

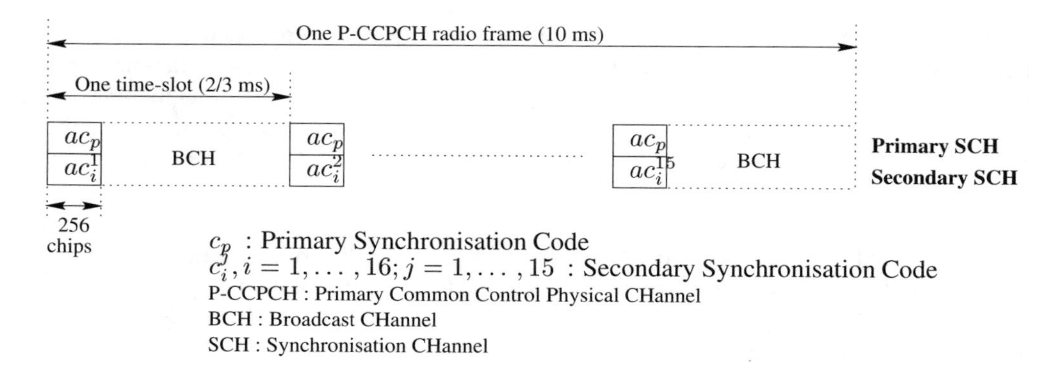

Figure 1.40: Frame structure of the UTRA DL synchronisation channel (SCH), which is mapped to the first 256 chips of the P-CCPCH of Figure 1.23. The primary and secondary SCH are transmitted in parallel using code multiplexing. The parameter a is a gain factor used to indicate the presence ($a = +1$) or absence ($a = -1$) of STTD encoding in the P-CCPCH.

constraints, one could construct

$$\begin{aligned} c_{i,j}^{\text{repeated}} &= \binom{i+j-1}{j} \\ &= \frac{(i+j-1)!}{j!(i-1)!} \\ &= \frac{30!}{15! \cdot 15!} \\ &= 155,117,520 \end{aligned} \tag{1.40}$$

different such sequences, where $i = 16$ and $j = 15$. However, the 15 different 256-chip SSCs of Figure 1.40 must be constructed so that their cyclic shifts are also unique, since these sequences have to be uniquely recognised before synchronisation. In other words, none of the cyclic shifts of the 64 required $15 \times 256 = 3840$-chip sequences can be identical to any of the other sequences' cyclic shifts. Provided that these conditions are satisfied, the 15 specific 256-chip secondary SCH sequences can be recognised within one 10 ms-radio frame-duration of 15 slots. Thus, both slot and frame synchronisation can be established within the particular 10 ms frame received. Using this technique, initial cell identification and synchronisation can be carried out in the following three basic steps.

Step 1: The MS uses the 256-chip PSC of Figure 1.40 to perform cross-correlation with all the received Primary SCHs of the BSs in its vicinity. The BS with the highest correlator output is then chosen, which constitutes the best cell site associated with the lowest path-loss. Several periodic correlator output peaks have to be identified in order to achieve a high BS detection reliability, despite the presence of high-level interference. *Slot synchronisation is also achieved* in this step by recognising the 15 consecutive c_p sequences, providing 15 periodic correlation peaks.

Step 2: Once the best cell site is identified, the primary scrambling code subset of that cell site is found by cross-correlating the Secondary SCH with the 16 possible SSCs in each of the 15 time-slots of Figure 1.40. This can be easily implemented using 16 correlators, since the timing of the SSCs is known from Step 1. Hence, there are a total of $15 \times 16 = 240$ correlator outputs. From these outputs, a total of $64 \times 15 = 960$ decision variables corresponding to the 64 possible secondary SCH sequences and 15 cyclic shifts of each $15 \times 256 = 3840$-chip sequence are obtained. The highest decision variable determines the primary scrambling code subset. Consequently, *frame synchronisation is also achieved.*

Step 3: With the primary scrambling code subset identified and frame synchronisation achieved, the primary scrambling code itself is acquired in UTRA by cross-correlating the received CPICH signal — which is transmitted synchronously with the P-CCPCH — on a symbol-by-symbol basis with the eight possible primary scrambling codes belonging to the identified primary scrambling code subset. Note that the CPICH is used in this case, because it is scrambled by the same primary scrambling code as the P-CCPCH and also uses a predefined pilot sequence and so it can be detected more reliably. By contrast, the P-CCPCH carries the unknown BCH information. Once the exact primary scrambling code is identified, the BCH information of Table 1.5, which is conveyed by the P-CCPCH of Figure 1.23, can be detected.

1.3.2.9.2 Cell Identification in the TDD Mode The procedure of cell identification in the TDD mode is somewhat different from that in FDD mode. In the TDD mode, a combination of three 256-chip SSCs out of 16 unique SSCs are used to identify one of 32 SSC code groups allocated to that cell. Each code group contains four different scrambling codes and four corresponding long (for Type 1 burst) and short (for Type 2 burst) basic midamble codes, which were introduced in the context of Figure 1.27. Each code group is also associated with a specific time offset, t_{offset}. The three SSCs, c_i^1, c_i^2, and c_i^3, are transmitted in parallel with the PSC, c_p, at a time offset t_{offset} measured from the start of a time-slot, as shown in Figure 1.41. Similarly to the FDD mode, the PSC is based on a so-called generalised hierarchical Golay sequence [118], which is common to all the cells in the system. Initial cell identification and synchronisation in the TDD mode can also be carried out in three basic steps.

Step 1: The MS uses the 256-chip PSC of Figure 1.41 to perform cross-correlation with all the received PSC of the BSs in its vicinity. The BS associated with the highest correlator output is then chosen, which constitutes the best cell site exhibiting the lowest path-loss. Slot synchronisation is also achieved in this step. If only one time-slot per frame is used to transmit the SCH as outlined in the context of Figure 1.27, then frame synchronisation is also achieved.

Step 2: Once the PSC of the best cell site is identified, the three SSCs transmitted in parallel with the PSC in Figure 1.41 can be identified by cross-correlating the received signal with the 16 possible prestored SSCs. The specific combination of the three SSCs will identify the code group used by the corresponding cell. The specific frame timing of that cell also becomes known from the time offset t_{offset} associated with that code group.

If two time-slots per frame are used to transmit the SCH as outlined in the context of Figure 1.27, then the second PSC must be detected at an offset of seven or eight time-slots with respect to the first one in order to achieve frame synchronisation.

Step 3: As mentioned in Section 1.3.2.3, each basic midamble code defined in the context of Figure 1.27 is associated with a midamble code set. The P-CCPCH of Table 1.5 is always associated with the first midamble of that set. Hence, with the code group identified and frame synchronisation achieved, the cell-specific scrambling code and the associated basic

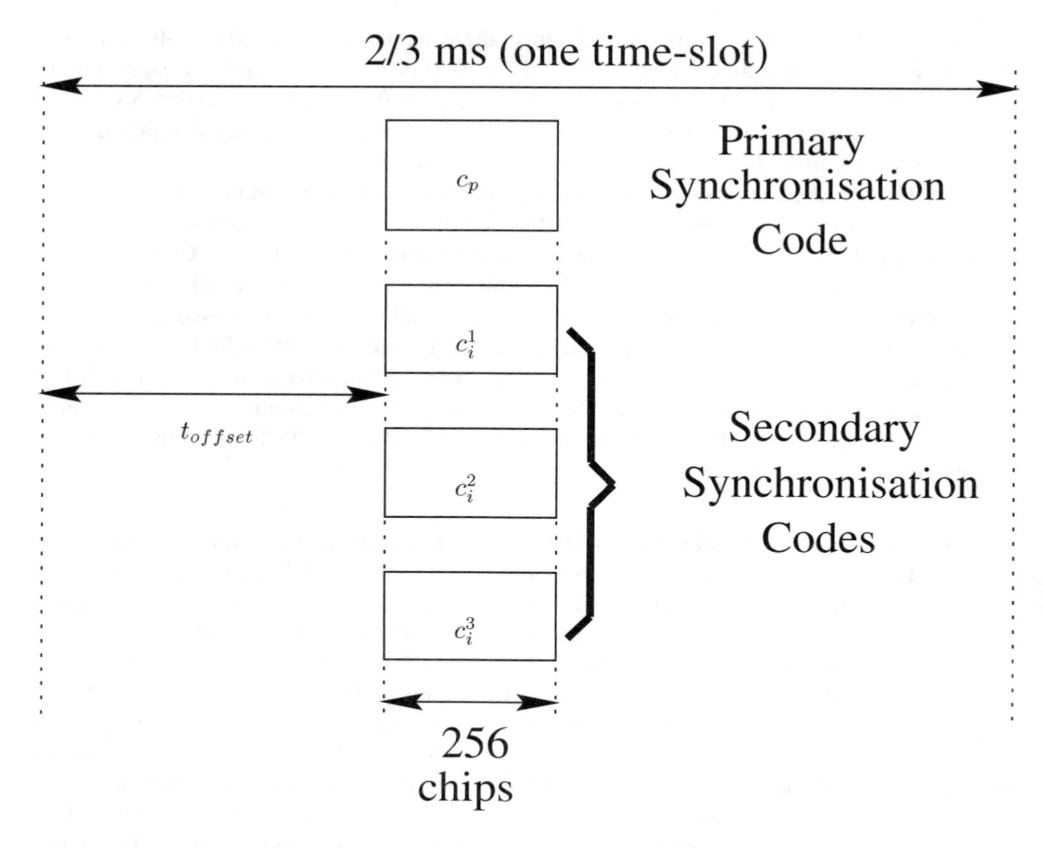

Figure 1.41: Time-slot structure of the UTRA TDD DL synchronisation channel (SCH), which obeys the format of Figure 1.19. The primary and three secondary synchronisation codes are transmitted in parallel at a time offset t_{offset} from the start of a time-slot.

midamble code are acquired in the TDD mode of UTRA by cross-correlating the four possible midamble codes with the P-CCPCH. Once the exact basic midamble code is identified, the associated scrambling code will be known, and the BCH information of Table 1.5, which is conveyed by the P-CCPCH of Figure 1.23, can be detected. Having highlighted the FDD and TDD UTRA cell-selection and synchronisation solutions, let us now consider some of the associated handover issues.

1.3.2.10 Handover

In this section, we consider the handover issues in the context of the FDD mode, since the associated procedures become simpler in the TDD mode, where the operations can be carried out during the unused time-slots. Theoretically, DS-CDMA has a frequency reuse factor of one [120]. This implies that neighbouring cells can use the same carrier frequency without interfering with each other, unlike in TDMA or FDMA. Hence, seamless uninterrupted handover can be achieved when mobile users move between cells, since no switching of car-

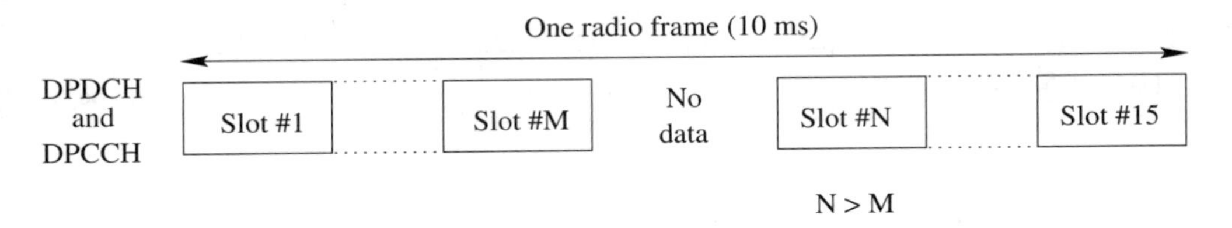

Figure 1.42: uplink frame structure in compressed mode operation during UTRA handovers.

rier frequency and synthesiser retuning is required. However, in *hierarchical cell structures* (HCS)[9] catering, for example, for high-speed mobiles with the aid of a macrocell oversailing a number of microcells, using a different carrier frequency is necessary in order to reduce the intercell interference. In this case, inter-frequency handover is required. Furthermore, because the various operational GSM systems used different carrier frequencies, handover from UTRA systems to GSM systems will have to be supported during the transitory migration phase, while these systems will coexist. Thus, handovers in terrestrial UMTSs can be classified into inter-frequency and intra-frequency handovers.

1.3.2.10.1 Intra-Frequency Handover or Soft Handover Soft handover [121, 122] involves no frequency switching because the new and old cell use the same carrier frequency. The MS will continuously monitor the received signal levels from the neighbouring cells and compares them against a set of thresholds. This information is fed back to the network. Based on this information, if a weak or strong cell is detected, the network will instruct the MS to drop or add the cell from/to its active BS set. In order to ensure a seamless handover, a new link will be established before relinquishing the old link, using the *make before break* approach.

1.3.2.10.2 Inter-Frequency Handover or Hard Handover In order to achieve handovers between different carrier frequencies without affecting the data flow, a technique known as *compressed mode* can be used [123]. With this technique, the UL data, which normally occupies the entire 10 ms frame of Figure 1.19 is time-compressed, so that it only occupies a portion of the frame, that is, slot#1-slot#M and slot#N-slot#15, while no data is transmitted during the remaining portion, that is, slot#(M+1)-slot#(N-1). The latter interval is known as the idle period, as shown in Figure 1.42. There are two types of frame structures for the DL compressed mode, as shown in Figure 1.43. In the Type A structure, shown at the top of Figure 1.43, no data is transmitted after the pilot field of slot#M until the start of the pilot field of slot#(N-1) in order to maximise the transmission gap length. By contrast, in the Type B structure shown at the bottom of Figure 1.43, a TPC command is transmitted in slot#(M+1) during the idle period in order to optimise the power control.

The idle period has a variable duration, but the maximum period allowable within a 15-slot, 10 ms radio frame is seven slots. The idle period can occur either at the centre of a 10 ms frame or at the end and the beginning of two consecutive 10 ms frames, such that the idle period spans over two frames. However, in order to maintain the seamless operation of all MSs occupying the uncompressed 15-slot, 10 ms frame, the duration of all time-slots has

[9]Microcells overlaid by a macrocell.

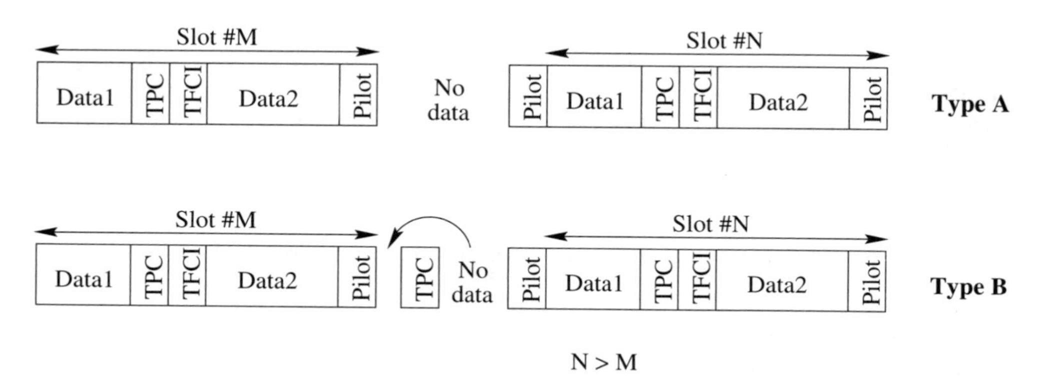

Figure 1.43: downlink frame structure in compressed mode operation during UTRA handovers using the transmission formats of Figure 1.19.

to be shortened by 'compressing' their data. The compression of data can be achieved by channel-code puncturing, a procedure that obliterates some of the coded parity bits, thereby slightly reducing the code's error correcting power, or by adjusting the spreading factor. In order to maintain the quality of the link, the instantaneous power is also increased during the compressed mode operation. After receiving the data, the MS can use this idle period in the 10 ms frame, to switch to other carrier frequencies of other cells and to perform the necessary link-quality measurements for handover.

Alternatively, a twin-receiver can be used in order to perform inter-frequency handovers. One receiver can be tuned to the desired carrier frequency for reception, while the other receiver can be used to perform handover link-quality measurements at other carrier frequencies. This method, however, results in a higher hardware complexity at the MS.

The 10 ms frame length of UTRA was chosen so that it is compatible with the multiframe length of 120 ms in GSM. Hence, the MS is capable of receiving the Frequency Correction Channel (FCCH) and Synchronisation Channel (SCH) messages in the GSM [28] frame using compressed mode transmission and to perform the necessary handover link-quality measurements [90].

1.3.2.11 Intercell Time Synchronisation in the UTRA TDD Mode

Time synchronisation between BSs is required when operating in the TDD mode in order to support seamless handovers. A simple method of maintaining intercell synchronisation is by periodically broadcasting a reference signal from a source to all the BSs. The propagation delay can be easily calculated, and hence compensated, from the fixed distance between the source and the receiving BSs. There are three possible ways of transmitting this reference signal, namely, via the terrestrial radio link, via the physical wired network, or via the Global Positioning System (GPS).

Global time synchronisation in 3G mobile radio systems is achieved by dividing the synchronous coverage region into three areas, namely, the so-called subarea, main area and coverage area, as shown in Figure 1.44. Intercell synchronisation within a sub-area is provided by a subarea reference BS. Since the subarea of Figure 1.44 is smaller than the main area, transmitting the reference signal via the terrestrial radio link or the physical wired network

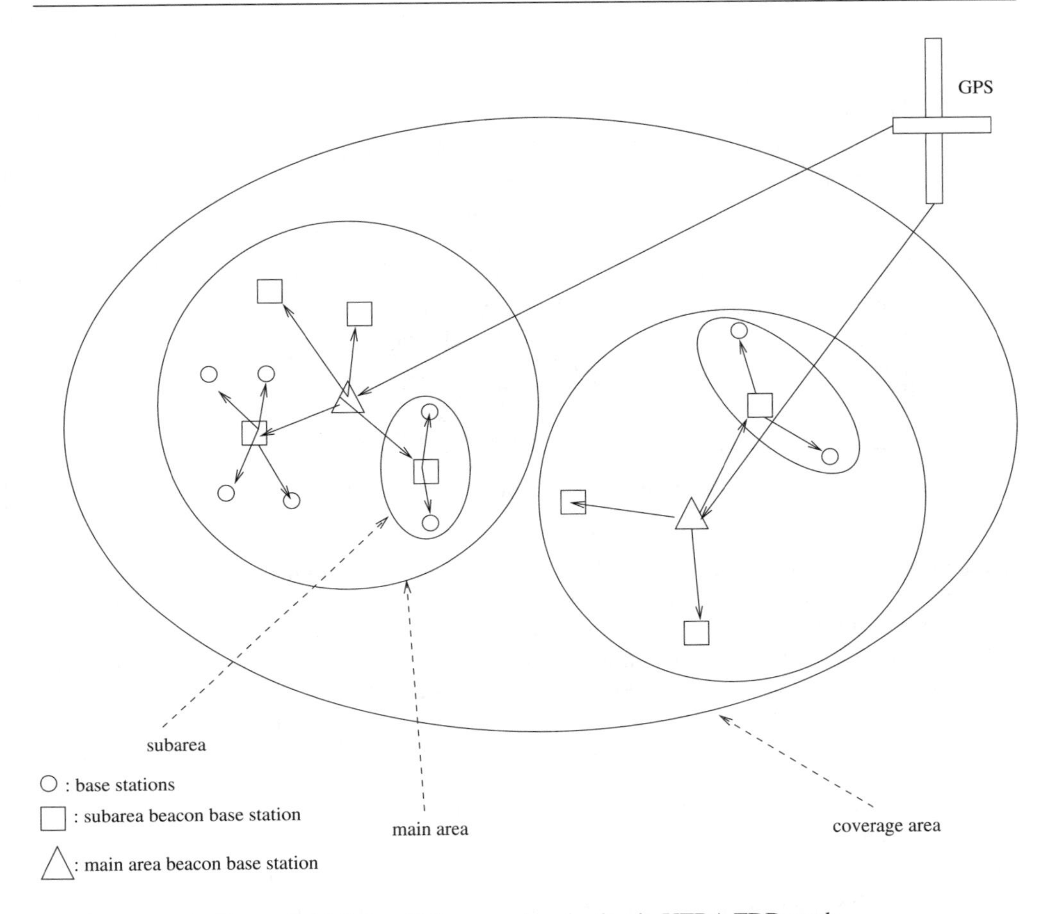

Figure 1.44: Intercell time synchronisation in UTRA TDD mode.

is more feasible. All the subarea reference BSs in a main area are in turn synchronised by a main-area reference BS. Similarly, the reference signal can be transmitted via the terrestrial radio link or the physical wired network. Finally, all the main-area reference BSs are synchronised using the GPS. The main advantage of dividing the coverage regions into smaller areas is that each lower hierarchical area can still operate on its own, even if the synchronisation link with the higher hierarchical areas is lost.

1.3.3 The cdma2000 Terrestrial Radio Access [124–126]

The current 2G mobile radio systems standardised by TIA in the United States are IS-95-A and IS-95-B [124]. The radio access technology of both systems is based on narrowband DS-CDMA with a chip rate of 1.2288 Mcps, which gives a bandwidth of 1.25 MHz. IS-95-A was commercially launched in 1995, supporting circuit and packet mode transmissions at a maximum bit rate of only 14.4 kbps [124]. An enhancement to the IS-95-A standards, known as IS-95-B, was developed and introduced in 1998 in order to provide higher data

Radio Access Technology	DS-CDMA, Multicarrier CDMA
Operating environments	Indoor/Outdoor to indoor/Vehicular
Chip rate (Mcps)	1.2288/3.6864/7.3728/11.0592/14.7456
Channel bandwidth (MHz)	1.25/3.75/7.5/11.25/15
Duplex modes	FDD and TDD
Frame length	5 and 20 ms
Spreading factor	variable, 4 to 256
Detection scheme	Coherent with common pilot channel
Intercell operation	FDD : Synchronous TDD : Synchronous
Power control	Open and closed loop
Handover	Soft-handover Inter-frequency handover

Table 1.10: The cdma2000 basic parameters.

rates, on the order of 115.2 kbps [31]. This was feasible without changing the physical layer of IS-95-A. However, this still falls short of the 3G mobile radio system requirements. Hence, the technical committee TR45.5 within TIA has proposed cdma2000, a 3G mobile radio system that is capable of meeting all the requirements laid down by ITU. One of the problems faced by TIA is that the frequency bands allocated for the 3G mobile radio system, identified during WARC'92 to be 1885–2025 MHz and 2110–2200 MHz, have already been allocated for Personal Communications Services (PCS) in the United States from 1.8 GHz to 2.2 GHz. In particular, the CDMA PCS based on the IS-95 standards has been allocated the frequency bands of 1850–1910 MHz and 1930–1990 GHz. Hence, the 3G mobile radio systems have to fit into the allocated bandwidth without imposing significant interference on the existing applications. Thus, the framework for cdma2000 was designed so that it can be overlaid on IS-95 and it is backwards compatible with IS-95. Most of this section is based on [124–126].

1.3.3.1 Characteristics of cdma2000

The basic parameters of cdma2000 are shown in Table 1.10. The cdma2000 system has a basic chip rate of 3.6864 Mcps, which is accommodated in a bandwidth of 3.75 MHz. This chip rate is in fact three times the chip rate used in the IS-95 standards, which is 1.2288 Mcps. Accordingly, the bandwidth was also trebled. Hence, the existing IS-95 networks can also be used to support the operation of cdma2000. Higher chip rates on the order of $N \times 1.2288$ Mcps, $N = 6, 9, 12$ are also supported. These are used to enable higher bit rate transmission. The value of N is an important parameter in determining the channel-coding rate and the channel bit rate. In order to transmit the high chip-rate signals ($N > 1$), two modulation techniques are employed. In the *direct-spread modulation mode*, the symbols are spread according to the chip rate and transmitted using a single carrier, giving a bandwidth of $N \times 1.25$ MHz. This method is used on both the uplink and downlink. In *multicarrier (MC) modulation*, the symbols to be transmitted are de-multiplexed into separate signals, each of which is then spread at a chip rate of 1.2288 Mcps. N different carrier frequencies are used to transmit these spread signals, each of which has a bandwidth of 1.25 MHz. This method

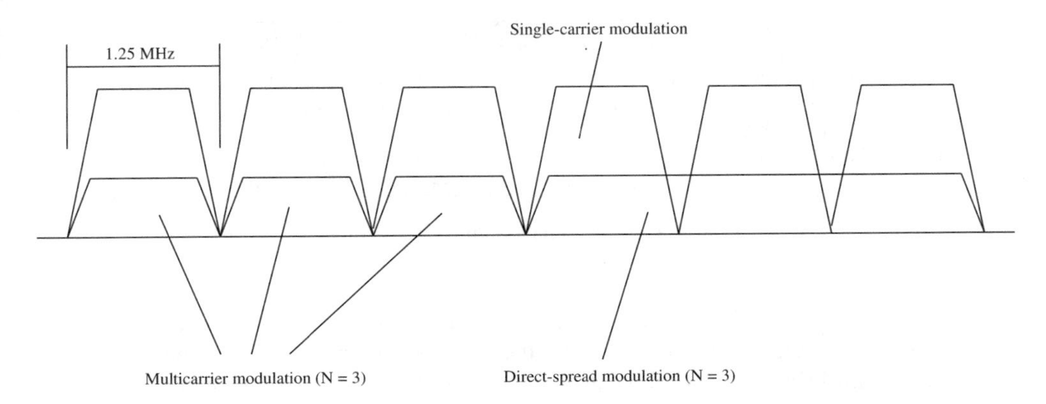

Figure 1.45: Example of an overlay deployment in cdma2000. The multicarrier mode is only used in the downlink.

is used for the downlink only, because in this case, transmit diversity can be achieved by transmitting the different carrier frequencies over spatially separated antennas.

By using multiple carriers, cdma2000 is capable of overlaying its signals on the existing IS-95 1.25 MHz channels and its own channels, while maintaining orthogonality. An example of an overlay scenario is shown in Figure 1.45. Higher chip rates are transmitted at a lower power than lower chip rates, thereby keeping the interferences to a minimum.

Similarly to UTRA and IMT-2000, cdma2000 also supports TDD operation in unpaired frequency bands. In order to ease the implementation of a dual-mode FDD/TDD terminal, most of the techniques used for FDD operation can also be applied in TDD operation. The difference between these two modes is in the frame structure, whereby an additional guard time has to be included for TDD operation.

In contrast to UTRA and IMT-2000, where the pilot symbols of Figure 1.21 are time-multiplexed with the dedicated data channel on the downlink, cdma2000 employs a common code multiplexed continuous pilot channel on the downlink, as in the IS-95 system. The advantage of a common downlink pilot channel is that no additional overhead is incurred for each user. However, if adaptive antennas are used, then additional pilot channels have to be transmitted from each antenna.

Another difference with respect to UTRA and IMT-2000 is that the base stations are operated in synchronous mode in cdma2000. As a result, the same PN code but with different phase offsets can be used to distinguish the base stations. Using one common PN sequence can expedite cell acquisition as compared to a set of PN sequences, as we have seen in Section 1.3.2.9 for IMT-2000/UTRA. Let us now consider the cdma2000 physical channels.

1.3.3.2 Physical Channels in cdma2000

The physical channels (PHCH) in cdma2000 can be classified into two groups, namely Dedicated Physical Channels (DPHCH) and Common Physical Channels (CPHCH). DPHCHs carry information between the base station and a single mobile station, while CPHCHs carry information between the base station and several mobile stations. Table 1.11 shows the collection of physical channels in each group. These channels will be elaborated on during our

Dedicated Physical Channels (DPHCH)	Common Physical Channels (CPHCH)
Fundamental Channel (FCH) (UL/DL)	Pilot Channel (PICH) (DL)
Supplemental Channel (SCH) (UL/DL)	Common Auxiliary Pilot Channel (CAPICH) (DL)
Dedicated Control Channel (DCCH) (UL/DL)	Forward Paging Channel (PCH) (DL)
Dedicated Auxiliary Pilot Channel (DAPICH) (DL)	Sync Channel (SYNC) (DL)
Pilot Channel (PICH) (UL)	Access Channel (ACH) (UL)
	Common Control Channel (CCCH) (UL/DL)

Table 1.11: The cdma2000 physical channels.

further discourse. Typically, all physical channels are transmitted using a frame length of 20 ms. However, the control information on the so-called Fundamental Channel (FCH) and Dedicated Control Channel (DCCH) can also be transmitted in 5 ms frames.

Each base station transmits its own downlink Pilot Channel (PICH), which is shared by all the mobile stations within the coverage area of the base station. Mobile stations can use this common downlink PICH in order to perform channel estimation for coherent detection, soft handover, and fast acquisition of strong multipath rays for RAKE combining. The PICH is transmitted orthogonally along with all the other downlink physical channels from the base station by using a unique orthogonal code (Walsh code 0) as in the IS-95 system The optional Common Auxiliary Pilot Channels (CAPICH) and Dedicated Auxiliary Pilot Channels (DAPICH) are used to support the implementation of antenna arrays. CAPICHs provide spot coverage shared among a group of mobile stations, while a DAPICH is directed toward a particular mobile station. Every mobile station also transmits an orthogonal code-multiplexed uplink pilot channel (PICH), which enables the base station to perform coherent detection in the uplink as well as to detect strong multipaths and to invoke power control measurements. This differs from IS-95, which supports only noncoherent detection in the uplink due to the absence of a coherent uplink reference. In addition to the pilot symbols, the uplink PICH also contains time-multiplexed power control bits assisting in downlink power control. A power control bit is multiplexed onto the 20 ms frame every 1.25 ms, giving a total of 16 power control bits per 20 ms frame or 800 power updates per second, implying a very agile, fast response power control regime. Each 1.25 ms duration is referred to as a Power Control Group, as shown in Figure 1.46.

The use of two dedicated data physical channels, namely, the so-called Fundamental (FCH) and Supplemental (SCH) channels, optimises the system during multiple simultaneous service transmissions. Each channel carries a different type of service and is coded and interleaved independently. However, in any connection, there can be only one FCH, but several SCHs can be supported. For a FCH transmitted in a 20 ms frame, two sets of uncoded data rates, denoted as Rate Set 1 (RS1) and Rate Set 2 (RS2), are supported. The data rates in RS1 and RS2 are 9.6/4.8/2.7/1.5 kbps and 14.4/7.2/3.6/1.8 kbps, respectively. Regardless of the uncoded data rates, the coded data rate is 19.2 kbps and 38.4 kbps for RS1 and RS2, respectively, when the rate-control parameter is $N = 1$. The 5 ms frame only supports one data rate, which is 9.6 kbps. The SCH is capable of transmitting higher data rates than the FCH. The SCH supports variable data rates ranging from 1.5 kbps for N=1 to as high as 2073.6 kbps, when N=12. Blind rate detection [127] is used for SCHs not exceeding 14.4 kbps, while rate information is explicitly provided for higher data rates. The dedicated control physical channel has a fixed uncoded data rate of 9.6 kbps on both 5 ms and 20 ms frames. This control channel rate is more than an order of magnitude higher than that of the IS-95 system hence it

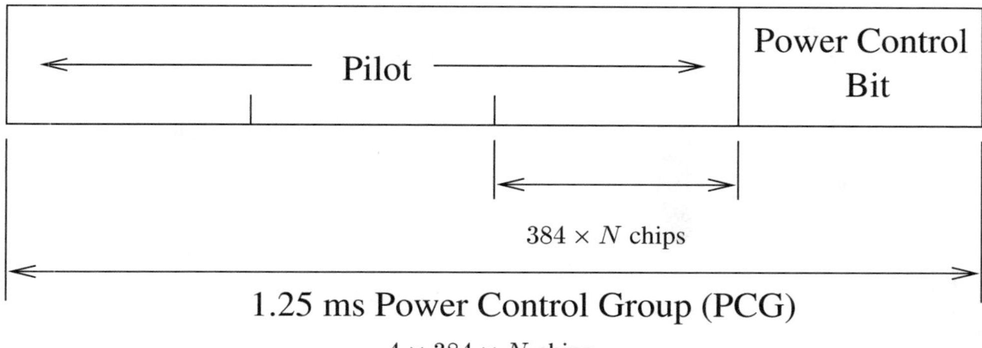

Figure 1.46: Uplink pilot channel structure in cdma2000 for a 1.25 ms duration PCG, where $N = 1, 3, 6, 9, 12$ is the rate-control parameter.

supports a substantially enhanced system control.

The Sync Channel (SYCH) — note the different acronym in comparison to the SCH abbreviation in UTRA/IMT-2000 — is used to aid the initial synchronisation of a mobile station to the base station and to provide the mobile station with system-related information, including the Pseudo Noise (PN) sequence offset, which is used to identify the base stations and the long code mask, which will be defined explicitly in Section 1.3.3.4. The SYCH has an uncoded data rate of 1.2 kbps and a coded data rate of 4.8 kbps.

Paging functions and packet data transmission are handled by the downlink Paging Channel (PCH) and the downlink Common Control Channel (CCCH). The uncoded data rate of the PCH can be either 4.8 kbps or 9.6 kbps. The CCCH is an improved version of the PCH, which can support additional higher data rates, such as 19.2 and 38.4 kbps. In this case, a 5 ms or 10 ms frame length will be used. The PCH is included in cdma2000 in order to provide IS-95-B functionality.

In TDD mode, the 20 ms and 5 ms frames are divided into 16 and 4 time-slots, respectively. This gives a duration of 1.25 ms per time-slot, as shown in Figure 1.47. A guard time of 52.08 μs and 67.44 μs is used for the downlink in multicarrier modulation and for direct-spread modulation, respectively. In the uplink, the guard time is 52.08 μs. Having described the cdma2000 physical channels of Table 1.11, let us now consider the service multiplexing and channel-coding aspects.

1.3.3.3 Service Multiplexing and Channel Coding

Services of different data rates and different QoS requirements are carried by different physical channels, namely, by the FCH and SCH of Table 1.11. This differs from UTRA and IMT-2000, whereby different services were time-multiplexed onto one or more physical channels, as highlighted in Section 1.3.2.4. These channels in cdma2000 are code-multiplexed using Walsh codes. Two types of coding schemes are used in cdma2000, as shown in Table 1.12. Basically, all channels use convolutional codes for forward error correction. However, for SCHs at rates higher than 14.4 kbps, turbo coding [108] is preferable. The rate of the input data stream is matched to the given channel rate by either adjusting the coding rate or using

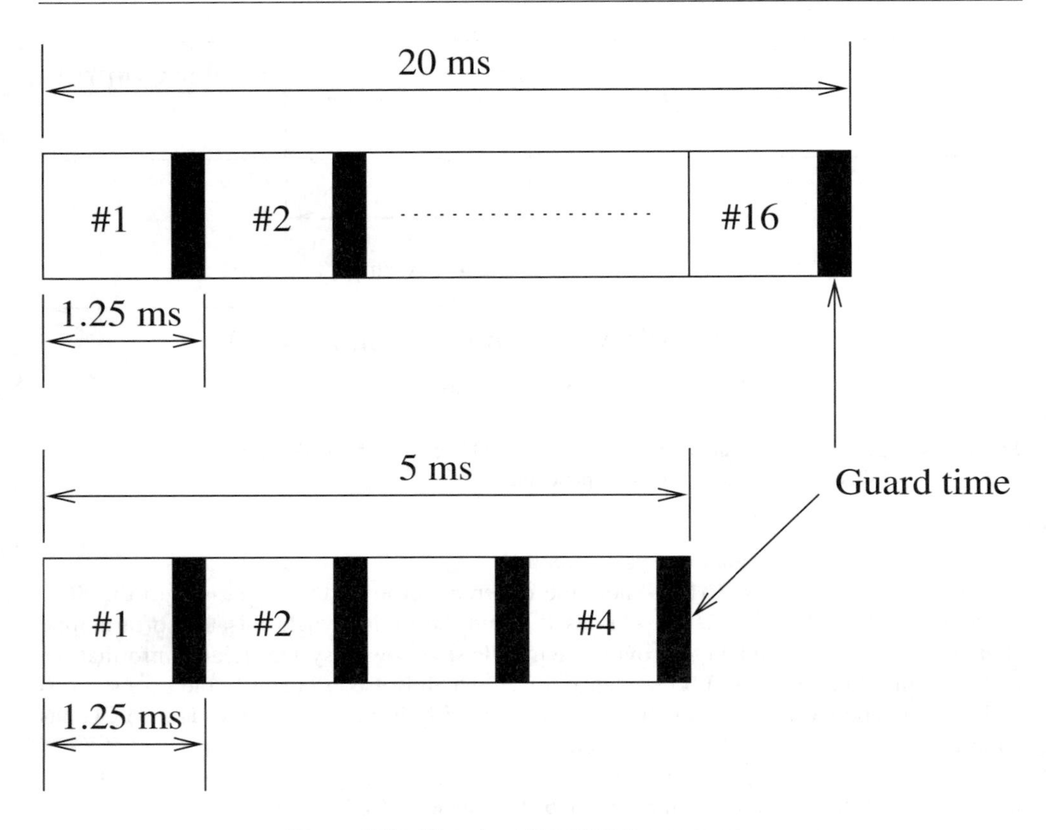

Figure 1.47: The cdma2000 TDD frame structure.

	Convolutional	Turbo
Rate	1/2 or 1/3 or 1/4	1/2 or 1/3 or 1/4
Constraint length	9	4

Table 1.12: The cdma2000 channel-coding parameters.

symbol repetition with and without symbol puncturing, or alternatively, by sequence repetition. Tables 1.13 and 1.14 show the coding rate and the associated rate matching procedures for the various downlink and uplink physical channels, respectively, when $N = 1$. Following the above brief notes on the cdma2000 channel coding and service multiplexing issues, let us now turn to the spreading and modulation processes.

1.3.3.4 Spreading and Modulation

There are generally three layers of spreading in cdma2000, as shown in Table 1.15. Each user's uplink signal is identified by different offsets of a long code, a procedure that is similar to that of the IS-95 system portrayed in [128]. As seen in Table 1.15, this long code is an m-sequence with a period of $2^{42} - 1$ chips. The construction of m-sequences was highlighted

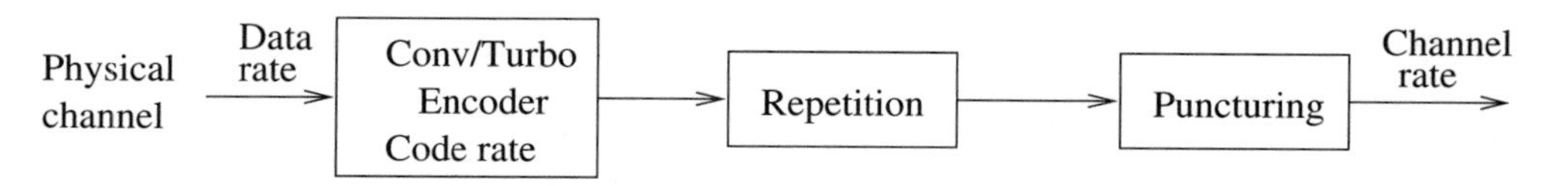

Physical Channel	Data Rate	Code Rate	Repetition	Puncturing	Channel Rate
SYCH	1.2 kbps	1/2	×2	0	4.8 ksps
PCH	4.8 kbps	1/2	×2	0	19.2 ksps
	9.6 kbps	1/2	×1	0	19.2 ksps
CCCH	9.6 kbps	1/2	×1	0	19.2 ksps
	19.2 kbps	1/2	×1	0	38.4 ksps
	38.4 kbps	1/2	×1	0	76.8 ksps
FCH	1.5 kbps	1/2	×8	1 of 5	19.2 ksps
	2.7 kbps	1/2	×4	1 of 9	19.2 ksps
	4.8 kbps	1/2	×2	0	19.2 ksps
	9.6 kbps	1/2	×1	0	19.2 ksps
	1.8 kbps	1/3	×8	1 of 9	38.4 ksps
	3.6 kbps	1/3	×4	1 of 9	38.4 ksps
	7.2 kbps	1/3	×2	1 of 9	38.4 ksps
	14.4 kbps	1/3	×1	1 of 9	38.4 ksps
SCH	9.6 kbps	1/2	×1	0	19.2 ksps
	19.2 kbps	1/2	×1	0	38.4 ksps
	38.4 kbps	1/2	×1	0	76.8 ksps
	76.8 kbps	1/2	×1	0	153.6 ksps
	153.6 kbps	1/2	×1	0	307.2 ksps
	307.2 kbps	1/2	×1	0	614.4 ksps
	14.4 kbps	1/3	×1	1 of 9	38.4 ksps
	28.8 kbps	1/3	×1	1 of 9	76.8 ksps
	57.6 kbps	1/3	×1	1 of 9	153.6 ksps
	115.2 kbps	1/3	×1	1 of 9	307.2 ksps
	230.4 kbps	1/3	×1	1 of 9	614.4 ksps
DCCH	9.6 kbps	1/2	×1	0	19.2 ksps

Table 1.13: The cdma2000 downlink physical channel (see Table 1.11) coding parameters for $N = 1$, Where repetition × 2 implies transmitting a total of two copies.

by proakis [5]. Different user offsets are obtained using a long code mask. Orthogonality between the different physical channels of the same user belonging to the same connection in the uplink is maintained by spreading using Walsh codes.

In contrast to the IS-95 downlink of Figure 1.42 of [128], whereby Walsh code spreading is performed prior to QPSK modulation, the data in cdma2000 is first QPSK modulated before spreading the resultant I and Q branches with the same Walsh code. In this way, the number of Walsh codes available is increased twofold due to the orthogonality of the I and Q carriers. The length of the uplink/downlink (UL/DL) channelisation Walsh codes of Table 1.15 varies according to the data rates. All the base stations in the system are

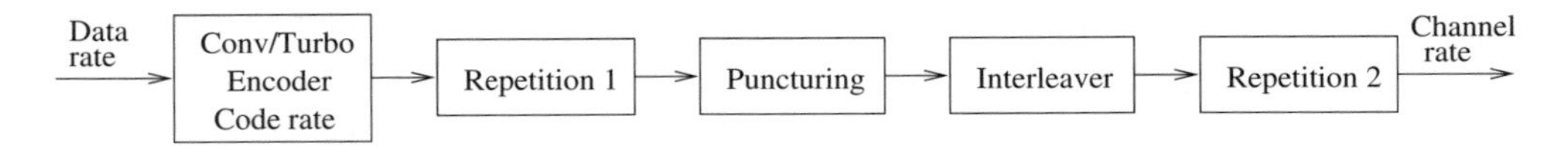

Physical Channel	Data Rate	Code Rate	Repetition 1	Puncturing	Repetition 2	Channel Rate
CCCH	19.2 kbps	1/4	×1	0	×4	307.2 ksps
	38.4 kbps	1/4	×1	0	×2	307.2 ksps
FCH	1.5 kbps	1/4	×8	1 of 5	×8	307.2 ksps
	2.7 kbps	1/4	×4	1 of 9	×8	307.2 ksps
	4.8 kbps	1/4	×2	0	×8	307.2 ksps
	9.6 kbps	1/4	×1	0	×8	307.2 ksps
	1.8 kbps	1/4	×16	1 of 3	×4	307.2 ksps
	3.6 kbps	1/4	×8	1 of 3	×4	307.2 ksps
	7.2 kbps	1/4	×4	1 of 3	×4	307.2 ksps
	14.4 kbps	1/4	×2	1 of 3	×4	307.2 ksps
SCH	9.6 kbps	1/4	×1	0	×16	614.4 ksps
	19.2 kbps	1/4	×1	0	×8	614.4 ksps
	38.4 kbps	1/4	×1	0	×4	614.4 ksps
	76.8 kbps	1/4	×1	0	×2	614.4 ksps
	153.6 kbps	1/4	×1	0	×1	614.4 ksps
	307.2 kbps	1/2	×1	0	×1	614.4 ksps
ACH	4.8 kbps	1/4	×1	0	×8	307.2 ksps
	9.6 kbps	1/4	×1	0	×4	307.2 ksps
DCCH	9.6 kbps	1/4	×1	0	×4	307.2 ksps

Table 1.14: The cdma2000 uplink physical channel (see Table 1.11) coding parameters for $N = 1$, where repetition $\times$ 2 implies transmitting a total of two copies.

	Channelisation Codes (UL/DL)	User-specific Scrambling Codes (UL)	Cell-specific Scrambling Codes (DL)
Type of codes	Walsh codes	Different offsets of a real m-sequence	Different offsets of a complex m-sequence
Code length	Variable	$2^{42} - 1$ chips	2^{15} chips
Type of Spreading	BPSK	BPSK	QPSK
Data Modulation	DL : QPSK UL : BPSK		

Table 1.15: Spreading parameters in cdma2000.

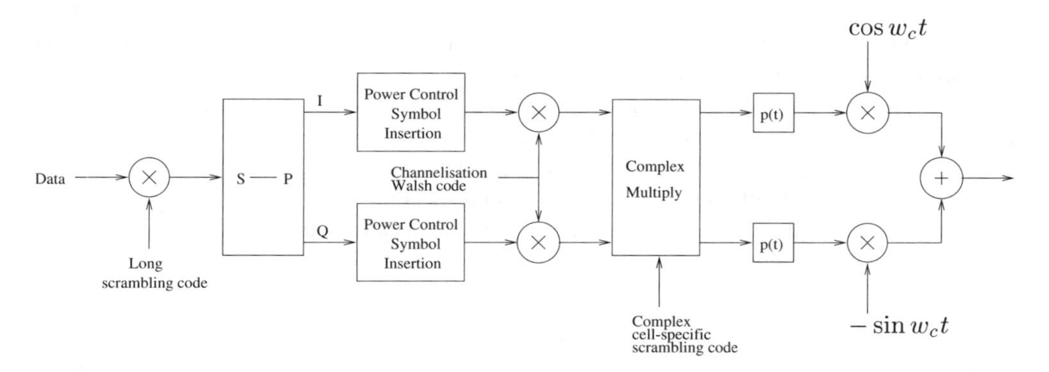

Figure 1.48: The cdma2000 downlink transmitter. The long scrambling code is used for the purpose of improving user privacy. Hence, only the paging channels and the traffic channels are scrambled with the long code. The common pilot channel and the SYNC channel are not scrambled by this long code (the terminology of Table 1.15 is used).

distinguished by different offsets of the same complex downlink m-sequence, as indicated by Table 1.15. This downlink m-sequence code is the same as that used in IS-95, which has a period of $2^{15} = 32768$, and it is derived from m-sequences. The feedback polynomials of the shift registers for the I and Q sequences are $X^{15} + X^{13} + X^9 + X^8 + X^7 + X^5 + 1$ and $X^{15} + X^{12} + X^{11} + X^{10} + X^6 + X^5 + X^4 + X^3 + 1$, respectively. The offset of these codes must satisfy a minimum value, which is equal to $N \times 64 \times$Pilot_Inc, where Pilot_Inc is a code reuse parameter, which depends on the topology of the system, analogously to the frequency reuse factor in FDMA. Let us now focus on downlink spreading issues more closely.

1.3.3.4.1 Downlink Spreading and Modulation Figure 1.48 shows the structure of a downlink transmitter for a physical channel. In contrast to the IS-95 downlink transmitter shown in [128], the data in the cdma2000 downlink transmitter shown in Figure 1.48 are first QPSK modulated before spreading using Walsh codes. As a result, the number of Walsh codes available is increased twofold due to the orthogonality of the I and Q carriers, as mentioned previously. The user data are first scrambled by the long scrambling code by assigning a different offset to different users for the purpose of improving user privacy, which is then mapped to the I and Q channels. This long, scrambling code is identical to the uplink user-specific scrambling code given in Table 1.15. The downlink pilot channels of Table 1.11 (PICH, CAPICH, DAPICH) and the SYNC channel are not scrambled with a long code since there is no need for user-specificity. The uplink power control symbols are inserted into the FCH at a rate of 80 Hz, as shown in Figure 1.48. The I and Q channels are then spread using a Walsh code and complex multiplied with the cell-specific complex PN sequence of Table 1.15, as portrayed in Figure 1.48. Each base station's downlink channel is assigned a different Walsh code in order to eliminate any intracell interference since all Walsh codes transmitted by the serving base station are received synchronously. The length of the downlink channelisation Walsh code of Table 1.15 is determined by the type of physical channel and its data rate. Typically for $N = 1$, downlink FCHs with data rates belonging to RS1, that is, those transmitting at 9.6/4.8/2.7/1.5 kbps, use a 128-chip Walsh code, and those in RS2, transmitting at 14.4/7.2/3.6/1.8 kbps, use a 64-chip Walsh code. Walsh codes for downlink SCHs can range from 4-chip to 128-chip Walsh codes. The downlink PICH is an unmodulated sequence (all 0 s) spread by Walsh code 0. Finally, the complex spread data in Figure 1.48 are baseband filtered using the Nyquist filter impulse responses $p(t)$ in Figure 1.48 and modulated on a carrier frequency.

For the case of multicarrier modulation, the data is split into N branches immediately after the long code scrambling of Figure 1.48 which was omitted in the figure for the sake of simplicity. Each of the N branches is then treated as a separate transmitter and modulated using different carrier frequencies.

1.3.3.4.2 Uplink Spreading and Modulation The uplink cdma2000 transmitter is shown in Figure 1.49. The uplink PICH and DCCH of Table 1.11 are mapped to the I data channel, while the uplink FCH and SCH of Table 1.11 are mapped to the Q channel in Figure 1.49. Each of these uplink physical channels belonging to the same user is assigned different Walsh channelisation codes in order to maintain orthogonality, with higher rate channels using shorter Walsh codes. The I and Q data channels are then spread by complex multiplication with the user-specifically offset real m-sequence based scrambling code of Table 1.15 and a complex scrambling code, which is the same for all the mobile stations in the system, as seen at the top of Figure 1.49. However, this latter complex scrambling code is not explicitly shown in Table 1.15, since it is identical to the downlink cell-specific scrambling code. This complex scrambling code is only used for the purpose of quadrature spreading. Thus, in order to reduce the complexity of the base station receiver, this complex scrambling code is identical to the cell-specific scrambling code of Table 1.15 used on the downlink by all the base stations.

1.3.3.5 Random Access

The mobile station initiates an access request to the network by repeatedly transmitting a so-called access probe until a request acknowledgement is received. This entire process of sending a request is known as an access attempt. Within a single access attempt, the request may be sent to several base stations. An access attempt addressed to a specific base station is known as a subattempt. Within a subattempt, several access probes with increasing power can be sent. Figure 1.50 shows an example of an access attempt. The access probe transmission follows the slotted ALOHA algorithm, which is a relative of PRMA. An access probe can be divided into two parts, as shown in Figure 1.51. The access preamble carries a nondata-bearing pilot channel at an increased power level. The so-called access channel message capsule carries the data-bearing Access Channel (ACH) or uplink Common Control Channel (CCCH) messages of Table 1.11 and the associated nondata-bearing pilot channel. The structure of the pilot channel is similar to that of the uplink pilot channel (PICH) of Figure 1.46 except that in this case there are no time-multiplexed power control bits. The preamble length in Figure 1.51 is an integer multiple of the 1.25 ms slot intervals. The specific access preamble length is indicated by the base station, which depends on how fast the base station can search the PN code space in order to recognise an access attempt. The ACH is transmitted at a fixed rate of either 9.6 or 4.8 kbps, as seen in Table 1.14. This rate is constant for the duration of the access probe of Figure 1.50. The ACH or CCCH and their associated pilot channel are spread by the spreading codes of Table 1.15, as shown in Figure 1.52. Different ACHs or CCCHs and their associated pilot channels are spread by different long codes.

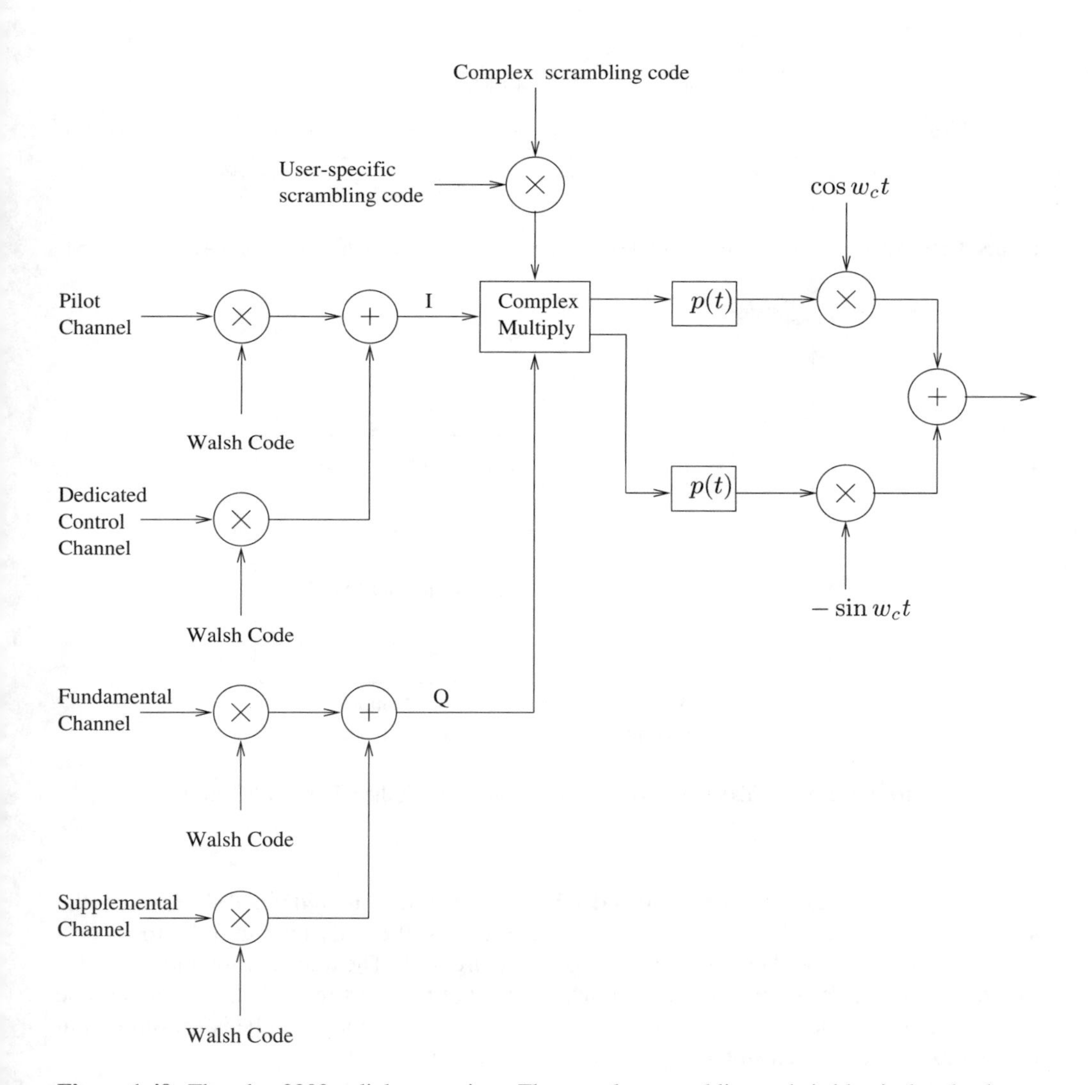

Figure 1.49: The cdma2000 uplink transmitter. The complex scrambling code is identical to the downlink cell-specific complex scrambling code of Table 1.15 used by all the base stations in the system (the terminology of Table 1.15 is used).

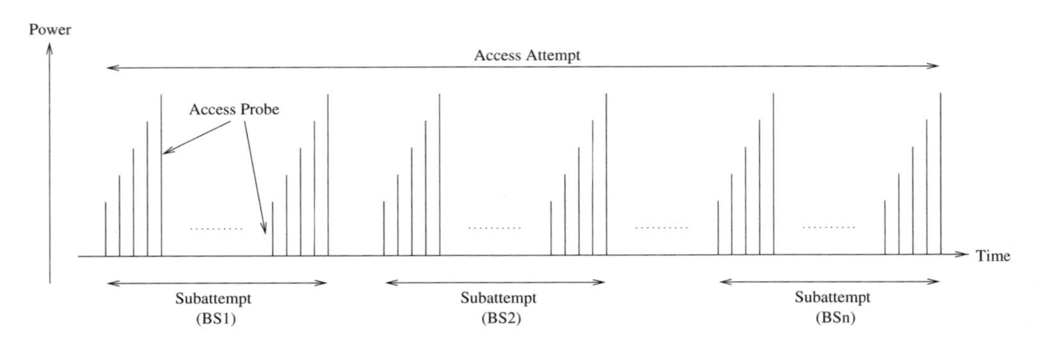

Figure 1.50: An access attempt by a mobile station in cdma2000 using the access probe of Figure 1.51.

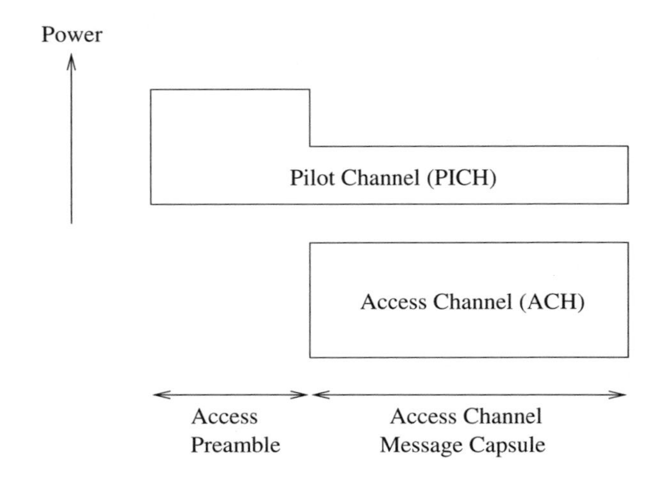

Figure 1.51: A cdma2000 access probe transmitted using the regime of Figure 1.50.

The access probes of Figures 1.50 and 1.51 are transmitted in predefined slots, where the slot length is indicated by the base station. Each slot is sufficiently long in order to accommodate the preamble and the longest message of Figure 1.51. The transmission must begin at the start of each 1.25 ms slot. If an acknowledgement of the most recently transmitted probe is not received by the mobile station after a time-out period, another probe is transmitted in another randomly chosen slot, obeying the regime of Figure 1.50.

Within a subattempt of Figure 1.50, a sequence of access probes is transmitted until an acknowledgement is received from the base station. Each successive access probe is transmitted at a higher power compared to the previous access probe, as shown in Figure 1.53. The initial power (IP) of the first probe is determined by the open-loop power control plus a nominal offset power that corrects for the open-loop power control imbalance between uplink and downlink. Subsequent probes are transmitted at a power level higher than the previous probe. This increased level is indicated by the Power Increment (PI). Let us now highlight some of the cdma2000 handover issues.

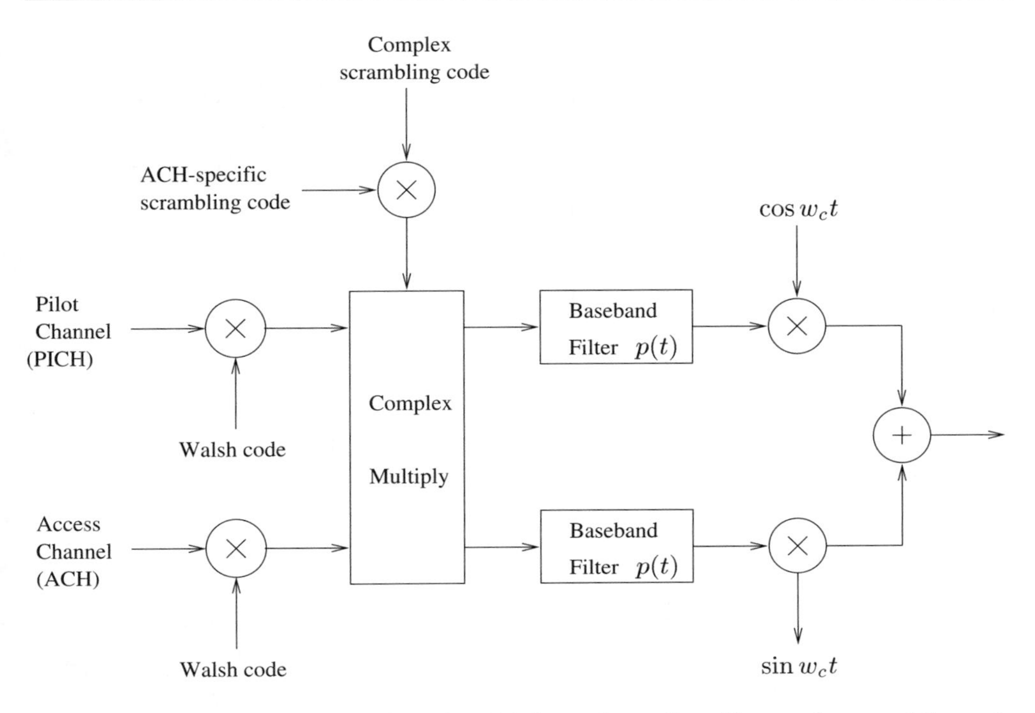

Figure 1.52: The cdma2000 access channel modulation and spreading. The complex scrambling code is identical to the downlink cell-specific complex scrambling code of Table 1.15 used by all the base stations in the system (the terminology of Table 1.15 is used).

1.3.3.6 Handover

Intra-frequency or soft-handover is initiated by the mobile station. While communicating, the mobile station may receive the same signal from several base stations. These base stations constitute the Active Set of the mobile station. The mobile station will continuously monitor the power level of the received pilot channels (PICH) transmitted from neighbouring base stations, including those from the mobile station's active set. The power levels of these base stations are then compared to a set of thresholds according to an algorithm, which will be highlighted later in this chapter. The set of thresholds consists of the static thresholds, which are maintained at a fixed level, and the dynamic thresholds, which are dynamically adjusted based on the total received power. Subsequently, the mobile station will inform the network when any of the monitored power levels exceed the thresholds.

Whenever the mobile station detects a PICH, whose power level exceeds a given static threshold, denoted as T_1, this PICH will be moved to a candidate set and will be searched and compared more frequently against a dynamically adjusted threshold denoted as T_2. This value of T_2 is a function of the received power levels of the PICHs of the base stations in the active set. This process will determine whether the candidate base station is worth adding to the active set. If the overall power level in the active set is weak, then adding a base station of higher power will improve the reception. By contrast, if the overall power level in the active set is relatively high, then adding another high-powered base station may not only be

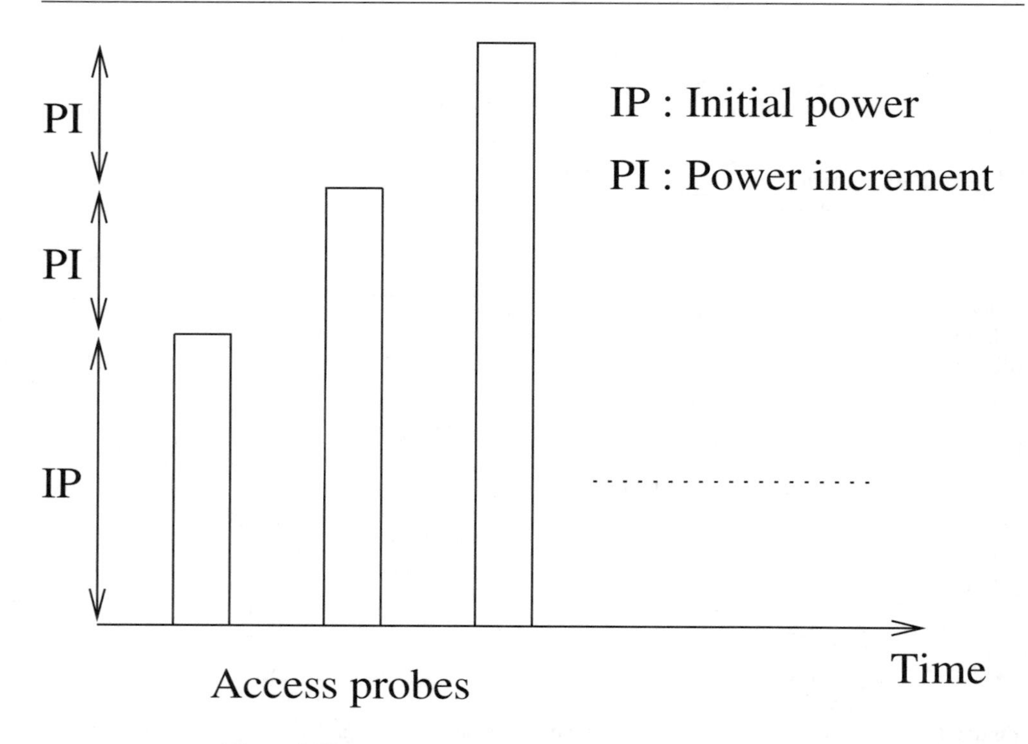

Figure 1.53: Access probes within a subattempt of Figure 1.50.

unnecessary, but may actually utilise more network resources.

For the base stations that are already in the active set, the power level of their corresponding PICH is compared to a dynamically adjusted threshold, denoted as T_3, which is also a function of the total power of the PICH in the active set, similar to T_2. This is to ensure that each base station in the active set is contributing sufficiently to the overall power level. If any of the PICH's power level dropped below T_3 after a specified period of time allowed in order to eliminate any uncertainties due to fading which may have caused fluctuations in the power level, the base station will again be moved to the candidate set where it will be compared with a static threshold T_4. At the same time, the mobile station will report to the network the identity of the low-powered base station in order to allow the corresponding base station to increase its transmit power. If the power level decreases further below a static threshold, denoted as T_4, then the mobile station will again report this to the network and the base station will subsequently be dropped from the candidate set.

Inter-frequency or hard-handovers can be supported between cells having different carrier frequencies. Here we conclude our discussions on the cdma2000 features and provide some rudimentary notes on a number of advanced techniques, which can be invoked in order to improve the performance of the 3G W-CDMA systems.

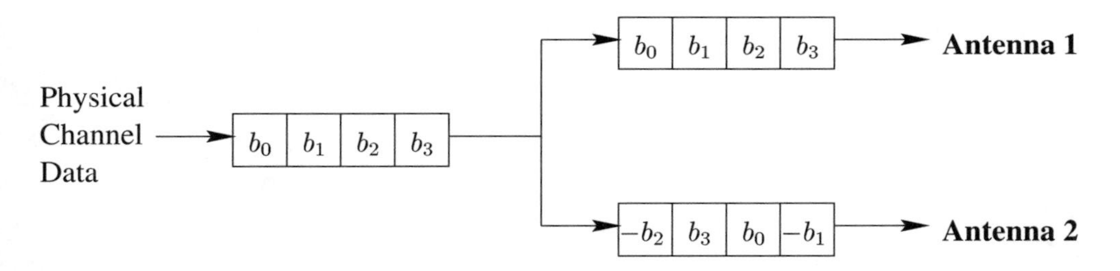

Figure 1.54: Transmission of a physical channel using Space Time block coding Transmit Diversity (STTD).

1.3.4 Performance-Enhancement Features

The treatment of adaptive antennas, multi-user detection, interference cancellation, or the portrayal of transmit diversity techniques is beyond the scope of this chapter. Here we simply provide a few pointers to the associated literature.

1.3.4.1 Downlink Transmit Diversity Techniques

1.3.4.1.1 Space Time Block Coding-Based Transmit Diversity Further diversity gain can be provided for the mobile stations by upgrading the base station with the aid of Space Time block coding assisted Transmit Diversity (STTD) [119], which can be applied to all the DL physical channels with the exception of the SCH. Typically the data of physical channels are encoded and transmitted using two antennas, as shown in Figure 1.54.

1.3.4.1.2 Time-Switched Transmit Diversity Time-Switched Transmit Diversity (TSTD) [129] is only applicable to the SCH, and its operation becomes explicit in Figure 1.55.

1.3.4.1.3 Closed-Loop Transmit Diversity Closed-loop transmit diversity is only applicable to the DPCH and PDSCH messages of Table 1.5 on the DL, which is illustrated in Figure 1.56. The weights w_1 and w_2 are related to the DL channel's estimated phase and attenuation information, which are determined and transmitted by the MS to the BS using the FBI D field, as portrayed in Figure 1.20. The weights for each antenna are independently measured by the MS using the corresponding pilot channels CPICH1 and CPICH2.

1.3.4.2 Adaptive Antennas

The transmission of time-multiplexed user-specific pilot symbols on both the UL and DL as seen for UTRA in Figures 1.20–1.24 facilitates the employment of adaptive antennas. Adaptive antennas are known to enhance the capacity and coverage of the system [130, 131].

1.3.4.3 Multi-User Detection/Interference Cancellation

Following Verdú's seminal paper [66], extensive research has shown that Multi-user Detection (MUD) [65, 68, 132–137] and Interference Cancellation techniques [64, 138–148] can

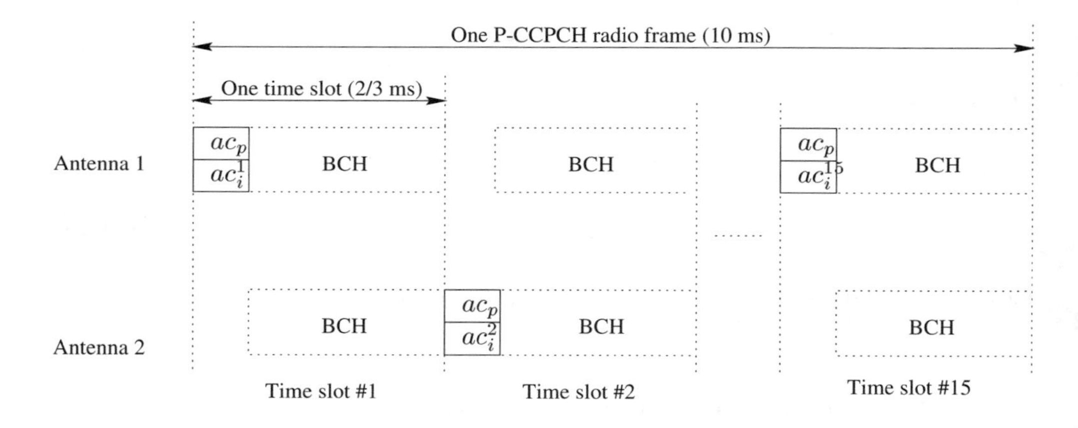

Figure 1.55: Frame structure of the UTRA DL synchronisation channel (SCH), transmitted by a TSTD scheme. The primary and secondary SCH are transmitted alternatively from Antennas 1 and 2. The parameter a is a binary flag used to indicate the presence ($a = +1$) or absence ($a = -1$) of STTD encoding in the P-CCPCH.

substantially improve the performance of the CDMA link in comparison to conventional RAKE receivers. However, using long scrambling codes increases the complexity of the MUD [31]. As a result, UTRA introduced an optional short scrambling code, namely, the S(2) code of Table 1.9, as mentioned in Section 1.3.2.6.4, in order to reduce the complexity of MUD [91]. Another powerful technique is invoking burst-by-burst adaptive CDMA [69, 105] in conjunction with MUD.

However, interference cancellation and MUD schemes require accurate channel estimation, in order to reproduce and deduct or cancel the interference. Several stages of cancellation are required in order to achieve a good performance, which in turn increases the canceller's complexity. It was shown that recursive channel estimation in a multistage interference canceller improved the accuracy of the channel estimation and hence gave improved BER performance [84].

Because of the complexity of the multi-user or interference canceller detectors, they were originally proposed for the UL. However, recently reduced-complexity DL MUD techniques have also been proposed [149].

1.3.5 Summary of 3G Systems

We have presented an overview of the terrestrial radio transmission technology of 3G mobile radio systems proposed by ETSI, ARIB, and TIA. All three proposed systems are based on Wideband-CDMA. Despite the call for a common global standard, there are some differences in the proposed technologies, notably, the chip rates and intercell operation. These differences

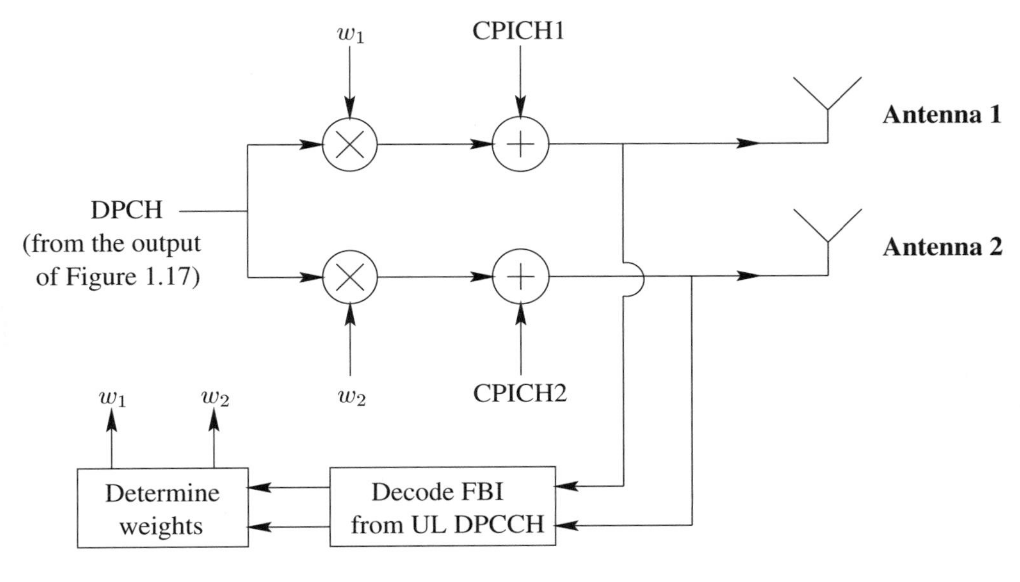

Figure 1.56: Transmission of the DL DPCH using a closed-loop transmit diversity technique.

are partly due to the existing 2G infrastructure already in use all over the world, and are specifically due to the heritage of the GSM and the IS-95 systems. Huge capital has been invested in these current 2G mobile radio systems. Therefore, the respective regional standard bodies have endeavoured to ensure that the 3G systems are compatible with the 2G systems. Because of the diversified nature of these 2G mobile radio systems, it is not an easy task to reach a common 3G standard that can maintain perfect backwards compatibility. Non-coherent M-ary orthogonal CDMA is described in the next chapter.

1.4 Summary and Conclusions

Following the rudimentary introduction of Sections 1.1–1.2.6, Section 1.3 reviewed the 3G WB-CDMA standard proposals. The 3G systems are more amenable to the transmission of interactive video signals than their more rigid 2G counterparts. This is due partly to the higher supported bit rate and partly to the higher variety of available transmission integrities and bit rates. During our further discourse we will rely on this chapter and quantify the network performance of various joint-detection-based CDMA systems.

Chapter 2

Burst-by-Burst Adaptive Wireless Transceivers

L. Hanzo, P.J. Cherriman, C.H. Wong, E.L. Kuan, T. Keller[1]

2.1 Motivation

In recent years the concept of intelligent multi-mode, multimedia transceivers (IMMT) has emerged in the context of wireless systems [67, 150–152] and the range of various existing solutions that have found favour in existing standard systems was summarised in the excellent overview by Nanda *et al.* [153]. *The aim of these adaptive transceivers is to provide mobile users with the best possible compromise amongst a number of contradicting design factors, such as the power consumption of the hand-held portable station (PS), robustness against transmission errors, spectral efficiency, teletraffic capacity, audio/video quality and so forth [152].* In this introductory chapter we have to limit our discourse to a small subset of the associated wireless transceiver design issues, referring the reader for a deeper exposure to the literature cited [151]. A further advantage of the IMMTs of the near future is that due to their flexibility they are likely to be able to reconfigure themselves in various operational modes in order to ensure backwards compatibility with existing, so-called second generation standard wireless systems, such as the Japanese Digital Cellular [154], the Pan-American IS-54 [155] and IS-95 [156] systems, as well as the Global System of Mobile Communications (GSM) [11] standards.

The fundamental advantage of burst-by-burst adaptive IMMTs is that - regardless of the propagation environment encountered - when the mobile roams across different environments subject to pathloss, shadow- and fast-fading, co-channel-, intersymbol- and multi-user in-

[1] This chapter is based on L. Hanzo, C.H. Wong, P.J. Cherriman: Channel-adaptive wideband wireless video telephony, ©IEEE Signal Processing Magazine, July 2000; Vol. 17., No. 4, pp 10-30 and on L. Hanzo, P.J. Cherriman, Ee Lin Kuan: Interactive cellular and cordless video telephony: State-of-the-art, system design principles and expected performance, ©IEEE Proceedings of the IEEE, Sept. 2000, pp 1388-1413.

terference, while experiencing power control errors, the system will always be able to configure itself in the highest possible throughput mode, whilst maintaining the required transmission integrity. Furthermore, whilst powering up under degrading channel conditions may disadvantage other users in the system, invoking a more robust - although lower throughput - transmission mode will not. The employment of the above burst-by-burst adaptive modems in the context of Code Division Multiple Access (CDMA) is fairly natural and it is motivated by the fact that all three third-generation mobile radio system proposals employ CDMA [11, 124, 157].

2.2 Narrowband Burst-by-Burst Adaptive Modulation

In burst-by-burst Adaptive Quadrature Amplitude Modulation (BbB-AQAM) a high-order, high-throughput modulation mode is invoked, when the instantaneous channel quality is favourable [13]. By contrast, a more robust lower order BbB-AQAM mode is employed, when the channel exhibits inferior quality, for improving the average BER performance. In order to support the operation of the BbB-AQAM modem, a high-integrity, low-delay feedback path has to be invoked between the transmitter and receiver for signalling the estimated channel quality perceived by the receiver to the remote transmitter. This strongly protected message can be for example superimposed on the reverse-direction messages of a duplex interactive channel. The transmitter then adjusts its AQAM mode according to the instructions of the receiver in order to be able to meet its BER target.

A salient feature of the proposed BbB-AQAM technique is that regardless of the channel conditions, the transceiver achieves always the best possible multi-media source-signal representation quality - such as video, speech or audio quality - by automatically adjusting the achievable bitrate and the associated multimedia source-signal representation quality in order to match the channel quality experienced. The AQAM modes are adjusted on a near-instantaneous basis under given propagation conditions in order to cater for the effects of pathloss, fast-fading, slow-fading, dispersion, co-channel interference (CCI), multi-user interference, etc. Furthermore, when the mobile is roaming in a hostile outdoor - or even hilly terrain - propagation environment, typically low-order, low-rate modem modes are invoked, while in benign indoor environments predominantly the high-rate, high source-signal representation quality modes are employed.

BbB-AQAM has been originally suggested by Webb and Steele [158], stimulating further research in the wireless community for example by Sampei *et al.* [159], showing promising advantages, when compared to fixed modulation in terms of spectral efficiency, BER performance and robustness against channel delay spread. Various systems employing AQAM were also characterised in [13]. The numerical upper bound performance of narrow-band BbB-AQAM over slow Rayleigh flat-fading channels was evaluated by Torrance and Hanzo [160], while over wide-band channels by Wong and Hanzo [161]. Following these developments, the optimisation of the BbB-AQAM switching thresholds was carried employing Powell-optimisation using a cost-function, which was based on the combination of the target BER and target Bit Per Symbol (BPS) performance [162]. Adaptive modulation was also studied in conjunction with channel coding and power control techniques by Matsuoka *et al.* [163] as well as Goldsmith and Chua [164].

In the early phase of research more emphasis was dedicated to the system aspects of adaptive modulation in a narrow-band environment. A reliable method of transmitting the modulation control parameters was proposed by Otsuki *et al.* [165], where the parameters were embedded in the transmission frame's mid-amble using Walsh codes. Subsequently, at the receiver the Walsh sequences were decoded using maximum likelihood detection. Another technique of estimating the required modulation mode used was proposed by Torrance and Hanzo [166], where the modulation control symbols were represented by unequal error protection 5-PSK symbols. The adaptive modulation philosophy was then extended to wide-band multi-path environments by Kamio *et al.* [167] by utilising a bi-directional Decision Feedback Equaliser (DFE) in a micro- and macro-cellular environment. This equalisation technique employed both forward and backward oriented channel estimation based on the pre-amble and post-amble symbols in the transmitted frame. Equaliser tap gain interpolation across the transmitted frame was also utilised, in order to reduce the complexity in conjunction with space diversity [167]. The authors concluded that the cell radius could be enlarged in a macro-cellular system and a higher area-spectral efficiency could be attained for micro-cellular environments by utilising adaptive modulation. The latency effect, which occurred when the input data rate was higher than the instantaneous transmission throughput was studied and solutions were formulated using frequency hopping [168] and statistical multiplexing, where the number of slots allocated to a user was adaptively controlled.

In reference [169] symbol rate adaptive modulation was applied, where the symbol rate or the number of modulation levels was adapted by using $\frac{1}{8}$-rate 16QAM, $\frac{1}{4}$-rate 16QAM, $\frac{1}{2}$-rate 16QAM as well as full-rate 16QAM and the criterion used to adapt the modem modes was based on the instantaneous received signal-to-noise ratio and channel delay spread. The slowly varying channel quality of the uplink (UL) and downlink (DL) was rendered similar by utilising short frame duration Time Division Duplex (TDD) and the maximum normalised delay spread simulated was 0.1. A variable channel coding rate was then introduced by Matsuoka *et al.* in conjunction with adaptive modulation in reference [163], where the transmitted burst incorporated an outer Reed Solomon code and an inner convolutional code in order to achieve high-quality data transmission. The coding rate was varied according to the prevalent channel quality using the same method, as in adaptive modulation in order to achieve a certain target BER performance. A so-called channel margin was introduced in this contribution, which adjusted the switching thresholds in order to incorporate the effects of channel quality estimation errors. As mentioned above, the performance of channel coding in conjunction with adaptive modulation in a narrow-band environment was also characterised by Goldsmith and Chua [164]. In this contribution, trellis and lattice codes were used without channel interleaving, invoking a feedback path between the transmitter and receiver for modem mode control purposes. The effects of the delay in the feedback path on the adaptive modem's performance were studied and this scheme exhibited a higher spectral efficiency, when compared to the non-adaptive trellis coded performance.

Subsequent contributions by Suzuki *et al.* [170] incorporated space-diversity and power-adaptation in conjunction with adaptive modulation, for example in order to combat the effects of the multi-path channel environment at a 10Mbits/s transmission rate. The maximum tolerable delay-spread was deemed to be one symbol duration for a target mean BER performance of 0.1%. This was achieved in a Time Division Multiple Access (TDMA) scenario, where the channel estimates were predicted based on the extrapolation of previous channel quality estimates. Variable transmitted power was then applied in combination with adaptive modulation in reference [164], where the transmission rate and power adaptation was optimised in order to achieve an increased spectral efficiency. In this treatise, a slowly varying channel was assumed and the instantaneous received power required in order to achieve a certain upper bound performance was assumed to be known prior to transmission. Power control in conjunction with a pre-distortion type non-linear power amplifier compensator was studied in the context of adaptive modulation in reference [171]. This method was used to mitigate the non-linearity effects associated with the power amplifier, when QAM modulators were used.

Results were also recorded concerning the performance of adaptive modulation in conjunction with different multiple access schemes in a narrow-band channel environment. In a TDMA system, dynamic channel assignment was employed by Ikeda *et al.*, where in addition to assigning a different modulation mode to a different channel quality, priority was always given to those users in reserving time-slots, which benefitted from the best channel quality [172]. The performance was compared to fixed channel assignment systems, where substantial gains were achieved in terms of system capacity. Furthermore, a lower call termination probability was recorded. However, the probability of intra-cell hand-off increased as a result of the associated dynamic channel assignment (DCA) scheme, which constantly searched for a high-quality, high-throughput time-slot for the existing active users. The application of adaptive modulation in packet transmission was introduced by Ue, Sampei and Morinaga [173], where the results showed improved data throughput. Recently, the performance of adaptive modulation was characterised in conjunction with an automatic repeat request (ARQ) system in reference [174], where the transmitted bits were encoded using a cyclic redundant code (CRC) and a convolutional punctured code in order to increase the data throughput.

A recent treatise was published by Sampei, Morinaga and Hamaguchi [175] on laboratory test results concerning the utilisation of adaptive modulation in a TDD scenario, where the modem mode switching criterion was based on the signal-to-noise ratio and on the normalised delay-spread. In these experimental results, the channel quality estimation errors degraded the performance and consequently a channel estimation error margin was devised, in order to mitigate this degradation. Explicitly, the channel estimation error margin was defined as the measure of how much extra protection margin must be added to the switching threshold levels, in order to minimise the effects of the channel estimation errors. The delay-spread also degraded the performance due to the associated irreducible BER, which was not compensated by the receiver. However, the performance of the adaptive scheme in a delay-spread impaired channel environment was better than that of a fixed modulation scheme. Lastly, the experiment also concluded that the AQAM scheme can be operated for a Doppler frequency of $f_d = 10$ Hz with a normalised delay spread of 0.1 or for $f_d = 14$ Hz with a normalised delay spread of 0.02, which produced a mean BER of 0.1% at a transmission rate of 1 Mbits/s.

Lastly, the latency and interference aspects of AQAM modems were investigated in [168, 176]. Specifically, the latency associated with storing the information to be transmitted during severely degraded channel conditions was mitigated by frequency hopping or statistical multiplexing. As expected, the latency is increased, when either the mobile speed or the channel SNR are reduced, since both of these result in prolonged low instantaneous SNR intervals. It was demonstrated that as a result of the proposed measures, typically more than 4 dB SNR reduction was achieved by the proposed adaptive modems in comparison to the conventional fixed-mode benchmark modems employed. However, the achievable gains depend strongly on the prevalant co-channel interference levels and hence interference cancellation was invoked in [176] on the basis of adjusting the demodulation decision boundaries after estimating the interfering channel's magnitude and phase.

Having reviewed the developments in the field of narrowband AQAM, let us now consider wideband AQAM modems in the next section.

2.3 Wideband Burst-by-Burst Adaptive Modulation

In the above narrow-band channel environment, the quality of the channel was determined by the short-term SNR of the received burst, which was then used as a criterion in order to choose the appropriate modulation mode for the transmitter, based on a list of switching threshold levels, l_n [158–160]. However, in a wideband environment, this criterion is not an accurate measure for judging the quality of the channel, where the existence of multi-path components produces not only power attenuation of the transmission burst, but also inter-symbol interference. Consequently, appropriate channel quality criteria have to be defined, in order to estimate the wideband channel quality for invoking the most appropriate modulation mode.

2.3.1 Channel quality metrics

The most reliable channel quality estimate is the BER, since it reflects the channel quality, irrespective of the source or the nature of the quality degradation. The BER can be estimated with a certain granularity or accuracy, provided that the system entails a channel decoder or - synonymously - Forward Error Correction (FEC) decoder employing algebraic decoding [11, 177]. If the system contains a so-called soft-in-soft-out (SISO) channel decoder, such as a turbo decoder [107], the BER can be estimated with the aid of the Logarithmic Likelihood Ratio (LLR), evaluated either at the input or the output of the channel decoder. Hence a particularly attractive way of invoking LLRs is employing powerful turbo codecs, which provide a reliable indication of the confidence associated with a particular bit decision. The LLR is defined as the logarithm of the ratio of the probabilities associated with a specific bit being binary zero or one. Again, this measure can be evaluated at both the input and the output of the turbo channel codecs and both of them can be used for channel quality estimation.

In the event that no channel encoder / decoder (codec) is used in the system, the channel quality expressed in terms of the BER can be estimated with the aid of the mean-squared error (MSE) at the output of the channel equaliser or the closely related metric, the Pseudo-Signal-to-Noise-Ratio (Pseudo-SNR) [161]. The MSE or pseudo-SNR at the output of the channel equaliser have the important advantage that they are capable of quantifying the severity of the Inter-Symbol-Interference (ISI) and/or CCI experienced, in other words quantifying the Signal-to-Interference-plus-Noise-Ratio (SINR).

In our proposed systems the wideband channel-induced degradation is combated not only by the employment of adaptive modulation but also by equalisation. In following this line of thought, we can formulate a two-step methodology in mitigating the effects of the dispersive wideband channel. In the first step, the equalisation process will eliminate most of the inter-symbol interference based on a Channel Impulse Response (CIR) estimate derived using the channel sounding midamble and consequently, the signal-to-noise and residual interference ratio at the output of the equaliser is calculated.

We found that the residual channel-induced ISI at the output of the DFE is near-Gaussian distributed and that if there are no decision feedback errors, the pseudo-SNR at the output of the DFE, γ_{dfe} can be calculated as [67, 161, 178]:

$$\begin{aligned} \gamma_{dfe} &= \frac{\text{Wanted Signal Power}}{\text{Residual ISI Power + Effective Noise Power}} \\ &= \frac{E\left[|S_k \sum_{m=0}^{N_f-1} C_m h_m|^2\right]}{\sum_{q=-(N_f-1)}^{-1} E\left[|f_q S_{k-q}|^2\right] + N_o \sum_{m=0}^{N_f-1} |C_m|^2}, \end{aligned} \tag{2.1}$$

where C_m and h_m denotes the DFE's feed-forward coefficients and the channel impulse response, respectively. The transmitted signal and the noise spectral density is represented by S_k and N_o. Lastly, the number of DFE feed-forward coefficients is denoted by N_f. By utilising the pseudo-SNR at the output of the equaliser, we are ensuring that the system performance is optimised by employing equalisation and AQAM [13] in a wideband environment according to the following switching regime:

$$\text{Modulation Mode} = \begin{cases} NoTX & \text{if } \gamma_{DFE} < f_0 \\ BPSK & \text{if } f_0 < \gamma_{DFE} < f_1 \\ 4QAM & \text{if } f_1 < \gamma_{DFE} < f_2 \\ 16QAM & \text{if } f_2 < \gamma_{DFE} < f_3 \\ 64QAM & \text{if } \gamma_{DFE} > f_3, \end{cases} \tag{2.2}$$

where $f_n, n = 0...3$ are the pseudo-SNR thresholds levels, which are set according to the system's integrity requirements and the modem modes may assume $0 \ldots 6$ bits/symbol transmissions corresponding to no transmissions (No TX), Binary Phase Shift Keying (BPSK), as well as 4- 16- and 64QAM [13]. We note, however that in the context of the interactive BbB-AQAM videophone schemes introduced during our later discourse for quantifying the service-related benefits of such adaptive transceivers we refrained from employing the No Tx mode. This allowed us to avoid the associated latency of the buffering required for storing the information, until the channel quality improved sufficiently for allowing transmission of the buffered bits.

In references [179, 180] a range of novel Radial Basis Function (RBF) assisted BbB-AQAM channel equalisers have been proposed, which exhibit a close relationship with the so-called Bayesian schemes. Decision feedback was introduced in the design of the RBF equaliser in order to reduce its computational complexity. The RBF DFE was found to give similar performance to the conventional DFE over Gaussian channels using various BbB-AQAM schemes, while requiring a lower feedforward and feedback order. Over Rayleigh-fading channels similar findings were valid for binary modulation, while for higher order modems the RBF-based DFE required increased feedforward and feedback orders in order to outperform the conventional MSE DFE scheme. Then turbo BCH codes were invoked [179] for improving the associated BER and BPS performance of the scheme, which was shown to give a significant improvement in terms of the mean BPS performance compared to that of the uncoded RBF equaliser assisted adaptive modem. Finally, a novel turbo equalisation scheme

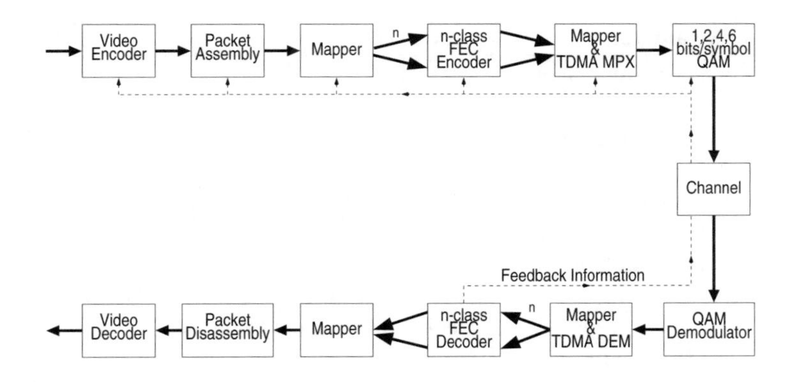

Figure 2.1: Reconfigurable transceiver schematic diagram.

was presented in [180], which employed an RBF DFE instead of the conventional trellis-based equaliser, which was advocated in most turbo equaliser implementations. The so-called Jacobian logarithmic complexity reduction technique was proposed, which was shown to achieve an identical BER performance to the conventional trellis-based turbo equaliser, while incurring a factor 4.4 lower 'per-iteration' complexity in the context of 4QAM.

In summary, in contrast to the narrowband, statically reconfigured multimode systems of [151], in this section wideband, near-instantaneously reconfigured or burst-by-burst adaptive modulation was invoked, in order to quantify the achievable service-related benefits, as perceived by users of such systems. More specifically, the achievable video performance benefits of wireless BbB-AQAM video transceivers will be quantified in this section, when using the H.263 video encoder [151]. Similar BbB-AQAM speech and audio transceivers were portrayed in [181].

It is an important element of the system that when the binary BCH [11, 177] or turbo codes [107, 177] protecting the video stream are overwhelmed by the plethora of transmission errors, the systems refrains from decoding the video packet in order to prevent error propagation through the reconstructed frame buffer [151]. Instead, these corrupted packets are dropped and the reconstructed frame buffer will not be updated, until the next packet replenishing the specific video frame area arrives. The associated video performance degradation is fairly minor for packet dropping or frame error rates (FER) below about 5%. These packet dropping events are signalled to the remote decoder by superimposing a strongly protected one-bit packet acknowledgement flag on the reverse-direction packet, as outlined in [151]. In the proposed scheme we also invoked the adaptive rate control and packetisation algorithm of [151], supporting constant Baud-rate operation.

Having reviewed the basic features of adaptive modulation, in the forthcoming section we will characterise the achievable service-related benefits of BbB-AQAM video transceivers, as perceived by the users of such systems.

Parameter	Value
Carrier Frequency	1.9 GHz
Vehicular Speed	30 mph
Doppler frequency	85 Hz
Norm. Doppler fr.	3.27×10^{-5}
Channel type	COST 207 Typ. Urban (Figure 2.2)
No. of channel paths	4
Data modulation	Adaptive QAM (BPSK, 4-QAM, 16-QAM, 64-QAM)
Receiver type	Decision Feedback Equaliser No. of Forward Filter Taps = 35 No. of Backward Filter Taps = 7

Table 2.1: Modulation and channel parameters.

2.4 Wideband BbB-AQAM Video Transceivers

Again, in this section we set out to demonstrate the service-quality related benefits of a wideband BbB-AQAM in the context of a wireless videophone system employing the programmable H.263 video codec in conjunction with an adaptive packetiser. The system's schematic diagram is shown in Figure 2.1, which will be referred to in more depth during our further discourse.

In these investigations 176x144 pixel QCIF-resolution, 30 frames/s video sequences were transmitted, which were encoded by the H.263 video codec [151, 182] at bitrates resulting in high perceptual video quality. Table 2.1 shows the modulation- and channel parameters employed. The COST207 [50] four-path typical urban (TU) channel model was used, which is characterised by its CIR in Figure 2.2. We used the Pan-European FRAMES proposal [183] as the basis for our wideband transmission system, invoking the frame structure shown in Figure 2.3. Employing the FRAMES Mode A1 (FMA1) so-called non-spread data burst mode required a system bandwidth of 3.9 MHz, when assuming a modulation excess bandwidth of 50% [13]. A range of other system parameters are shown in Table 2.2. Again, it is important to note that the proposed AQAM transceiver of Figure 2.1 requires a duplex system, since the AQAM mode required by the receiver during the next received video packet has to be signalled to the transmitter. In this system we employed TDD and the feedback path is indicated by the dashed line in the schematic diagram of Figure 2.1.

Again, the proposed video transceiver of Figure 2.1 is based on the H.263 video codec [182]. The video coded bitstream was protected by near-half-rate binary BCH coding [11] or by half-rate turbo coding [107] in all of the burst-by-burst adaptive wideband AQAM modes [13]. The AQAM modem can be configured either under network control on a more static basis, or under transceiver control on a near-instantaneous basis, in order to operate as a 1, 2, 4 and 6 bits/symbol scheme, while maintaining a constant signalling rate. This allowed us to support an increased throughput expressed in terms of the average number of bits per symbol (BPS, when the instantaneous channel quality was high, leading ultimately to an increased video quality in a constant bandwidth.

The transmitted bitrate for all four modes of operation is shown in Table 2.3. The un-

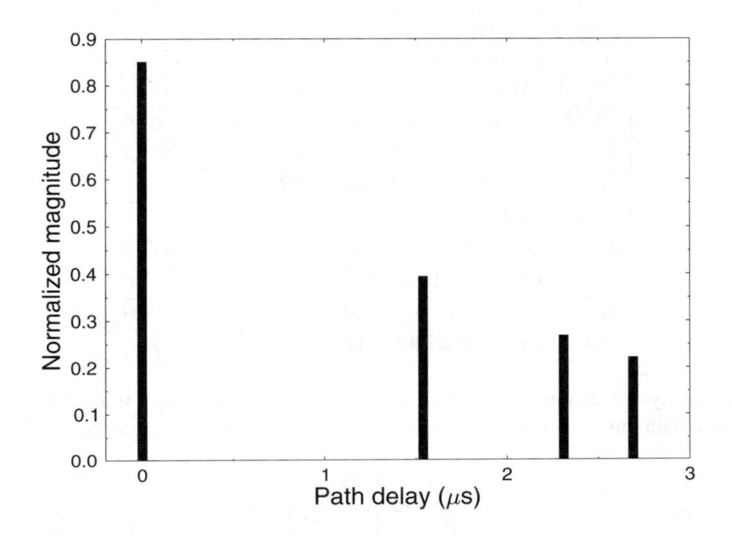

Figure 2.2: Normalised channel impulse response for the COST 207 [50] four-path Typical Urban (TU) channel.

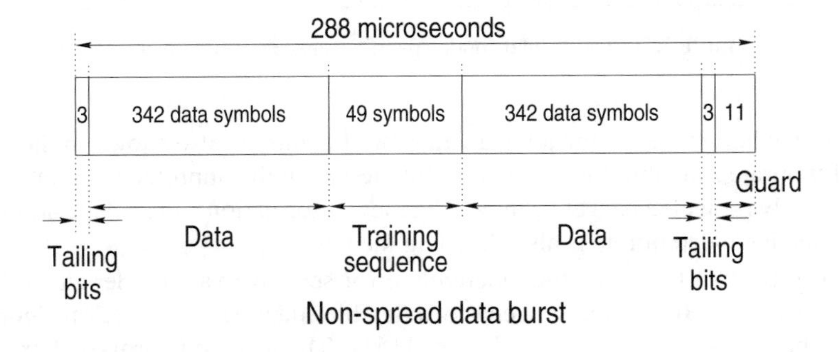

Figure 2.3: Transmission burst structure of the FMA1 non-spread data burst mode of the FRAMES proposal [183].

Features	**Value**
Multiple access	TDMA
Duplexing	TDD
No. of Slots/Frame	16
TDMA frame length	4.615 ms
TDMA slot length	288μs
Data Symbols/TDMA slot	684
User Data Symbol Rate (KBd)	148.2
System Data Symbol Rate (MBd)	2.37
Symbols/TDMA slot	750
User Symbol Rate (KBd)	162.5
System Symbol Rate (MBd)	2.6
System Bandwidth (MHz)	3.9
Eff. User Bandwidth (kHz)	244

Table 2.2: Generic system features of the reconfigurable multi-mode video transceiver, using the non-spread data burst mode of the FRAMES proposal [183] shown in Figure 2.3.

Features	Multi-rate System			
Mode	BPSK	4QAM	16QAM	64QAM
Bits/Symbol	1	2	4	6
FEC	Near Half-rate BCH			
Transmission bitrate (kbit/s)	148.2	296.4	592.8	889.3
Unprotected bitrate (kbit/s)	75.8	151.7	303.4	456.1
Effective Video-rate (kbit/s)	67.0	141.7	292.1	446.4
Video fr. rate (Hz)	30			

Table 2.3: Operational-mode specific transceiver parameters.

protected bitrate before approximately half-rate BCH coding is also shown in the table. The actual useful bitrate available for video is slightly less than the unprotected bitrate due to the required strongly protected packet acknowledgement information and packetisation information. The effective video bitrate is also shown in the table.

In order to be able to invoke the inherently error-sensitive variable-length coded H.263 video codec in a high-BER wireless scenario, a flexible adaptive packetisation algorithm was necessary, which was highlighted in reference [151]. The technique proposed exhibits high flexibility, allowing us to drop corrupted video packets, rather than allowing errorneous bits to contaminate the reconstructed frame buffer of the H.263 codec. This measure prevents the propagation of errors to future video frames through the reconstructed frame buffer of the H.263 codec. More explicitly, corrupted video packets cannot be used by either the local or the remote H.236 decoder, since that would result in unacceptable video degradation over a prolonged period of time due to the error propagation inflicted by the associated motion vectors and run-length coding. Upon dropping the erroneous video packets, both the local and remote H.263 reconstruction frame buffers are updated by a blank packet, which corresponds to assuming that the video block concerned was identical to the previous one.

A key feature of our proposed adaptive packetisation regime is therefore the provision of a strongly error protected binary transmission packet acknowledgement flag [151], which instructs the remote decoder not to update the local and remote video reconstruction buffers in the event of a corrupted packet. This flag can be for example conveniently repetition-coded, in order to invoke Majority Logic Decision (MLD) at the decoder. Explicitly, the binary flag is repeated an odd number of times and at the receiver the MLD scheme counts the number of binary ones and zeros and opts for the logical value, constituting the majority of the received bits. These packet acknowledgement flags are then superimposed on the forthcoming reverse-direction packet in our advocated Time Division Duplex (TDD) regime [151] of Table 2.2, as seen in the schematic diagram of Figure 2.1.

The proposed BbB-AQAM modem maximises the system capacity available by using the most appropriate modulation mode for the current instantaneous channel conditions. As stated before, we found that the pseudo-SNR at the output of the channel equaliser was an adequate channel quality measure in our burst-by-burst adaptive wide-band modem. A more explicit representation of the wideband AQAM regime is shown in Figure 2.4, which displays the variation of the modulation mode with respect to the pseudo SNR at channel SNRs of 10 and 20 dB. In these figures, it can be seen explicitly that the lower-order modulation modes were chosen, when the pseudo SNR was low. In contrast, when the pseudo SNR was high, the higher-order modulation modes were selected in order to increase the transmission throughput. These figures can also be used to exemplify the application of wideband AQAM in an indoor and outdoor environment. In this respect, Figure 2.4(a) can be used to characterise a hostile outdoor environment, where the perceived channel quality was low. This resulted in the utilisation of predominantly more robust modulation modes, such as BPSK and 4QAM. Conversely, a less hostile indoor environment is exemplified by Figure 2.4(b), where the perceived channel quality was high. As a result, the wideband AQAM regime can adapt suitably by invoking higher-order modulation modes, as evidenced by Figure 2.4(b). Again, this simple example demonstrated that wideband AQAM can be utilised, in order to provide a seamless, near-instantaneous reconfiguration between for example indoor and outdoor environments.

2.5 BbB-AQAM Performance

The mean BER and BPS performances were numerically calculated [161] for two different target BER systems, namely for the **High-BER** and **Low-BER** schemes, respectively. The results are shown in Figure 2.5 over the COST207 TU Rayleigh fading channel of Figure 2.2. The targeted mean BERs of the **High-BER** and **Low-BER** regime of 1% and 0.01% was achieved for all average channel SNRs investigated, since this scheme also invoked a no-transmission mode, when the channel quality was extremely hostile. In this mode only dummy data was transmitted, in order to facilitate monitoring the channel's quality.

At average channel SNRs below 20 dB the lower-order modulation modes were dominant, producing a robust system in order to achieve the targeted BER. Similarly, at high average channel SNRs the higher-order modulation mode of 64QAM dominated the transmission

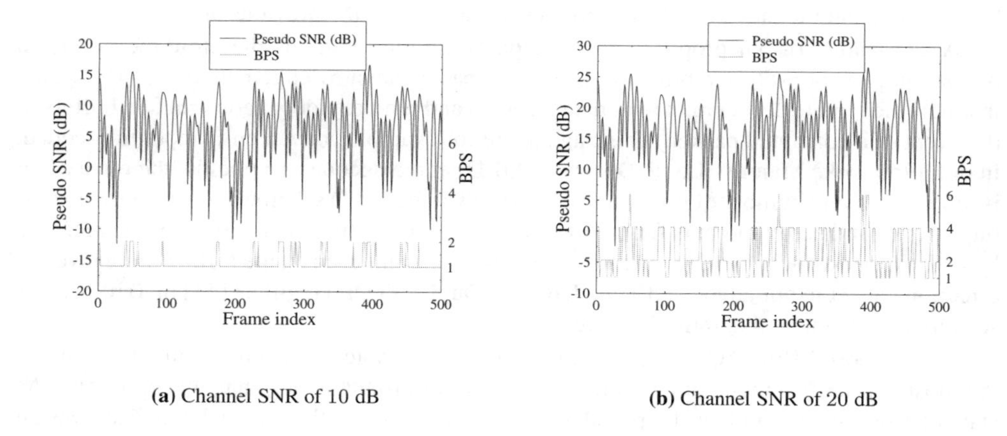

(a) Channel SNR of 10 dB

(b) Channel SNR of 20 dB

Figure 2.4: Modulation mode variation with respect to the pseudo SNR defined by Equation 2.1 over the **TU Rayleigh fading channel**. The BPS throughputs of 1, 2, 4 and 6 represent BPSK, 4QAM, 16QAM and 64QAM, respectively.

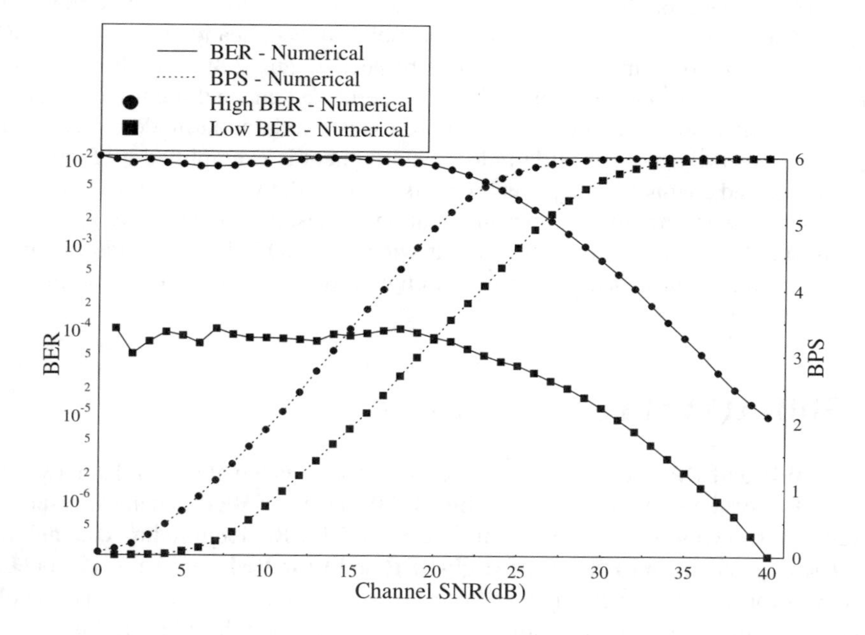

Figure 2.5: Numerical mean BER and BPS performance of the wideband equalised AQAM scheme for the **High-BER** and **Low-BER** regime over the COST207 TU Rayleigh fading channel.

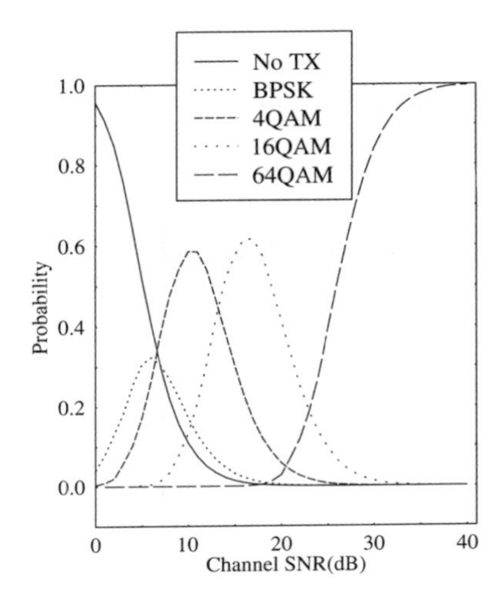

(a) High-BER transmission regime over the TU Rayleigh fading channel

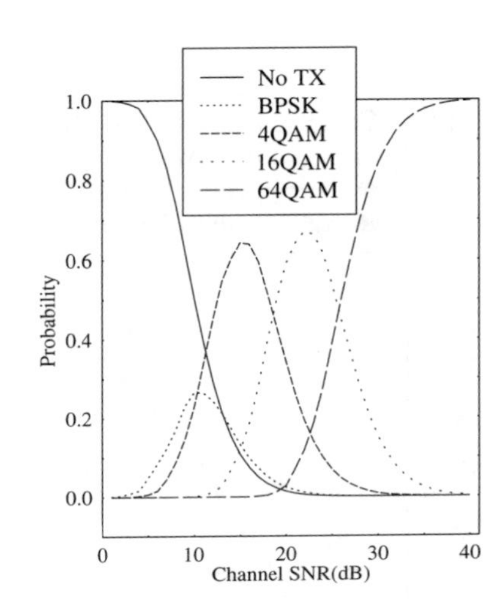

(b) Low-BER transmission regime over the TU Rayleigh fading channel

Figure 2.6: Numerical probabilities of each modulation mode utilised for the wideband AQAM and DFE scheme over the **TU Rayleigh Fading channel** for the **(a) Low-BER Transmission** regime and **(b) Low-BER Transmission** regime.

regime, yielding a lower mean BER than the target, since no higher-order modulation mode could be legitimately invoked. This is evidenced by the modulation mode probability results shown in Figure 2.6 for the COST207 TU Rayleigh fading channel of Figure 2.2. The targeted mean BPS values for the **High-BER** and **Low-BER** regime of 4.5 and 3 were achieved at approximately 19 dB channel SNR for the COST207 TU Rayleigh fading channels. However, at average channel SNRs below 3 dB the above-mentioned no-transmission or transmission blocking mode was dominant in the **Low-BER** system and thus the mean BER performance was not recorded for that range of average channel SNRs.

The transmission throughput achieved for the **High-BER** and **Low-BER** transmission regimes is shown in Figure 2.7. The transmission throughput for the **High-BER** transmission regime was higher than that of the **Low-BER** transmission regime for the same transmitted signal energy due to the more relaxed BER requirement of the **High-BER** transmission regime, as evidenced by Figure 2.7. The achieved transmission throughput of the wideband AQAM scheme was higher than that of the BPSK, 4QAM and 16QAM schemes for the same average channel SNR. However, at higher average channel SNRs the throughput performance of both schemes converged, since 64QAM became the dominant modulation mode for the wideband AQAM scheme. SNR gains of $1 - 3$ dB and $7 - 9$ dB were recorded for the **High-BER** and **Low-BER** transmission schemes, respectively. These gains were

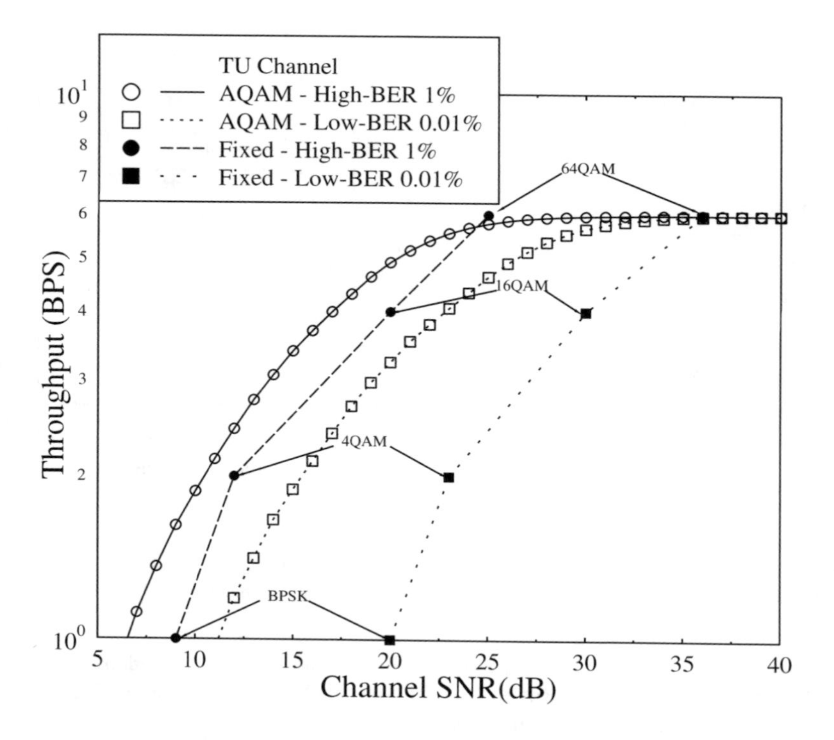

Figure 2.7: Transmission throughput of the wideband AQAM and DFE scheme and fixed modulation modes over the **TU Rayleigh Fading channel** for both the **High-BER** and **Low-BER** transmission regimes.

considerably lower than those associated with narrow-band AQAM, where 5 - 7 dB and 10 - 18 dB of gains were reported for the **High-BER** and **Low-BER** transmission scheme, respectively [168, 176]. This was expected, since in the narrow-band environment the fluctuation of the instantaneous SNR was more severe, resulting in increased utilisation of the modulation switching mechanism. Consequently, the instantaneous transmission throughput increased, whenever the fluctuations yielded a high received instantaneous SNR. Conversely, in a wide-band channel environment the channel quality fluctuations perceived by the DFE were less severe due to the associated multi-path diversity, which was exploited by the equaliser.

Having characterised the wideband BbB-AQAM modem's performance, let us now consider the entire video transceiver of Figure 2.1 and Tables 2.1-2.3 in the next section.

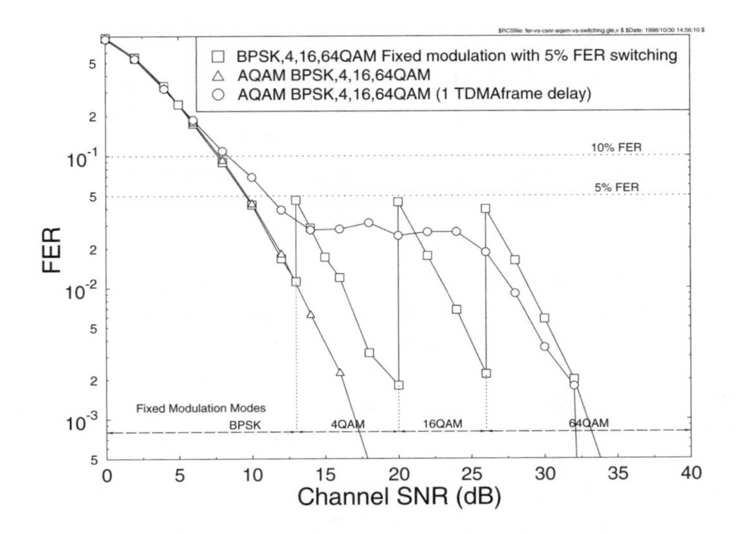

Figure 2.8: Transmission FER (or packet loss ratio) versus Channel SNR comparison of the four fixed modulation modes (BPSK, 4QAM, 16QAM, 64QAM) with 5% FER switching and adaptive burst-by-burst modem (AQAM). AQAM is shown with a realistic one TDMA frame delay between channel estimation and mode switching, and a zero delay version is included as an upper bound. The channel parameters were defined in Table 2.1 and near-half-rate BCH coding was employed [184] Cherriman, Wong, Hanzo, 2000 ©IEEE.

2.6 Wideband BbB-AQAM Video Performance

As a benchmarker, the statically reconfigured modems of reference [151] were invoked in Figure 2.8, in order to indicate how a system would perform, which cannot act on the basis of the near-instantaneously varying channel quality. As it can be inferred from Figure 2.8, such a statically reconfigured transceiver switches its mode of operation from a lower-order modem mode, such as for example BPSK to a higher-order mode, such as 4QAM, when the channel quality has improved sufficiently for the 4QAM mode's FER to become lower than 5 % after reconfiguring the transceiver in this more long-term 4QAM mode.

In order to assess the effects of imperfect channel estimation on BbB-AQAM we considered two scenarios. In the first scheme the adaptive modem always chose the perfectly estimated AQAM modulation mode, in order to provide a maximum upper bound performance. In the second scenario the modulation mode was based upon the perfectly estimated AQAM modulation mode for the previous burst, which corresponded to a delay of one TDMA frame duration of 4.615 ms. This second scenario represents a practical burst-by-burst adaptive modem, where the one-frame channel quality estimation latency is due to superimposing the receiver's required AQAM mode on a reverse-direction packet, for informing the transmitter concerning the best mode to be used for maintaining the target performance.

Figure 2.8 demonstrates on a logarithmic scale that the 'one-frame channel estimation delay' AQAM modem manages to maintain a similar FER performance to the fixed rate

	BPSK	4QAM	16QAM	64QAM
Standard	<10 dB	≥ 10 dB	≥ 18 dB	≥ 24 dB
Conservative	<13 dB	≥ 13 dB	≥ 20 dB	≥ 26 dB
Aggressive	<9 dB	≥ 9 dB	≥ 17 dB	≥ 23 dB

Table 2.4: SINR estimate at output of the equaliser required for each modulation mode in Burst-by-Burst Adaptive modem, ie. switching thresholds

BPSK modem at low SNRs, although we will see during our further discourse that AQAM provides increasingly higher bitrates, reaching six times higher values than BPSK for high channel SNRs, where the employment of 64QAM is predominant. In this high-SNR region the FER curve asymptotically approaches the 64QAM FER curve for both the realistic and the ideal AQAM scheme, although this is not visible in the figure for the ideal scheme, since this occurs at SNRs outside the range of Figure 2.8. Again, the reason for this performance discrepancy is the occasionally misjudged channel quality estimates of the realistic AQAM scheme. Additionally, Figure 2.8 indicates that the realistic AQAM modem exhibits a near-constant 3% FER at medium SNRs. The issue of adjusting the switching thresholds in order to achieve the target FER will be addressed in detail at a later stage in this section and the thresholds invoked will be detailed with reference to Table 2.4. Suffice to say at this stage that the average number of bits per symbol - and potentially also the associated video quality - can be increased upon using more 'aggressive' switching thresholds. However, this results in an increased FER, which tends to decrease the video quality, as it will be discussed later in this section. Having shown the effect of the BbB-AQAM modem on the transmission FER, let us now demonstrate the effects of the AQAM switching thresholds on the system's performance in terms of the associated FER performance.

2.6.1 AQAM Switching Thresholds

The set of switching thresholds used in all the previous graphs was the 'standard' set shown in Table 2.4, which was determined on the basis of the required channel SINR for maintaining the specific target video FER. In order to investigate the effect of different sets of switching thresholds, we defined two new sets of thresholds, a more 'conservative' set, and a more 'aggressive' set, employing less robust, but more bandwidth-efficient modem modes at lower SNRs. The more conservative switching thresholds reduced the transmission FER at the expense of a lower effective video bitrate. By contrast, the more aggressive set of thresholds increased the effective video bitrate at the expense of a higher transmission FER. The transmission FER performance of the realistic burst-by-burst adaptive modem, which has a one TDMA frame delay between channel quality estimation and mode switching is shown in Figure 2.9 for the three sets of switching thresholds of Table 2.4. It can be seen that the more 'conservative' switching thresholds reduce the transmission FER from about 3% to about 1% for medium channel SNRs, while the more 'aggressive' thresholds increase the transmission FER from about 3% to 4-5%. However, since FERs below 5% are not objectionable in video quality terms, this FER increase is an acceptable compromise for attaining a higher effective video bitrate.

The effective video bitrate for the realistic adaptive modem with the three sets of switching thresholds is shown in Figure 2.10. The more conservative set of switching thresholds

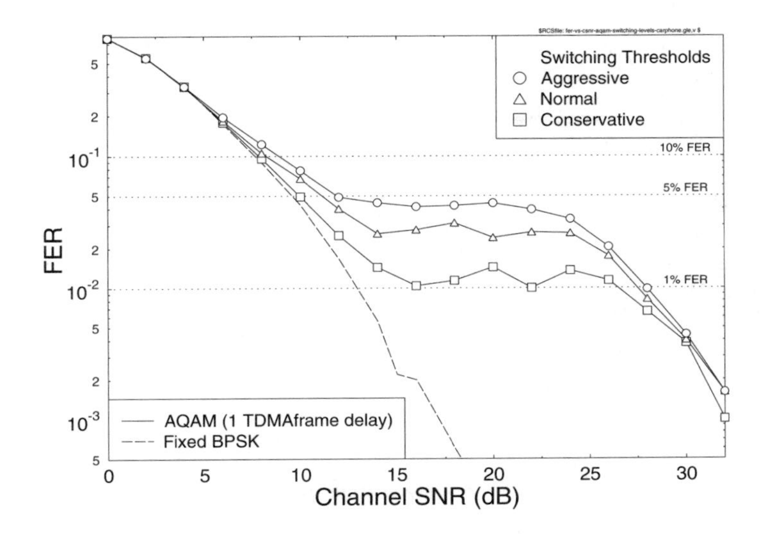

Figure 2.9: Transmission FER (or packet loss ratio) versus Channel SNR comparison of the fixed BPSK modulation mode and the adaptive burst-by-burst modem (AQAM) for the three sets of switching thresholds described in Table 2.4. AQAM is shown with a realistic one TDMA frame delay between channel estimation and mode switching. The channel parameters were defined in Table 2.1 [184] Cherriman, Wong, Hanzo, 2000 ©IEEE.

reduces the effective video bitrate but also reduces the transmission FER. The aggressive switching thresholds increase the effective video bitrate, but also increase the transmission FER. Therefore the optimal switching thresholds should be set such that the transmission FER is deemed acceptable in the range of channel SNRs considered. Let us now consider the performance improvements achievable, when employing powerful turbo codecs.

2.6.2 Turbo-coded AQAM videophone performance

Let us now demonstrate the additional performance gains that are achievable when a somewhat more complex turbo codec [107] is used in comparison to similar-rate algebraically decoded binary BCH codecs [11]. The generic system parameters of the turbo-coded reconfigurable multi-mode video transceiver are the same as those used in the BCH-coded version summarised in Table 2.2. Turbo-coding schemes are known to perform best in conjunction with square-shaped turbo interleaver arrays and their performance is improved upon extending the associated interleaving depth, since then the two constituent encoders are fed with more independent data. This ensures that the turbo decoder can rely on two quasi-independent data streams in its efforts to make as reliable bit decisions as possible. A turbo interleaver size of 18×18 bits was chosen, requiring 324 bits for filling the interleaver. The required so-called recursive systematic convolutional (RSC) component codes had a coding rate of 1/2 and a constraint length of $K = 3$. After channel coding the transmission burst length became 648 bits, which facilitated the decoding of all AQAM transmission bursts

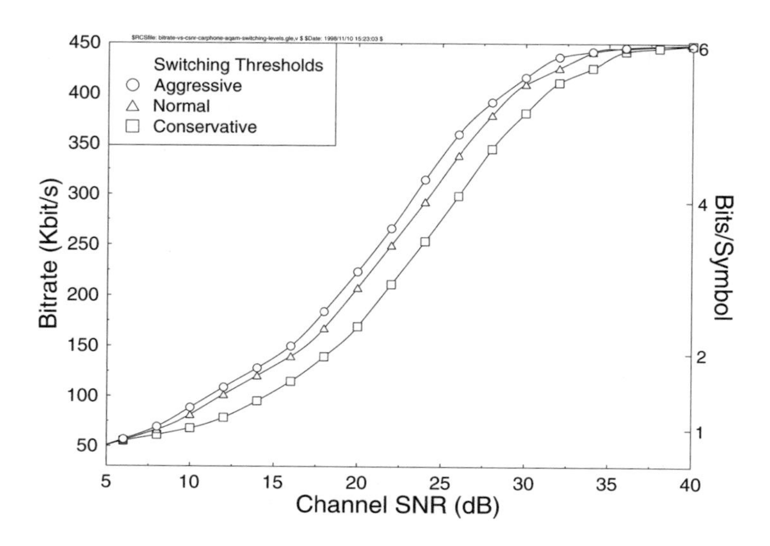

Figure 2.10: Video bitrate versus channel SNR comparison for the adaptive burst-by-burst modem (AQAM) with a realistic one TDMA frame delay between channel estimation and mode switching for the three sets of switching thresholds as described in Table 2.4. The channel parameters were defined in Table 2.1 [184] Cherriman, Wong, Hanzo, 2000 ©IEEE.

Features	Multi-rate System			
Mode	BPSK	4QAM	16QAM	64QAM
Bits/Symbol	1	2	4	6
FEC	Half-Rate Turbo coding with CRC			
Transmission bitrate (kbit/s)	140.4	280.8	561.6	842.5
Unprotected bitrate (kbit/s)	66.3	136.1	275.6	415.2
Effective Video-rate (kbit/s)	60.9	130.4	270.0	409.3
Video fr. rate (Hz)	30			

Table 2.5: Operational-mode specific turbo-coded transceiver parameters.

independently. The operational-mode specific turbo transceiver parameter are shown in Table 2.5, which should be compared to the corresponding BCH-coded parameters of Table 2.3. The turbo-coded parameters result in a 10% lower effective throughput bitrate compared to the similar-rate BCH-codecs under error-free conditions. However, Figure 2.11 demonstrates that the PSNR video quality versus channel SNR performance of the turbo-coded AQAM modem becomes better than that of the BCH-coded scenario, when the channel quality degrades. Having highlighted the operation of wideband single-carrier burst-by-burst AQAM modems, let us now consider briefly in the next two sections how the above burst-by-burst

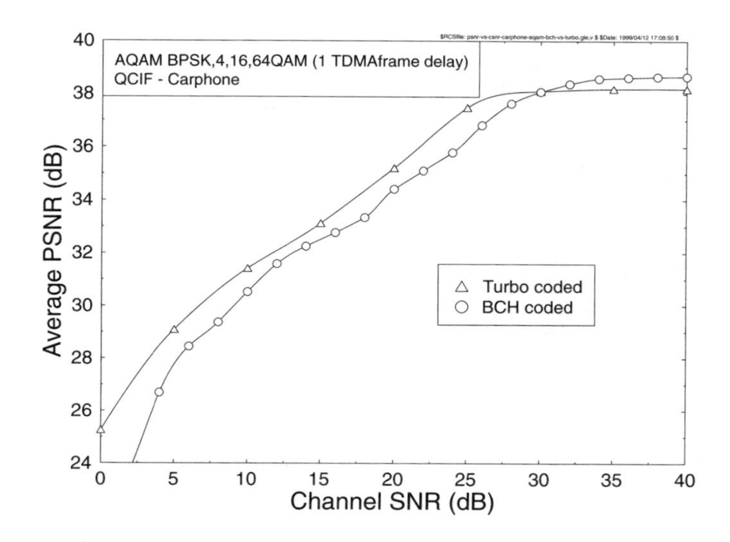

Figure 2.11: Decoded video quality (PSNR) versus transmission FER (or packet loss ratio) comparison of the realistic adaptive burst-by-burst modems (AQAM) using either BCH or turbo coding. The channel parameters were defined in Table 2.1 [184] Cherriman, Wong, Hanzo, 2000 ©IEEE.

adaptive principles can be extended to CDMA and Orthogonal Frequency Division Multiplex (OFDM) systems [13, 185].

2.7 Burst-by-burst Adaptive Joint-detection CDMA Video Transceiver

2.7.1 Multi-user Detection for CDMA

In the previous chapter a simple conceptual introduction was provided to CDMA, assuming the employment of simple single-user receivers. Then the most recent family of CDMA-based third-generation standards was reviewed. In this chapter we introduce a number of advanced near-instantaneouly adaptive transceiver concepts, which may find their way into future standards, in order to enhance the performance of the existing systems. We also introduce the concept of multi-user detection in an effort to maintain a near-single-user performance, whilst supporting a multiplicity of users. These adaptive system concepts are discussed in significantly more depth in [67, 151].

The effects of multi-user interference (MAI) are similar to those of the Intersymbol Interference (ISI) inflicted by the multipath propagation channel. More specifically, each user in a K-user system will suffer from MAI due to the other $(K - 1)$ users. This MAI can also be viewed as a single user's signal contaminated by the ISI due to $(K - 1)$ propagation paths in a multipath channel. Therefore, conventional equalisation techniques used to

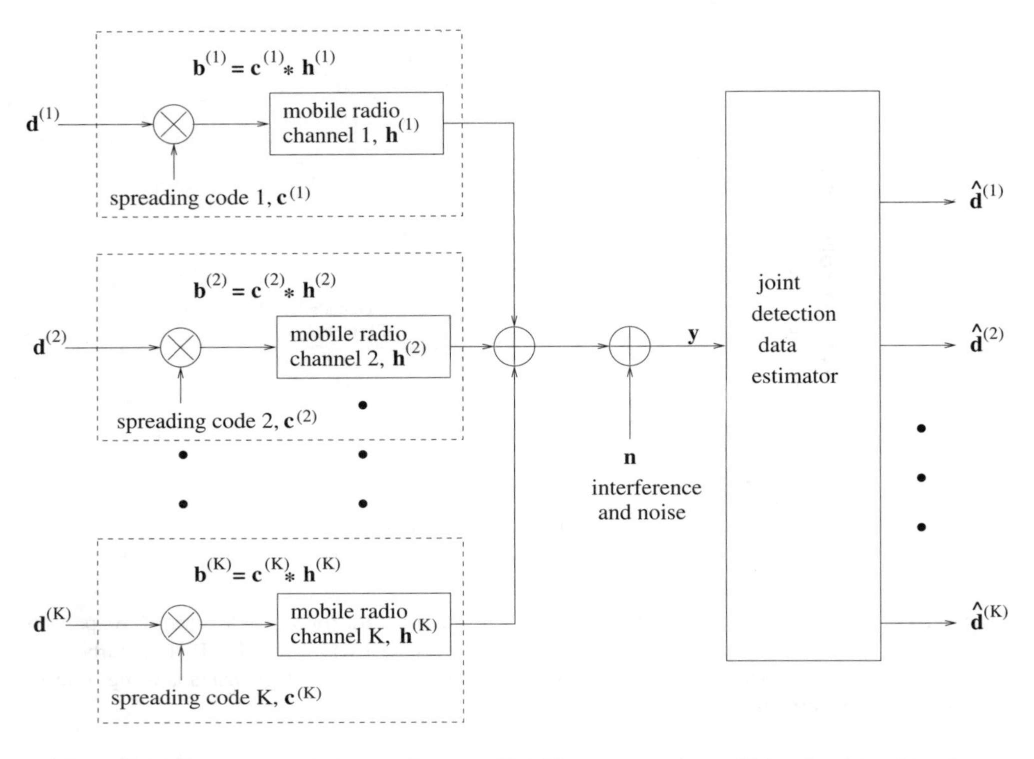

Figure 2.12: System model of a synchronous CDMA system on the up-link using joint detection.

mitigate the effects of ISI can be modified for employment in multi-user detection assisted CDMA systems. The so-called joint detection (JD) receivers constitute a category of multi-user detectors developed for synchronous burst-based CDMA transmissions and they utilise these techniques.

Figure 2.12 depicts the block diagram of a synchronous joint-detection assisted CDMA system model for up-link transmissions. There are a total of K users in the system, where the information is transmitted in bursts. Each user transmits N data symbols per burst and the data vector for user k is represented as $\mathbf{d}^{(\mathbf{k})}$. Each data symbol is spread with a user-specific spreading sequence, $\mathbf{c}^{(\mathbf{k})}$, which has a length of Q chips. In the uplink, the signal of each user passes through a different mobile channel characterised by its time-varying complex impulse response, $\mathbf{h}^{(\mathbf{k})}$. By sampling at the chip rate of $1/T_c$, the impulse response can be represented by W complex samples. Following the approach of Klein and Baier [186], the received burst can be represented as $\mathbf{y} = \mathbf{A}\mathbf{d} + \mathbf{n}$, where $\mathbf{y}$ is the received vector and consists of the synchronous sum of the transmitted signals of all the K users, corrupted by a noise sequence, $\mathbf{n}$. The matrix $\mathbf{A}$ is referred to as the system matrix and it defines the system's response, representing the effects of MAI and the mobile channels. Each column in the matrix represents the combined impulse response obtained by convolving the spreading sequence of a user with its channel impulse response, $\mathbf{b}^{(\mathbf{k})} = \mathbf{c}^{(\mathbf{k})} * \mathbf{h}^{(\mathbf{k})}$. This is the impulse response experienced by a transmitted data symbol. Upon neglecting the effects of the noise the joint detection formulation is simply based on inverting the system matrix $\mathbf{A}$, in order to recover the data vector constituted by the superimposed transmitted information of all the K CDMA users.

2.7.2 JD-ACDMA Modem Mode Adaptation and Signalling

In mobile communications systems typically power control techniques are used to mitigate the effects of pathloss and slow fading. However, in order to counteract the problem of fast fading and co-channel interference, agile and tight-specification power control algorithms are required [187]. Another technique that can be used to overcome the problems due to fading is adaptive-rate transmission [158, 188], where the information rate is varied according to the quality of the channel.

Different methods of multi-rate transmission have been proposed by Ottosson and Svensson [189]. According to the multi-code method, multiple codes are assigned to a user requiring a higher bit rate [189]. Multiple data rates can also be provided by a multiple processing-gain scheme, where the chip rate is kept constant but the data rates are varied by changing the processing gain of the spreading codes assigned to the users. Performance comparisons for both of these schemes have been carried out by Ottosson and Svensson [189] and Ramakrishna and Holtzman [190], demonstrating that both schemes achieved similar performance. Saquib and Yates [191] and Johansson and Svensson [192] have also investigated the employment of the so-called decorrelating detector and the successive interference cancellation receiver for multi-rate CDMA systems.

Adaptive rate transmission schemes, where the transmission rate is adapted according to the channel quality have also been proposed. Abeta *et al.* [193] have conducted investigations into an adaptive CDMA scheme, where the transmission rate is modified by varying the channel code rate and the processing gain of the CDMA user, employing the carrier to interference and noise ratio (CINR) as the switching metric. In their investigations, the overall packet rate was kept constant by transmitting in shorter bursts, when the transmission bit rate was high and lengthening the burst when the bit rate was low. This resulted in a decrease in interference power, which translated to an increase in system capacity. Hashimoto *et al.* [194] extended this work to show that the proposed system was capable of achieving a higher capacity with a smaller hand-off margin and lower average transmitter power. In these schemes, the conventional RAKE receiver was used for the detection of the data symbols. Kim [188] analysed the performance of two different methods of combatting the mobile channel's variations, which were the adaptation of the transmitter power to compensate for channel variations or the switching of the information rate to suit the channel conditions. Using a RAKE receiver, it was demonstrated that rate adaptation provided a higher average information rate than power adaptation for a given average transmit power and a given BER.

In our design example here we also propose to vary the information rate in accordance with the channel quality. However, in comparison to conventional power control techniques - which may disadvantage other users by increasing their transmitted powers in an effort to maintain the quality of their own links - the JD-AQAM scheme employed does not disadvantage other users. This is achieved by 'non-destructively' adjusting the modulation mode of the user supported accoding to the near-instantaneous channel quality experienced. Additionally, burst-by-burst adaptive transceivers are capable of increasing the network capacity, as we will demonstrate in the book. This is because conventional transceivers would drop a call, when the interference levels become excessive. By contrast, adaptive transceivers reconfigure themselves in a more robust coding/modulation mode.

In this section we will quantify the expected video performance of a range of intelligent multi-mode CDMA transceivers, employing joint detection (JD) multi-user reception CDMA techniques at the BS, which are optional in the 3G system proposals due to their high implementational complexity and hence are likely to be employed only in future implementations of the 3G standards. As a potential further future enhancement, we will also invoke the powerful principle of burst-by-burst adaptive JD-CDMA (JD-ACDMA) transmissions, which was discussed in some depth in Section 2.7. Burst-by-burst adaptive transmissions can be readily accommodated by JD-CDMA receivers, as it will be augmented in more detail below. The duplex JD-ACDMA video transceiver used in our system design example operates on the basis of the following philosophy:

- The channel quality estimation is based on evaluating the Mean Squared Error (MSE) at the output of the JD-CDMA multi-user equaliser at the receiver, as suggested for wideband single-carrier Kalman-filtered DFE-based modems by Liew *et al.* in [195].
- The decision concerning the modem mode to be used by the local transmitter for the forthcoming CDMA transmission burst is based on the prediction of the expected channel quality.
- Specifically, if the channel quality can be considered predictable, then the channel quality estimate for the uplink can be extracted from the received signal and *the receiver instructs the local transmitter as to what modem mode to use in its next transmission burst. We refer to this regime as open–loop adaptation.* In this case, the transmitter has to explicitly signal the modem modes to the receiver.
- By contrast, if the channel cannot be considered reciprocal, then the channel quality estimation is still performed at the receiver, but *the receiver has to instruct the remote transmitter as to what modem modes have to be used at the transmitter, in order to meet the target integrity requirements of the receiver. We refer to this mode as closed–loop adaptation.*

2.7.3 The JD-ACDMA Video Transceiver

In this JD-CDMA system performance study we transmitted 176x144 pixel Quarter Common Intermediate Format (QCIF) and 128x96 pixel Sub-QCIF (SQCIF) video sequences at 10 frames/s using a reconfigurable Time Division Multiple Access / Code Division Multiple Access (TDMA / CDMA) transceiver, which can be configured as a 1, 2 or 4 bit/symbol scheme. The H.263 video codec [196] extensively employs variable-length compression techniques and hence achieves a high compression ratio. However, as all entropy- and variable-length coded bit streams, its bits are extremely sensitive to transmission errors.

This error sensitivity was counteracted in our system by invoking the adaptive video packetisation and video packet dropping regime of [106], when the channel codec protecting the video stream became incapable of removing all channel errors. Specifically, we refrained from decoding the corrupted video packets in order to prevent error propagation through the reconstructed video frame buffer [106, 196]. Hence - similarly to our AQAM / TDD-based system design example - these corrupted video packets were dropped at both the transmitter and receiver and the reconstructed video frame buffer was not updated, until the next

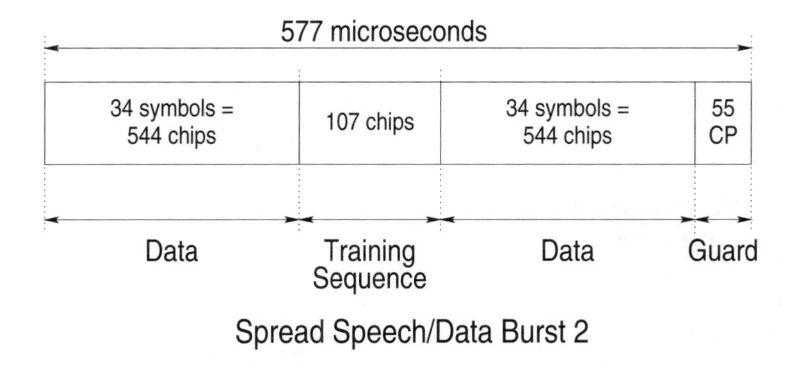

Figure 2.13: Transmission burst structure of the FMA1 spread speech/data mode 2 of the FRAMES proposal [183].

video packet replenishing the specific video frame area was received. This required a low-delay, strongly protected video packet acknowledgement flag, which was superimposed on the transmitted payload packets [106]. As in the system design example of the previous section, the associated video performance degradation was found perceptually unobjectionable for transmission burst error rates below about 5%.

The associated JD-ACDMA video system parameters are summarised in Table 2.7.3, which will be addressed in more depth during our further discourse. Employing a low spreading factor of 16 allowed us to improve the system's multi-user performance with the aid of joint-detection techniques [68], whilst imposing a realistic implementational complexity. This is because the JD operation is based on inverting the system matrix, which is constructed from the convolution of the channel's impulse response (CIR) and the spreading codes. Hence maintaining a low spreading factor (SF) is critical as to the implementational complexity. We note furthermore that the implementation of the joint detection receivers is independent of the number of bits per symbol associated with the modulation mode used, since the receiver simply inverts the associated system matrix and invokes a decision concerning the received symbol, irrespective of how many bits per symbol were used. *Therefore, joint detection receivers are amenable to amalgamation with the above 1, 2 and 4 bit/symbol CDMA modem, since they do not have to be reconfigured each time the modulation mode is switched.*

In this performance study we used the Pan-European FRAMES proposal [183] as the basis for our CDMA system. The associated transmission burst structure is shown in Figure 2.13, while a range of generic system parameters are summarised in Table 2.7.3. In our performance studies we used the COST207 [50] seven-path bad urban (BU) channel model, whose impulse response is portrayed in Figure 2.14.

Again, the remaining generic system parameters are defined in Table 2.7.3. In our JD-ACDMA design example we investigated the performance of a multi-mode convolutionally coded video system employing joint detection, while supporting two users. The associated convolutional codec parameters are summarised in Table 2.7.3 along with the operational-mode specific transceiver parameters of the multi-mode JD-ACDMA system. As seen in Table 2.7.3, when the channel is benign, the unprotected video bit rate will be approximately 26.9 kbit/s in the 16QAM/JD-CDMA mode. However, as the channel quality degrades, the

Parameter	
Multiple access	TDMA/CDMA
Channel type	COST 207 Bad Urban
Number of paths in channel	7
Normalised Doppler frequency	3.7×10^{-5}
CDMA spreading factor	16
Spreading sequence	Random
Tx. Frame duration	4.615 ms
Tx. Slot duration	577 μs
Joint detection CDMA receiver	Whitening matched filter (WMF) or Minimum mean square error block decision feedback equaliser (MMSE-BDFE)
No. of Slots/Frame	8
TDMA slots/Video packet	3
Chip Periods/TDMA slot	1250
Data Symbols/TDMA slot	68
User Data Symbol Rate (kBd)	14.7
System Data Symbol Rate (kBd)	117.9

Table 2.6: Generic system parameters using the FRAMES spread speech/data mode 2 proposal [183].

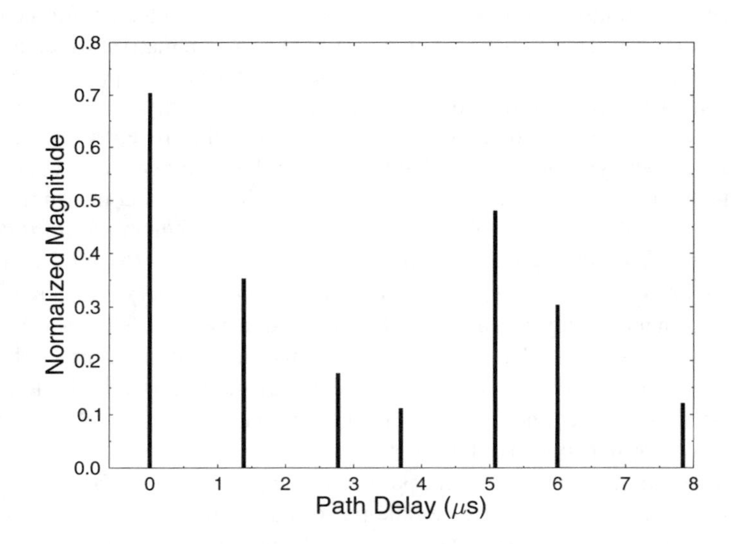

Figure 2.14: Normalised channel impulse response for the COST 207 [50] seven-path Bad Urban channel.

Features	Multi-rate System		
Mode	BPSK	4QAM	16QAM
Bits/Symbol	1	2	4
FEC	Convolutional Coding		
Octal Gen. Pol.	561; 753		
Coding-rate	$R = 1/2$		
Constraint-length	$K = 9$		
Transmitted bits/packet	204	408	816
Total bit rate (kbit/s)	14.7	29.5	58.9
FEC-coded bits/packet	102	204	408
Assigned to FEC-coding (kbit/s)	7.4	14.7	29.5
Error detection per packet	16 bit CRC		
Feedback bits / packet	9		
Video packet size	77	179	383
Packet header bits	8	9	10
Video bits/packet	69	170	373
Unprotected video-rate (kbit/s)	5.0	12.3	26.9
Video framerate (Hz)	10		

Table 2.7: Operational-mode specific JD-ACDMA video transceiver parameters used in our design example.

modem will switch to the BPSK mode of operation, where the video bit rate drops to 5 kbit/s and for maintaining a reasonable video quality, the video resolution has to be reduced to SQCIF (128x96 pels).

2.7.4 JD-ACDMA Video Transceiver Performance

The burst-by-burst adaptive JD-ACDMA scheme of our design example maximises the system's throughput expressed in terms of the number of bits per transmitted non-binary symbol by allocating the highest possible number of bits to a symbol based on the receiver's perception concerning the instantaneous channel quality. When the instantaneous channel conditions degrade, the number of bits per symbol (BPS) is reduced in order to maintain the required target transmission burst error rate. Figure 2.15 provides a snap-shot of the JD-ACDMA system's mode switching dynamics, which is based on the fluctuating channel conditions determined by all factors influencing the channel's quality, such as pathloss, fast-fading, slow-fading, dispersion, co-channel interference, etc. The adaptive modem uses the SINR estimate at the output of the joint-detector, in order to estimate the instantaneous channel quality, and hence to set the modulation mode. The probability density function (PDF) of the JD-ACDMA scheme using each modulation mode for a particular average channel SNR is portrayed in Figure 2.16. It can be seen at high channel SNRs that the modem predominantly uses the 16QAM/JD-ACDMA modulation mode, while at low channel SNRs the BPSK mode is most prevalent. However, the PDF is widely spread, indicating that often the

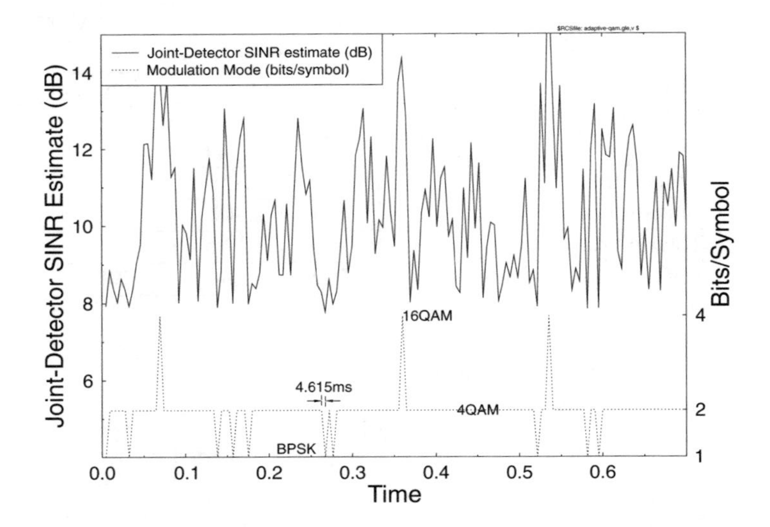

Figure 2.15: Example of modem mode switching in a dynamically reconfigured burst-by-burst modem in operation, where the modulation mode switching is based upon the SINR estimate at the output of the joint-detector over the channel model of Figure 2.14.

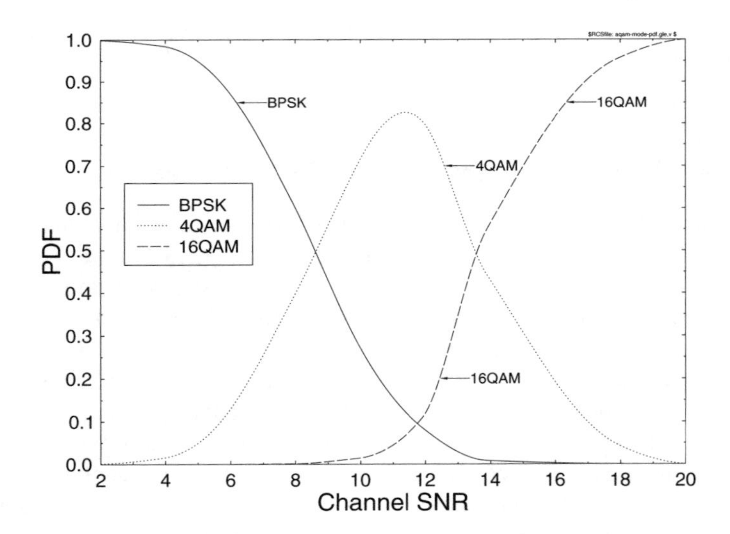

Figure 2.16: PDF of the various adaptive modem modes versus channel SNR over the channel model of Figure 2.14.

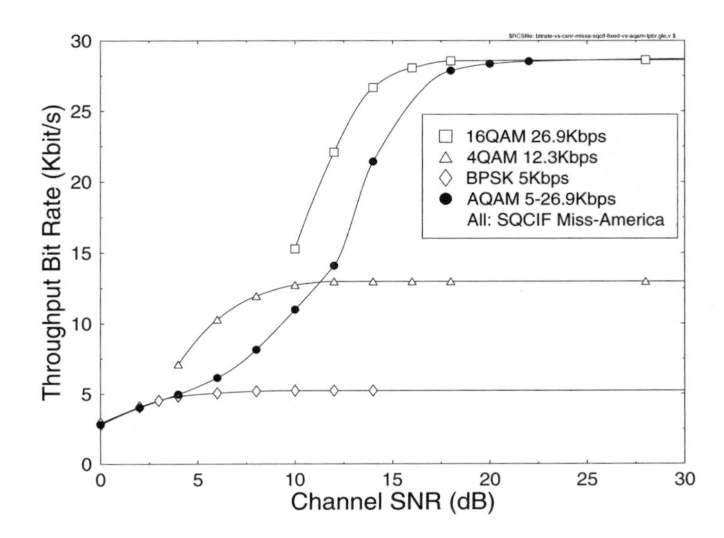

Figure 2.17: Throughput bit rate versus channel SNR comparision of the three fixed modulation modes (BPSK, 4QAM, 16QAM) and the adaptive burst-by-burst modem (AQAM), both supporting two users with the aid of joint detection over the channel model of Figure 2.14.

channel quality is misjudged by the receiver due to unpredictable channel quality fluctuations caused by a high doppler frequency or co-channel interference, etc. Hence in certain cases BPSK is used under high channel quality conditions or 16QAM is employed under hostile channel conditions.

The advantage of the dynamically reconfigured burst-by-adaptive JD-ACDMA modem over a statically reconfigured system, which would be incapable of near-instantaneous channel quality estimation and modem mode switching is that the video quality is smoothly - rather than abruptly - degraded, as the channel conditions deteriorate and vice versa. By contrast, a less 'agile' statically switched or reconfigured multi-mode system results in more visible reductions in video quality, when the modem switches to a more robust modulation mode, as it is demonstrated in Figure 2.17. Explicitly, Figure 2.17 shows the throughput bit rate of the dynamically reconfigured burst-by-burst adaptive modem, compared to the three modes of a less agile, statically switched multi-mode system. The reduction of the fixed modem modes' effective throughput at low SNRs is due to the fact that under such channel conditions an increased fraction of the transmitted packets have to be dropped, reducing the effective throughput, since dropped packets do not contribute towards the system's effective throughput. The figure shows the smooth reduction of the throughput bit rate, as the channel quality deteriorates. The burst-by-burst modem matches the BPSK mode's bit rate at low channel SNRs, and the 16QAM mode's bit rate at high SNRs. In this example the dynamically reconfigured burst-by-burst adaptive modem characterised in the figure perfectly estimates the prevalent channel conditions although in practice the estimate of channel quality is not perfect and it is inherently delayed. Hence our results constitute the best-case performance.

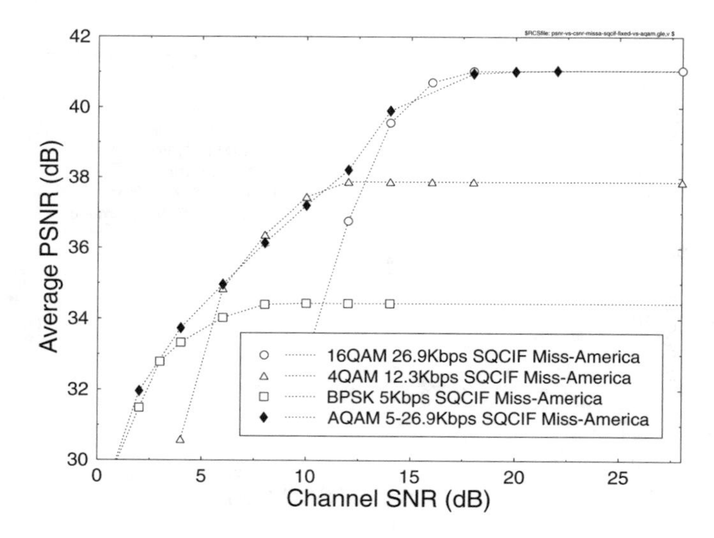

Figure 2.18: Average decoded video quality (PSNR) versus channel SNR comparision of the fixed modulation modes of BPSK, 4QAM and 16QAM, and the burst-by-burst adaptive modem. Both supporting two-users with the aid of joint detection. These results were recorded for the Miss-America video sequence at SQCIF resolution (128x96 pels) over the channel model of Figure 2.14

The smoothly varying effective throughput bit rate of the burst-by-burst adaptive modem translates into a smoothly varying video quality, as the channel conditions change. The video quality measured in terms of the average PSNR is shown versus the channel SNR in Figure 2.18 in contrast to that of the individual modem modes. The figure demonstrates that the burst-by-burst adaptive modem provides equal or better video quality over a large proportion of the SNR range shown than the individual modes. However, even at channel SNRs, where the adaptive modem has a slightly reduced PSNR, the perceived video quality of the adaptive modem is better, since the video packet loss rate is far lower than that of the fixed modem modes.

2.8 Subband-Adaptive OFDM Video Transceivers

In order to demonstrate the benefits of the proposed near-instantaneously adaptive video transceivers also in the context of OFDM schemes [13, 185], in this section we compare the performance of a subband-adaptive OFDM video scheme [151] to that of a fixed modulation mode transceiver under identical propagation conditions, while having the same transmission bitrate. The subband-adaptive modem is capable of achieving a lower BER, since it can disable transmissions over low quality sub-carriers and compensate for the lost throughput by invoking a higher-order modulation mode, than that of the fixed-mode transceiver over the high-quality sub-carriers.

	BPSK mode	QPSK mode
Packet rate	4687.5 Packets/s	
FFT length	512	
OFDM symbols/packet	3	
OFDM symbol duration	$2.6667\mu s$	
OFDM time frame	80 Timeslots = $213\mu s$	
Normalised Doppler frequency, f'_d	1.235×10^{-4}	
OFDM symbol normalised Doppler frequency, F_D	7.41×10^{-2}	
FEC coded bits/packet	1536	3072
FEC-coded video bitrate	7.2 Mbps	14.4 Mbps
Unprotected Bits/Packet	766	1534
Unprotected bitrate	3.6 Mbps	7.2 Mbps
Error detection CRC (bits)	16	16
Feedback error flag bits	9	9
Packet header bits/packet	11	12
Effective video bits/packet	730	1497
Effective video bitrate	3.4 Mbps	7.0 Mbps

Table 2.8: System parameters for the fixed QPSK and BPSK transceivers, as well as for the corresponding subband-adaptive OFDM (AOFDM) transceivers for Wireless Local Area Networks (WLANs).

Table 2.8 shows the system parameters for the fixed-mode BPSK and QPSK transceivers, as well as for the corresponding AOFDM transceivers. The system employs constraint length three, half-rate turbo coding, using octal generator polynomials of 5 and 7 as well as random turbo interleavers, where the channel- and turbo-interleaver depth was adjusted for each AOFDM transmission burst, in order to facilitate burst-by-burst or symbol-by-symbol based OFDM demodulation and turbo decoding. Therefore the unprotected bitrate is approximately half the channel coded bitrate. The protected to unprotected video bitrate ratio is not exactly half, since two tailing bits are required to reset the convolutional encoders' memory to their default state in each transmission burst. In both the BPSK and QPSK modes 16-bit Cyclic Redundancy Checking (CRC) is used for error detection and 9 bits are used to encode the reverse link feedback acknowledgement information by simple repetition coding. The packet acknowledgement flag decoding ensues using majority logic decisions. The packetisation [151] requires a small amount of header information added to each transmitted packet, which is 11 and 12 bits per packet for BPSK and QPSK, respectively. The effective or useful video bitrates for the fixed BPSK and QPSK modes are then 3.4 and 7.0 Mbps.

The fixed-mode BPSK and QPSK transceivers are limited to one and two bits per symbol, respectively. By contrast, the proposed AOFDM transceivers operate at the same bitrate as their corresponding fixed modem mode counterparts, although they can vary their modulation mode on a subband by subband basis between 0, 1, 2 and 4 bits per symbol. Zero bits per symbol implies that transmissions are disabled for the subband concerned.

The 'micro-adaptive' nature of the subband-adaptive modem is characterised by Figure 2.19, portraying at the top a contour plot of the channel Signal-to-Noise Ratio (SNR) for each subcarrier versus time. This channel SNR fluctuation was recorded here for the short indoor WLAN channel impulse response of Figure 2.20 having a maximum dispersion of about 60ns, which was referred to as the short Wireless Asynchronous Transfer Mode (WATM) channel in [13].

At the centre and bottom of the figure the modulation mode chosen for each 32-subcarrier subband is shown versus time for the 3.4 and 7.0 Mbps target-rate subband-adaptive modems, respectively. Again, this was recorded for the short WATM channel impulse response of Figure 2.20. It can be seen that when the channel is of high quality – like for example at about frame 1080 – the subband-adaptive modem used the same modulation mode as the equivalent fixed rate modem in all subcarriers. When the channel is hostile – like around frame 1060 – the subband-adaptive modem used a lower-order modulation mode in some subbands, than the equivalent fixed mode scheme, or in extreme cases disabled transmission for that subband. In order to compensate for the loss of throughput in this subband a higher-order modulation mode was used in the highest quality subbands.

One video packet is transmitted per OFDM symbol, therefore the video packet loss ratio is the same as the OFDM symbol error ratio. The video packet loss ratio is plotted versus the channel SNR in Figure 2.21. It is shown in the graph that the subband-adaptive transceivers – or synonymously termed as microscopic-adaptive (μAOFDM), in contrast to OFDM symbol-by-symbol adaptive transceivers – have a lower packet loss ratio (PLR) at the same SNR compared to the fixed modulation mode transceiver. Note in Figure 2.21 that the subband-adaptive transceivers can operate at lower channel SNRs than the fixed modem mode transceivers, while maintaining the same required video packet loss ratio. Again, the figure labels the subband-adaptive OFDM transceivers as μAOFDM, implying that the adaptation is not noticeable from the upper layers of the system. A macro-adaption could be applied in addition to the microscopic adaption by switching between different target bitrates on an OFDM symbol-by-symbol basis, as the longer-term channel quality improves and degrades. This issue was further investigated in [151].

The figure shows that when the channel quality is high, the throughput bitrate of the fixed and adaptive transceivers is identical. However, as the channel degrades, the loss of packets due to channel impairments results in a lower throughput bitrate. The lower packet loss ratio of the subband-adaptive transceiver results in a higher throughput bitrate than that of the fixed modulation mode transceiver. Finally, these improved throughput bitrate results translate to the enhanced decoded video quality performance results evaluated in terms of Peak Signal-to-Noise Ratio (PSNR) in Figure 2.22. Again, for high channel SNRs the performance of the fixed and adaptive OFDM transceivers is identical. However, as the channel quality degrades, the video quality of the subband-adaptive transceiver degrades less dramatically than that of the corresponding fixed modulation mode transceiver.

2.9 Summary and Conclusions

In contrast to the statically reconfigured narrow-band multimode video transceivers [151], in this chapter we have advocated BbB-AQAM based wireless transceivers [151]. We justified their service-related benefits in terms of the video quality improvements perceived by the users of such systems. As an example, the channel quality perceived by the channel equaliser

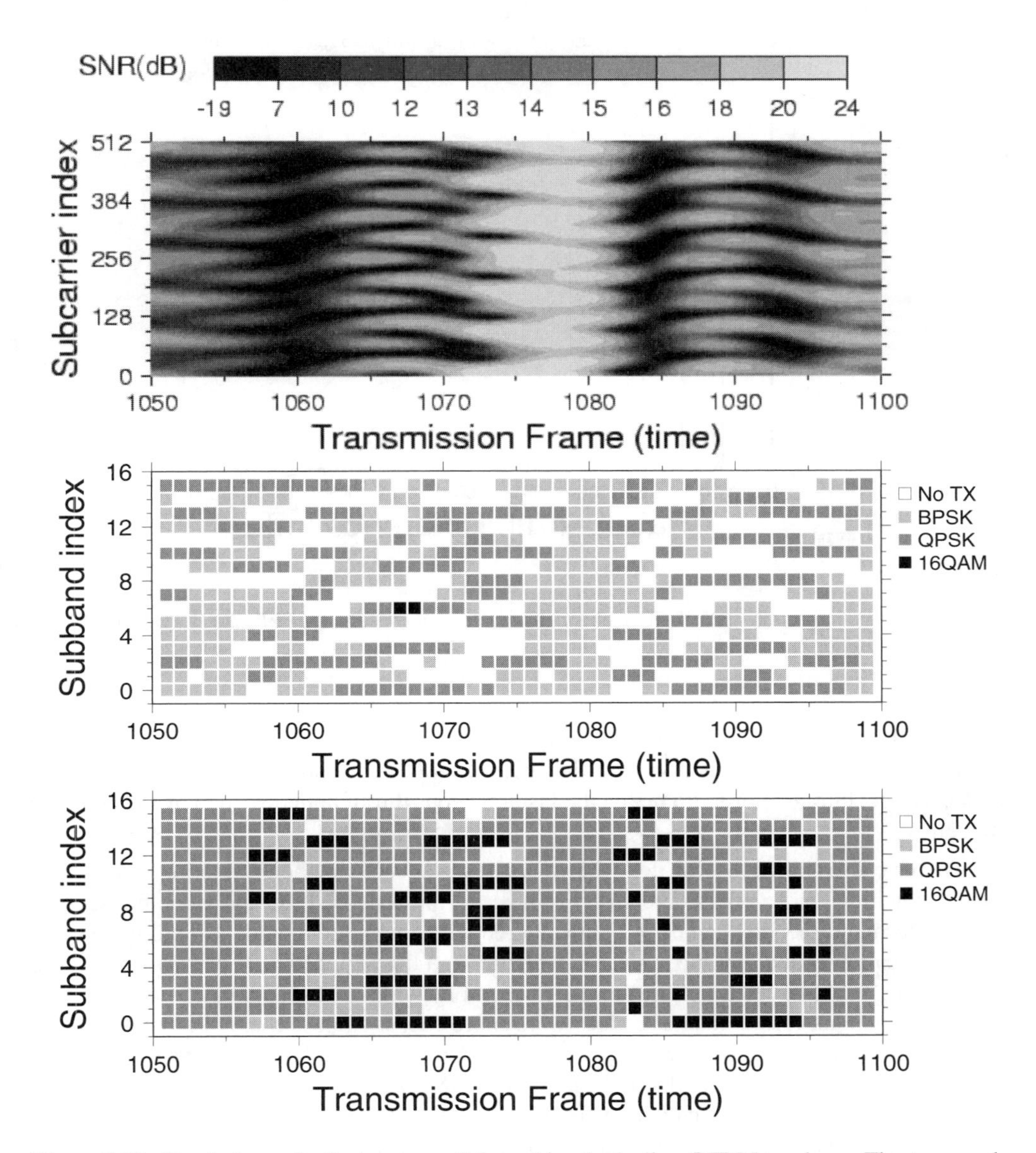

Figure 2.19: The 'micro-adaptive' nature of the subband-adaptive OFDM modem. The top graph is a contour plot of the channel SNR for all 512 subcarriers versus time. The bottom two graphs show the modulation modes chosen for all 16 32-subcarrier subbands for the same period of time. The middle graph shows the performance of the 3.4Mbps subband-adaptive modem, which operates at the same bitrate as a fixed BPSK modem. The bottom graph represents the 7.0Mbps subband-adaptive modem, which operated at the same bitrate as a fixed QPSK modem. The average channel SNR was 16dB. ©IEEE, 2001, Hanzo, Cherriman, Streit [151]

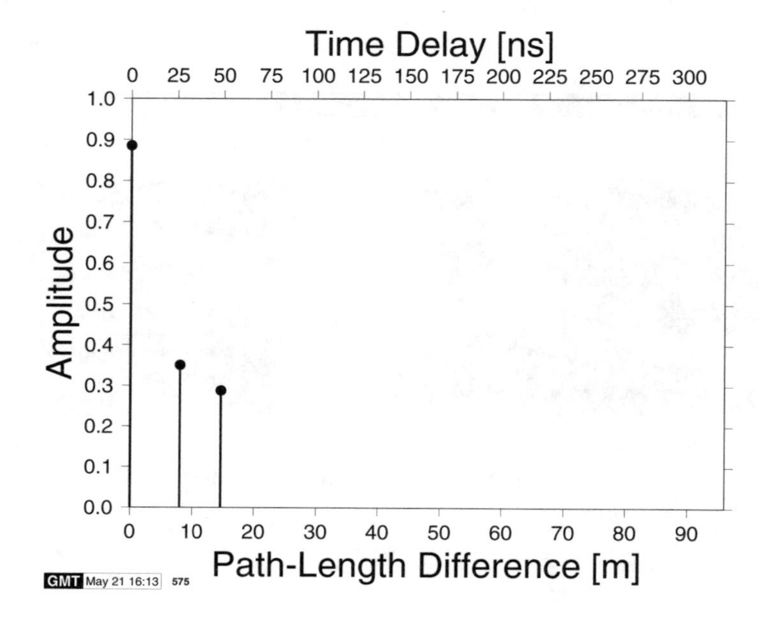

Figure 2.20: Indoor three-path WATM channel impulse response. ©IEEE, 2001, Hanzo, Webb, Keller [13].

or the multi-user equaliser was used for controlling the AQAM modes. When an adaptive packetiser is used in conjunction with the AQAM modem, it continually adjusts the video codec's target bitrate in order to match the instantaneous throughput provided by the adaptive modem.

We have also shown that the delay between the instants of channel estimation and AQAM mode switching has an effect on the performance of the proposed AQAM video transceiver. This performance penalty can be mitigated by reducing the modem mode signalling delay. It was also demonstrated that the system can be tuned to the required FER performance using appropriate AQAM switching thresholds. In harmony with our expectations, we found that the more complex turbo channel codecs were more robust against channel effects than the lower-complexity binary BCH codecs. Lastly, the AQAM principles were extended to joint-detection assisted AQAM/CDMA and adaptive OFDM systems, where similar findings were confirmed to those found in the context of unspread AQAM.

It is a natural thought to combine these adaptive transceivers [67, 150–152] with diversity aided Multiple Input, Multiple Output (MIMO) systems and space-time coding [177, 197–200] in a further effort towards mitigating the effects of fading and rendering the channel more Gaussian-like. **A vital question in this context is, whether adaptive transceivers retain their performance advantages in conjunction with MIMOs?** As expected, no significant joint benefits accrue, since both of these regimes aim at mitigating the effects of fading and once the fading is mitigated sufficiently for it to become near-Gaussian, no further fading counter-measures are necessary. It is worth noting, however that MIMOs have been predominantly studied in the context of narrowband or non-dispersive fading channels or in

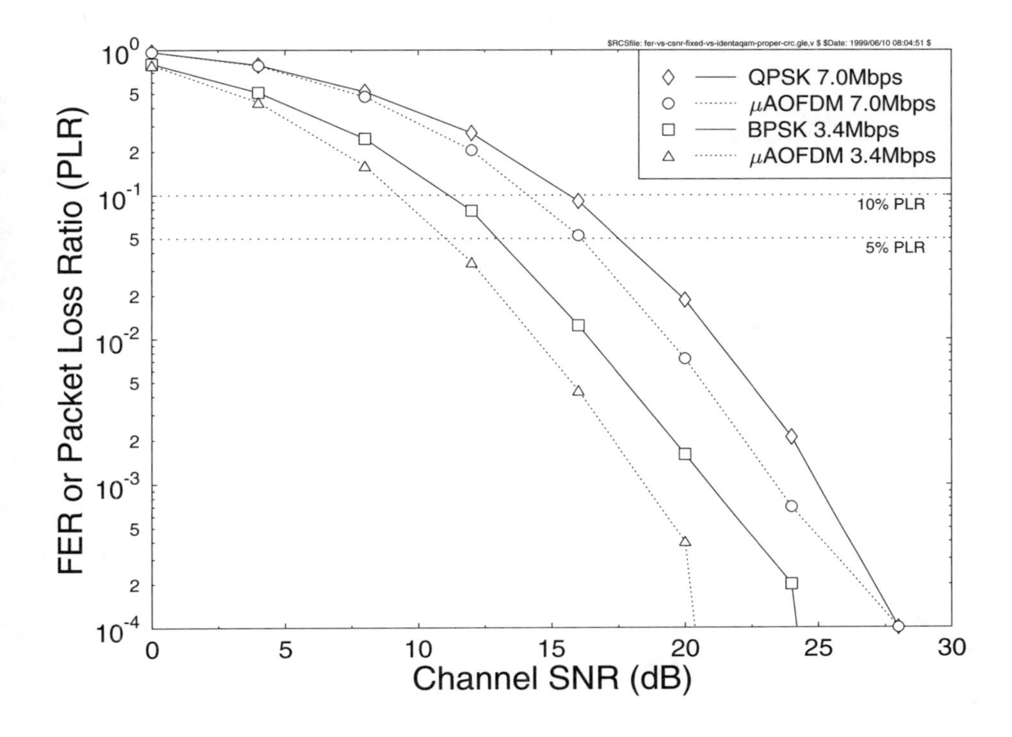

Figure 2.21: Frame Error Rate (FER) or video packet loss ratio (PLR) versus channel SNR for the BPSK and QPSK fixed modulation mode OFDM transceivers and for the corresponding subband-adaptive μAOFDM transceiver, operating at identical effective video bitrates, namely at 3.4 and 7.0 Mbps, over the channel model of Figure 2.20 at a normalised Doppler frequency of $F_D = 7.41 \times 10^{-2}$. ©IEEE, 2001, Hanzo, Cherriman, Streit [151].

conjunction with OFDM - a scheme that decomposes a high-rate bit stream into a high number of low-rate bit streams - thereby rendering the dispersive channel non-dispersive for each of the low-rate composite streams.

A further problem, when invoking high-order receiver diversity in an effort to mitigate the effects of fading and hence rendering the wireless channel Gaussian-like is that the receiver complexity increases. It is a more attractive proposition to employ complex, transmit diversity assisted base stations, which allows us to aim for low-complexity terminals. In this context in recent years space-time codecs have found favour and have also been proposed for the IMT2000 system and for multi-user HIPERLAN 2 type systems.

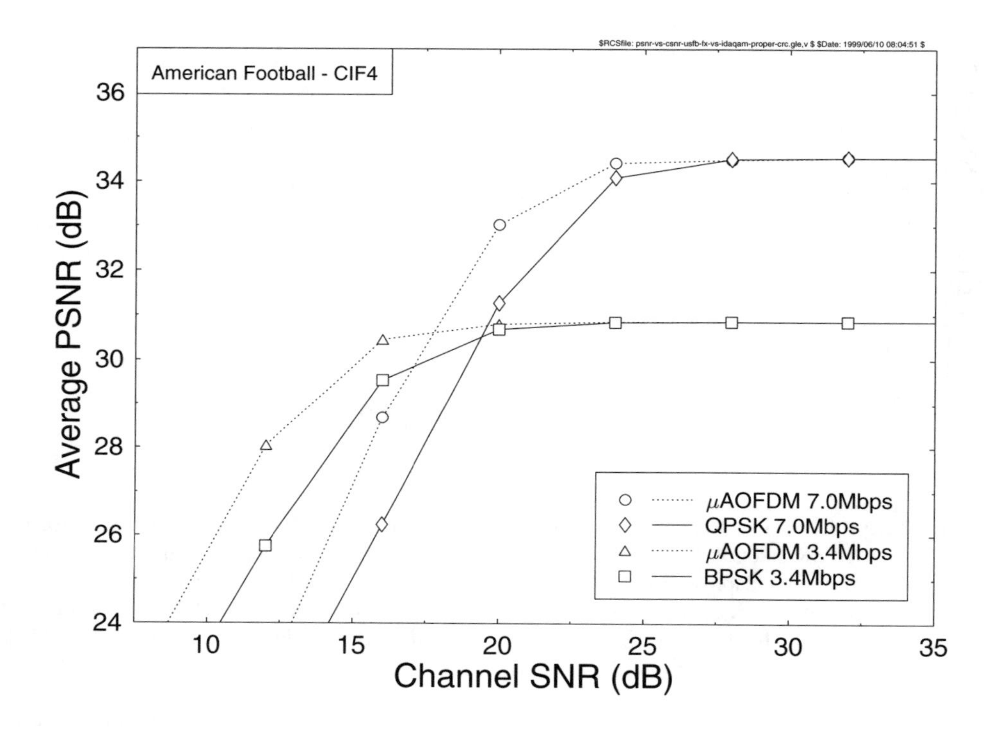

Figure 2.22: Average video quality expressed in PSNR versus channel SNR for the BPSK and QPSK fixed modulation mode OFDM transceivers and for the corresponding μAOFDM transceiver operating at identical channel SNRs over the channel model of Figure 2.20 at a normalised Doppler frequency of $F_D = 7.41 \times 10^{-2}$. ©IEEE, 2001, Hanzo, Cherriman, Streit [151].

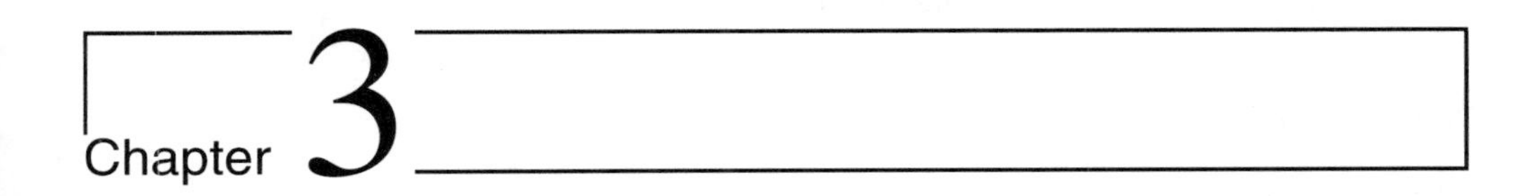

Intelligent Antenna Arrays and Beamforming

3.1 Introduction

Adaptive beamforming was initially developed in the 1960s for the military applications of sonar and radar, in order to remove unwanted noise and jamming from the output. The related literature of the past 40 years is extremely rich [201–237] and since this book is mainly concerned with the networking aspects of wireless systems, rather than with specific antenna array designs, here we will restrict our discussions on the topic to a rudimentary overview.

The first fully adaptive array was conceived in 1965 by Applebaum [238], which was designed to maximise the Signal-to-Noise Ratio (SNR) at the array's output. An alternative approach to cancelling unwanted interference is the Least Mean Squares (LMS) error algorithm of Widrow [239]. While a simple idea, satisfactory performance can be achieved under specific conditions. Further work on the LMS algorithm, by Frost [240] and Griffiths [241], introduced constraints to ensure that the desired signals were not filtered out along with the unwanted signals. The optimisation process takes place as before, but the antenna gain is maintained constant in the desired direction. For stationary signals, both algorithms converge to the optimum Wiener solution [3, 240, 242]. A different technique was proposed in 1969 by Capon [243] using a Minimum-Variance Distortionless Response (MVDR) or the Maximum Likelihood Method (MLM). In 1974, Reed *et al.* demonstrated the power of the Sample-Matrix Inversion (SMI) technique, which determines the adaptive antenna array weights directly [244]. Unlike the algorithms of Applebaum [238] and Widrow [239], which may suffer from slow convergence if the eigenvalue spread of the received sample correlation matrix is relatively large, the performance of the SMI technique is virtually independent of the eigenvalue spread.

In recent years the tight frequency reuse of cellular systems has stimulated renewed research interests in the field [3, 6, 242, 245]. In this book we will attempt to review the recent literature and highlight the most important research issues for UMTS, HiperLAN and WATM applications, while providing some performance results. We commence in Section 3.2 by reviewing beamforming and its potential benefits, then we provide a generic signal model in Section 3.2.3 and we describe the processes of element and beam space beamforming. In Section 3.3 we highlight a range of adaptive beamforming algorithms and consider the less commonly examined downlink scenario in Section 3.3.5. Lastly in Section 3.3.6 we provide some performance results and outline our future work.

3.2 Beamforming

The signals induced in different elements of an antenna array are combined to form a single output of the array. This process of combining the signals from the different elements is known as beamforming. This section describes the basic characteristics of an antenna, the advantages of using beamforming techniques in a mobile radio environment [3,6], and a generic signal model for use in beamforming calculations. For further details on the associated issues the reader is referred to [3, 6, 8, 238–242, 244–250].

3.2.1 Antenna Array Parameters

Below we provide a few definitions used throughout this report in order to describe antenna systems:

Radiation Pattern The radiation pattern of an antenna is the relative distribution of the radiated power as a function of direction in space. The radiation pattern of an antenna array is the product of the element pattern and the array factor, both of which are defined below. If $f(\theta, \phi)$ is the radiation pattern of each antenna element and $F(\theta, \phi)$ is the array factor, then the array's radiation pattern, $G(\theta, \phi)$, which is also referred to as the beam pattern, is given by

$$G(\theta, \phi) = f(\theta, \phi)F(\theta, \phi). \tag{3.1}$$

Figure 3.1 gives an example of a stylised antenna element response, an array factor of an 8 element linear array with an element spacing of $\lambda/2$ steered at 0° and the radiation pattern, which results from combining the two.

Array Factor The array factor, $F(\theta, \phi)$, is the far-field radiation pattern of an array of isotropically radiating elements, where θ is the azimuth angle and ϕ is the elevation angle.

Main Lobe The main lobe of an antenna radiation pattern is the lobe containing the direction of maximum radiated power.

Sidelobes Sidelobes are lobes of the antenna radiation pattern, which do not constitute the mainlobe. They allow signals to be received in directions other than that of the main lobe and hence they are undesirable, but they are also unavoidable.

Beamwidth The beamwidth of an antenna is the angular width of the main lobe. The 3 dB beamwidth is the angular width between the points on the main lobe that are 3 dB below the peak of the main lobe. A smaller beamwidth results from an array of a greater aperture size, which is the distance between the two farthest elements of the array.

Antenna Efficiency Antenna efficiency is the ratio of the total power radiated by the antenna to the total power input to the antenna.

Grating Lobes When the distance between the antenna array elements, d, exceeds $\lambda/2$, spatial under-sampling of the received radio frequency carrier wave takes place, causing secondary maxima [2, 247], referred to as grating lobes, to appear in the radiation pattern, which

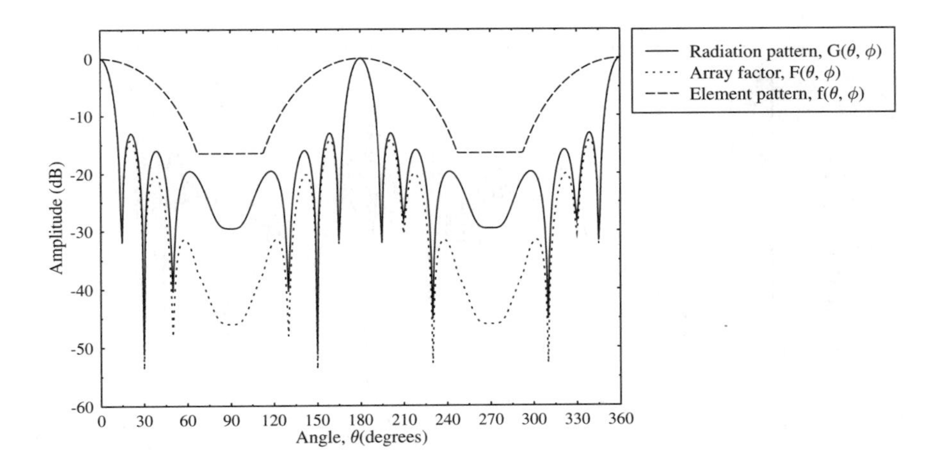

Figure 3.1: The array factor of an eight element linear array with an element spacing of $\lambda/2$ steered at $0°$, the response of each antenna element and the radiation pattern resulting from combining the two.

can be clearly seen in Figure 3.2. The spatial under-sampling results in ambiguities in the directions of the arriving signals, which manifests itself as copies of the main lobe in unwanted directions. The grating lobe phenomenon in spatial sampling is analogous to the well known aliasing effect in temporal sampling [247]. Therefore, the distance, d, between adjacent sensors in the array must be chosen to be less than or equal to $\lambda/2$, if grating lobes are to be avoided [247, 251]. However, an inter-element spacing of greater than $\lambda/2$ improves the spatial resolution of the array [2], i.e. reduces the 3 dB beamwidth as shown in Figure 3.2, and reduces the correlation between the signals arriving at adjacent antenna elements.

3.2.2 Potential Benefits of Antenna Arrays in Mobile Communications

3.2.2.1 Multiple Beams [6]

The formation of multiple beams, or sectorisation, uses multiple antennae at the base station in order to form beams that cover the whole cell site [251]. For example, three beams, each with a beamwidth of 120° may cover the entire 360° as seen in Figure 3.3. The coverage area of each beam may be regarded as a separate cell, with frequency assignment and handovers between beams performed in the usual manner [252]. No intelligence is required to locate a subscriber within a beam and to connect that beam to a radio channel unit. The use of multiple beams results in a reduction of the co-channel interference. In the uplink scenario, the signal received from the mobile station constitutes interference at only two base stations, and additionally in only one sector. In the downlink, the situation is similar, only now the sectors which can interfere with the user in the central cell are the images of the interfering

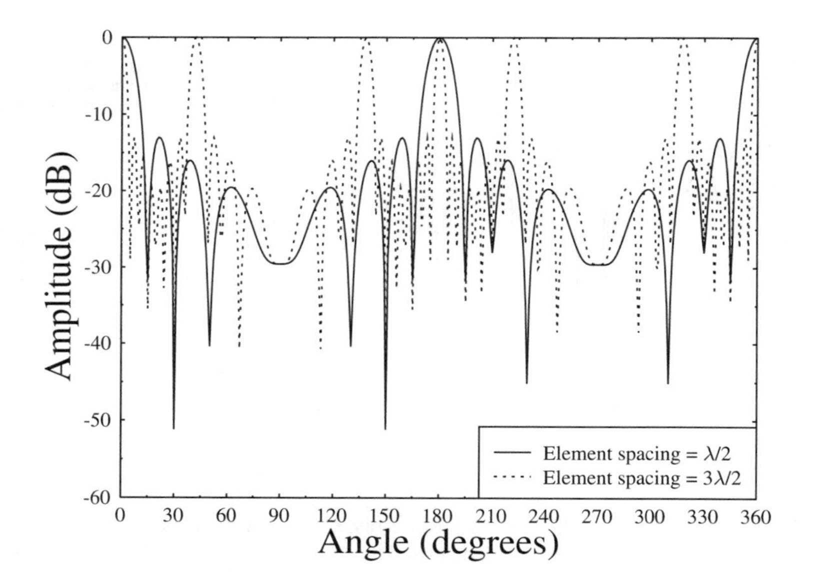

Figure 3.2: The array factor of an eight element uniform linear array with element spacing of $\lambda/2$ and $3\lambda/2$. The grating lobes associated with the spatial under-sampling-induced secondary maxima of the radiated carrier wave are clearly visible for the case when the element spacing is $3\lambda/2$.

sectors on the uplink [19], again, as shown in Figure 3.3.

3.2.2.2 Adaptive Beams [6]

The combined antenna array is used to find the location of each mobile, and then beams are formed, in order to cover different mobiles or groups of mobiles [20,253]. Each beam having its own coverage area may be considered as a co-channel cell, and thus be able to use the same carrier frequency [7,251]. In conventional sectorisation the location of the beams is fixed, while the adaptive system allows the beams to cover specific areas of the cell within which users are located [17]. In intelligent near-future systems the beams may follow the mobiles, which benefit from the concentrated transmission power, with inter-beam handovers occurring as necessary.

3.2.2.3 Null Steering [6,254]

In contrast to steering beams towards mobiles, null steering creates spatial radiation nulls towards co-channel mobiles [38]. The realisation of true nulls or zero response is not possible due to practical considerations, such as the isolation of the radio frequency components.

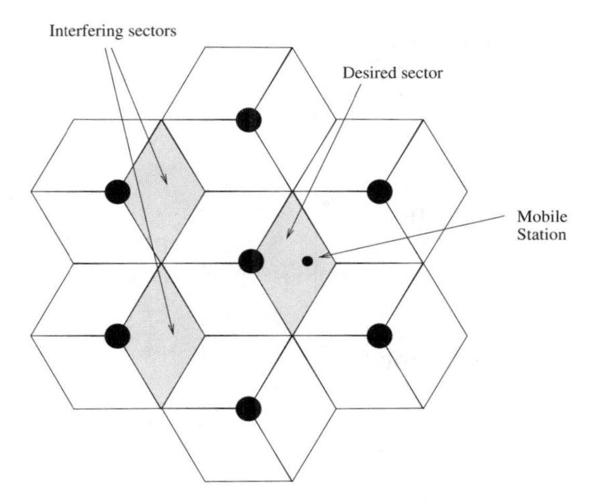

Figure 3.3: An example of sectorisation, using three sectors per base station, showing the reduced levels of interference with respect to an omni-directional base station antenna scenario.

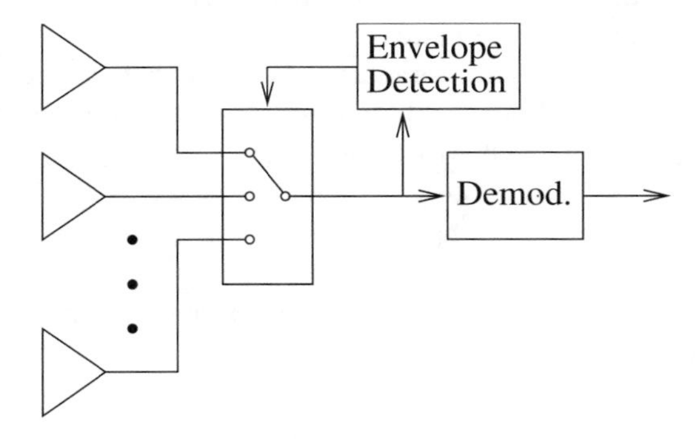

Figure 3.4: Switched-diversity combining.

The formation of spatial radiation nulls in the antenna response towards co-channel mobiles reduces the co-channel interference both on the uplink and the downlink [2, 253].

3.2.2.4 Diversity Schemes [6, 255]

The simplest and most commonly used diversity scheme is *switched diversity*. In this scheme the system switches between antennae, such that only one is in use at any one time [1, 256], as shown in Figure 3.4. The switching criterion is often the loss of received signal level at the antenna being used. The switching may be performed at the Radio Frequency (RF) stage, avoiding the need for a down-converter for each antenna.

Selection diversity is a more sophisticated version of switched diversity, where the system can monitor the signal level on all of the antennae simultaneously, and select the specific

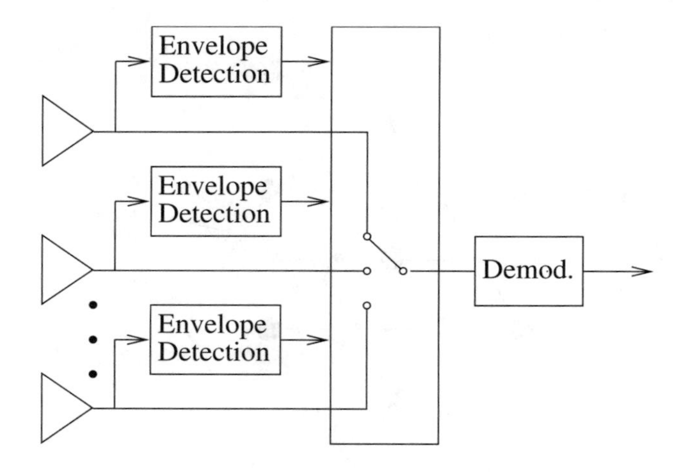

Figure 3.5: Selective-diversity combining.

branch exhibiting the highest SNR at any given time, thus requiring an RF front-end for each antenna in the system [1], as seen in Figure 3.5.

In a Rayleigh fading environment, the fading at each branch can be assumed to be independent provided that the branches are sufficiently far apart. If each branch has an instantaneous SNR of γ_l, the probability density function of γ_l is given by [3]

$$p(\gamma_l) = \frac{1}{\Gamma} e^{\frac{-\gamma_l}{\Gamma}} \tag{3.2}$$

where Γ denotes the mean SNR at each branch. The probability that a single branch has a SNR less than some threshold γ is given by [3]

$$P[\gamma_l \leq \gamma] = \int_0^\infty p(\gamma_l) d\gamma_l = 1 - e^{-\frac{\gamma}{\Gamma}}. \tag{3.3}$$

Therefore, the probability that all the branches fail to achieve an SNR higher than γ is [3]:

$$P_L(\gamma) = P[\gamma_1, \gamma_2, \ldots, \gamma_L \leq \gamma] = \left(1 - e^{-\frac{\gamma}{\Gamma}}\right)^L, \tag{3.4}$$

from which the probability density function of the fading magnitude in conjunction with selection diversity can be obtained,

$$p_L(\gamma) = \frac{d}{d\gamma} P_L(\gamma) = \frac{L}{\Gamma}\left(1 - e^{-\frac{\gamma}{\Gamma}}\right)^{L-1} e^{-\frac{\gamma}{\Gamma}}, \tag{3.5}$$

leading to the average SNR, $\bar{\gamma}$, of selection diversity assisted Rayleigh fading channels as [3]:

$$\bar{\gamma} = \int_0^\infty \gamma p_L(\gamma) d\gamma = \Gamma \sum_{l=1}^{L} \frac{1}{l}. \tag{3.6}$$

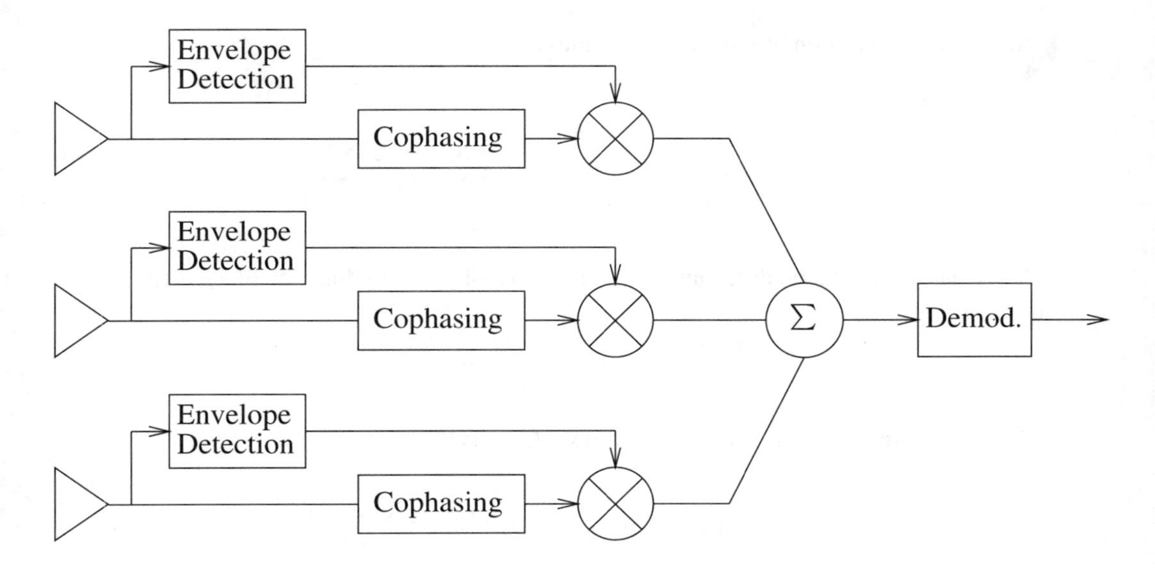

Figure 3.6: Optimal Combining.

In maximal ratio combining, which is also often referred to as optimal diversity combining, the signal of each antenna is weighted by its instantaneous Signal-to-Noise Ratio (SNR). The weighted signals are then combined for forming a single output, as shown in Figure 3.6. It has been shown that the maximal ratio combining technique is optimal, if the diversity branch signals are uncorrelated and follow a Rayleigh distribution [21], provided that the noise has a Gaussian distribution and a zero mean. If each branch has a gain, g_l, the output of the combiner is [3]

$$s_L = \sum_{l=1}^{L} g_l s_l, \tag{3.7}$$

and if each branch has the noise power, σ_n^2, the total noise power at the output of the combiner is [3]:

$$\sigma_N^2 = \sigma_n^2 \sum_{l=1}^{L} g_l^2. \tag{3.8}$$

Therefore, the SNR at the output of the combiner is given by

$$\gamma_L = \frac{s_l^2}{2\sigma_N^2}. \tag{3.9}$$

It can be easily shown that γ_L is maximised, when $g_l = s_l^2/\sigma_n^2$, which is the SNR in each branch. The expansion of Equation 3.9 is thus

$$\gamma_L = \frac{1}{2}\frac{\left(\sum_{l=1}^{L}\frac{s_l^2}{\sigma_n^2}s_l\right)^2}{\sigma_n^2\sum_{l=1}^{L}\left(\frac{s_l^2}{\sigma_n^2}\right)^2} = \frac{1}{2}\sum_{l=1}^{L}\frac{s_l^2}{\sigma_n^2} = \sum_{l=1}^{L}\gamma_l. \tag{3.10}$$

As γ_L has a chi-squared distribution [3], the probability density function of γ_L is [3]:

$$p(\gamma_L) = \frac{\gamma_L^{L-1}e^{-\frac{\gamma_L}{\Gamma}}}{\Gamma^L(L-1)!}. \tag{3.11}$$

The probability that γ_L is less than the threshold, γ, is [3]

$$P[\gamma_L \leq \gamma] = \int_0^{\gamma} p(\gamma_L)d\gamma_L = 1 - e^{-\frac{\gamma}{\Gamma}}\sum_{l=1}^{L}\frac{(\frac{\gamma}{\Gamma})^{l-1}}{(l-1)!}. \tag{3.12}$$

The expectation of Equation 3.12, $\bar{\gamma}_L$, is the average SNR at the output of the combiner:

$$\bar{\gamma}_L = \sum_{l=1}^{L}\bar{\Gamma} = L\Gamma, \tag{3.13}$$

where Γ is the mean SNR at each branch.

Optimal combining processes the signals received from an antenna array such that the contribution from unwanted co-channel sources is reduced, whilst enhancing that of the desired signal. The explicit knowledge of the directions of the interferences is not necessary, but some characteristics of the desired signal are required in order to protect it from cancellation as if it were an unwanted co-channel source [6]. A popular technique is to use a reference signal, such as a channel sounding sequence, which must be correlated with the desired signal. The scheme then phase-coherently combines all the signals that are correlated with the reference signal, whilst simultaneously cancelling the waveforms that are not correlated with this signal, resulting in the removal of co-channel interferences.

A base station using an optimal combining antenna array may adjust the array weights during the receive cycle, in order to enhance the signal arriving from a desired mobile. A system using the same frequency for receiving and transmitting the signals in different time slots, such as in the Time Division Duplex (TDD) Digital European Cordless Telephone (DECT) [257,258] system may be able to use the complex conjugate of these weights during the transmit cycle in order to pre-process the transmit signal and to enhance the signal received at the desired mobile, whilst suppressing this signal at the other mobiles. This process relies on the fact that the weights were adjusted during the receive cycle to reduce co-channel interference, thus placing nulls in the directions of co-channel mobiles [6]. Therefore, by employing the complex conjugate of these weights during the transmit cycle, the same antenna pattern may be produced, resulting in no energy transmitted towards the co-channel mobiles [6].

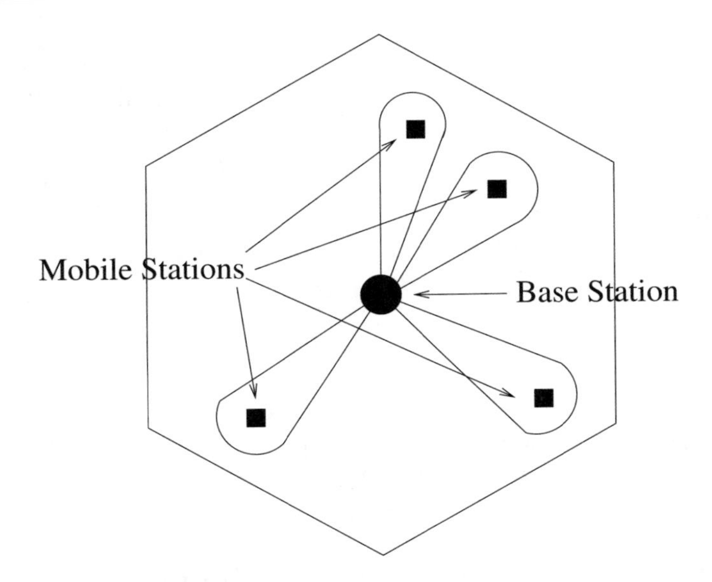

Figure 3.7: A cell layout showing how an antenna array can support many users on the same carrier frequency and timeslot with the advent of spatial filtering or Space Division Multiple Access (SDMA).

3.2.2.5 Reduction in Delay Spread and Multipath Fading

Delay spread is caused by multipath propagation, where a desired signal arriving from different directions is delayed due to the different distances travelled [17]. In transmit mode an intelligent antenna is able to focus the energy in the required direction, assisting in reducing the multipath reflections and thus delay spread. In receive mode the antenna array is able to perform optimal combining after delay compensation of the multipath signals incident upon it [1]. Those signals whose delays cannot be compensated for may be cancelled by the formation of nulls in their directions [18].

The directive nature of an antenna array also results in a smaller spread of Doppler frequencies encountered at the mobile [259]. For an omni-directional antenna at both the base station, and at the mobile the Direction-Of-Arrival (DOA) at the mobile is uniformly distributed. Hence the Doppler spectrum is given by Clarke's model [21] as:

$$S_r(f) = \frac{A_o^2}{\pi f_m \sqrt{1-(f/f_m)^2}}, \qquad |f| < f_m. \tag{3.14}$$

where A_o is the mean power transmitted and $f_m = v/\lambda$ is the maximum Doppler shift, where v is the velocity of the mobile and λ is the carrier wavelength. However, if a directional antenna is used at the base station then the Doppler power spectral density is given by [259]:

$$\begin{aligned} S_r(f) &= \frac{A_o^2}{f_m \sqrt{1-(f/f_m)^2}} \\ &\cdot [f_\theta(\phi_v + |\cos^{-1}(f/f_m)|) + f_\theta(\phi_v - |\cos^{-1}(f/f_m)|)],\ |f| < f_m, \end{aligned} \tag{3.15}$$

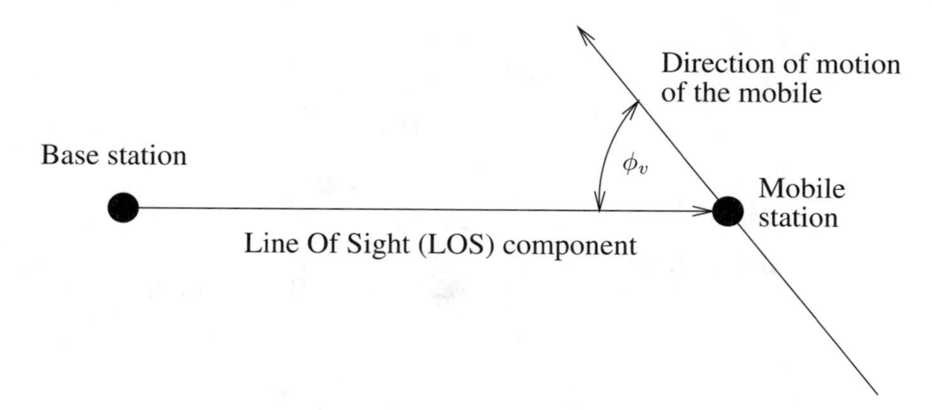

Figure 3.8: Illustration of the Line Of Sight (LOS) component arriving at the mobile from the base station showing the direction of motion of the mobile, ϕ_v.

where ϕ_v, as shown in Figure 3.8, is the direction of motion of the mobile with respect to the direction of the base station from the mobile and $f_\theta()$ is the PDF of the DOA of the multipath components at the mobile, as given by [259]:

$$f_\theta(\theta) = \begin{cases} \frac{R^2}{I}, & -\theta_1 < \theta \leq \theta_1 \\ \frac{(D\tan(\alpha))^2}{I(\sin(\theta)+\cos(\theta)\tan(\alpha))^2}, & \theta_1 < |\theta| \leq \theta_2 \\ \frac{R^2}{I}, & \theta_2 < \theta \leq -\theta_2 \end{cases} \tag{3.16}$$

where

$$I = 2R^2(\pi + \theta_1 - \theta_2) + 4D\sin(\alpha)\sqrt{R^2 - D^2\sin^2(\alpha)}. \tag{3.17}$$

Furthermore, 2α is the beamwidth of the so-called idealised 'flat-top' directional antenna, which has zero gain except over the angular spread of 2α, where the gain is 1, R is the radius of the circular area containing all the scatters and D is the separation distance between the base station and the mobile. Finally, θ_1 and θ_2 are constants calculated using $\theta = \cos^{-1}\left[\frac{D}{R}\sin^2(\alpha) \pm \frac{\cos(\alpha)}{R}\sqrt{R^2 - D^2\sin^2(\alpha)}\right]$. Figure 3.9 shows examples of the Doppler spectra for beamwidths of 2, 10 and 20 degrees for a mobile moving at angles of 0, 45 and 90 degrees with respect to the main LOS component, with a base station to mobile distance of 3 km, where the scatterers are all located within a circle of 1 km radius of the mobile.

3.2.2.6 Reduction in Co-channel Interference

An antenna array allows the implementation of spatial filtering, as shown in Figure 3.7, which may be exploited in both transmitting as well as receiving modes in order to reduce co-channel interferences [1, 2, 14, 15]. When transmitting, the antenna is used to focus the radiated energy in order to form a directive beam in the area, where the receiver is likely to

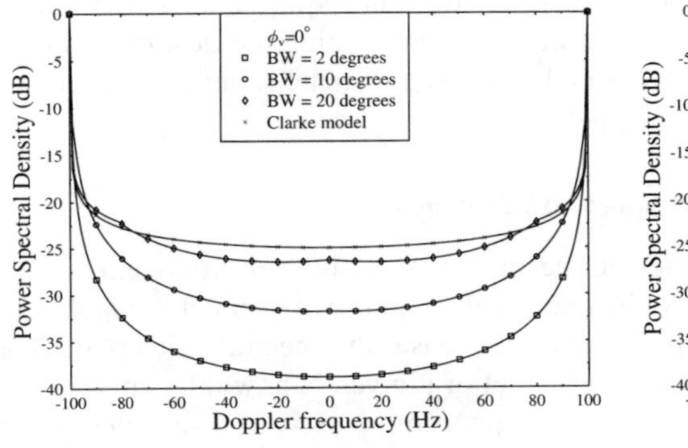

(a) Mobile's direction, $\phi_v = 0°$.

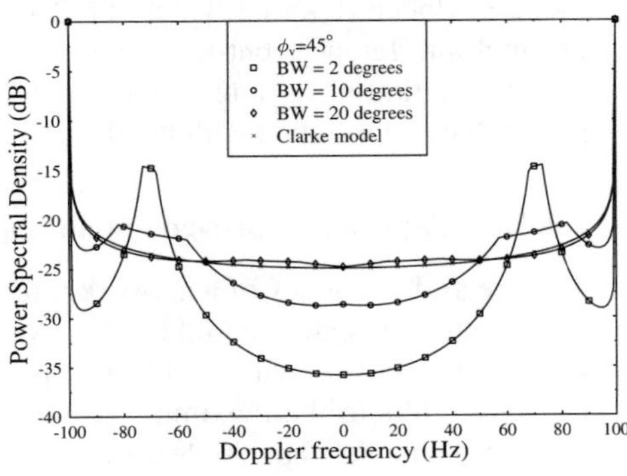

(b) Mobile's direction, $\phi_v = 45°$.

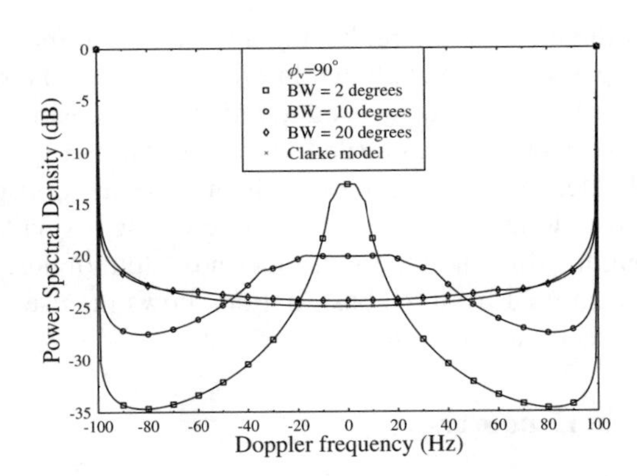

(c) Mobile's direction, $\phi_v = 90°$.

Figure 3.9: Doppler spectra at the mobile, when using a directional antenna at the base station, and an omnidirectional antenna at the mobile, is compared with Clarke's model. $R = 1$km, $D = 3$km, $f_m = 100$ Hz.

be. This in turn means that there is less interference in the other directions, where the beam is not pointing. The co-channel interference generated in transmit mode may be further reduced by forming beams exhibiting nulls in the directions of other receivers [6, 16]. This scheme deliberately reduces the transmitted energy in the direction of co-channel receivers and hence requires prior knowledge of their positions.

The employment of antenna arrays for reducing co-channel interference in the receive mode has been reported widely [1, 2, 6, 16–18]. It does not require knowledge of the co-channel interference, but must have some information concerning the desired signal, such as the direction of its source, a reference signal, such as a channel sounding sequence, or a signal that is correlated with the desired signal.

3.2.2.7 Capacity Improvement and Spectral Efficiency

The spectral efficiency of a network refers to the amount of traffic a given system with a certain spectral allocation could handle. An increase in the number of users of the mobile communications system without a loss of performance increases the spectral efficiency. Channel capacity refers to the maximum data rate a channel of a given bandwidth can sustain. An improved channel capacity leads to an ability to support more users of a specified data rate, implying a better spectral efficiency. The increased quality of service that results from the reduced co-channel interference and reduced multipath fading [18, 19] upon using smart antennae may be exchanged for an increased number of users [2, 20].

3.2.2.8 Increase in Transmission Efficiency

An antenna array is directive in nature, having a high gain in the direction where the beam is pointing. This property may be exploited in order to extend the range of the base station, resulting in a larger cell size or may be used to reduce the transmitted power of the mobiles. The employment of a directive antenna allows the base station to receive weaker signals than an omni-directional antenna. This implies that the mobile can transmit at a lower power and its battery life becomes longer, or it would be able to use a smaller battery, resulting in a smaller size and weight, which is important for hand-held mobiles. A corresponding reduction in the power transmitted from the base station allows the use of electronic components having lower power ratings and therefore, lower cost.

3.2.2.9 Reduction in Handovers

When the amount of traffic in a cell exceeds the cell's capacity, cell splitting is often used in order to create new cells [2], each with its own base station and frequency assignment. The reduction in cell size leads to an increase in the number of handovers performed. By using antenna arrays to increase the capacity of a cell [1] the number of handovers required may actually be reduced. Since each beam tracks a mobile [2], no handover is necessary, unless different beams using the same frequency cross each other.

3.2.3 Signal Model

Consider an array of L omni-directional antenna elements situated in the far field of a sinusoidal point source, as shown in Figure 3.10. Given that the array element separation is d and

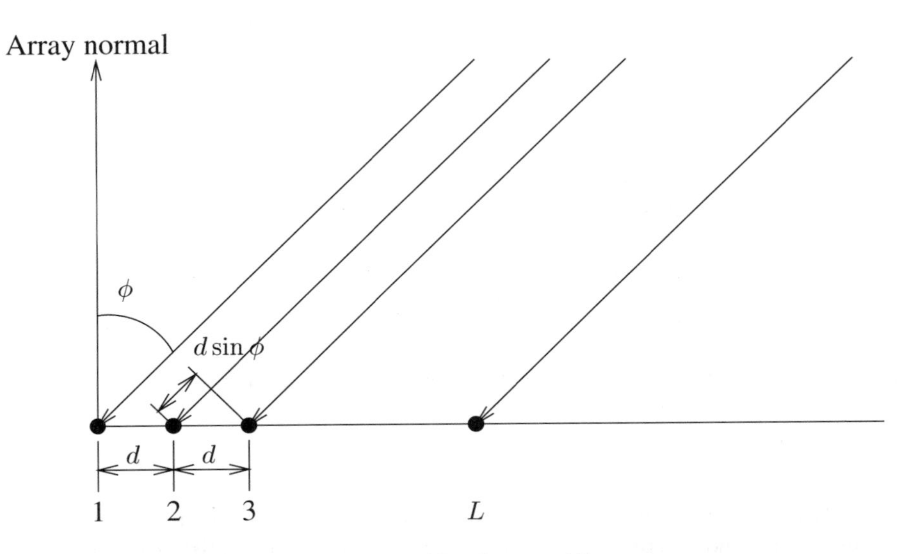

Figure 3.10: Reception by a uniformly spaced linear antenna array.

the plane wavefront is impinging upon the array at an angle of θ with respect to the array normal, the wavefront arrives at the $l+1^{th}$ element before arriving at the lth element. Again, as seen in Figure 3.10, the extra distance that the wavefront must travel to reach the lth element relative to the $l+1^{th}$ element is $d\sin\theta$. However, for an arbitrary array of L elements the relative delays, assuming that the point of zero delay is the origin, are given by

$$t_l(\theta) = \frac{x_l \sin\theta + y_l \cos\theta}{c}, \qquad l = 1, \ldots, L \tag{3.18}$$

where c is the speed of wave propagation, i.e. the speed of light, while x_l and y_l are the x and y-coordinates of the lth element with respect to the origin located at (0,0). The extra cosine term is due to the potential y-offset from the x-axis of the array elements which is zero, and thus omitted, from the example shown in Figure 3.10. The signal, $x_{l,i}(t)$, induced in the lth element due to the ith source can be expressed as

$$x_{l,i}(t) = m_i(t)e^{j\omega t_l(\theta)}, \tag{3.19}$$

with $m_i(t)$ denoting the complex modulating function. This expression is based upon the narrow-band assumption for array signal processing, which assumes that the bandwidth of the signal is sufficiently small, so that the weighting co-efficients maintain a constant phase variation across all of the antenna array elements.

Assuming M directional sources and isotropic background noise, the total signal at the lth element is

$$x_l = \sum_{i=1}^{M} m_i(t)e^{j\omega t_l(\theta)} + n_l(t), \tag{3.20}$$

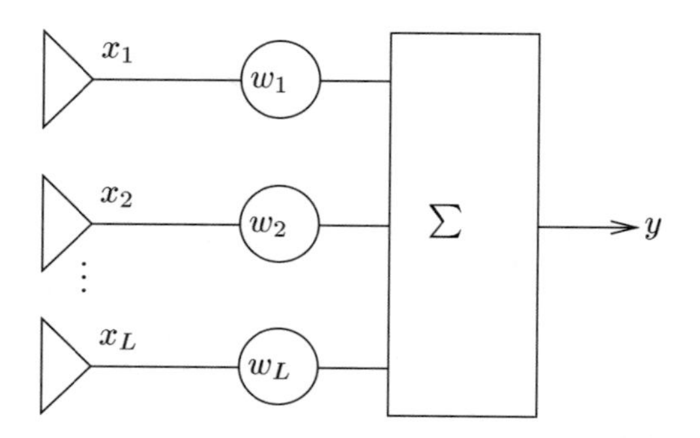

Figure 3.11: A beamformer sums the weighted antenna element signals, yielding the received signal $y(t) = \sum_{l=1}^{L} w_l^* x_l(t)$.

where $n_l(t)$ is a random noise component on the lth antenna array element, which includes background noise and electronic noise. It is assumed to be white noise with a mean of zero and a variance of σ_n^2.

The array factor, $F(\theta)$ which was introduced in Section 3.2.1 may be calculated thus as:

$$F(\theta) = \sum_{l=1}^{L} w_l e^{-j\omega t_l(\theta)}, \tag{3.21}$$

where w_l is the complex weighting applied to the lth element to steer the antenna beam in the direction of θ_0. The maximum value of $F(\theta)$ will occur when $\theta = \theta_0$, as shown previously in Figure 3.1.

Consider the narrow-band receiving beamformer, shown in Figure 3.11, where signals from each element are multiplied by a complex weight, $w_l, l = 1, \ldots, L$ and summed, in order to form the array output. The array output, $y(t)$ in Figure 3.11, at time t is given by

$$y(t) = \sum_{l=1}^{L} w_l^* x_l(t), \tag{3.22}$$

where * denotes the complex conjugate, $x_l(t)$ is the signal arriving from the lth element of the array, and w_l is the weight applied to the lth element. Representing the weights of the beamformer of Figure 3.11 as:

$$\underline{w} = [w_1, w_2, \ldots, w_L]^T, \tag{3.23}$$

and the signals induced in all elements as

$$\underline{x} = [x_1(t), x_2(t), \ldots, x_L(t)]^T, \tag{3.24}$$

the output of the beamformer receiver in Figure 3.11 becomes

$$y(t) = \underline{w}^T \underline{x}(t), \tag{3.25}$$

where the superscripts T and H, respectively, denote the transpose and complex conjugate transpose of a vector or matrix.

Let R define the L-by-L correlation matrix of the signal received by the L elements:

$$R = E[\underline{x}(t)\underline{x}^H(t)] = E\left\{\begin{bmatrix} x_1(t) \\ x_2(t) \\ \vdots \\ x_L(t) \end{bmatrix} \begin{bmatrix} x_1^*(t) & x_2^*(t) & \dots & x_L^*(t) \end{bmatrix}\right\}, \tag{3.26}$$

where the superscript H denotes Hermitian transposition (i.e., transposition combined with complex conjugation).

The correlation matrix R may be expressed in the expanded form:

$$R = \begin{bmatrix} r(0) & r(1) & \dots & r(L-1) \\ r(-1) & r(0) & \dots & r(L-2) \\ \vdots & \vdots & \ddots & \vdots \\ r(-L+1) & r(-L+2) & \cdots & r(0) \end{bmatrix}. \tag{3.27}$$

The element $r(0)$ on the main diagonal is always real-valued. For complex-valued data, the remaining elements of R assume complex values. The correlation matrix of a stationary discrete-time stochastic process is Hermitian [247], i.e. $R^H = R$. Alternatively, this may be written as $r(-k) = r^*(k)$, where $r(k)$ is the autocorrelation function of the stochastic process for a lag of k. Therefore, Equation 3.27 may be rewritten as

$$R = \begin{bmatrix} r(0) & r(1) & \dots & r(L-1) \\ r^*(1) & r(0) & \dots & r(L-2) \\ \vdots & \vdots & \ddots & \vdots \\ r^*(L-1) & r^*(L-2) & \cdots & r(0) \end{bmatrix}. \tag{3.28}$$

The elements of the matrix, R, denote the correlation between the output signals of the various antenna elements of Figure 3.11. For example, R_{ij} denotes the correlation between the ith and the jth elements of the array. Given that the steering vector associated with the direction θ_i, or the i_{th} source, can be described by an L-dimensional complex vector $\underline{s}_i$ as [242],

$$\underline{s}_i = [\exp(j\omega t_1(\theta_i)), \dots, \exp(j\omega t_L(\theta_i))]^T, \tag{3.29}$$

where L is the number of elements in the antenna array, and t_i is the time delay taken by a plane wave arriving from the ith source, located in the direction θ_i, and measured from the element at the origin, then the correlation matrix, R, of the array elements' outputs in Figure 3.11 may be expressed as [242]:

$$R = \sum_{i=1}^{M} p_i \underline{s}_i \underline{s}_i^H + \sigma_n^2 I, \tag{3.30}$$

where p_i is the power of the ith source, σ_n^2 is the noise power and I is the identity matrix.

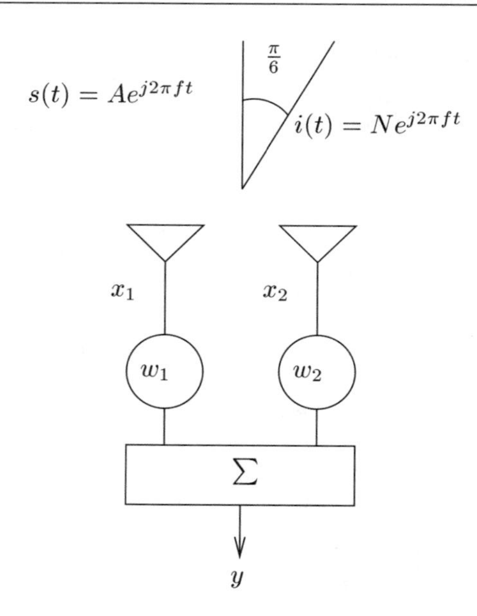

Figure 3.12: Example of a beamforming receiver problem with a wanted signal at 0° and interfering signal at 30° using an array element spacing of $\lambda/2$.

Using matrix notation, the correlation matrix, R, may be expressed in the following form [242, 260]:

$$R = ASA^H + \sigma_n^2 I = U\Lambda U^H, \tag{3.31}$$

where $S = E[\underline{s}_i \underline{s}_i^H]$ is the covariance matrix of the array elements' outputs in Figure 3.11, $A = [\underline{s}_1, \underline{s}_2, \ldots, \underline{s}_M]$ and is the $L \times M$ matrix of steering vectors. Furthermore, the diagonal matrix $\Lambda = \text{diag}[\lambda_1, \lambda_2, \ldots, \lambda_L]$ is constituted by the real eigenvalues of R, while U contains the corresponding unit-norm eigenvectors of R.

3.2.4 A Beamforming Example

Consider the antenna array shown in Figure 3.12, which consists of two omni-directional antenna elements having a spacing of $\frac{\lambda}{2}$. The desired unmodulated carrier signal, $s(t) = Ae^{j2\pi ft}$, arrives from the angle of θ_s=0 radians. The interfering signal, $i(t) = Ne^{j2\pi ft}$, arrives from the direction of θ_i=$\frac{\pi}{6}$ radians or 30°. Both signals have the same frequency, f. The signal arriving from each antenna array element is multiplied by a variable complex weight, and the weighted signals are then summed in order to form the array output. The array output due to the desired signal is

$$y_s(t) = Ae^{j2\pi ft}(w_1 + w_2). \tag{3.32}$$

For the array output, $y(t)$ in Figure 3.12, to be the desired signal $s(t)$, the following equation must be satisfied:

$$Ae^{j2\pi ft}(w_1 + w_2) = Ae^{j2\pi ft}, \tag{3.33}$$

which leads to

$$\begin{aligned} \Re[w_1] + \Re[w_2] &= 1 \\ \Im[w_1] + \Im[w_2] &= 0. \end{aligned} \tag{3.34}$$

The interfering signal arrives at the second array element with a phase lead of $\frac{\pi}{2}$ relative to the first element, since their spacing is $\lambda/2$ and the angle of incidence is 30°. Therefore, the array output due to the interfering signal is

$$y_i(t) = w_1 N e^{j2\pi ft} + w_2 N e^{j(2\pi ft + \pi/2)}. \tag{3.35}$$

For this to become zero we require that:

$$\begin{aligned} \Re[w_1] - \Re[w_2] &= 0 \\ \Im[w_1] + \Im[w_2] &= 0. \end{aligned} \tag{3.36}$$

Solving the simultaneous Equations 3.34 and 3.36 yields

$$w_1 = 0.5 - j0.5, w_2 = 0.5 + j0.5. \tag{3.37}$$

The beam pattern obtained using these weights is shown in Figure 3.13. The desired signal at 0° is attenuated by about 3 dB, but the unwanted interference at an angle of 30° is subjected to an attenuation of more than 30 dB. This example shows how beamforming and the cancellation of unwanted interferences may be accomplished. However, a practical beamformer does not require the information regarding the location, number and nature of the signal sources.

3.2.5 Analogue Beamforming

An antenna array consists of a number of antenna elements, the outputs of which are combined via an amplitude and phase control network, in order to form a desired antenna beam [20]. It is possible to perform analogue beamforming at the RF stage [20], using phase shifters and amplifiers, however, the high specification required of these devices renders them costly. An alternative solution is to down-convert the RF signal to an Intermediate Frequency (IF) and to perform the beamforming at the IF stage [3]. The disadvantage of this technique is that each antenna must have its own RF-to-IF receiver. Multiple beamformers must be used to form multiple beams, resulting in the distribution of the signal energy across all the formed beams. The output SNR is thus reduced, when the lower signal energy of the beams is combined with the increased noise injected by the increased number of RF and IF stages.

3.2.6 Digital Beamforming

The philosophy of digital beamforming is similar to that of analogue beamforming in that they both adjust the amplitude and phase of the signal arriving from each antenna element, but they

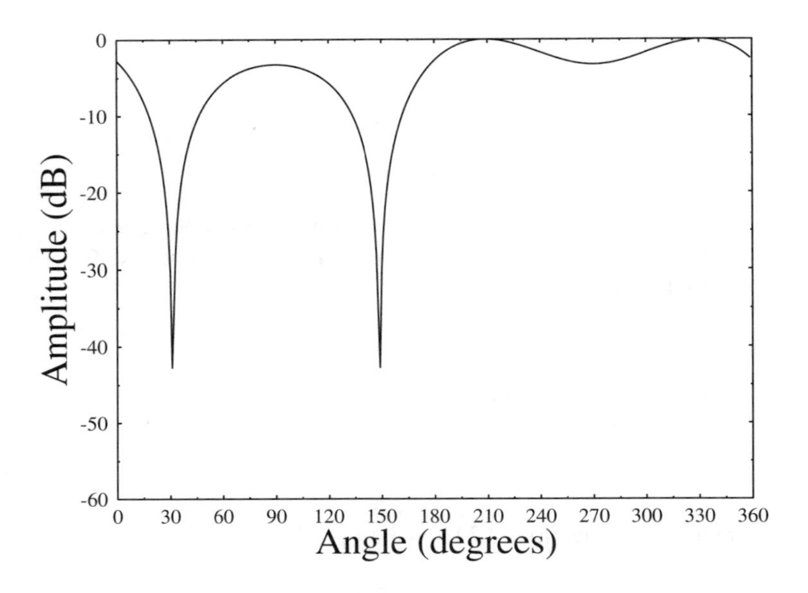

Figure 3.13: The beam pattern produced using Equation 3.21 for a two element array with an element spacing of $\lambda/2$ and element weights of $0.5 \pm j0.5$. The desired signal is at 0°, the interference is at 30°, while SNR=9.0 dB and INR=9.0 dB.

use different techniques to reach the same objective. The digitisation of the signal received at each antenna element ensures a higher information processing accuracy [254]. The RF signal received at each element is either digitised at RF or down-converted to IF and then digitised using an Analogue-to-Digital Convertor (ADC). The digital baseband signals then represent the amplitudes and phases of the signals received at each element of the array [254]. The process of beamforming weights these digital signals, thereby adjusting their amplitudes and phases, such that when added together they form the desired beam [20]. The receivers used in a digital beamforming system need not be as closely matched in phase and amplitude, as in an analogue network, since a calibration process can be performed by the controlling software, and any discrepancies can be removed by adjusting the weights appropriately [254].

3.2.7 Element-Space Beamforming

The beamforming process described in Sections 3.2.3-3.2.6 is referred to as element-space beamforming, where the digitised data signals, x_l, $l = 1, \cdots, L$, received from the array elements are directly multiplied by a set of weights, $w_l, l = 1, \ldots, L$, in order to form a beam at the desired angle, θ_k. By multiplying the received data signals, $x_1, \ldots, x_L$, by different sets of weights, w_l^k, where $l = 1, \cdots, L$, and $k = 1, \ldots, K$, *it is possible to*

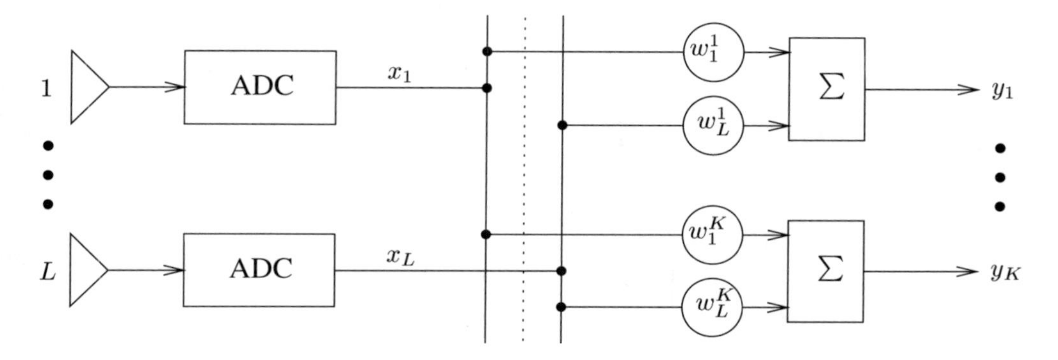

Figure 3.14: An element-space beamformer receiver with L antenna elements capable of forming K beams.

form beams steered in any direction, θ_k, where, again $k = 1, \ldots, K$. More explicitly, by multiplying the signal received at each antenna element by a given complex-valued weight, which may be different for each antenna element, the desired signal may be recovered. Each of the beamformers creates an independent beam, at an angle, θ_k, for receiving an arbitrary mobile's signal, by applying independent weights, w_l^k, $l = 1, \ldots, L$, $k = 1, \ldots, K$, to the array signals, yielding:

$$y(\theta_k) = \sum_{l=1}^{L} w_l^{k*} x_l, \qquad k = 1, \ldots, K \tag{3.38}$$

where $y(\theta_k)$ is the output of the beamformer in the direction of source k, $k = 1, \ldots, K$, which is located at the angle θ_k, $x_l(t)$ is a sample from the lth array element and $w_l^k, l = 1, \cdots, L$ represents the weights for forming a beam at angle θ_k. This equation is very similar to Equation 3.22, except for the addition of the superscript k, $k = 1, \ldots, K$ denoting the kth beam.

Figure 3.14 shows an element-space beamformer with L antenna elements, capable of forming K independent beams for receiving K mobiles' signals. Each of the K beams may independently reject sources of interference, whilst receiving the desired signal.

3.2.8 Beam-Space Beamforming

In contrast to the method of element-space beamforming, where the signals arriving from each of the L elements are weighted and summed to produce the desired output, the beam-space technique forms multiple fixed beams, using a fixed beamforming network, which may be spatially orthogonal. The output of each beam is then weighted and the resultant signals are combined to produce the desired output [3,242,246,247]. The signals from the beams, which are not used to supply the desired response may be used to cancel unknown interference [247].

Assuming that the outputs from each antenna element are equally weighted and have a uniform phase delay, the response of the array, the array factor $F(\Phi, \alpha)$ in Equation 3.21, produced by an incident plane wave arriving at the antenna array from direction θ, measured

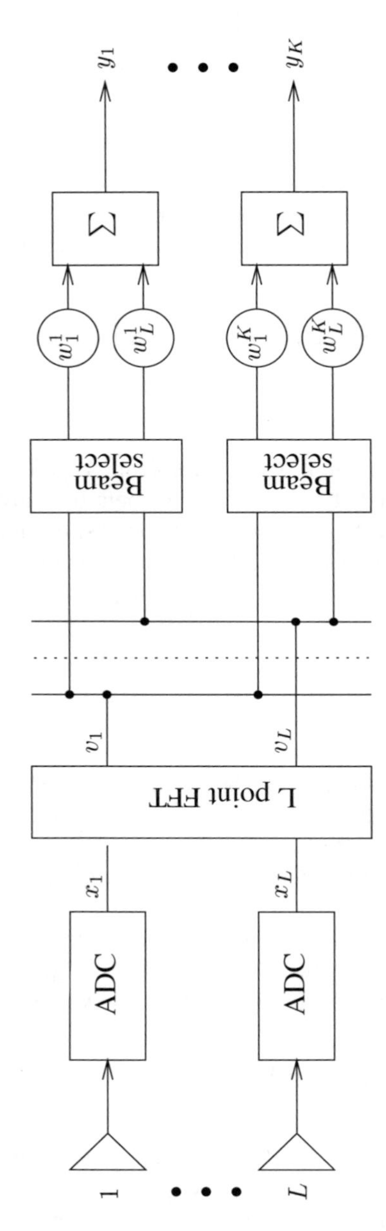

Figure 3.15: A beam-space beamformer receiver with L antenna elements capable of forming K beams [3].

with respect to the normal of the antenna array, is given by [247]

$$F(\Phi, \alpha) = \sum_{n=-N}^{N} e^{jn\Phi} e^{-jn\alpha}, \tag{3.39}$$

where $L = (2N + 1)$ is the total number of elements in the array, $\Phi = \frac{2\pi d}{\lambda} \sin\theta$ is the electrical angle, where d is the inter-elemental distance and α is a constant known as the uniform phase factor. Substituting Φ into Equation 3.39 leads to

$$F(\Phi, \alpha) = \sum_{n=-N}^{N} e^{j\omega t_n(\theta)} e^{-jn\alpha}, \tag{3.40}$$

where $t_n(\theta) = \frac{d \sin\theta}{c}$ and c is the propagation velocity of the received signal. This equation corresponds to Equation 3.21.

For $d = \lambda/2$, we have $\Phi = \pi \sin\theta$ [247]. Summing the geometric series in Equation 3.39, leads to [247]

$$F(\Phi, \alpha) = \frac{\sin[\frac{1}{2}(2N + 1)(\Phi - \alpha)]}{\sin[\frac{1}{2}(\Phi - \alpha)]}. \tag{3.41}$$

By assigning different values to α, the main beam of the antenna may be swept across the range, $-\pi \leq \Phi \leq \pi$. In order to generate an orthogonal set of $2N = L - 1$ beams, the uniform phase factor, α, may be assigned the following values [247]:

$$\alpha = \frac{\pi}{2N + 1} k, \qquad k = \pm 1, \pm 3, \cdots, \pm 2N - 1. \tag{3.42}$$

Figure 3.16 illustrates the variations in the magnitude of the array factor, $F(\Phi, \alpha)$, with $-\pi \leq \Phi \leq \pi$ for the case of $2N + 1 = 5$ elements and $\alpha = \pm\pi/5, \pm 3\pi/5$. The orthogonal beams generated by the beamforming network represent $2N$ independent directions, one per beam. Depending on the target direction of interest, a particular beam of the set is identified as the main beam and the remainder are viewed as auxiliary beams. From Figure 3.16 it can be seen that each of the auxiliary beams has a null in the direction of the main beam. Because of the fixed nature of these unweighted beams formed by the fixed beamformers of Figure 3.15, individual beam control requires interpolation between beams in order to fine-steer the resultant beam and linear combination of auxiliary beams to create nulls in the direction of interfering sources. Alternatively, beam-space beamforming requires a set of beam-space combiners to generate weighted outputs as shown in Figure 3.15. The Fast Fourier Transform (FFT) block in the diagram generates the orthogonal beams, the process by which this is done is analogous to the performance of an FFT in the time-domain, where it may be viewed as a bank of non-overlapping narrow-band filters whose passbands span the frequency of interest [247]. Hence, the L point FFT generates L spatially orthogonal beams.

3.3 Adaptive Beamforming

An antenna array uses an array of simple antennae, such as omni-directional antennae, and combines the signal induced in these antennae to form the array output. Each antenna forming part of the array is known as an element of the array. The direction where the maximum

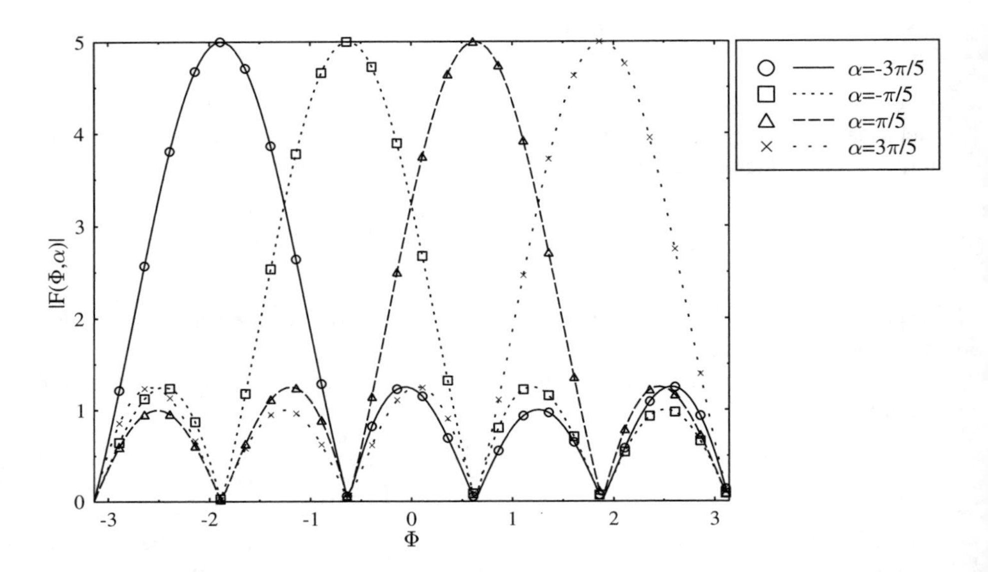

Figure 3.16: The array factor, $F(\Phi, \alpha)$, of a five element antenna array using beam-space beamforming showing the four spatially orthogonal beams that may be generated.

gain would appear is controlled by adjusting the phase between the different antenna elements. The phase and gain of the signals induced in each array element is adjusted such that the signals due to a source in the direction in which maximum gain is required are added in-phase. An adaptive antenna adjusts these phases and gains, known as weights, so that when the outputs from the antenna elements are combined, the desired output is achieved [6, 248]. The properties of the antenna array may be varied over time in order to optimise the system's performance with respect to different optimisation criteria. This criteria can include maximum power, maximum Signal to Noise Ratio (SNR), minimum interference and maximum Signal to Interference plus Noise Ratio (SINR) [246]. Depending upon the operational environment that the antenna is currently in, it can change its performance metric and control algorithm, in order to provide the best service for the users of the network [261]. For example, conventional beamforming/diversity may be used to give maximum received signal power, while a null steering algorithm results in minimum interference. Finally, maximising the SINR corresponds to optimum diversity combining. Given these examples and the generic optimisation criteria to maximise reliable information flow to users with minimum required resources such as power and bandwidth, it is plausible that using a range of different schemes may be necessary. The term intelligent antenna encompasses the technologies of diversity combining [1, 3, 6, 255, 256, 262], adaptive beamforming [3, 6, 8], optimum combining [3, 6], adaptive matching of the antenna's impedance to the receiver [263, 264], and space division multiple access [6, 8, 265, 266].

An adaptive antenna's parameters are automatically adjusted, in order to obtain an optimal or near-optimal array output. The optimisation cost-function and the method used to achieve this state are dependent upon the optimisation algorithm chosen. The need for an adaptive solution is obvious, once one considers that interference is seldom constant in either terms of either time or space and a fixed antenna response would be of little, if any, use.

3.3.1 Fixed Beams

The simplest technique of improving the system's performance is to use fixed multiple beams for both reception and transmission at the base station [251]. The strongest beam in the uplink will also be used for the downlink, since this is deemed to be the beam targeted at the desired user. On the uplink, the base station determines the direction of the path on which the strongest component of the desired signal arrives at the base station. On the downlink, the base station points a beam in the corresponding direction. Although this simple technique is not optimal, the SINR achievable at the mobile can be improved.

Leth-Espensen *et al.* [20] describe a system of array processing, where an algorithm searches through the 22 fixed beams that may be generated by the antenna array, in order to find the strongest receiver beam of the desired signal. More explicitly, an exhaustive search is performed over nine delay taps and the 22 directions until the tap and direction, which result in the maximum received power are obtained. The estimated Direction of Arrival (DoA) was compared to the actual DoA found using a Global Positioning System (GPS) receiver. When averaging the received signals over 21 GSM transmission bursts ($21 \times 8 \times 576\ \mu s \approx 100$ ms) the direction estimates occasionally indicated a direction quite different from that of the mobile. This was attributed to the received signal's lack of power due to undergoing a deep fade at that time. Increasing the number of bursts, over which the received signal was averaged, to 104 ($\approx$480 ms) gave significantly improved results. The performance of eight element arrays processing either 22 beams or eight beams as well as that of four element arrays processing eight or four beams were compared. The average performance gain of the eight element array using 22 beams over that of a single element was 9.8 dB. For the eight element, eight beam antenna the corresponding improvement was 8.8 dB and for the four element, eight beam array the gain was 8.7 dB. Finally, the gain offered by the four beam, four element array was 5.4 dB.

In a switched beam system [252] a mobile station is located within a specific antenna beam and the antenna is then switched in the required operational mode in order to communicate with the specific user supported by the selected beam. If one considers a cell split into three sectors, each of 120° coverage, the available channels are divided equally amongst the sectors. No intelligence is required to locate a mobile station within a sector and to initiate a call. In the event of the mobile station changing sector a handover is performed. An intelligent antenna system is able to switch from a given beam to a new beam without necessitating a handover, i.e. any of the beams can be assigned to one or more of the transceivers. Therefore, should all the users be located in one sector, then as many users as there are transceivers can be served. In contrast, using a conventionally sectorised base station the transceivers in the empty sectors would not be used, while calls in the high-traffic sectors would be blocked [252].

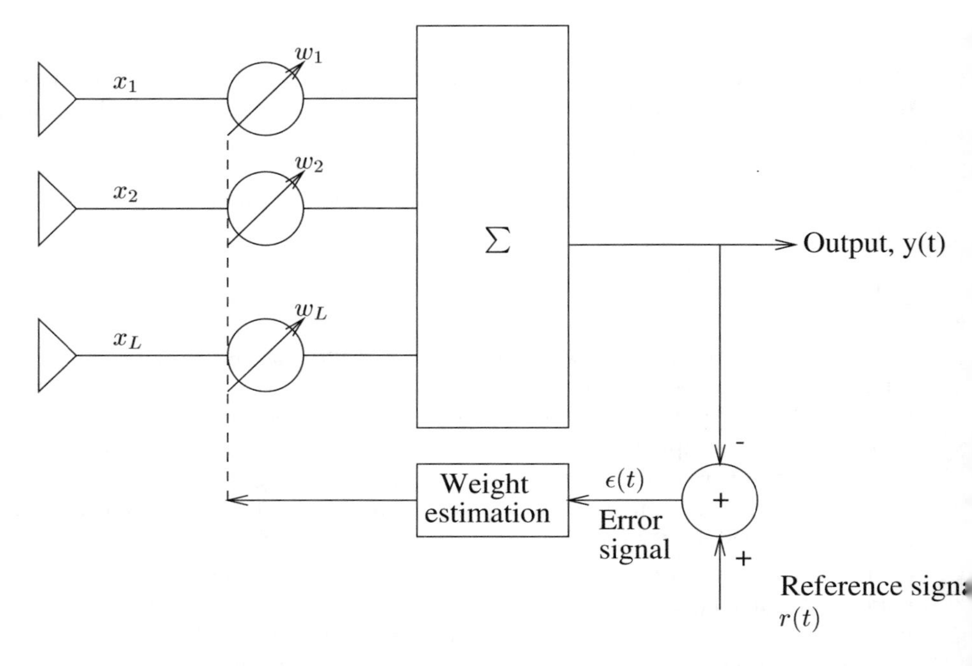

Figure 3.17: The structure of a temporal reference based beamformer with L antenna elements.

3.3.2 Temporal Reference Techniques

Temporal reference techniques refer to the design of array processors which optimise the receive antenna array weights, in order to be able to identify a known sequence at the output of the antenna array. This known desired sequence is termed the reference signal, which must be specifically designed so as to be easily identifiable, for example with the aid of a high auto-correlation peak, while being readily distinguishable from or uncorrelated with unwanted interferences and noise sources [3, 239, 249]. For example, in GSM [11] there are eight different channel sounding sequences used for identifying the eight co-channel base stations, therefore, inevitably, co-channel interferers will use identical sounding sequences to those used by the desired mobile user, hence the system may become unable to distinguish between the wanted signal and a co-channel interferer [1]. The spreading codes used in CDMA are inherently unique and they are therefore suitable for use as the user specific sequence. A significant advantage of the temporal reference technique is that, unlike the spatial reference approach, it does not need careful characterisation of the antenna array. Effects such as mutual coupling between the antenna array elements are readily handled by the adaptation routine, since the array weights are adjusted automatically, in order to cancel them [1].

Figure 3.17 shows the structure of a temporal reference based beamformer, where the array output is subtracted from the reference signal, $r(t)$ which assists in identifying the desired user, in order to generate the error signal $\epsilon(t) = r(t) - \underline{w}^H \underline{x}(t)$, which is then used to control the weights. The weights are adjusted such that the Mean Squared Error (MSE) between the array output and the reference signal is minimised, where the error is expressed as:

$$\epsilon^2(t) = [r(t) - \underline{w}^H \underline{x}(t)]^2. \tag{3.43}$$

Taking the expected values of both sides of Equation 3.43 we get

$$E[\epsilon^2(t)] = E[r^2(t)] - 2\underline{w}^H \underline{z} + \underline{w}^H R \underline{w}, \tag{3.44}$$

where $\underline{z} = E[\underline{x}(t) r^*(t)]$ is the cross-correlation between the reference signal and the array signal vector $\underline{x}(t)$ and $R = E[\underline{x}(t)\underline{x}^H(t)]$, as defined in Equations 3.26 and 3.27, is the correlation matrix of the array output signals.

The MSE surface is a quadratic function of the complex array weight vector $\underline{w}$ and it is minimised by setting its gradient with respect to $\underline{w}$ equal to zero:

$$\nabla_{\underline{w}}(E[\epsilon^2(t)]) = -2\underline{z} + 2R\underline{w} = 0, \tag{3.45}$$

yielding the well-known Wiener-Hopf equation for the optimal weight vector [3, 239, 242, 246, 247, 249] in the form of:

$$\underline{w}_{opt} = R^{-1}\underline{z}. \tag{3.46}$$

The Minimum Mean Square Error (MMSE) at the output of the array processor, also known as the Wiener filter, using these weights is given by [242]:

$$MMSE = E[|r(t)|^2] - \underline{z}^H R^{-1} \underline{z}. \tag{3.47}$$

In [267] a 16-bit reference signal was used in order to uniquely identify the mobiles. This contribution proposes an adaptive antenna algorithm suitable for GSM and the urban environment, since this is where the highest capacity is generally needed. More specifically, the 16-bit reference signal used in this system is the GSM equaliser's training sequence, which is one of the eight legitimate 16-bit codes exhibiting the highest main-peak to side-peak ratio in its auto-correlation function, which were found by exhaustive computer search of all 2^16 possible sequences. These 16-bit sequences were then extended to 26 bits by quasi-periodically repeating five bits at both ends of the sequence. Neighbouring base stations, and hence their mobiles, use a different one from the set of eight codes, as detailed in [11]. The algorithm described in this paper [267] calculates the initial weight vector using just the known training sequence. This weight vector is then applied to all the data in the burst and the result is passed to the GSM channel equaliser in order to detect the unknown bits. The detected bits are then input to the GSM modulator, in order to construct a modulated reference waveform for the entire burst and a new weight vector is calculated. This weight vector is applied to the whole data burst and the result is again passed to the GSM equaliser. Therefore, the Signal-to-Interference and Noise Ratio (SINR) is improved for the whole burst, rather than just for the training sequence. In the simulations carried out in [267] the process was repeated for a maximum of 20 iterations or until the same data bits were returned twice. It was found that the typical number of iterations required was three or four. The effect of varying the number of antenna elements was investigated. If the multipath components of the wanted signal are sufficiently delayed, so that they are uncorrelated with the reference signal, they are cancelled. These delayed paths can be exploited, if tapped delay-line filters are used in conjunction with amplitude and phase weighting of the antenna elements. The paper presents results for an eight element linear array with up to three taps.

Barrett and Arnott [1] describe a similar system, in which the modulated training sequence is compared to the signal at the array's output. After the training sequence has been received and the data detection begins, the system switches into decision directed mode, in which the demodulator decisions are remodulated in order to form the reference signal on the basis of the total received burst. Provided that the error rate is adequate (better than 10^{-2}), a reference signal generated by this method would allow the system to track interference changes in the propagation environment. Field trials were conducted for a system using an eight element adaptive antenna. The data received at each antenna was digitised and stored, in order to allow offline processing, enabling the comparison of different processing functions operating on the basis of the same recorded data. The results show a substantial improvement in terms of the demodulated SNR, when compared to that of a single element antenna. The optimum combining was implemented by updating the array weights every transmission burst (every 10 ms), and each update used 100 data snapshots taken from within the burst. The reference signal was obtained using decision directed operation (no training sequence was used) and the weights were updated using the Normalised Least Mean Squares (NLMS) algorithm. The amplitude resolution of the data and weights was eight bits. The results using optimum combining were found to be superior to those obtained using selection diversity.

3.3.2.1 Least Mean Squares

The Least Mean Squares (LMS) algorithm is the most common technique used for continuous adaptation [3, 239, 242, 247, 249]. It is based on the steepest-descent method, a well-known optimisation technique that recursively computes and updates the weight vector. The algorithm updates the weights at each iteration by estimating the gradient of the quadratic error surface and then changing the weights in the direction opposite to the gradient by a small amount in an attempt to minimise the Mean Square Error (MSE), as seen in Figure 3.18. The desired response, generated for example by inputting the reference sequence to the modulator is supplied to the algorithm, allowing the estimation error and thus the error surface, to be calculated. The constant that determines the amount by which the weights are adjusted during each iteration is referred to as the step size. When the step size is sufficiently small, the process leads these estimated weights to the near-optimal weights in Figure 3.18, whilst large step sizes allow faster convergence, but exhibit a larger residual MSE due to the non-optimal weights [247].

The updated value of the weight vector at time $n+1$ is computed using [3, 8, 242, 246–248, 268]:

$$\underline{w}(n+1) = \underline{w}(n) - \frac{1}{2}\mu\nabla(J(n)), \tag{3.48}$$

where $\underline{w}(n+1)$ denotes the new weights computed at the $(n+1)^{th}$ iteration; μ is the positive step size that controls the rate of convergence and hence determines how close the estimated weights approach the optimal weights and $\nabla(J(n))$ is an estimate of the gradient of the MSE, $J(n)$, where $J(n)$ is given by [242]:

$$J(n) = E[|r(n+1)|^2] + \underline{w}^H(n)R\underline{w}(n) - 2\underline{w}^H(n)\underline{z}, \tag{3.49}$$

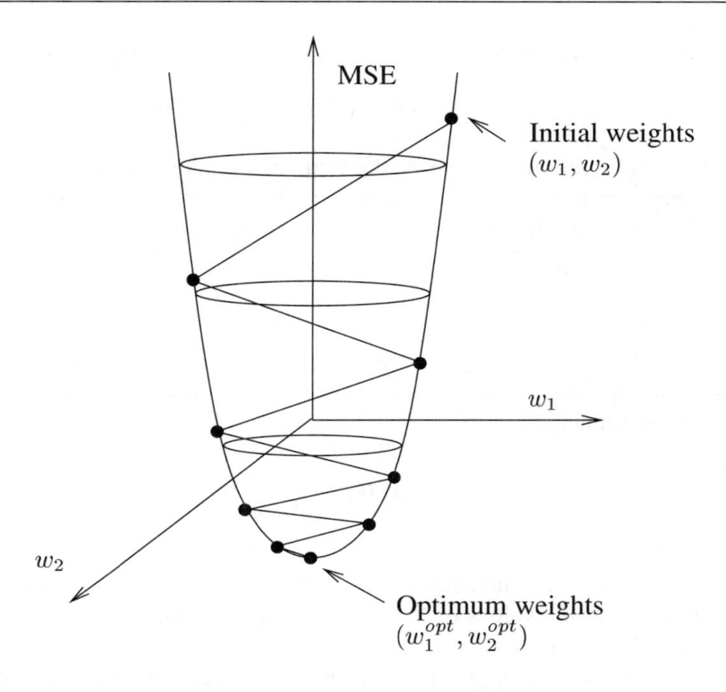

Figure 3.18: An example of the quadratic error surface and the weights of a two element system following the negative direction of the gradient in order to minimise the Mean Square Error (MSE).

where $r(n+1)$ is the reference signal at time $n+1$ and $\underline{z} = E[\underline{x}(t)r^*(t)]$ is the cross-correlation vector between the input vector $\underline{x}(n)$ and the desired response $r(n)$, while the correlation matrix, R, was defined in Equations 3.26 and 3.27.

Differentiating Equation 3.49 with respect to $\underline{w}(n)$ gives:

$$\nabla(J(n)) = 2R\underline{w}(n) - 2\underline{z}. \tag{3.50}$$

Therefore, the instantaneous estimate of the gradient vector becomes:

$$\begin{aligned}\hat{\nabla}(J(n)) &= 2\underline{x}(n)\underline{x}^H(n)\underline{w}(n) - 2\underline{x}(n)r^*(n) \\ &= 2\underline{x}(n)\epsilon^*(n),\end{aligned} \tag{3.51}$$

where $\epsilon^*(\underline{w}(n))$ is the error between the array output and the reference signal, which is formulated as:

$$\epsilon^*(n) = \underline{x}^H(n)\underline{w}(n) - r(n). \tag{3.52}$$

The array output in Figure 3.17 is given by:

$$y(n) = \underline{w}^H(n)\underline{x}(n). \tag{3.53}$$

Upon substituting Equation 3.52 in Equation 3.48 the weight adaptation equation becomes:

$$\underline{w}(n+1) = \underline{w}(n) + \mu \underline{x}(n)\epsilon^*(n). \tag{3.54}$$

Therefore, as Equation 3.52 shows, the estimated gradient, $\hat{\nabla}(J(n))$, is a function of the error, $\epsilon(n)$, between the array output, $y(n)$, and the reference signal, $r(n)$, and the received array signals, $\underline{x}(n)$, after the nth iteration. Convergence is guaranteed only, if [242, 248],

$$0 < \mu < \frac{1}{\lambda_{max}}, \tag{3.55}$$

where λ_{max} is the maximum eigenvalue of R, the correlation matrix of Equations 3.26 and 3.27. Therefore, the eigenvalue spread or ratio of the matrix R controls the rate of convergence [247] according to:

$$\chi(R) = \frac{\lambda_{max}}{\lambda_{min}}, \tag{3.56}$$

where $\chi(R) \geq 1$.

Under these conditions the algorithm is stable and the mean value of the estimated array weights converges to the values of the optimal weights. Within these bounds, the speed of adaptation and also the noise contaminating the weight vector are both determined by the size of μ. Since the trace of R is given by the sum of the diagonal elements of R [247], λ_{max} therefore cannot be greater than the trace of R, that is,

$$\lambda_{max} \leq tr[R] = \sum_{i=1}^{L} \lambda_i \tag{3.57}$$

where L is the number of antenna elements, and λ_i is the ith eigenvalue of R. Hence we have:

$$0 < \mu < \frac{1}{tr[R]}. \tag{3.58}$$

This is a more restrictive bound on μ than Equation 3.55, but it is much easier to apply, because the elements of R and the signal power can generally be more readily estimated than the eigenvalues of R. The efficiency of the LMS algorithm has been shown to approach a theoretical limit for adaptive algorithms, when the eigenvalues of R are equal or nearly equal [269]. When the eigenvalues of the correlation matrix R are widely spread, i.e. $\chi(R) = \frac{\lambda_{max}}{\lambda_{min}} \gg 1$, then, according to Haykin [247], the excess mean-squared error produced by the LMS algorithm with respect to the minimum is determined primarily by the largest eigenvalues [247], and the time taken for the average weight vector to converge is limited by the smallest eigenvalues. However, as the spread of the eigenvalues increases, the highest acceptable value of the stepsize μ required for maintaining stability decreases inevitably, resulting in slower convergence to the optimal weights. Selecting too small a value for μ results in a slow rate of convergence, and in a non-stationary environment may cause the estimated weights to lag behind the evolution of the optimal weights [254], a phenomena known as the weight vector lag. Alternatively, using too high a value for μ allows the vicinity of the solution point to be reached more rapidly, but the weights then wander around a larger region and cause a weight mis-adjustment error, as was demonstrated in Figure 3.18 [270]. This is due to μ being equivalent to the reciprocal of the memory of the system, where a large value of μ uses fewer samples to estimate R, and hence a degraded estimation is performed, resulting in an increase in the average excess mean-squared error after adaptation.

3.3.2.2 Normalised Least Mean Squares Algorithm

In the LMS algorithm, the correction $\mu\underline{x}(n)\epsilon^*(n)$ applied to the weight vector at time $n+1$ in Equation 3.54 is directly proportional to the input vector $\underline{x}(n)$. Therefore, when $\underline{x}(n)$ is large, the LMS algorithm experiences a gradient noise amplification problem [247]. Therefore an algorithm which normalises the weight vector correction with respect to the squared Euclidean norm of the input vector $\underline{x}(n)$ at time n can be invoked. At the nth iteration the step size is then given by [242, 247]:

$$\mu(n) = \frac{\mu_0}{\underline{x}^H(n)\underline{x}(n)} = \frac{\mu_0}{||\underline{x}(n)||^2}, \tag{3.59}$$

where μ_0 is a constant. The normalised LMS algorithm is convergent in the mean-square sense, if $0 < \mu_0 < 2$ [247]. However, if the input vector $\underline{x}(n)$ is small, then numerical problems may arise due to the associated division by a small number. Therefore Equation 3.59 may be modified to:

$$\mu(n) = \frac{\mu_0}{a + ||\underline{x}(n)||^2}, \tag{3.60}$$

where $a > 0$. Hence, the weight update formula of Equation 3.54 is modified to:

$$\underline{w}(n+1) = \underline{w}(n) + \frac{\mu_0}{a + ||\underline{x}(n)||^2}\underline{x}(n)\epsilon^*(n). \tag{3.61}$$

3.3.2.3 Sample Matrix Inversion

The Sample Matrix Inversion (SMI) algorithm is a method of directly calculating the antenna array weights based on an estimate of the correlation matrix, $R = E[\underline{x}(t)\underline{x}^H(t)]$ of the adaptive array output samples. The Wiener-Hopf solution for the optimal weights is repeated here from Equation 3.46, for convenience:

$$\underline{w}_{opt} = R^{-1}\underline{z}, \tag{3.62}$$

where $\underline{z} = E[\underline{x}(t)r(t)]$ is the cross-correlation between the reference signal, $r(t)$ and the array output signal, $\underline{x}(t)$. If the signal, noise and interference characteristics are stationary, then the correlation matrix can be evaluated and the optimal solution for the adaptive weights can be computed directly using the above equation, with the aid of matrix inversion. In practice however, due to the non-stationary mobile environments encountered, the adaptive processor must continually update the weight vector, in order to meet the new conditions imposed by the time-varying mobile environment. This need to regularly update the weight vector leads to the requirement of obtaining estimates of R and $\underline{z}$ in a finite observation interval, and thus to obtain a weight vector estimate. This approach is termed block-adaptive, where the statistics are estimated from a temporal block of data and are used in a periodic optimum weight calculation process. In the GSM system [11] it may be possible to use the synchronisation/channel sounding sequence in each burst to recompute the antenna array weights for each 4.615 ms burst.

If the cross-correlation vector $\underline{z} = E\underline{x}(t)\underline{x}^H(t)]$ is assumed to be known, then the optimal weight vector estimate, $\hat{\underline{w}}$ of Equation 3.62, for the situation when $\underline{x}(t)$ contains the reference-signal related desired signal, where $\hat{R}_{xx}$ is the block based estimate of the true correlation of the array's output samples, namely that of R_{xx}, may be determined using

$$\hat{\underline{w}}_1 = \hat{R}_{xx}^{-1}\underline{z}. \tag{3.63}$$

However, in the scenario when the received signal $\underline{x}(t)$ contains either noise of the interfering users' signals rather than the desired signal, the estimate of the correlation matrix R_{xx} is denoted by $\hat{R}_{nn}$, and the optimal antenna weights may be calculated thus according to:

$$\hat{\underline{w}}_2 = \hat{R}_{nn}^{-1}\underline{z}. \tag{3.64}$$

Therefore, the Signal-to-Noise Ratio (SNR) at the output of the combiner seen in Figure 3.17 may be written as [249]:

$$\left(\frac{s}{n}\right)_i = \frac{\hat{\underline{w}}_i^H \underline{s}\underline{s}^H \hat{\underline{w}}_i}{\hat{\underline{w}}_i^H R_{nn} \hat{\underline{w}}_i}, \tag{3.65}$$

where i assumes values of 1 or 2, according to the first of second scenarios above, and $\underline{s}$ denotes the reference-signal related desired signal component of the array output signal vector $\underline{x}$. The SNR $(s/n)_2$ is only defined during those time intervals, when a reference-signal related desired signal is actually present; the weight adjustment is assumed to take place when the desired signal is absent.

The estimate of the sample correlation matrix can be evaluated according to:

$$\hat{R}_{xx} = \frac{1}{N}\sum_{n=1}^{N} \underline{x}(n)\underline{x}^H(n), \tag{3.66}$$

where N is the size of the observation interval expressed in terms of the number of array output samples considered. Again, this approach is termed block-adaptive, where the statistics are estimated from a temporal block of data and used during the optimum weight calculation process. Given that each element of the matrix, $\hat{R}_{xx}$, is a random variable, the output SNR is also a random variable [244, 249]. The maximum achievable SNR at the output of the combiner seen in Figure 3.17 that may be obtained is:

$$SNR_{opt} = \underline{s}^H R_{nn}^{-1}\underline{s}. \tag{3.67}$$

The actual SNRs obtained using $\hat{\underline{w}}_1$ and $\hat{\underline{w}}_2$ may be normalised as follows [244, 249]:

$$\rho_i = \frac{(s/n)_i}{SNR_{opt}}. \tag{3.68}$$

Reed [244] examined the number of samples, N, required in order to achieve a high-quality estimate of the noise- or interference-related co-variance matrix, R_{nn}, and derived

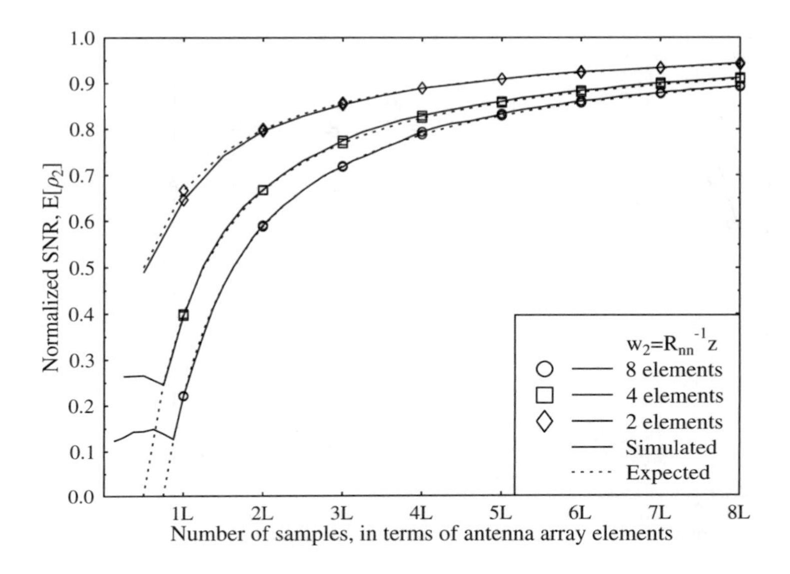

Figure 3.19: The expected normalised Signal-to-Noise Ratio (SNR), $E[\rho_2]$ evaluated from Equation 3.69, for various numbers of array output samples, in terms of the number of antenna array elements, used to construct the noise- or interference-only correlation matrix. Simulated results for identical scenarios are also presented for comparison. The SNR at each antenna array element was 12.0 dB.

the expected value of the normalised SNR at the output of the combiner seen in Figure 3.17, which was found to be:

$$E[\rho_2] = \frac{N + 2 - L}{N + 1}, \tag{3.69}$$

where L is the number of elements in the antenna array.

The expectation of the normalised SNR in Equation 3.69 employing the antenna weights calculated on the basis of the noise- or interference-only related co-variance matrix, is plotted in Figure 3.19 for two, four and eight element antenna arrays. Explicitly, Figure 3.19 suggests that as long as, N, the number of samples used to estimate the noise- or interference-related correlation matrix, R_{nn}, is greater than twice the number of antenna elements, the loss in $E[\rho_2]$ due to non-optimal weights is less than 3 dB. The expected values of $E[\rho_2]$ evaluated from Equation 3.69 are compared to values determined using simulations. The simulation based and theoretical SNRs were in good agreement. It is interesting to note that although both the normalised simulated and theoretical SNRs approach unity, implying approaching the optimum SNR in Equation 3.67, however the rate of convergence for both the theoretical and simulated values slows down, as the number of antenna elements used to form the antenna

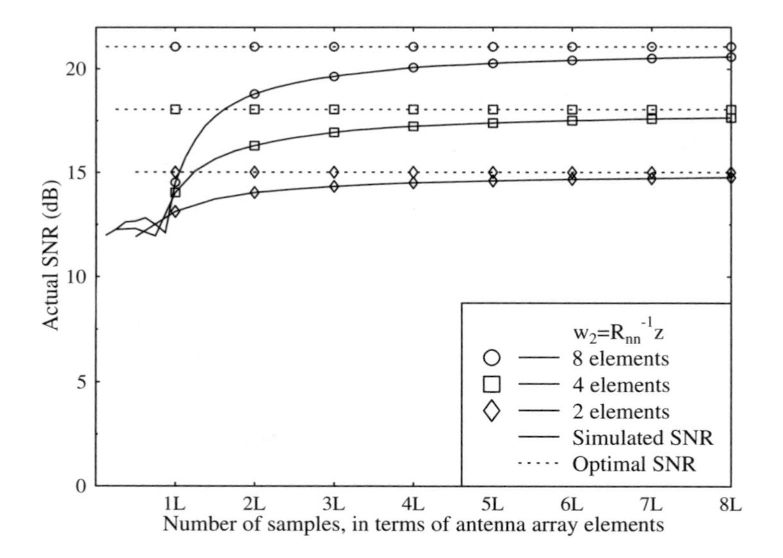

Figure 3.20: The SNR at the output of the array combiner determined by simulation and the optimal SNR according to Equation 3.67 for a varying number of array output samples, in terms of the number of antenna array elements, used to construct the noise- or interference-only correlation matrix. The SNR at each antenna array element was 12.0 dB.

array increases. This is expected, since as the number of antenna array elements increases, so does the optimum SNR that may be obtained according to Equation 3.67, as also seen in Figure 3.20.

Thus far we have assumed the knowledge of the cross-correlation vector $\underline{z}$, which is unrealistic in a practical system. Therefore, the optimal weight vector may be determined with the aid of the estimated cross-correlation vector $\hat{\underline{z}}$ according to :

$$\underline{\hat{w}}_2 = \hat{R}_{xx}^{-1}\,\hat{\underline{z}}, \tag{3.70}$$

where $\hat{\underline{z}}$ is the sample cross-correlation vector given by

$$\hat{\underline{z}} = \frac{1}{N}\sum_{n=1}^{N} \underline{x}(n)r^*(n), \tag{3.71}$$

and $r(n)$ is the reference signal.

The normalised SNR for a two element antenna array was determined by simulation using Equation 3.70, for estimating the optimum antenna array weights, is presented in Figure 3.21. This figure shows that the SNR of the received signal, using the antenna weights determined

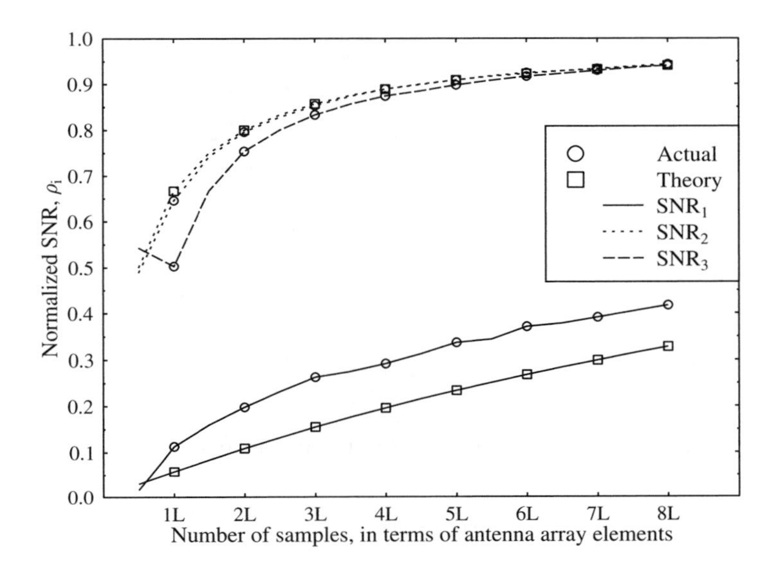

Figure 3.21: The normalised Signal-to-Noise Ratio (SNR), ρ_i, for various numbers of array output samples, in terms of the number of antenna array elements. Results are shown for $\underline{w}_1 = \hat{R}_{xx}^{-1}\underline{z}$, $\underline{w}_2 = \hat{R}_{nn}^{-1}\underline{z}$, and $\underline{w}_3 = \hat{R}_{xx}^{-1}\hat{\underline{z}}$ for both theory, according to Equations 3.63, 3.64 and 3.70, and simulation. The antenna array consisted of two antenna elements, separated by $\lambda/2$, and the SNR at each of which was 12.0 dB.

when the desired signal was present, is significantly lower than when using the weights obtained when the desired user's signal was absent. The simulated SNR, for the case of two antenna elements, when the desired signal was received is significantly higher than that predicted theoretically by Equation 3.68, although this phenomenon does not appear for the four and eight element antenna arrays characterised in Figure 3.22.

The SNR obtained using Equation 3.70 is shown to be comparable to the SNR obtained with the noise- or interference-only correlation matrix, R_{nn}, which is because the estimates $\hat{\underline{z}}$ and $\hat{R}_{xx}$ are highly correlated under strong desired signal conditions, and the errors in each estimate tend to compensate each other, thus yielding an improved weight estimate and faster convergence. Improvement of the transient response through careful selection of the initial weight vector is possible by invoking the following relationship [249]:

$$\hat{\underline{w}}_1 = \left[\frac{1}{N}\left(\sum_{n=1}^{N} \underline{x}(n)\underline{x}^H(n) + \alpha I\right)\right]^{-1} \underline{z} \tag{3.72}$$

where α is a scalar constant and I is the $N \times N$ identity matrix.

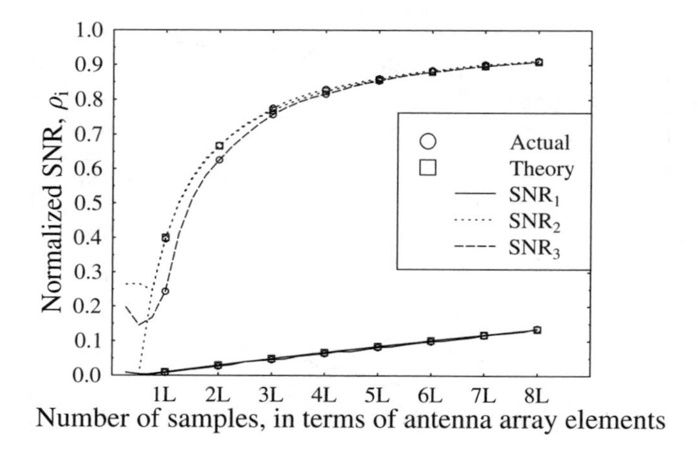

(a) Four elements

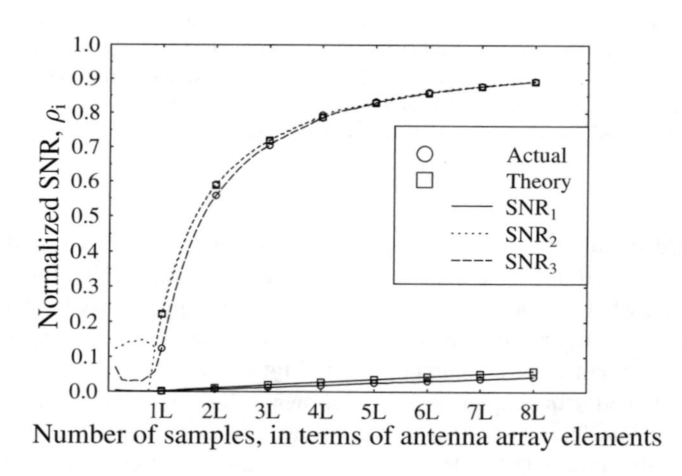

(b) Eight elements

Figure 3.22: The normalised Signal-to-Noise Ratio (SNR), ρ_i, for various numbers of samples, in terms of the number of antenna array elements. Results are shown for $\underline{w}_1 = \hat{R}_{xx}^{-1}\underline{z}$, $\underline{w}_2 = \hat{R}_{nn}^{-1}\underline{z}$, and and $\underline{w}_3 = \hat{R}_{xx}^{-1}\hat{\underline{z}}$ for both theory, according to Equations 3.63, 3.64 and 3.70, and simulation. The antenna elements were separated by $\lambda/2$. The SNR at each antenna element was 12.0 dB.

The estimate of R may be updated, when new samples arrive from the antenna, according to [242]:

$$\hat{R}(n+1) = \frac{n\hat{R}(n) + \underline{x}(n+1)\underline{x}^H(n+1)}{n+1}, \tag{3.73}$$

and a new estimate of the weights $\underline{\hat{w}}(n+1)$ at time instant $n+1$ may be made. The expression of the optimal weights in Equation 3.46 requires the inverse of R, and this process of estimating R and then its inverse may be combined to update the inverse of R from the array signal samples, $\underline{x}(n)$, using the Matrix Inversion Lemma [242, 246] which is given in its general form as:

$$(A + XX^H)^{-1} = A^{-1} - \frac{A^{-1}XX^HA^{-1}}{1 + X^HA^{-1}X}, \tag{3.74}$$

thus leading to:

$$\hat{R}^{-1}(n) = (n+1)\hat{R}^{-1}(n-1) - \frac{(1+\frac{1}{n})\hat{R}^{-1}(n-1)\underline{x}(n)\underline{x}^H(n)\hat{R}^{-1}(n-1)}{n + \underline{x}^H(n)\hat{R}^{-1}(n-1)\underline{x}(n)}, \tag{3.75}$$

with

$$\hat{R}^{-1}(0) = \frac{1}{\epsilon_0}I, \qquad \epsilon_0 > 0, \tag{3.76}$$

where I is the $N \times N$ identity matrix. This method of estimating the array weights using the inverse update technique is known as the Recursive Least Squares (RLS) algorithm.

Unlike for the LMS algorithms, the performance of the SMI algorithm is almost independent of the eigenvalue spread of R and it is similar to that of the steepest descent algorithm using a correlation matrix, R, of equal eigenvalues [246]. The matrix estimation in Equation 3.66 is only suitable for use in a stationary environment [246]. In a time varying environment a de-weighted matrix estimate may be more applicable [268], yielding:

$$\hat{R}(n) = \alpha\hat{R}(n-1) + (1-\alpha)\underline{x}(n)\underline{x}^H(n) \qquad 0 < \alpha < 1 \tag{3.77}$$

where α is the so-called 'forgetting factor'.

Hence, Equation 3.75 becomes

$$\hat{R}^{-1}(n) = \alpha^{-1}\hat{R}^{-1}(n-1) - \frac{(1-\alpha)\alpha^{-2}\hat{R}^{-1}(n-1)\underline{x}(n)\underline{x}^H(n)\hat{R}^{-1}(n-1)}{1 + (1-\alpha)\alpha^{-1}\underline{x}(n)\hat{R}^{-1}(n-1)\underline{x}^H(n)}. \tag{3.78}$$

The vector, $\underline{\hat{z}} = E[\underline{x}(n)r^*(n)]$, containing the correlation between the reference signal, $r(n)$, and the array output signals, $\underline{x}(n)$, must also be updated for each block of N received samples according to:

$$\underline{\hat{z}} = \frac{1}{N}\sum_{n=1}^{N}\underline{x}(n)r^*(n). \tag{3.79}$$

If an error term, $e = \hat{\underline{z}} - \underline{z}$, between the estimate of the correlation vector $\underline{z}$ and its actual value is used to represent the errors due to the estimation process, we may write

$$e = \hat{R}\underline{w}_{opt} - \underline{z}. \tag{3.80}$$

Therefore, the weight vector derived using the SMI method is a least squares solution. It can be shown theoretically that the array weights derived by the SMI approach converge more rapidly towards their final values than those generated by the LMS algorithm. However, there are practical difficulties associated with the employment of the SMI algorithm. Specifically, the inversion of the potentially large correlation matrix, R, requires a high complexity. Specifically, the complexity of the matrix inversion is proportional to L^3, where L is the matrix dimensionality, and it is thus very computationally expensive. However, the matrix inversion may be avoided by using the recursive techniques of Equation 3.75.

In [15] Strandell *et al.* investigated the performance of an adaptive antenna system using the SMI adaptation algorithm. The system was integrated into an existing DCS-1800 base station and used the 26-bit equaliser training sequence in each traffic burst as the reference signal. The performance of the adaptive antenna was evaluated in the laboratory initially, so as to avoid multipath propagation. It was shown that the algorithm was capable of suppressing an interferer, when the power of the interferer was within the dynamic range of the Analogue-to-Digital Converter (ADC) used to digitise the signals arriving at the antenna array elements. The ADC had an eight-bit resolution giving approximately a 48 dB dynamic range spanning from -32 dBm to -80 dBm. Consequently, below -80 dBm the interferer is buried in the noise and no suppression is possible. Therefore, stronger interferers are suppressed more effectively than weak ones. The adaptive antenna was found to improve the SIR by more than 30 dB in conjunction with an interferer power at -40 dBm and a desired input signal power between -70 dBm and -40 dBm. When either of the signal levels exceeded the dynamic range of the ADC, the SIR improvement was very low, even less than 0 dB in some circumstances.

The performance of the antenna was then evaluated in an open terrain environment, with no obstacles within 500 m of the antenna. It was found that even though there was some array pattern distortion, or angular pointing error in the direction of the main beam, the interfering signal located at an angle of 90° with respect to the desired signal was suppressed by about 25 dB relative to the main beam. The pointing error of the main beam was due to the relatively short, 26-bit training sequence used, leading to a poorly estimated array output correlation matrix, when the desired signal was present in the matrix [271]. A solution to this problem is the positive diagonal loading technique [271], where adding a small value to the diagonal elements of the matrix results in faster weight convergence. In conjunction with a perfectly estimated array output correlation matrix all the noise eigenvalues are identical and equal to the noise variance [15]. In contrast, a poor estimate of the array output correlation matrix gives non-identical eigenvalues, resulting in a distorted array pattern. If the loading value is larger than the noise eigenvalues, but smaller than the eigenvalues of the desired and interfering signal, then the overall noise level is increased, resulting in almost identical noise eigenvalues [272]. The loading value l was chosen so that $l/\sigma^2 \approx 10^2$ [271]. The diagonal loading decreases the SIR, but increases the SNR due to the lower sidelobe levels, leaving the SINR unchanged [271]. The SIR improvement achieved by the adaptive antenna was measured for Direction-Of-Arrival (DOA) separations ranging from 2.5° to 180° at a constant input SIR of 20 dB. The interference suppression capability varied from 31 dB for a 180° angular separation to 26 dB for a 2.5° separation. However, as a consequence of the limited array beamwidth, the SNR gain decreased upon decreasing the DOA separation, reaching a minimum of -10 dB at 5° separation.

3.3.2.4 Recursive Least Squares

The RLS algorithm exploits the matrix inversion lemma defined in Equation 3.74 for updating the antenna array element weights. As the RLS algorithm utilises information contained in the array's combiner output data as shown by Equations 3.74 and 3.75, extending back to the time when the algorithm was initiated, the rate of convergence is typically an order of magnitude higher than that of the LMS algorithm. This performance improvement, however, is achieved at the expense of a substantial increase in computational complexity.

The correlation matrix, R of the array output, at time n, may be updated thus according to [3, 242, 247]:

$$R(n) = \delta_0 R(n-1) + \underline{x}(n)\underline{x}^H(n), \tag{3.81}$$

where, similarly to Equation 3.77, the 'forgetting factor', δ_0, is used to de-emphasise old array output samples. The value $1/(1-\delta_0)$ is known as the memory of the algorithm, and for example when $\delta_0 = .99$, the memory of the algorithm is approximately 100 samples, while $R(n-1)$ is the previous value of the correlation matrix, R, at time $n-1$.

Similarly, the cross-correlation vector between the array output signal and the desired signal may be calculated as:

$$\underline{z}(n) = \delta_0 \underline{z}(n-1) + \underline{x}(n)r(n). \tag{3.82}$$

Equation 3.46 states how the optimal receive antenna weights may be obtained, which is repeated here for convenience:

$$\underline{w}_{opt} = R^{-1}\underline{z}, \tag{3.83}$$

leading to,

$$\begin{aligned} \hat{\underline{w}}(n) &= R^{-1}(n)\underline{z} \\ &= \delta_0 R^{-1}(n)\underline{z}(n-1) + R^{-1}(n)\underline{x}(n)r^*(n), \end{aligned} \tag{3.84}$$

where

$$R^{-1}(n) = \frac{1}{\delta_0}\left[R^{-1}(n-1) - \frac{R^{-1}(n-1)\underline{x}(n)\underline{x}^H(n)R^{-1}(n-1)}{\delta_0 + \underline{x}^H(n)R^{-1}(n-1)\underline{x}(n)}\right] \tag{3.85}$$

with

$$R^{-1}(0) = \frac{1}{\epsilon_0}I, \qquad \epsilon_0 > 0, \tag{3.86}$$

as in Equation 3.76, when using the SMI algorithm. Therefore, with the aid of:

$$R^{-1}(n) = \frac{1}{\delta_0}\left[R^{-1}(n-1) - q(n)\underline{x}^H(n)R^{-1}(n-1)\right], \tag{3.87}$$

where

$$q(n) = \frac{R^{-1}(n-1)\underline{x}(n)}{\delta_0 + \underline{x}^H(n)R^{-1}(n-1)\underline{x}(n)} \tag{3.88}$$

we arrive at [3,247],

$$\underline{w}(n) = \underline{w}(n-1) + q(n)[r^*(n) - \underline{w}^H(n-1)\underline{x}(n)], \tag{3.89}$$

where the square-bracketed term represents the error, $e(n) = r^*(n) - y(n)$ between the desired signal and the array output signal after processing. As can be seen from Equation 3.85, the inversion of the correlation matrix, $R(n)$ required by Equation 3.83, has been replaced by the simple update formula of Equation 3.87, requiring scalar division, thus significantly reducing the complexity imposed.

3.3.3 Spatial Reference Techniques

Spatial reference adaptation [1,3,8,238–242,260] relies on information regarding the direction of arrival of the desired signal and its multipath components. There are numerous different methods for obtaining estimates of the DOA information with the aid of the received antenna array signals [3,242,260]. Wave-number estimation techniques [3,242,243,260] are based on the decomposition of the array output correlation matrix, $R = E[\underline{x}(t)\underline{x}^H(t)]$, whose terms consist of estimates of the correlation between the signals at the elements of the antenna array in Figure 3.10. The so-called MUltiple SIgnal Classification (MUSIC) algorithm [3,242,260] and the Estimation of Signal Parameters by Rotational Invariance Techniques (ESPRIT) both use this approach [242,260]. However, these algorithms are not effective for detecting coherent signals [242,260]. The parametric estimation techniques [242,260] are mainly maximum likelihood estimation (MLE) based algorithms, where the ML estimates of desired parameters, such as the angles of arrival, are the ones for which the likelihood function is maximised. These techniques impose a high computational complexity and also require the antenna array to be accurately calibrated. Again, further information concerning these algorithms may be found in [7,9,242,273].

3.3.3.1 Antenna Calibration

Antenna calibrating procedures [7,253,274] can be readily incorporated in a digital beamforming array, facilitating the realisation of highly selective antenna patterns exhibiting ultralow sidelobes. The feature of self-calibration is an advantage, but may indeed also be an essential requirement for a system employing an array of elemental receivers constituted by multiple, cascaded active components [7,254]. Several techniques are available, such as the injection of precise radio frequency test signals at the receiver front-ends [15,253], focusing on a source at a known position in the near or far-field, or employment of a known, well defined scatterer of the transmitted signal.

In order to improve the SIR of the signal received by an adaptive antenna array, nulls can be created in the antenna array's radiation pattern in the direction of strong co-channel interferers. However, the depth and angular position of these nulls are very sensitive to phase and amplitude errors within the antenna array [7]. The performance of RF components generally varies over temperature, time and frequency. A study conducted by Tsoulos and Beach [7] found that a temperature variation of 14°C to 27°C resulted in a maximum amplitude variation of ±1.5 dB and a ±180° maximum phase error across the antenna array. Performing a calibration of phase and amplitude mis-matches between the antenna array elements at the time of manufacture would not take into account temperature variations and ageing effects [253]. Reference [7] noted that even under the same room temperature the amplitude and phase mismatches varied from day to day. Therefore, an online calibration procedure is required that can take place, whilst the base station continues to function normally. Only the active components have to be calibrated, the passive components are assumed to be less susceptible to temperature and time. After calibration the amplitude mismatch was limited to ±0.04 dB and the phase mismatch to ±0.4°.

The calibration process of an 8×8 element receiver antenna array developed for the pan-European TSUNAMI (II) SDMA Field Trial was described by Passman and Wixforth in [274]. The aim of the calibration procedure was to reduce the phase error to less than 3° and the amplitude error to less than 0.5 dB. The receive antenna array, as shown in Figure 3.23, consists of ten linearly spaced active subarrays, each of which consists of eight vertically separated single antenna elements. The 1st and 10th subarrays act as dummy elements in an attempt to maintain a consistent mutual coupling between subarrays across the entire array. The provision of circuitry to allow the reception of both vertically and horizontally polarised signals at each of the eight subarrays implies that the reception of 16 different polarisations is possible. The calibration of the antenna can be separated into two stages, namely the offline calibration after manufacture and the online calibration performed during operation. The offline calibration measures the characteristics of the passive components in the signal path and assumes that the 16:1 Wilkinson divider and the 20 dB directional couplers are stable over both time and temperature. More specifically, the online calibration procedure uses the Wilkinson divider and the directional couplers to inject a calibration signal into each of the eight signal paths dedicated to horizontal polarisation and the eight paths for vertical polarisation. The magnitude and phase response of these 16 signal paths is then measured in the baseband in order to characterise the entire antenna system. However, fully characterising this antenna array receiver at all of the frequencies of interest would generate vast amounts of data, and require an impractical length of time.

Fortunately, it is possible to use a reduced set of measurements [274]. Measurements of the antenna array's forward transfer function, S_{21}, between the central calibration port and the 16 receiver ports for both the vertically and horizontally polarised signals were found to be essential for characterising the calibration network itself. Phase differences of up to 20° and amplitude variations of 2 dB were measured between two seemingly identical calibration signal paths, despite the symmetrical layout of the Wilkinson divider [274]. Further measurements of S_{21} between each subarray port and all other subarray ports, in order to account for mutual coupling of the subarrays showed coupling levels of below -30 dB between all ports. Thus far, the characterisation of the calibration network has required 16 phase and magnitude values, while the mutual coupling between the subarrays necessitated a further $(2 \times 8)^2 = 256$ readings. Additionally, any imbalances between subarrays in the magnitude radiation patterns over all specified azimuth and elevation angles must also be measured, leading to a still significant amount of information that must be processed.

Simmonds and Beach [253] described how an 8 element adaptive antenna array can be calibrated, with no interruption to the network, for both transmission and reception. The aim of the scheme was to achieve a post-calibration accuracy of 3° phase and 0.5 dB magnitude

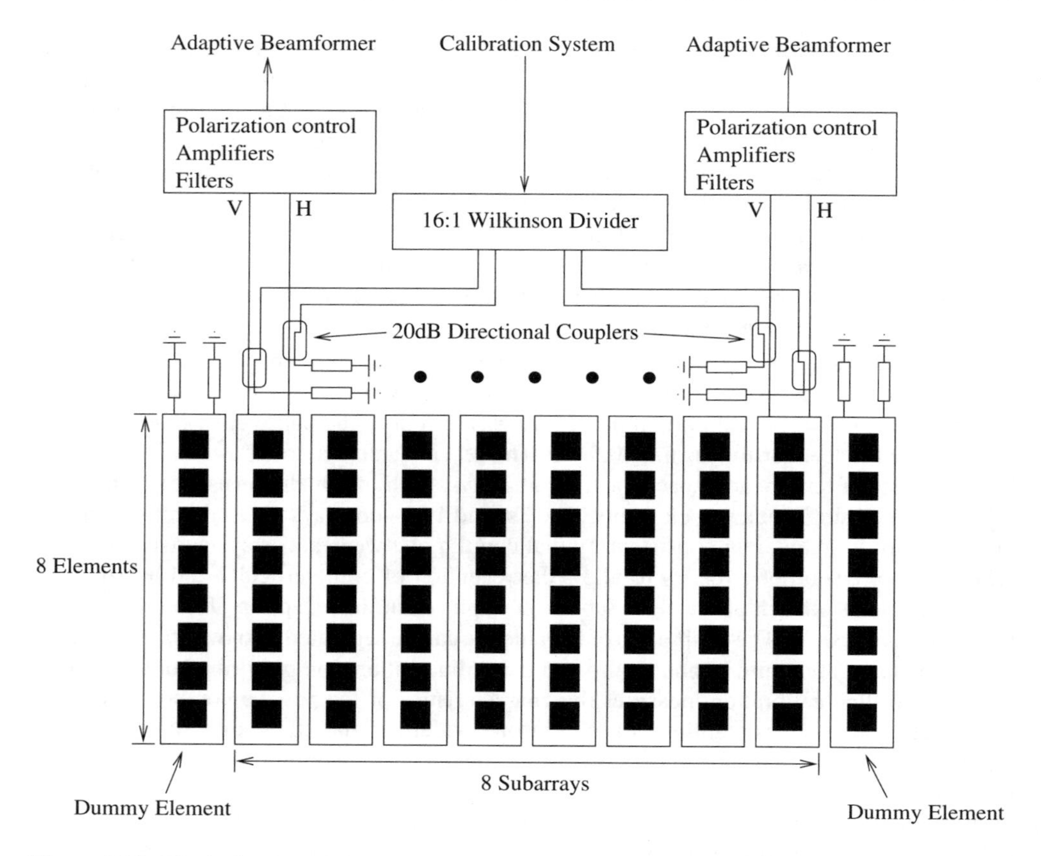

Figure 3.23: Block diagram of the 8×8 element antenna array receiver and in-built calibration system of Passman and Wixforth showing the horizontal and vertical polarisation ports [274].

error across the array. The design of the process allows the receive calibration to be performed during the unallocated timeslots within the DCS1800 frame structure. A Continuous Wave (CW) signal is injected simultaneously into each of the receiver antenna array elements via directional couplers and a power divider/combiner. Digital attenuators allow the injected signal strength to be varied over a range of 60 dB in 2 to 3 dB steps. The errors associated with the received signal phase and amplitude are measured in the baseband and the beamformer weights are adjusted appropriately, in order to produce the desired beam pattern. Moreover, the same technique cannot be used for transmitter calibration, since this would result in spurious RF transmission. In the proposed scheme the transmitter would be calibrated in the even timeslots, except for timeslot zero, which is used for the Broadcast Control CHannel (BCCH) in DCS1800 and GSM [253]. Each branch of the antenna array is sampled using the directional couplers and the resulting signals are down-converted to baseband. These 8-bit quantised I and Q samples are compared to the baseband digital beamformer outputs, in order to obtain the correction factor required for each array path.

3.3.4 Blind Adaptation

Blind adaptation [3, 242, 260] of the array weights has several advantages over both the spatial [6, 7, 9, 242, 260, 273] and temporal reference [1, 3, 6, 239, 249, 267] based systems of Sections 3.3.3 and 3.3.2. Temporal reference assisted systems must achieve synchronisation and perform demodulation, before weight adaptation can commence, whereas spatial reference aided systems require very strictly calibrated hardware and rely on DoA information. However, the typically large angular spread of the incoming signals in small picocells makes this difficult to attain. In contrast, a blind adaptation scheme [3, 242, 260, 266] does not require training sequences or any information concerning the antenna array's geometry. Dispensing with the reference or training sequence results in potentially increased data rates. For example, a capacity increase of 17% can be achieved in the uplink for GSM [266] upon invoking blind joint space-time equalisation. However, using for example the so-called constant modulus adaptive algorithm [275] can lead to the capture of interfering signals instead of the wanted signal, an issue argued more explicitly in [276, 277].

3.3.4.1 Constant Modulus Algorithm

The Constant Modulus (CM) algorithm [275] operates on the principle that the amplitude of the receive antenna array output should remain constant, unless the interference causes fluctuations. If the transmitted signal, $s(n)$, has a constant envelope, then the combiner output, $y(n)$ in Figure 3.17, should also have a constant envelope. However, if multipath fading occurs, then the combiner output, $y(n)$, will have a fluctuating envelope. The objective of CM beamforming [3, 242, 260] is to restore the array output to a constant envelope signal, on average. This can be accomplished by adjusting the array weight vector, $\underline{w}$, in such a way, so as to minimise a certain cost function.

In the classic paper by Godard [275], who used the CM property in order to carry out blind channel equalisation, the criterion was to minimise the functions $D^{(p)}$, referred to as the dispersion of order p ($p > 0$ integer), defined by

$$D^{(p)}(n) = E\left[(|y(n)|^p - R_p(n))^2\right], \tag{3.90}$$

with R_p being real positive constants given by:

$$R_p(n) = \frac{E[|a(n)|^{2p}]}{E[|a(n)|^p]}, \tag{3.91}$$

where $a(n)$ is the transmitted data symbol. The standard cost function of [3, 242, 275]

$$G^{(p)}(n) = E\left[(|y(n)|^p - |a(n)|^p)^2\right], \tag{3.92}$$

which is not used in blind array weight adaptation, is independent of the carrier phase but depends on the magnitude of the antenna array's output signal, $|y(n)|$, and that of the transmitted signal, $|a(n)|$. In contrast, the function $D^{(p)}$, used in the CM algorithm is independent of both the carrier phase and the data symbol's magnitude [275].

The most often used practical case is that of $p = 2$, where

$$D^{(2)}(n) = E[(|y(n)|^2 - R_2(n))^2], \tag{3.93}$$

with

$$R_2 = \frac{E[|a(n)|^4]}{E[|a(n)|^2]}, \tag{3.94}$$

and the cost function, $D^2(n)$, is effectively the mean squared error between the magnitude of the antenna array's output signal squared and the constant $R_2(n)$. Hence, again, the main difference between the conventional cost function of Equation 3.92 and that of the constant modulus algorithm in Equation 3.90 is that the constant modulus algorithm does not assume the knowledge of the data sequence's magnitude, $|a(n)|$, it rather attempts to minimise the difference with respect to the constant quantity $R_2(n)$, which is related to the moments of $|a(n)|$ by Equation 3.94. In other words, the CM beamforming algorithm directs the combiner's output to a constant envelope.

In [242] a cost function is given in the form of:

$$J(n) = \frac{1}{2}E[(|y(n)|^2 - y_0^2)^2], \tag{3.95}$$

where y_0 is the desired amplitude in the absence of interference.

The objective is to find a set of values for the array weight vector, $\underline{w}$, that will minimise the given cost function. This may be accomplished using the following equation [242] :

$$\underline{w}(n+1) = \underline{w}(n) - 2\mu(|y(n)|^2 - y_0^2)y(n)\underline{x}(n+1), \tag{3.96}$$

or employing the update formula of [3]:

$$\underline{w}(n+1) = \underline{w}(n) + \mu[R_p(n) - |y(n)|^2]y(n)\underline{x}(n), \tag{3.97}$$

which are used in a steepest descent fashion to update the array weights and are essentially identical, apart from Equation 3.96 using the current sample, $\underline{x}(n+1)$, of the array's output, while Equation 3.97 using the previous sample, $\underline{x}(n)$. These equations are identical to the update regime of Equation 3.54 used in the LMS algorithm, with the only difference being the error term.

There are two conditions, which may lead to a zero-gradient situation, where the algorithm stops adapting. The first is the condition of $|y(n)| = 1$, which represents the desired convergence optimum. The second is $y(n) = 0$, which also forces the gradient to become zero. However, fortunately this is not a practical problem, since the point $y(n) = 0$ is not a stable equilibrium and the system noise moves the weight vector from this zero-gradient point. A further problem in a hostile fading environment is that the beamformer may incorrectly select the interference as the signal to process, so as to maintain a constant modulus, rather than the desired signal.

In [266] a blind array weight adaptation technique was described by Laurila and Bonek, which performs joint space-time equalisation, separation and detection of multiple unsynchronised co-channel digital signals. The scheme exploits the facts that the signals are of fixed symbol rate, have a CM and a Finite Alphabet (FA) of symbols. Simulations were conducted for an eight-element Uniform Linear Array (ULA) with an element spacing of $\lambda/2$ [266]. The equaliser order was five. Although the simulation parameters were not optimised, the system gave results demonstrating that comparable BER can be achieved, when compared to reference-assisted adaptation methods.

3.3.5 Adaptive Arrays in the Downlink

Adaptive arrays have been more often studied for receiving uplink data at the base station. However, they are equally suitable for transmitting data by the base station in the downlink. It is possible to steer a transmitting array in the same way as one used for reception, so as to minimise the downlink interference inflicted upon co-channel mobiles. The wide frequency separation between the uplink and downlink frequency bands used in the Frequency Division Duplexing (FDD) GSM system, for example, results in uncorrelated fading between the up- and the down-link. Therefore, the weights calculated for reception are typically unsuitable for employment in transmit mode. In contrast, in a Time Division Duplexing (TDD) system, such as UTRA [11] it may be possible to re-use the receive mode weights, provided that the location of the mobile has not changed significantly between timeslots, i.e. if the duration of the timeslots is sufficiently short.

In the uplink scenario, the receive array can adapt to changes in the propagation medium by observing its own outputs and modifying its own processing, since there is an in-built feedback mechanism, as was shown in Figure 3.17. When used in the transmit mode, an adaptive antenna array at the base station needs an additional feedback signal from the mobile receivers, in order to give the base station a means of measuring its own beam patterns. The array, by directing a mainlobe towards a mobile, could nonetheless produce a spurious fade in the desired signal or inflict interference upon other mobiles.

The scheme proposed by Gerlach and Paulraj [278] uses feedback of the signals received at the mobiles, in order to calculate the transmitter antenna weights to employ. The paper describes a system where data transmission is temporarily halted in for the transmission of probing signals. Each probing signal is sent on an orthogonal channel in the time, frequency or code domain so that the receivers may measure the response of each probing signal. The responses to each of the probing signals at each of the receivers are fed back to the transmitter, allowing the channel responses to be estimated. Simulations were performed, by Gerlach and Paulraj [278], which showed that at a low mobile speed of 2.5 miles per hour (mph) adequate signal separation required a data feedback rate in the order of a few kbit/s, making the approach only viable for static or slow-moving receivers. It is worth noting here that the 3G UTRA system has a total control channel rate of about 10 kbit/s.

Further to this scheme, Gerlach and Paulraj [279] presented a method which reduces the feedback rate by exploiting that as the array's weight vector fluctuates due to the mobile receiver's motion, the weight vector's fluctuations will be confined to a certain subspace of its total vector space. In contrast to the channel weight vector itself, the channel vector's subspace is much more stable during the mobile's motion, and this fact can reduce the required feedback rates. The method is best suited to environments having either a low number of propagation paths, or for several paths approaching the base station from similar angles. This implies that there must be only a few scattering bodies near to the base station. As the mobile receiver moves, its array weight vector varies at the fast fading rate, but the fluctuations are confined to the subspace Ψ_k, where the subscript k denotes the kth mobile, which varies slowly. A beamformer based on this more stable subspace structure, rather than the array weight vector, will need a lower feedback rate. Hence, the subspace structure tends to be more useful when the subspace dimension, $\dim[\Psi_k]$, is small. The subspace dimension, $\dim[\Psi_k]$, will only be small however, if the number of propagation paths is low or if all of the paths have approximately the same angle of departure from the array. The paper derives a subspace beamformer and presents results obtained using simulations. The required feedback rate for a mobile moving at 35 mph was estimated to be 250 bits/s. While this is a best-case estimate, it is significantly reduced in comparison to the rate in [278] and it is also less than the feedback rate used for power control in Qualcomm's IS-95 cellular system [279]. Hence, such a regime could realistically be used in a UTRA-type system.

Martin and Gaspard [251] presented a system based on the Discrete Fourier Transform (DFT) Beamspace technique. Each user's signal was transmitted on the particular DFT beam, which offered the largest mean power level during the uplink reception. With a four-element linear array the system provided a 175% radio capacity gain over a conventional base station. An eight-element array resulted in a gain of 200% in radio capacity. However, the downlink capacity using this method was not matched to the uplink capacity. Similarly enhanced downlink capacity was achieved using exact DOA information, where the downlink's transmission beam was steered in the direction of the strongest multipath component received at the uplink. This provided an estimated 350% increase in radio capacity.

Monot *et al.* [16] also used a DOA based system. Their prototype implemented the Capon [243] and the MUSIC [3, 242, 260] algorithms using a five element antenna array, and it was reported to have successfully estimated the DOA of the different paths, in an environment consisting of one main path and a set of spatially dispersed other paths.

3.3.6 Adaptive Beamforming Performance Results

The performance of the SMI algorithm of Section 3.3.2.3 using the direct matrix inversion formula of Equation 3.70 and the iterative matrix inversion lemma in Equation 3.75 as well as that of the ULMS and NLMS algorithms of Sections 3.3.2.1 and 3.3.2.2 was compared for identical scenarios. The effects of varying the reference signal lengths and the SNR as well as INR on the level of interference rejection were measured. For the situations exposed to different SNRs and INRs, the eigenvalue spread $\chi(R)$ of Equation 3.56 is summarised in Table 3.1.

SNR (dB)	INR (dB)	$\chi(R)$
3.0	3.0	4.4
3.0	9.0	8.3
3.0	27.0	402.2
9.0	3.0	8.3
9.0	9.0	5.4
9.0	27.0	120.6
27.0	3.0	403.3
27.0	9.0	120.6
27.0	27.0	5.8

Table 3.1: Eigenvalue spread, $\chi(R) = \frac{\lambda_{max}}{\lambda_{min}}$, of Equation 3.56 evaluated for the array output cross-correlation matrix, R, for different values of SNR and INR.

The effects of varying the reference signal length, the signal-to-noise, and the interference-to-noise ratios on the interference rejection achieved were evaluated and a complexity analysis was performed. The ability of the various beamforming algorithms to combine multipath signals, whilst rejecting interference was also investigated. The modulation scheme used in the simulations was BPSK. Our associated results are summarised in the forthcoming sections.

3.3.6.1 Two Element Adaptive Antenna Using Sample Matrix Inversion

Recall that the SMI algorithm of Section 3.3.2.3 directly inverts the sample correlation matrix, $\hat{R}_{xx} = E[\underline{x}(t)\underline{x}^H(t)]$, in order to find the optimal antenna element weights according to Equation 3.70. Specifically, we have $\hat{\underline{w}}_3 = \hat{R}_{xx}^{-1}\hat{\underline{z}}$, where $\hat{\underline{z}}$ is the sample cross-correlation vector between the array output vector, $\underline{x}$, and the reference signal, r. The iterative version of this technique, as described in Section 3.3.2.3, forms the inverse of the sample correlation matrix, $\hat{R}_{xx}^{-1}$, based on the received signal samples using Equation 3.75, and iteratively updates it according to:

$$\hat{R}^{-1}(n) = \hat{R}^{-1}(n-1) - \frac{\hat{R}^{-1}(n-1)\underline{x}(n)\underline{x}^H(n)\hat{R}^{-1}(n-1)}{1+\underline{x}^H(n)\hat{R}^{-1}(n-1)\underline{x}(n)}, \tag{3.98}$$

with $\hat{R}^{-1}(0) = \frac{1}{\epsilon_0}I$ where ϵ_0 is a scalar value greater than zero.

The interference rejection achieved using the SMI algorithm as a function of the reference signal length is shown in Figure 3.24(a) for equal values of SNR and INR, i.e. for equal signal and interferer powers. The graph also shows how the interference rejection increases, as the SNR and INR are increased. The performance of the direct inversion method of Equation 3.70 and the iterative method of Equation 3.75 was found to be identical using a value of $\epsilon_0 = 0.01$ in Equation 3.76 to initialise the estimate of R^{-1}. For a setting of $\epsilon_0 = 0.3$ the difference between the rejection levels was of the order of 0.01 dB, while a 0.1 dB interference rejection reduction resulted from $\epsilon_0 = 0.9$. As stated earlier in Section 3.3.2.3, an adequate performance can be achieved after processing only $2M$ data samples, where M is the number of sources present, which was two in this case. Figure 3.24(b) shows that the interference rejection is only affected by the INR and appears to be independent of the SNR.

The rate at which the interference rejection increases, as the SNR and INR improve is shown in Figure 3.25. The increased SNR and INR values allow for more accurate estimates of the array output cross-correlation matrix, R, thus resulting in improved interference rejection. The rate of increase of the interference rejection slows down as the SNR and INR increase, since the limit of the estimation accuracy is approached. As expected, the longer reference lengths allow for a better estimate of R and hence exhibit higher interference rejection levels for sufficiently high SNR and INR values. In contrast, for low SNR and INR the estimation quality of R is poor, resulting in marginal performance improvements due to extending the reference sequence length.

3.3.6.2 Two Element Adaptive Antenna Using Unconstrained Least Mean Squares

The Unconstrained Least Mean Squares (ULMS) technique [3,239–242] of beamforming was described in more detail in Section 3.3.2.1 but is based around the weight update formula of Equation 3.54, i.e. $\underline{w}(n+1) = \underline{w}(n) - \mu\underline{x}(n)\epsilon^*(n)$, where μ is a constant controlling the rate of convergence and $\epsilon(n)$ is the error between the combiner output, $y(n)$, and the reference signal, $r(n)$. For each array output sample, $x(n)$, the new antenna element weights

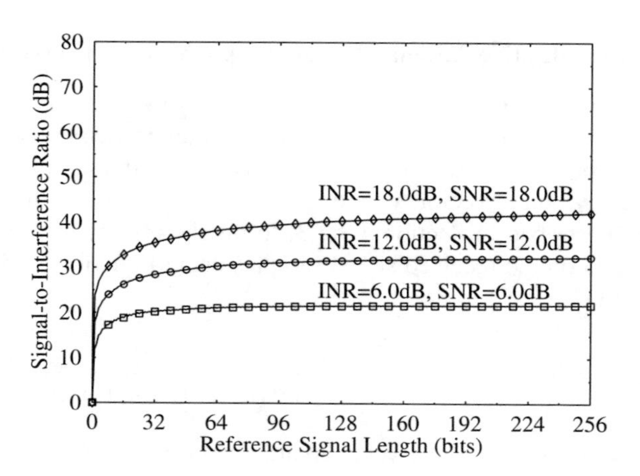

(a) Equal SNR and INR

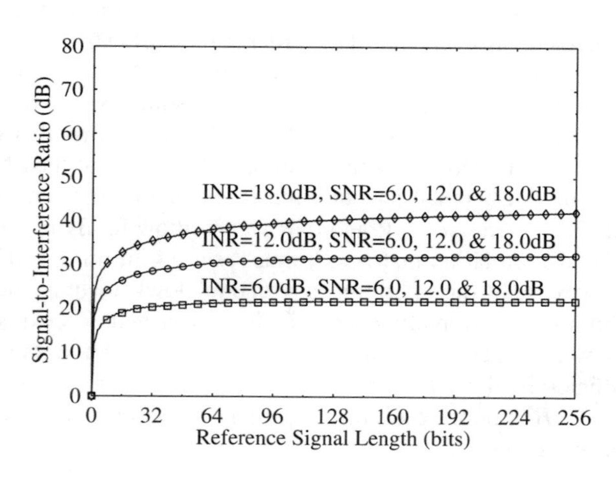

(b) Unequal SNR and INR

Figure 3.24: The interference rejection achieved using **SMI** beamforming upon varying the reference signal lengths for a two element antenna array using an element spacing of $\lambda/2$. The source was at $0°$ and the interferer at $30°$, whilst $\epsilon_0 = \mathbf{0.01}$ evaluating 10000 averaged runs over a Gaussian channel.

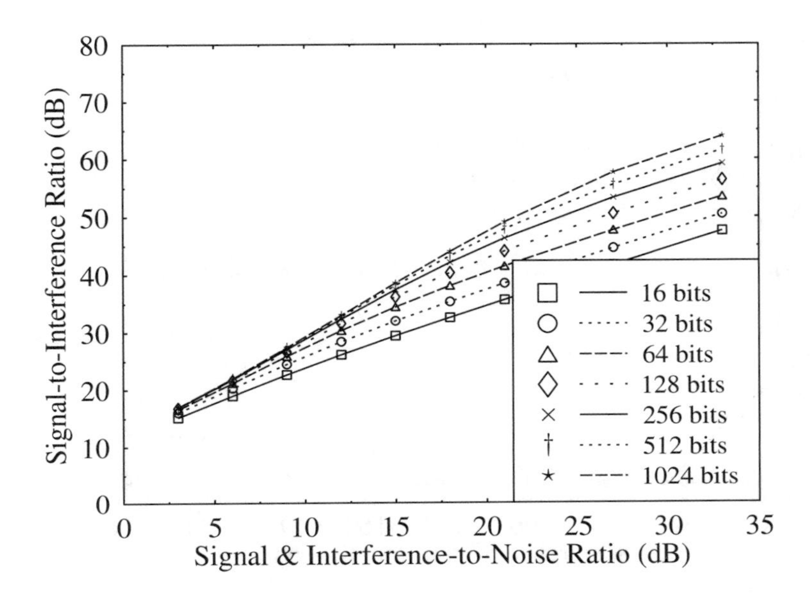

Figure 3.25: The interference rejection achieved versus SNR and INR using **SMI** beamforming upon varying the reference signal lengths for a two element antenna array using an element spacing of $\lambda/2$, at **equal SNR and INR**. The source was at 0° and the interferer at 30°, whilst $\epsilon_0 = \mathbf{0.01}$ evaluating 10000 averaged runs over a Gaussian channel.

are calculated, in order to minimise the mean square error between the measured array output and the desired array output.

The performance of the ULMS algorithm of Section 3.3.2.1 was studied using $\mu = 0.0000005$, $\mu = 0.00005$, $\mu = 0.0005$ in Equation 3.54 and varying the prevalent SNR and INR. It was found that convergence was extremely slow using $\mu = 0.0000005$, and a reasonable level of interference rejection required an SNR and INR of 33.0 dB in conjunction with a reference length of 1024 bits. This shows the dependence of the ULMS algorithm upon the received signal strength, which is evidenced by Figures 3.26, 3.27(a) and 3.27(b). Additionally, Figure 3.29(a) shows that step size is insufficient to allow convergence to an acceptable level of interference rejection regardless of the reference length or the number of iterations. In contrast, using a value of $\mu = 0.00005$ in Figure 3.26 results in significantly faster convergence for all SNRs and INRs, where best performance was achieved by the stronger signals. However, the step size is excessive for SNRs and INRs in excess of about 20 dB and the phenomenon of weight jitter can be seen becoming apparent. Figure 3.28 illustrates this further since it can be seen that the interference rejection achieved actually decreases for high SNRs and INRs upon increasing the number of iterations. Again, this phenomenon is due to weight jitter around the optimal solution for high values of SNR as well as INR and it becomes more

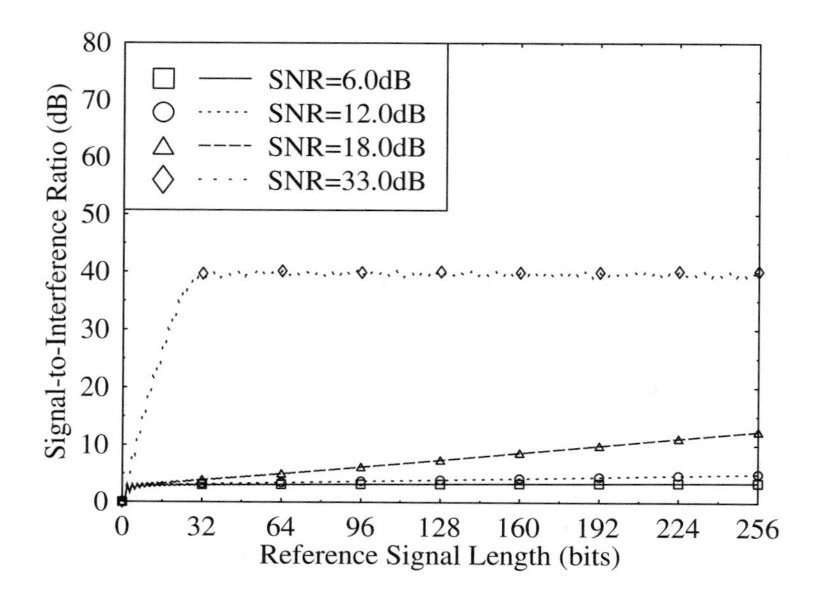

Figure 3.26: The interference rejection achieved using **ULMS** beamforming upon varying the reference signal lengths, a two element antenna array, using an element spacing of $\lambda/2$, at **equal SNR and INR**. The source was at 0° and the interferer at 30°, whilst $\mu = \mathbf{0.00005}$ evaluating 10000 averaged runs over a Gaussian channel.

prevalent for a large step size of 0.0005, which may be seen in Figure 3.29(b). Increasing the step size to 0.0005 results in a levelling off or even a reduction in the interference rejection achieved, as shown in Figure 3.27(b). Therefore, if the step size, μ, is chosen to be small, weak signals associated with low SNRs and INRs limit the convergence speed and may not be of much practical use, while strong signals allow for rapid convergence, as displayed in Figure 3.26. However, if μ is large then the convergence is rapid even for weak signals, but the algorithm exhibits weight jitter, resulting in poor performance and potential instability.

3.3.6.3 Two Element Adaptive Antenna Using Normalised Least Mean Squares

The Normalised Least Mean Squares (NLMS) algorithm [242, 247] of Section 3.3.2.2 uses a data dependent step size calculated using Equation (3.59), namely $\mu(n) = \frac{\mu_0}{\|\underline{x}(n)\|^2}$, in order to eliminate the deficiencies of the ULMS method of Section 3.3.2.1. Figure 3.30 characterises the algorithm's convergence, when $\mu_0 = 0.2$ in Equation 3.59.

When compared to the performance of the algorithm using the larger step sizes of $\mu_0 = 0.5$ and $\mu_0 = 1.0$ in Figures 3.31 and 3.32, it can be seen that for a small reference signal length the level of interference rejection is increased in conjunction with the larger step sizes,

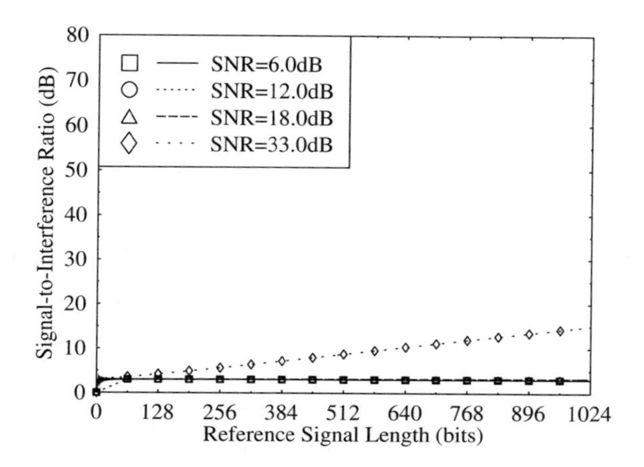

(a) $\mu = 0.0000005$

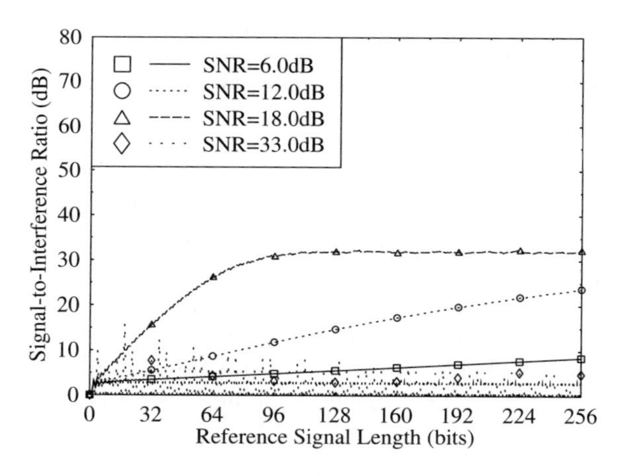

(b) $\mu = 0.0005$

Figure 3.27: The interference rejection achieved using **ULMS** beamforming upon varying the reference signal lengths for a two element antenna array using an element spacing of $\lambda/2$, at **equal SNR and INR**. The source was at 0° and the interferer at 30° evaluating 10000 averaged runs over a Gaussian channel.

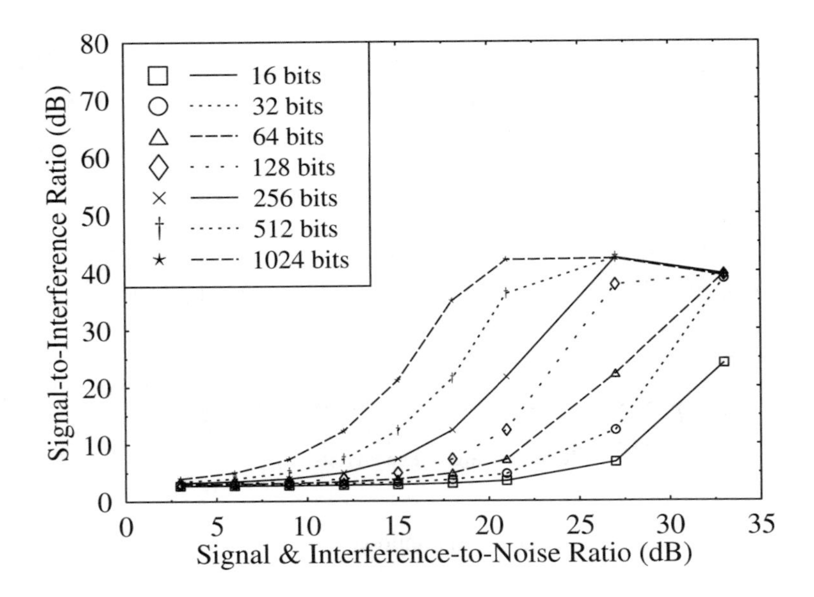

Figure 3.28: The interference rejection achieved using **ULMS** beamforming upon varying the reference signal lengths, SNR and INR. A two element antenna array was used with an element spacing of $\lambda/2$, at **equal SNR and INR**. The source was at 0° and the interferer at 30° while $\mu = \mathbf{0.00005}$ evaluating 10000 averaged runs over a Gaussian channel.

due to their faster rates of convergence. However, after the final interference rejection level has been reached, the algorithm performs better for smaller step sizes, attaining a higher level of interference rejection at the end of the convergence phase, and significantly lower weight jitter. For example, using $\mu_0 = 1.0$ when the SNR and INR was 6.0 dB, the interference rejection became approximately 15 dB exhibiting a jitter of ± 2.5 dB. In the case of $\mu_0 = 0.2$, the interference rejection was 20 dB exhibiting virtually no jitter effects. The performance difference became even more marked for higher SNR and INR levels.

Figure 3.33 demonstrates how the interference rejection increases, as the SNR and INR improve. When $\mu_0 = 0.2$, the rate of convergence is too slow for the optimal solution to be reached for reference signal lengths of 16 and 32 bits. For a reference signal length of 64 bits, a near optimal solution is obtained at low values of SNR and INR but as the SNR and INR increase, the performance of the algorithm does not improve beyond a certain point. This performance limitation experience for short reference signal lengths is due to the limited estimation quality of the mean of the received signal. Using a larger step size, hence allowing for faster convergence, resulted in shorter reference signal lengths converging to the optimal weights, although the final value of interference rejection reached did not match that of the smaller step sizes.

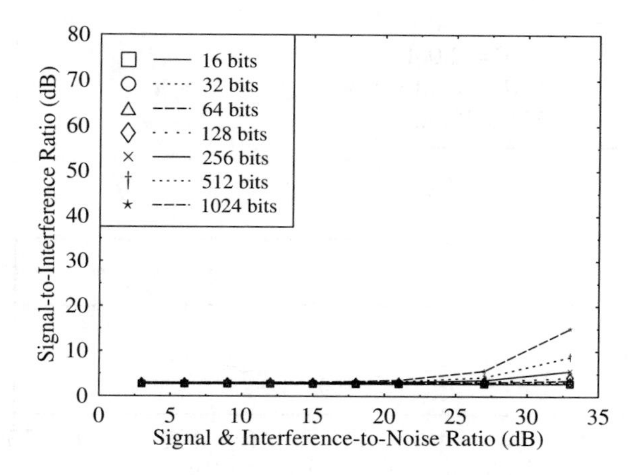

(a) $\mu = 0.0000005$

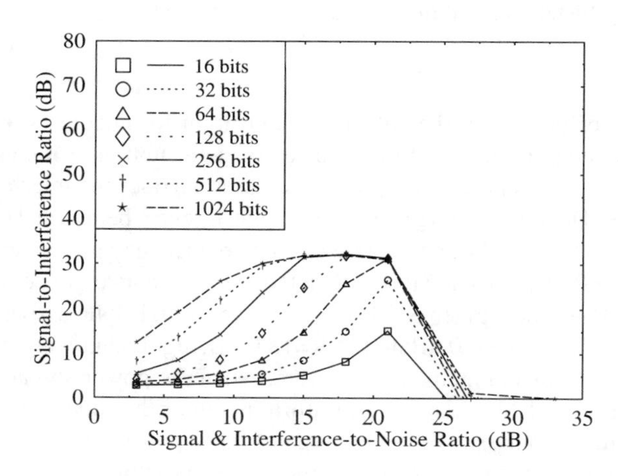

(b) $\mu = 0.0005$

Figure 3.29: The interference rejection achieved using **ULMS** beamforming upon varying the reference signal lengths, SNR and INR. A two element antenna array was used with an element spacing of $\lambda/2$, at **equal SNR and INR**. The source was at $0°$ and the interferer at $30°$ evaluating 10000 averaged runs over a Gaussian channel.

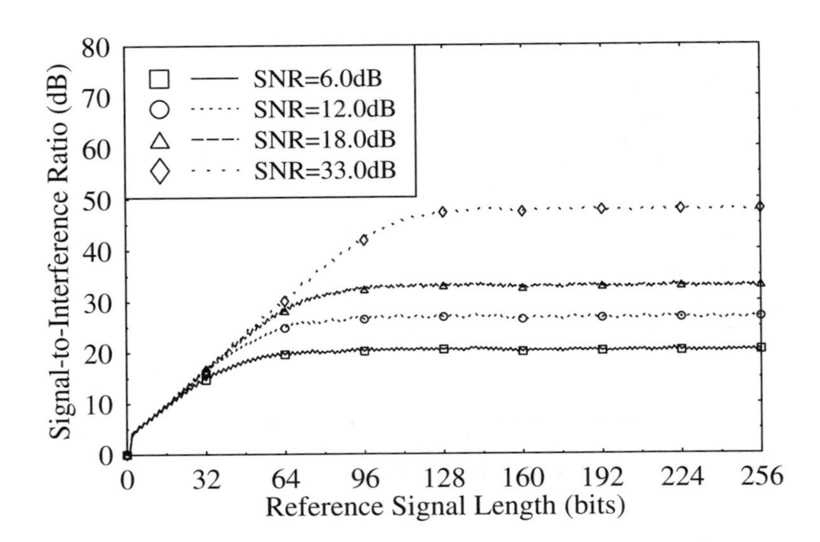

Figure 3.30: The interference rejection achieved using **NLMS** beamforming upon varying the reference signal lengths for a two element antenna array with an element spacing of $\lambda/2$, at **equal SNR and INR**. The source was at 0° and the interferer at 30° whilst $\mu_0 = \mathbf{0.2}$ evaluating 10000 averaged runs over a Gaussian channel.

The performance of the NLMS beamforming algorithm for unequal values of the SNR and the INR is portrayed in Figure 3.34 From the associated subfigures it can be seen that as the INR improves, i.e. as the interference power increases, so does the interference rejection, regardless of the SNR. However, for a given level of interference, better interference rejection is achieved for a higher SNR, although the rate of convergence may be slower, as seen for the case when we have SNR=18.0 dB and the INR=6.0 dB. Faster convergence was observed for higher values of the INR, for a given SNR. However, for a high INR associated with a low SNR, i.e. for example for SNR=6.0 dB and INR=18.0 dB, significant weight jitter occurred, whilst fast convergence was maintained. Therefore, when the power spread of the received signals is substantial, the NLMS adaptive beamforming algorithm does not perform as well as the SMI algorithm. In contrast, when the range of input powers is smaller, the algorithm performs well and for more than six antenna elements, this is achieved at a lower complexity than that of the SMI algorithm, as will be shown in Section 3.3.6.5.

3.3.6.4 Performance of a Three Element Adaptive Antenna Array

The interference rejection capabilities of a three element uniformly spaced linear adaptive array were investigated upon increasing the number of interference sources. The purpose of these experiments was to determine how the array behaved, when the total number of sources and interferers exceeded the degrees of freedom of the array, which was defined as

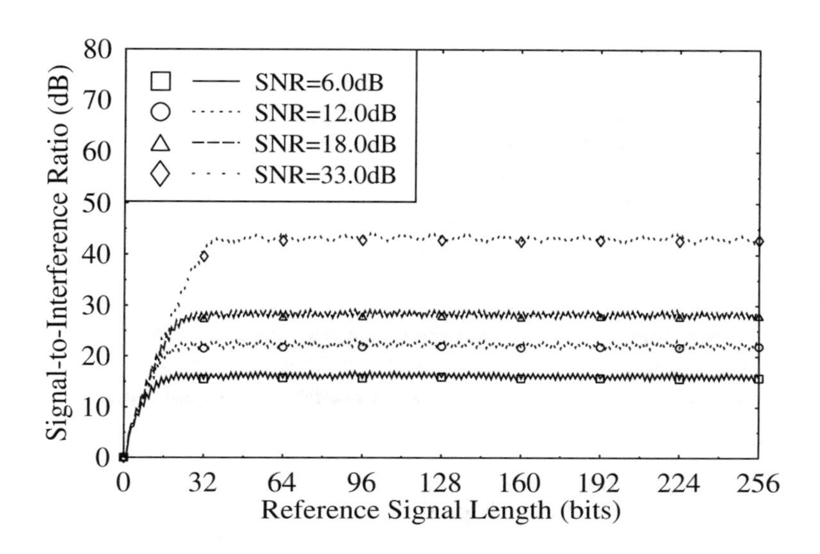

Figure 3.31: The interference rejection achieved using **NLMS** beamforming upon varying the reference signal lengths for a two element antenna array with an element spacing of $\lambda/2$, at **equal SNR and INR**. The source was at 0° and the interferer at 30° whilst $\mu_0 = \mathbf{0.5}$ evaluating 10000 averaged runs over a Gaussian channel.

the number of sources and/or interferences that may simultaneously be steered towards or nulled. The source was located at 15°, interferer 1 was at -30°, interferer 2 at 60°, interferer 3 was located at 80° and lastly, interference source 4 at -70°. It was assumed that the sources were point sources located in the far-field of the antenna array, benefiting from pure line of sight propagation without multipaths. Figure 3.35 shows the locations of the desired source and the interfering sources graphically. The simulations were carried out in conjunction with a 256-bit reference signal using the SMI and NLMS algorithms.

From the antenna array beam patterns portrayed in Figure 3.36 it can be observed that successful nulling of the interference source was accomplished for all the scenarios considered. A minimum interference rejection of 40 dB was attained for an INR of 9 dB, and when the INR was increased to 21 dB, an even higher rejection was achieved.

Figure 3.37 shows the array response for the situation where two interferers are incident upon the antenna array, having equal signal strengths to that of the desired signal. For the cases illustrated in Figure 3.37(a), where one of the sources of interference is at a -30° angle with respect to the array, good rejection of both sources of interference is achieved, whilst maintaining a perfect response in the direction of the desired source. Even for the situation, where the interference sources are located fairly close to each other, i.e. at -30° and -70°, strong nulling is maintained. Placing the interferers closer together, at angles of 60° and 80°, resulted in an interference rejection of over 45 dB, albeit exhibiting some beam and null mis-alignment. Spreading the interferers further apart, with each one tending to 'end-fire'

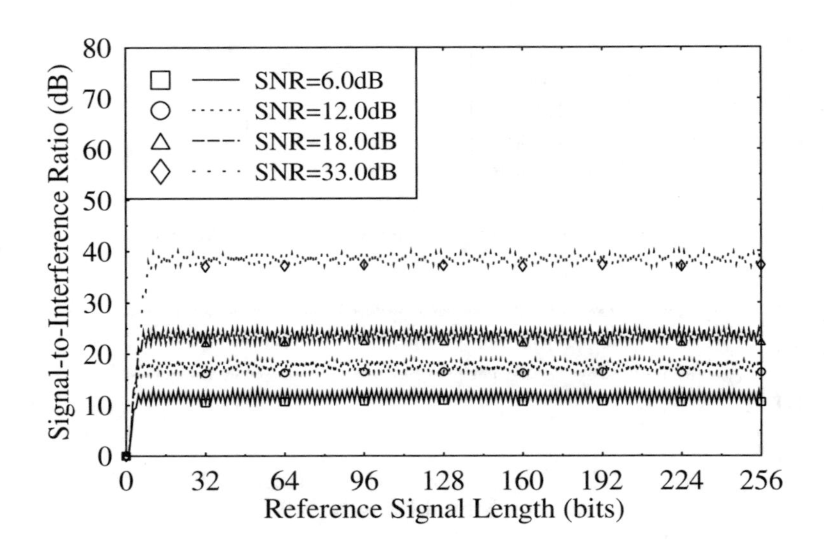

Figure 3.32: The interference rejection achieved using **NLMS** beamforming upon varying the reference signal lengths for a two element antenna array with an element spacing of $\lambda/2$, at **equal SNR and INR**. The source was at 0° and the interferer at 30° whilst $\mu_0 = \mathbf{1.0}$ evaluating 10000 averaged runs over a Gaussian channel.

at opposite ends of the array leads to some beam mis-steering, but nevertheless, maintaining good rejection of the sources of interference. Separating the interferers further so that they were located at -70° and 80° yielded significantly poorer results with an average interference rejection of about 25 dB. However, this is still perfectly acceptable and levels significantly higher than this would be unrealisable due to hardware limitations.

From Figure 3.38 it can be seen that, if two sources of interference are present, and one of them is weaker than the other, then the stronger one will be nulled more effectively than the weaker one. The SNR of the desired signal does not appear to affect the interference rejection.

When three sources of interference and one desired signal source are incident upon a three element antenna array, the performance of the array is reduced compared to the situation, when fewer sources impinge upon the array concurrently. In Figure 3.39(a) it can be seen that an interference rejection ratio of at least 15 dB is achieved for all of the interference sources simultaneously, where greater than 20 dB rejection ratios are also frequently obtained. The results presented in Figure 3.39(b) are better than those in Figure 3.39(a), exhibiting a minimum interference rejection of 25 dB. Therefore, the interference rejection obtainable when the number of sources equals the number of antenna elements appears to be dependent upon the location of the sources, but on average a good interference rejection performance is observed. Reducing the SNR from 21 dB to 9.0 dB, whilst keeping the INR at 21 dB produced the results depicted in Figure 3.40. The beam patterns in this figure are

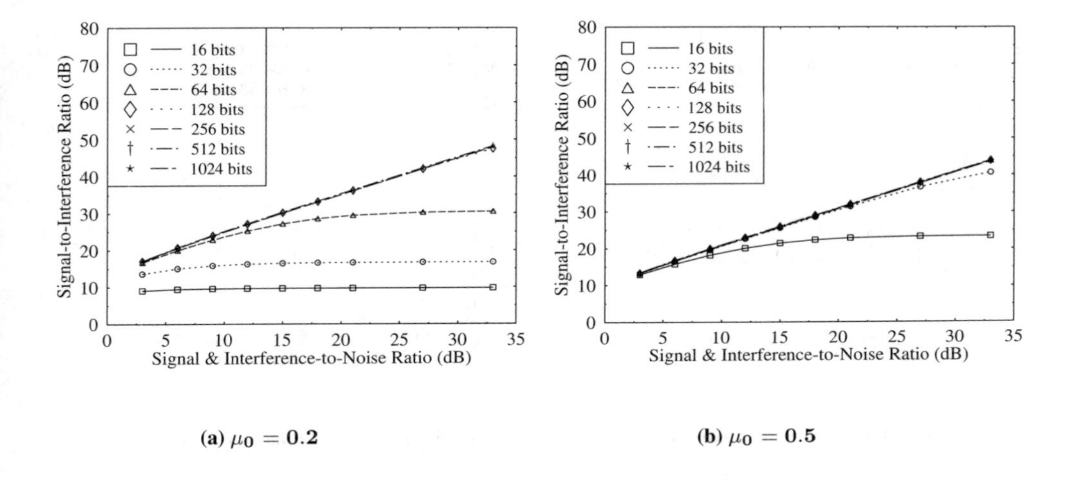

(a) $\mu_0 = 0.2$

(b) $\mu_0 = 0.5$

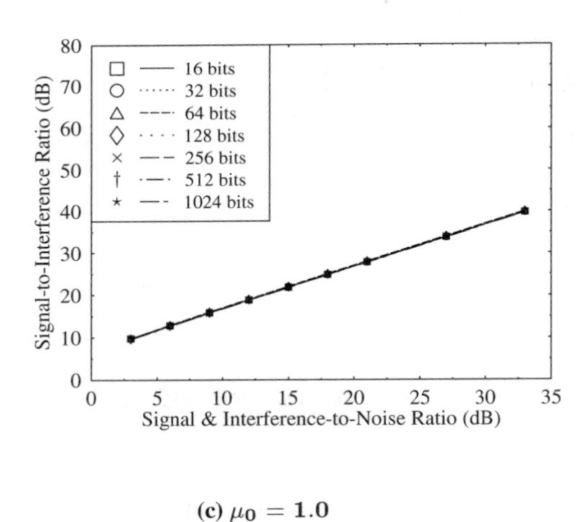

(c) $\mu_0 = 1.0$

Figure 3.33: The interference rejection achieved using **NLMS** beamforming upon varying the reference signal lengths, and SNR and INR, for a two element antenna array with an element spacing $\lambda/2$, at **equal SNR and INR**. The source was at $0°$ and the interferer at $30°$ evaluating 10000 averaged runs over a Gaussian channel.

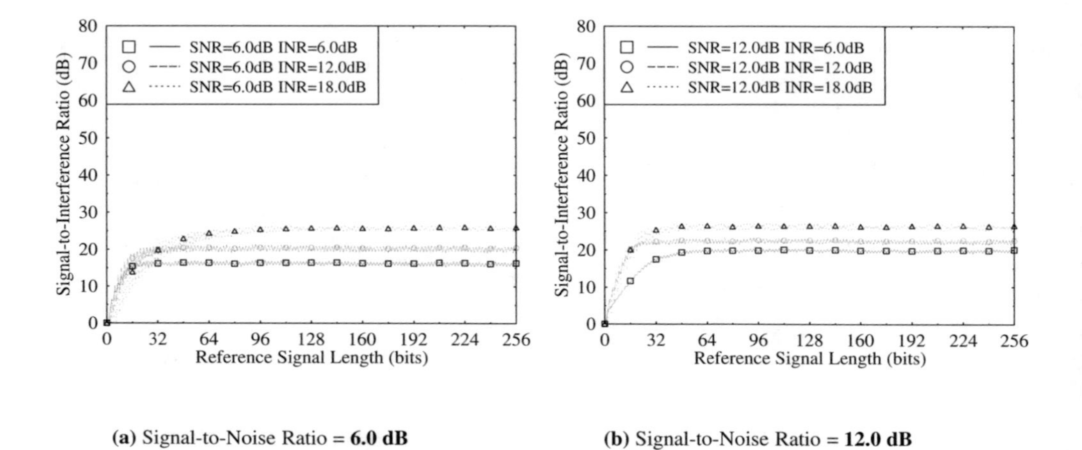

(a) Signal-to-Noise Ratio = **6.0 dB**

(b) Signal-to-Noise Ratio = **12.0 dB**

SNR=18.0dB INR=6.0dB
SNR=18.0dB INR=12.0dB
SNR=18.0dB INR=18.0dB
Signal-to-Interference Ratio (dB)
Reference Signal Length (bits)

(c) Signal-to-Noise Ratio = **18.0 dB**

Figure 3.34: The interference rejection achieved using **NLMS** beamforming upon varying the reference signal lengths, and SNR and INR, for a two element antenna array with an element spacing $\lambda/2$, at **unequal SNR and INR**. The source was at 0° and the interferer at 30° whilst $\mu_0 = \mathbf{0.5}$ evaluating 10000 averaged runs over a Gaussian channel.

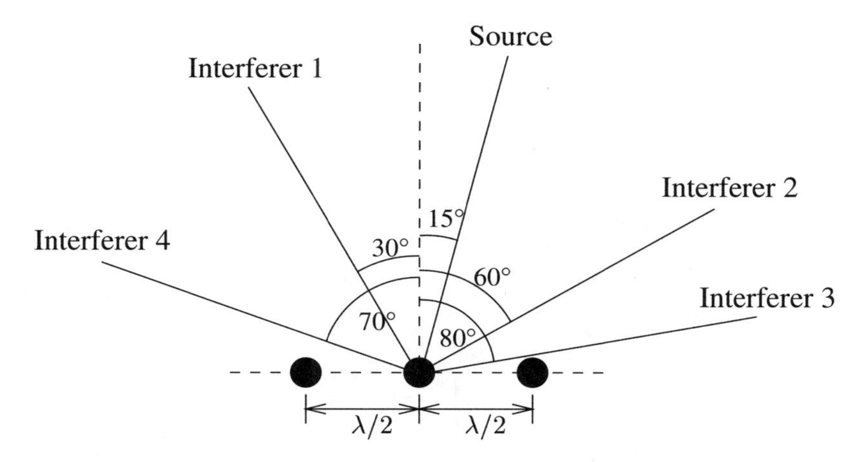

Figure 3.35: Locations of the desired source and the interferers with respect to the three element linear array with $\lambda/2$ element spacing.

similar in form to those of Figure 3.39, where the Interference-to-Noise Ratios was 21 dB, but the depths of the nulls are shallower. Although the nulls are less deep, the INRs are not as high, so the resultant SIR should not be any higher. Furthermore, the nulls are generally still more than 15 to 20 dB deep, which should be sufficient for effective interference rejection.

The performance of the three element antenna array when the desired source and the four interfering sources, all exhibiting equal signal power, are incident upon it, is shown in Figure 3.41(a). The antenna array response is virtually identical for the scenario when all the sources have SNRs of 21 dB, to that when the SNRs are equal to 9 dB. The array succeeds in suppressing all of the interference sources by at least 15 dB, where one of the interferers is nulled by more than 40 dB. In the situation when one of the interference sources has an INR of 9 dB, as in Figure 3.41(b), it is nulled less strongly than in the case of an INR of 21 dB. Although the associated null-depth was reduced from 43 dB to 29 dB, due to the associated 21-9=12 dB decrease in the power of the interferer, the SIR only fell by 2 dB to 20 dB. However, the rejection of the other interference sources increased slightly.

The beam patterns obtained for exactly the same scenarios, except using the NLMS beamforming algorithm along with $\mu_0 = 0.5$, are presented in Figures 3.42 to 3.46. From the graphs in Figure 3.42 it can be observed that the nulls formed by the NLMS adaptive beamforming algorithm are not as deep as those of the SMI algorithm. As for the SMI algorithm, the null depths are also shallower when the INRs are lower.

In the case of two sources of interference, as shown in Figure 3.43, the algorithm has again successfully nulled the sources, albeit with a lower attenuation than that achieved by the SMI algorithm as may be seen in Figure 3.37. This is, however of purely academic interest, since null depths of 50 dB would be unrealisable. For three interferers, all having the same power as the desired source, this phenomenon persists, as it does when the interference sources are of lower power. The corresponding results for three interferers are portrayed in Figure 3.44 for an SNR and INR value of 21.0 dB. Observe, however, in Figure 3.45(b) that the interference rejection for the source at an angle of 60° is significantly lower at 20 dB than

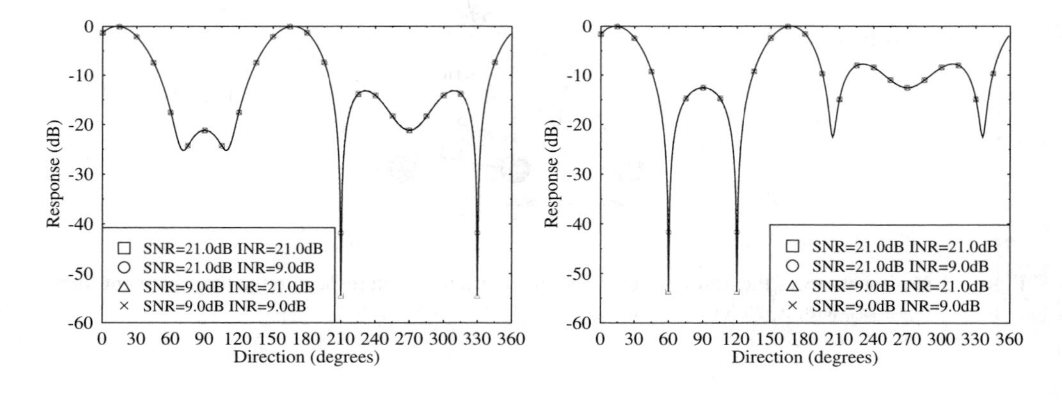

(a) Source located at 15°, interference at −30°.

(b) Source located at 15°, interference at 60°.

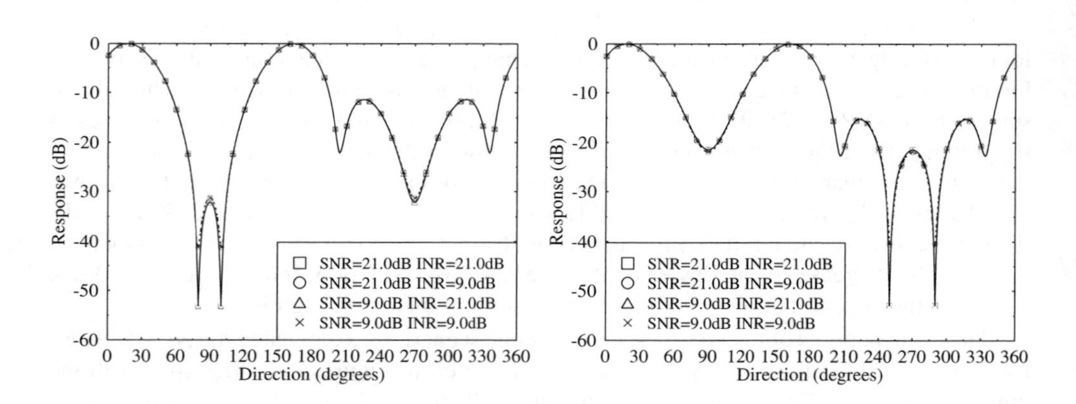

(c) Source located at 15°, interference at 80°.

(d) Source located at 15°, interference at −70°.

Figure 3.36: Beam patterns of a three element uniformly spaced linear array with an inter-element spacing of $\lambda/2$ with one desired source and **one source of interference**. The **SMI** beamforming algorithm was used with a reference length of 256 bits.

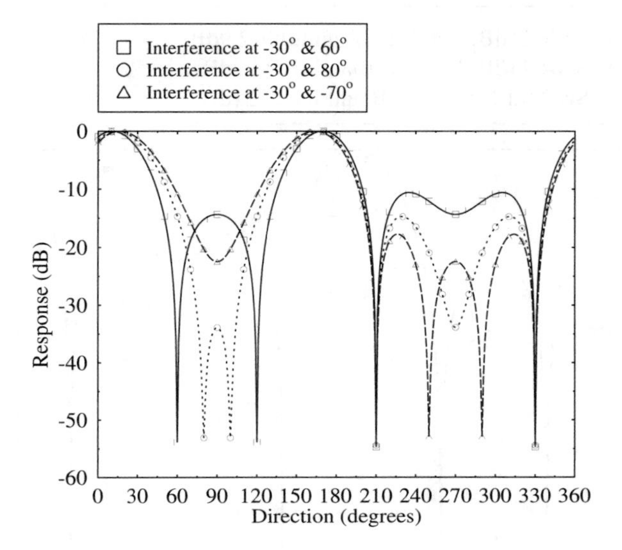

(a) Source located at 15°, interferers located at **-30°** and, **80°** or **-30°** and **-70°**.

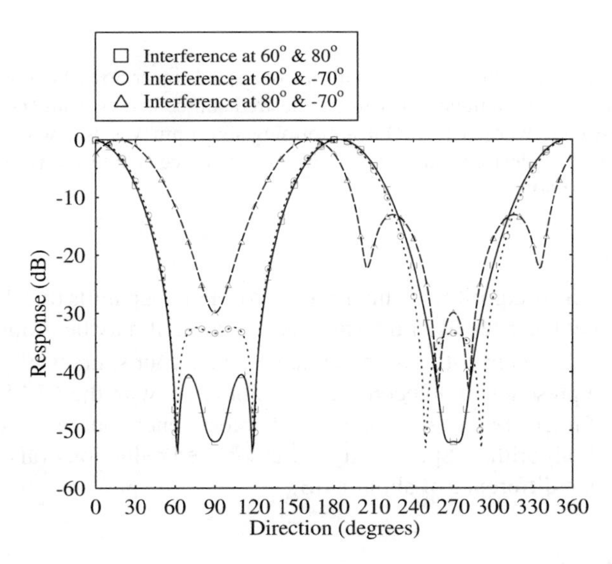

(b) Source located at 15°, interferers located at **60°** and **80°**, or **60°** and **-70°**, or **80°** and **-70°**.

Figure 3.37: Beam patterns of a three element uniformly spaced linear array having an inter-element spacing of $\lambda/2$ in conjunction with one desired source and **two sources of interference**. The **SMI** beamforming algorithm was used with a reference length of 256 bits.

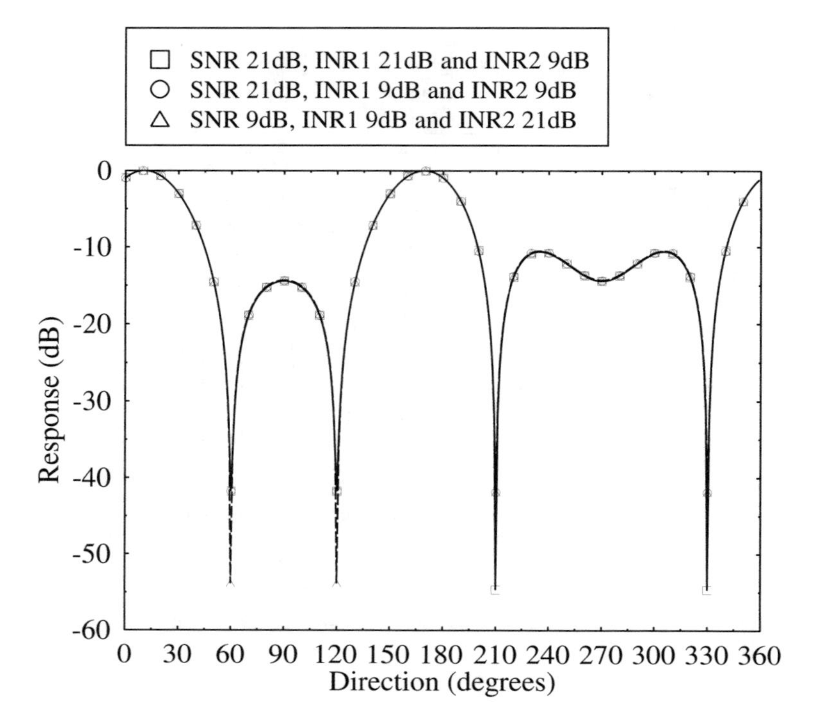

Figure 3.38: Beam patterns of a three element uniformly spaced linear array having an inter-element spacing of $\lambda/2$ in conjunction with one desired source and **two sources of interference** with **unequal powers**. The **SMI** beamforming algorithm was used with a reference length of 256 bits. The desired source was at 15°, interference source 1 was located at -30° and interferer 2 at 60°.

that obtained using the SMI algorithm, which was 27 dB. For deep nulls this difference would have little impact, but at these levels of interference rejection, it may be problematic.

Figure 3.46 shows the beam patterns encountered, when four sources of interference and one desired source are present simultaneously. In conjunction, with the NLMS beamforming algorithm the levels of interference rejection for each interference source are lower than those obtained using the SMI algorithm. Specifically, the associated reductions vary from only 2 dB to 17 dB, having a mean difference of about 8 dB.

3.3.6.5 Complexity analysis

The relative complexities of the DMI, ULMS and NLMS beamforming algorithms for a reference signal length of 16 symbols are portrayed in Figure 3.47. The direct matrix inversion algorithm requires the average of the cross-correlation matrix, R, which is a square-shaped matrix of size L, where L is the number of antenna elements. In order to calculate each

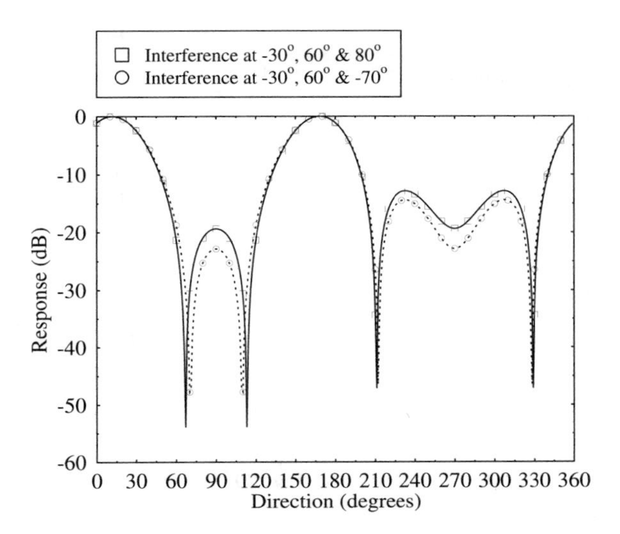

(a) Source located at 15°, interferers located at **-30°**, **60°** and **80°** or **-30°**, **60°** and **-70°**.

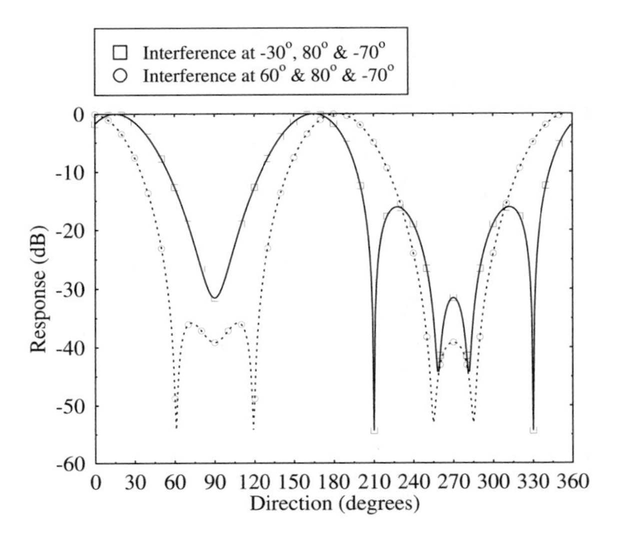

(b) Source located at 15°, interferers located at **-30°**, **80°** and **-70°**, or **60°**, **80°** and **-70°**.

Figure 3.39: Beam patterns of a three element uniformly spaced linear array having an inter-element spacing of $\lambda/2$ in conjunction with one desired source and **three sources of interference**. The **SMI** beamforming algorithm was used with a reference length of 256 bits. The SNR and INRs were **21.0 dB**

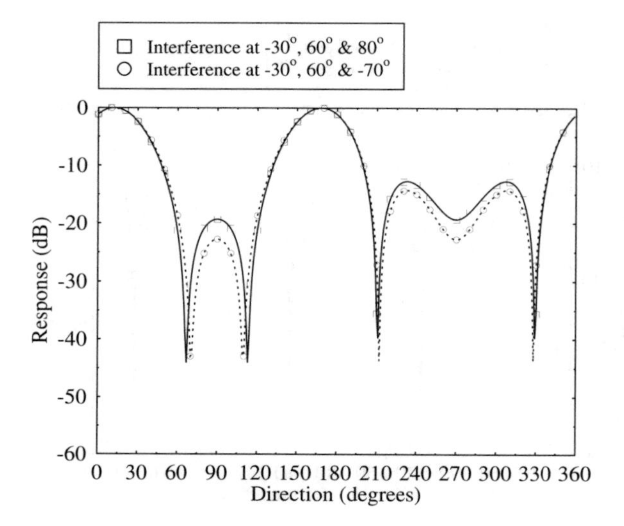

(a) Source located at 15°, interferers located at **-30°**, **60°** and **80°** or **-30°**, **60°** and **-70°**.

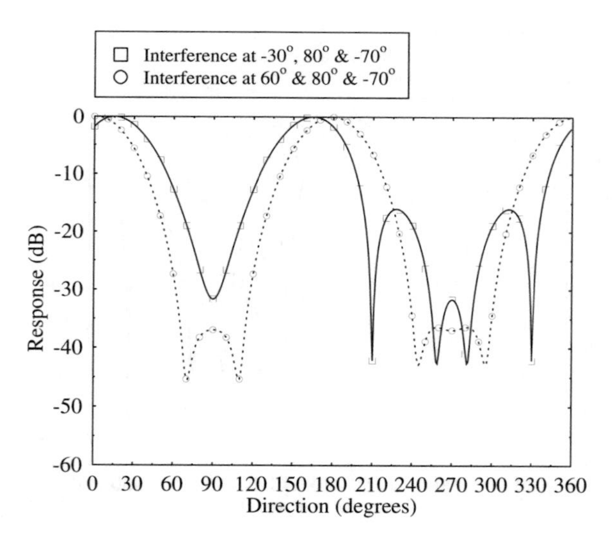

(b) Source located at 15°, interferers located at **-30°**, **80°** and **-70°**, or **60°**, **80°** and **-70°**.

Figure 3.40: Beam patterns of a three element uniformly spaced linear array having an inter-element spacing of $\lambda/2$ in conjunction with one desired source and **three sources of interference**. The **SMI** beamforming algorithm was used with a reference length of 256 bits. The SNR was **21.0 dB** whilst the INRs were **9.0 dB**

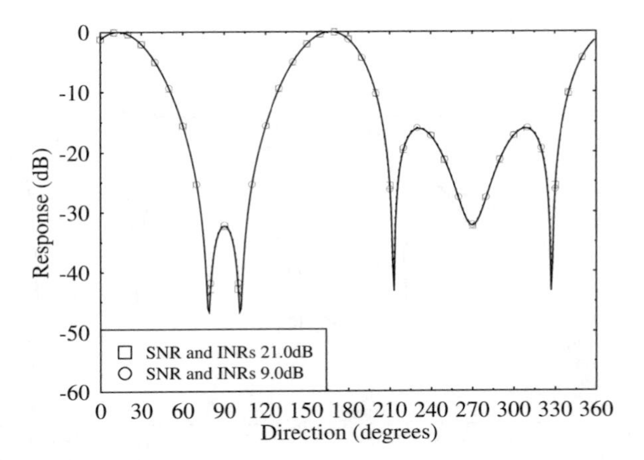

(a) Equal SNR and INR of 21.0 dB.

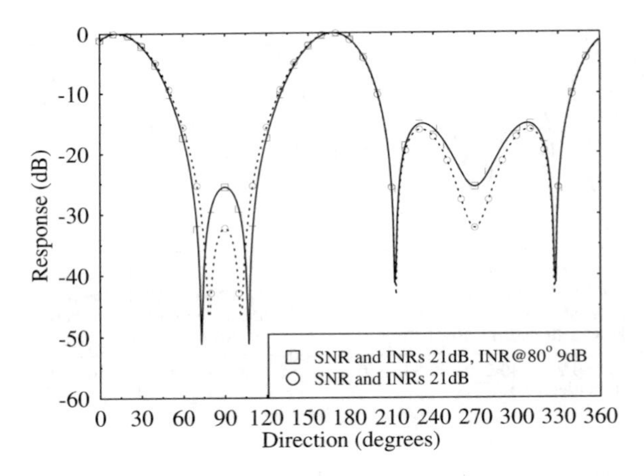

(b) Comparison between all SNRs and INRs of 21.0 dB, and all at 21.0 dB except the interferer at 80° which has an INR of 9.0 dB.

Figure 3.41: Beam patterns of a three element uniformly spaced linear array having an inter-element spacing of $\lambda/2$ in conjunction with one desired source located at 15°, and **four sources of interference** located at **-30°**, **60°**, **80°** and **-70°**. The **SMI** beamforming algorithm was used with a reference length of 256 bits.

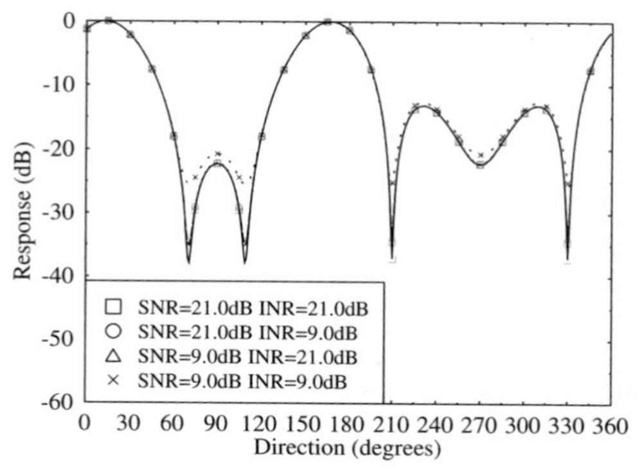

(a) Source located at 15°, interference at −30°.

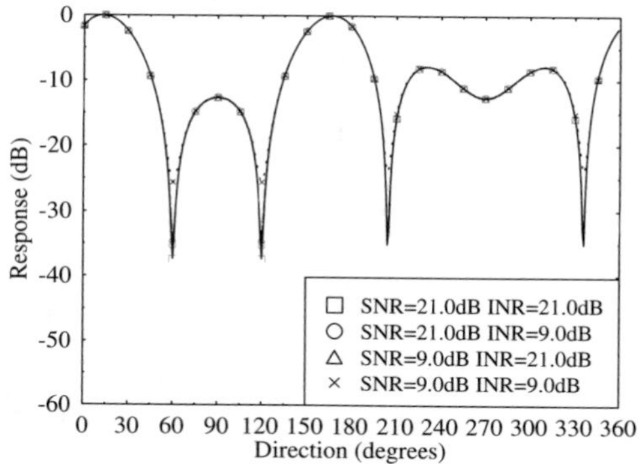

(b) Source located at 15°, interference at 60°.

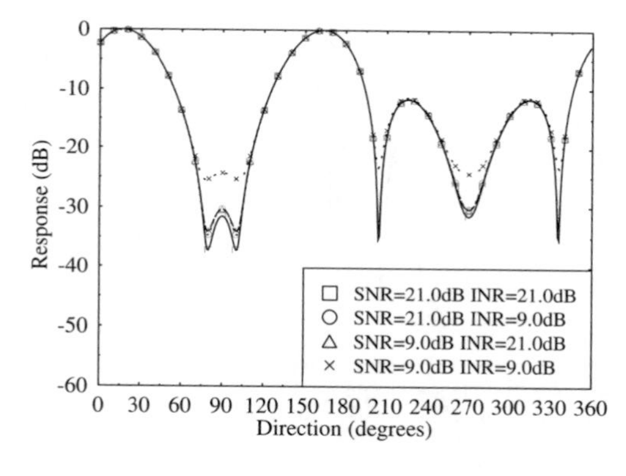

(c) Source located at 15°, interference at 80°.

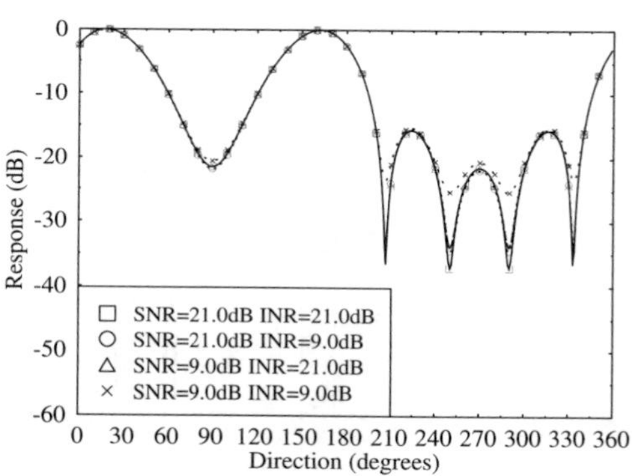

(d) Source located at 15°, interference at −70°.

Figure 3.42: Beam patterns of a three element uniformly spaced linear array with an inter-element spacing of $\lambda/2$ with one desired source and **one source of interference**. The **NLMS** beamforming algorithm was used with a reference length of 256 bits.

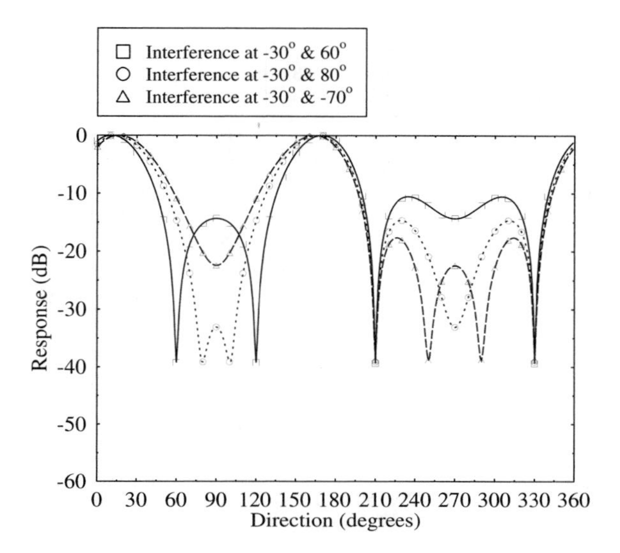

(a) Source located at 15°, interferers located at **-30°** and, **80°** or **-30°** and **-70°**.

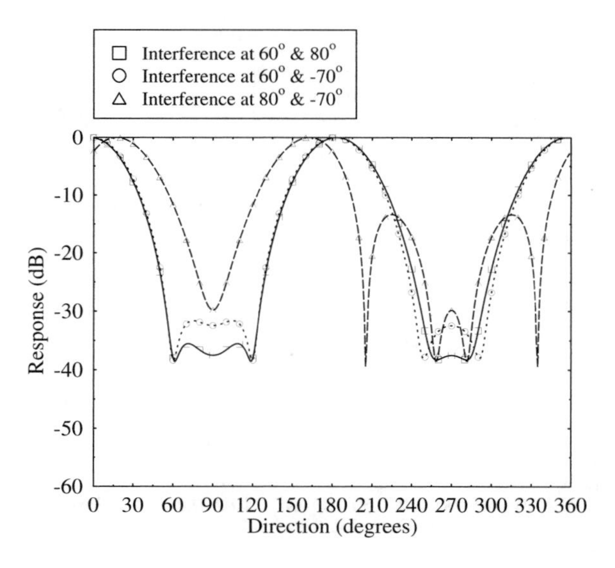

(b) Source located at 15°, interferers located at **60°** and **80°**, or **60°** and **-70°**, or **80°** and **-70°**.

Figure 3.43: Beam patterns of a three element uniformly spaced linear array with an inter-element spacing of $\lambda/2$ with one desired source and **two sources of interference**. The **NLMS** beamforming algorithm was used with a reference length of 256 bits.

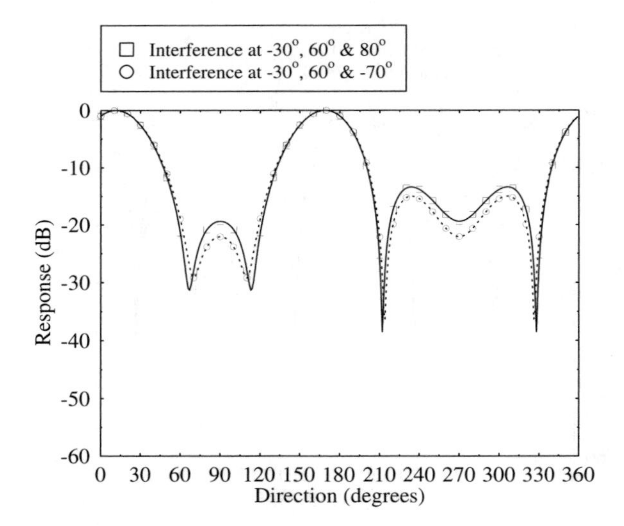

(a) Source located at 15°, interferers located at **-30°**, **60°** and **80°** or **-30°**, **60°** and **-70°**.

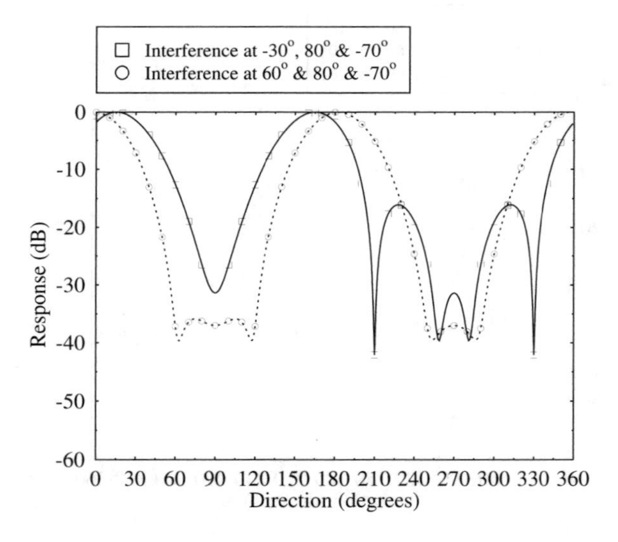

(b) Source located at 15°, interferers located at **-30°**, **80°** and **-70°**, or **60°**, **80°** and **-70°**.

Figure 3.44: Beam patterns of a three element uniformly spaced linear array with an inter-element spacing of $\lambda/2$ with one desired source and **three sources of interference**. The **NLMS** beamforming algorithm was used with a reference length of 256 bits. The SNR and INRs were **21.0 dB**

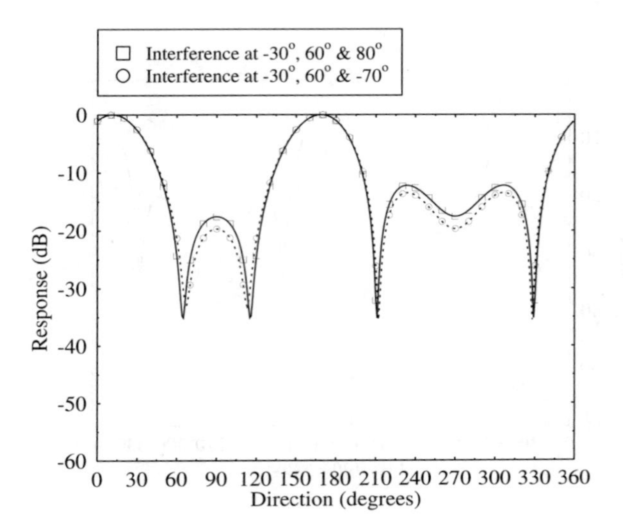

(a) Source located at 15°, interferers located at **-30°**, **60°** and **80°** or **-30°**, **60°** and **-70°**.

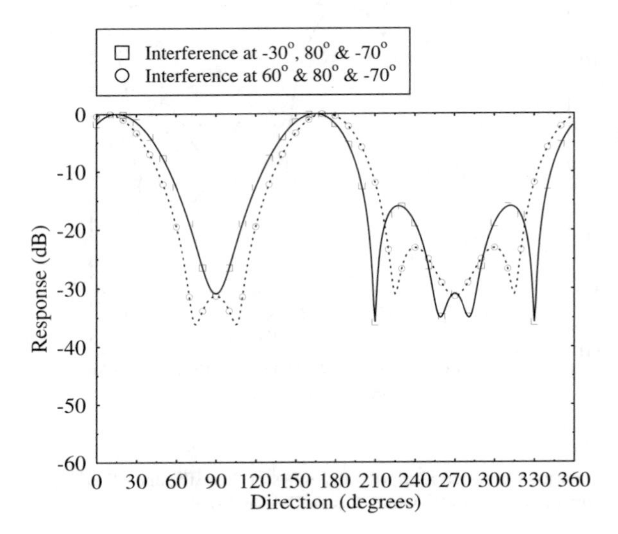

(b) Source located at 15°, interferers located at **-30°**, **80°** and **-70°**, or **60°**, **80°** and **-70°**.

Figure 3.45: Beam patterns of a three element uniformly spaced linear array with an inter-element spacing of $\lambda/2$ with one desired source and **three sources of interference**. The **NLMS** beamforming algorithm was used with a reference length of 256 bits. The SNR was **21.0 dB** whilst the INRs was **9.0 dB**

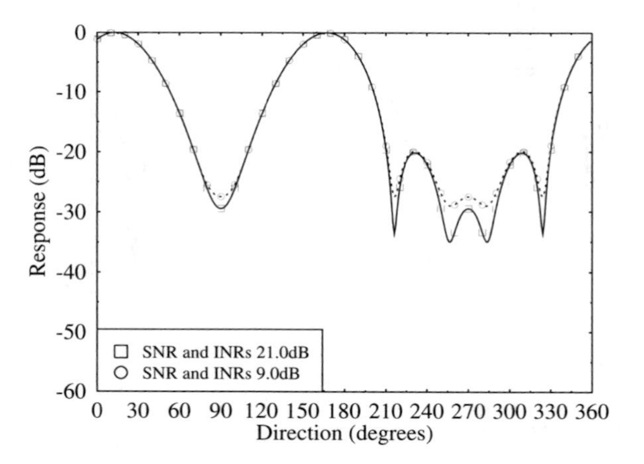

(a) Equal SNR and INR of 21.0 dB.

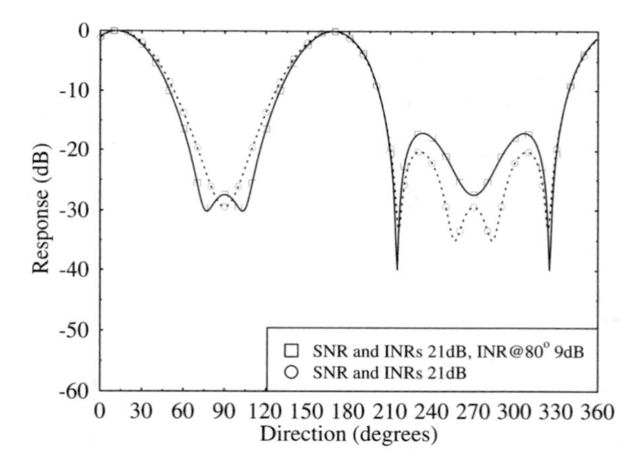

(b) Comparison between all SNRs and INRs of 21.0 dB, and all at 21.0 dB except the interferer at 80° which has an INR of 9.0 dB.

Figure 3.46: Beam patterns of a three element uniformly spaced linear array with an inter-element spacing of $\lambda/2$ with one desired source located at 15°, and **four sources of interference** located at **-30°**, **60°**, **80°** and **-70°**. The **NLMS** beamforming algorithm was used with a reference length of 256 bits.

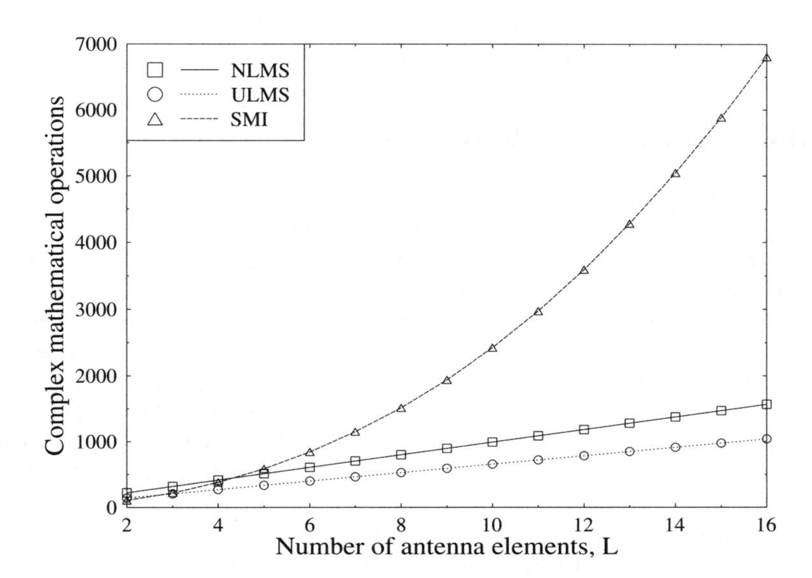

Figure 3.47: The relative complexities of the DMI, ULMS and NLMS beamforming algorithms for a reference signal length of 16 symbols.

element of the matrix, R, N complex multiplications and $N - 1$ complex additions must be performed, where N is the sample size, in bits, used to generate the cross-correlation matrix, R. Due to the Hermitian nature of the matrix, R, it is only necessary to execute these instructions $L(L+1)/2$ times, rather than L^2 times, as would be expected.

Therefore, $NL(L+1)/2$ complex multiplications and $L(L+1)(N-1)/2$ complex additions are required for forming the matrix, leading to a total of $L(L+1)(2N-1)/2$ complex operations. However, upon assuming that a Multiply-and-ACcumulate (MAC) instruction exists in the implementation, this complexity figure reduces to $NL(L+1)$. The Hermitian cross-correlation matrix, R, must then be inverted requiring $L^3/2 + L^2$ complex operations [244], rather than the usual L^3 operations required for a non-Hermitian matrix. In order to calculate the correlation between the reference signal and the array output vector requires a further L complex multiplications and $L - 1$ complex additions, reducing to L complex operations assuming a MAC instruction. Then, from the inverted matrix, R^{-1}, and the correlation vector, $\underline{z}$, the weight vector, $\underline{w}$, may be obtained after L^2 complex operations. Therefore, the total complexity of the SMI beamforming algorithm is $L(L+1)(2N-1)/2+L^3/2+2L^2+2L-1$ complex operations.

The ULMS adaptive beamformer requires only $2L + 1$ complex multiplications and $2L$ complex additions per iteration, rendering it the least complex algorithm. However, the NLMS technique is more practical, since its performance is less dependent upon the input power. The additional complexity associated with this algorithm is the $L + 1$ complex multiplications required for calculating the current value of μ. Therefore, the final complexity of the NLMS algorithm is equivalent to $3L + 2$ complex multiplications and $3L$ complex additions per bit received. Hence the total number of complex operations required by the NLMS beamforming algorithm is $N(3L + 2 + 3L) = N(6L + 2)$.

3.4 Summary and Conclusions

In this chapter we commenced in Section 3.2.2 by considering the possible applications of antenna arrays and their related benefits. A signal model was then described in Section 3.2.3 and a rudimentary example of how beamforming operates was presented. Section 3.3 highlighted the process of adaptive beamforming in conjunction with several different temporal reference techniques detailed, along with the approaches of spatial reference techniques and the associated process of antenna array calibration. The challenges that must be overcome before beamforming for the downlink becomes feasible were also discussed in Section 3.3.5.

In Section 3.3.6 results were presented showing how the SMI, ULMS and NLMS beamforming algorithms of Sections 3.3.2.3, 3.3.2.1 and 3.3.2.2 behaved for a two element adaptive antenna having varying eigenvalue spread and reference signal length. The SMI algorithm was shown to converge very rapidly, irrespective of the eigenvalue spread, and the level of interference rejection was found to be purely dependent upon the interference power, regardless of the desired signal power. However, in Section 3.3.6.2 the convergence characteristics of the ULMS adaptive beamforming algorithm were shown to be heavily dependent upon both the desired signal power and the interfering signal powers. The NLMS algorithm, in contrast, was found to be far superior in this respect, and considering its significantly lower complexity than that of the SMI technique, offered good performance.

The performance of the SMI and NLMS algorithms was then compared in Section 3.3.6.4 for a three element antenna array with one desired source and between one and four sources of interference. The results obtained in Section 3.3.6.4 further evidenced the better performance of the SMI algorithm, but as was shown in Section 3.3.6.5, this was achieved at a significantly higher complexity, when the number of array elements was higher than four. For a low number of elements, the SMI algorithm was found to have a lower complexity than both the ULMS and the NLMS techniques. However, as the number of antenna elements used in the array increased, the complexity of the SMI technique exponentially increased, while that of the LMS routines increased only linearly. Therefore, for about ten array elements the complexity of the SMI algorithm was about twice that of the NLMS technique.

In the next chapter we consider the performance benefits that may be obtained with the advent of adaptive antenna arrays in a cellular radio network.

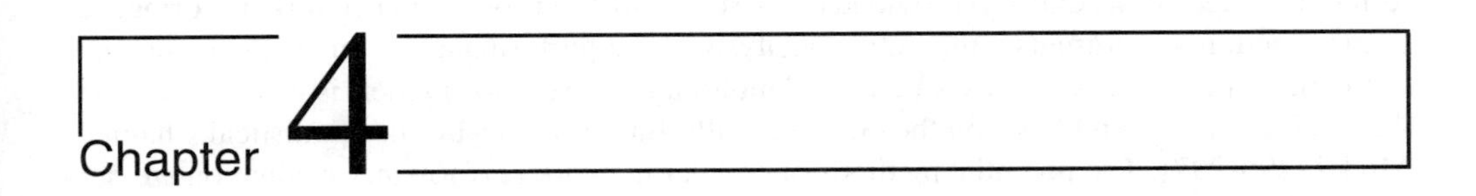

Adaptive Arrays in an FDMA/TDMA Cellular Network

4.1 Introduction

Cellular networks are typically interference limited, with co-channel interference arising from cellular frequency reuse, ultimately limiting the quality and capacity of wireless networks [280, 281]. However, Adaptive Antenna Arrays (AAAs) are capable of exploiting the spatial dimension in order to mitigate this co-channel interference and thus to increase the achievable network capacity [3, 6, 38, 242, 250, 282]. Since an AAA may receive signals with a high gain from one direction, whilst nulling signals arriving from other directions, it is inherently suited to a CCI-limited cellular network. Thus a beam may be formed to communicate with the desired mobile, whilst nulling interfering mobiles [6]. Assuming that each mobile station is uniquely identifiable, it is a relatively simple task to calculate the antenna array's receiver weights, so as to maximise the received SINR. The use of adaptive antenna arrays in a cellular network is an area of intensive research and adaptive antenna array's have been studied widely in the context of both interference rejection and in single-cell situations [1, 15, 18, 261, 267, 268]. More recently, work has been expanded to cover the analysis and performance benefits of using base stations equipped with adaptive antenna arrays across the whole of a cellular network [2, 265, 283].

A further approach to improving the network performance is the employment of Dynamic Channel Allocation (DCA) techniques [284–292], which offer substantially improved call-blocking, packet dropping, and grade-of-service performance in comparison to Fixed Channel Allocation (FCA). A range of so-called distributed DCA algorithms were investigated by Cheng and Chuang [290] where a given physical channel could be invoked anywhere in the network, provided that the associated channel quality was sufficiently high. As compromise schemes, locally optimised distributed DCA algorithms were proposed, for example, by Delli Priscoli *et al.* [293, 294], where the system imposed an exclusion zone for reusing a given physical channel around the locality, where it was already assigned.

In Sections 4.2.1-4.2.3 we briefly consider how an adaptive antenna array may be modelled for employment in a network level simulator, followed by a short overview of a variety of channel allocation schemes in Section 4.3. This section also provides a brief performance summary of the various channel allocation schemes based on our previous work [23, 295], which suggested for the scenarios considered [23, 295] that the Locally Optimised Least Interference Algorithm (LOLIA) provided the best overall compromise in network performance terms. Section 4.4 presents a theoretical analysis of the performance of an adaptive antenna in a cellular network. A summary of several multipath propagation models is given in Section 4.5, with particular emphasis on the Geometrically Based Single-Bounce Statistical Channel Model [296, 297]. The potential methods of cellular network performance evaluation are described in Section 4.3.3.4, as are the parameters of the network simulated in later sections. Simulation results for Fixed Channel Allocation (FCA) and two Dynamic Channel Allocation (DCA) schemes using single element antennas, as well as two- and four-element adaptive antenna arrays for Line-Of-Sight (LOS) scenarios are presented and analysed in Section 4.6.2.1. Furthermore, simulation-specific details of the multipath model are given in Section 4.6.1, with the associated results obtained for the FCA and the LOLIA in the context of two, four and eight element adaptive antenna arrays presented in Section 4.6.2.2. Performance results for a network using power control over a multipath channel in conjunction with two and four element adaptive antenna arrays are provided in Section 4.6.2.3, followed by the description of a network using Adaptive Quadrature Amplitude Modulation (AQAM) in Section 4.6.2.4.

Performance results were also obtained for AQAM and the FCA algorithm as well as the LOLIA, with both two- and four-element adaptive antenna arrays. Results using the 'wraparound' technique, described in Section 4.6.1, which removes the cellular edge effects observed at the simulation area perimeter of a 'desert-island' scenario, are then presented in Sections 4.6.3.1-4.6.3.4. Finally, a performance summary of the investigated networks is given in Section 4.7.

4.2 Modelling Adaptive Antenna Arrays

The interference rejection cability of an antenna array is determined by both the direction of arrival of the interference and the angle of arrival of the desired signal and therefore ultimately by the angular separation between the two. The direction of arrival and angle of arrival may be used interchangably throughout our discussions. The number of interferers and their signal strengths also affects the achievable attenuation of each of the interferers. This section attempts to derive a simple relationship between these factors for low-complexity modelling of an adaptive antenna array.

4.2.1 Algebraic Manipulation with Optimal Beamforming

Given that the steering vector associated with the direction θ_i of the ith source can be described by an L-dimensional complex vector $\underline{s}_i$ as [242],

$$\underline{s}_i = [\exp(j\omega t_1(\theta_i)), \ldots, \exp(j\omega t_L(\theta_i))]^T, \quad (4.1)$$

where L is the number of elements in the antenna array, and t_i is the time delay experienced by a plane wave arriving from the ith source direction, θ_i, and measured from the antenna element at the origin. Then the correlation matrix, R, of the steering vector $\underline{s}_i$, may be expressed as [242]:

$$R = \sum_{i=1}^{M} p_i \underline{s}_i \underline{s}_i^H + \sigma_n^2 I, \tag{4.2}$$

where p_i is the power of the ith source, σ_n^2 is the noise power and I is the identity matrix.

Assuming optimal beamforming under the constraint of a unit response in the wanted user's direction, then the weight vector of the AAA is [242]:

$$\underline{w} = \frac{R^{-1}\underline{s}_0}{\underline{s}_0^H R^{-1} \underline{s}_0}. \tag{4.3}$$

The array factor, $F(\theta)$, in the direction θ may be formulated as [38]:

$$F(\theta) = \sum_{l=1}^{L} w_l e^{-j\omega t_l(\theta)}. \tag{4.4}$$

Therefore, given that the desired signal arrives from the direction θ_0, and an interfering signal arrives from the angle θ_1, the corresponding array responses are $F(\theta_0)$ and $F(\theta_1)$, respectively. Hence, the level of interference rejection, $F(\theta_0) - F(\theta_1)$, when one desired signal and one interfering signal are received at a two-element antenna array, may be calculated using Equation 4.4 to be:

$$F(\theta_0) - F(\theta_1) = \frac{(2p_1 + \sigma_n^2)e^{\frac{j\omega\lambda \sin\theta_0}{2c}} - (p_1 + \sigma_n^2)e^{\frac{j\omega\lambda(2\sin\theta_0 - \sin\theta_1)}{2c}} - p_1 e^{\frac{j\omega\lambda\sin\theta_1}{2c}}}{(2p_1 + 2\sigma_n^2)e^{\frac{j\omega\lambda\sin\theta_0}{2c}} - p_1 e^{\frac{j\omega\lambda\sin\theta_1}{2c}} - p_1 e^{\frac{j\omega\lambda(2\sin\theta_0 - \sin\theta_1)}{2c}}}, \tag{4.5}$$

where the terms interference rejection is defined as the difference between the array response in the direction of the desired signal source and that in the directions of the interfering source.

As can be seen from this equation, there is a non-linear relationship between the two angles of arrival and the achievable interference rejection. Furthermore, the achievable interference rejection is independent of the desired signal's received power, p_0, and it is solely dependent upon the power of the interfering signal, p_1. Expanding this technique to either an antenna array having more elements or to catering for more interfering sources, or to multiple incident beams, led to overly complicated expressions which would be too complex to evaluate in real-time. In order to avoid the associated complexity, the quantities required for interference rejection in a given scenario could be stored in lookup tables. However, the size of the table required to store all of the information would be impractical. For example, for the desired source, one dimension would be required for the angle of arrival and then another one for every interference source. Two further table dimensions would be required to store the angle of arrival and interference power. Therefore, the simple situation involving just one interferer, with a received power dynamic range of 40 dB, would require an array of $180 \times 180 \times 40 = 1,296,000$ elements, at an angular resolution of $1°$, and an interferer power resolution of 1 dB. For two interference sources this figure increases to $180 \times 180 \times 40 \times 180 \times 40 = 0.3312 \times 10^9$ elements, which is clearly excessive.

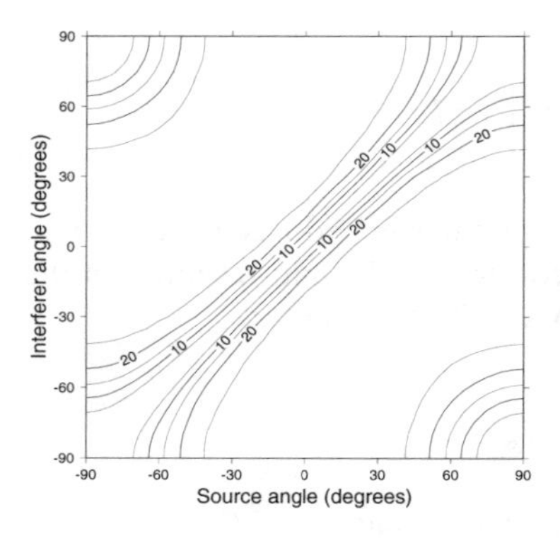

(a) Desired signal SNR = 3.0 dB, Interference SNR = 3.0 dB

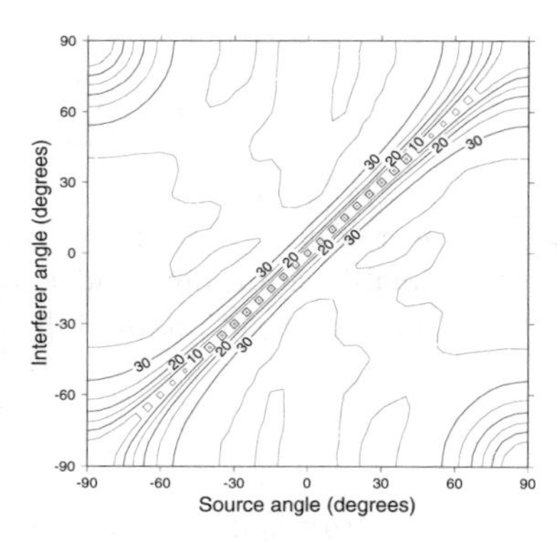

(b) Desired signal SNR = 3.0 dB, Interference SNR = 12.0 dB

Figure 4.1: Contour plots of interference rejection achieved using a four element antenna array with an inter-element spacing of $\lambda/2$ using SMI beamforming with a reference signal length of 16 bits. The angles of arrival of the signals from the desired source and the interfering source were swept over the range, -90 degrees to +90 degrees.

4.2.2 Using Probability Density Functions

Due to the inherent complexities of performing large-scale network simulations, whilst invoking the required beamforming operations, we conducted an investigation into the probability distribution of the interference rejection ratio achieved by an adaptive antenna array. For our initial studies a two element antenna array with the elements located $\lambda/2$ apart was considered, with one desired source and one interfering source. Therefore, the average interference rejection achieved in decibels, for a given source-direction and power as well as interferer-direction and power could be determined. Unfortunately, as it can be seen from Figure 4.1(a), the achievable interference rejection was not based upon a linear relationship between the two angles of arrival. Furthermore, Figure 4.1(b) illustrates that the interference rejection achieved was also related to the power, or the Signal-to-Noise Ratio (SNR), of the undesired interference source, which was 3 dB or 12 dB. As it was found in Section 4.2.1, attempting to construct a model or probability density function to cater for these parameters was not easily achievable. Rather than attempting to find the Probability Density Function (PDF) relating the two angles of arrival and interference power to the interference rejection achieved, a brief study was initiated for determining the PDF of the interference rejection achieved with respect to the angular separation between the desired signal and interfering signal. Figure 4.2 shows the probability density function of interference rejection achieved for one interference source and one desired source versus their angular separation. As this figure shows, the distribution of the interference rejection varies significantly, as the separation between the sources changes. As a consequence of the PDF's dependence on the angular separation encountered, modelling the achievable interference rejection expressed in decibels

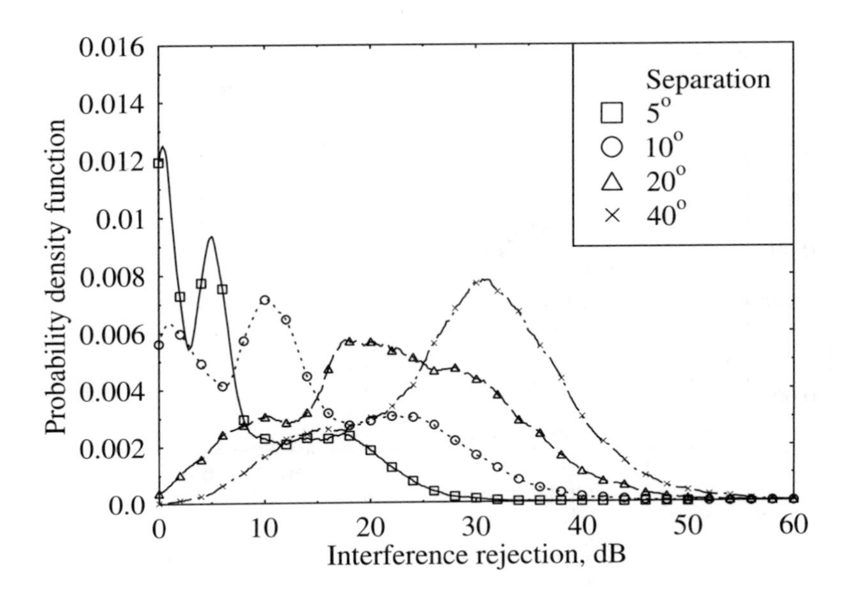

Figure 4.2: The PDF of the interference rejection (dB) achieved for various angular separations of the desired signal and the interfering signal. The angles of arrival of both signals were varied over the range of -90 to +90 degrees and were of equal power. The antenna array consisted of two elements separated by $\lambda/2$.

is an arduous task. Due to the complex nature of the PDF illustrated in Figure 4.2, an analysis of a smaller range of angles of arrival was conducted, in order to construct a piecewise valid model. The results are displayed in Figures 4.3(a) and 4.3(b) for angle of arrival spreads of $\pm 30°$ and $\pm 10°$, respectively. While these PDFs appear to be considerably simpler than that in Figure 4.2, it was not possible to match the PDFs to any commonly known distributions. Additionally, no information was available with regard to the correlation between successive interference rejection values. For these reasons, and due to the difficulties associated with adding multipath, it was decided to cease work on constructing a suitable interference rejection model and instead to implement an actual SMI beamformer within the simulation program as described in the following section.

.8

4.2.3 Sample Matrix Inversion Beamforming

The process of defining a suitable model of an adaptive antenna array was becoming increasingly complex, resulting in the decision to implement an SMI beamformer in the simulation software. The SMI beamforming algorithm of Section 3.3.2.3 was chosen due to its inde-

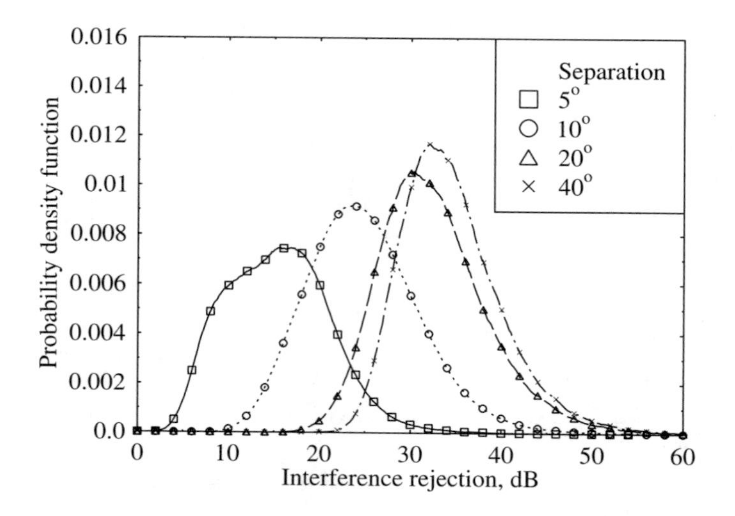

(a) Angular spread=$\pm30°$.

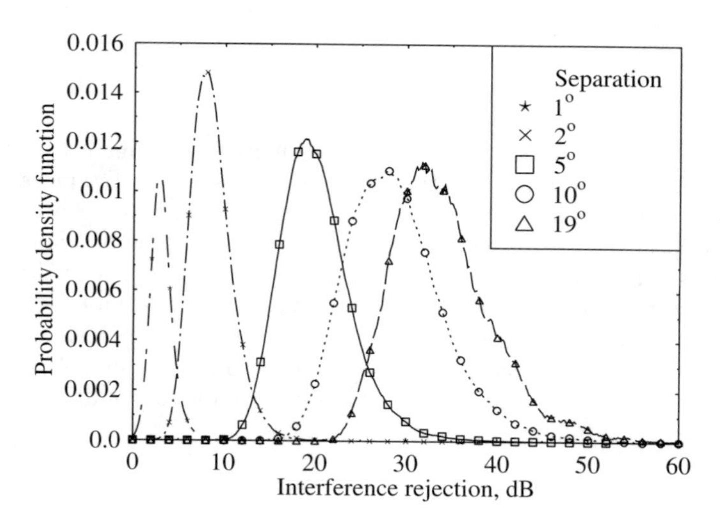

(b) Angular spread=$\pm10°$.

Figure 4.3: The PDF of the interference rejection achieved for the desired signal and the interfering signal angular separations of 5, 10 and 20 degrees. The desired signal and the interfering signal were of equal power. The antenna array consisted of two elements separated by $\lambda/2$.

pendence from the received signal strengths, as well as due to its fast convergence with the aid of few data samples and for the sake of its good overall performance in terms of its interference rejection capability. The reference signal was chosen to be eight bits in length as a compromise between the quality of the sample correlation matrix, R, and the computational complexity required. Since a cellular network is an interference limited system, the addition of noise to the received signal vector was neglected. A result of this was that occasionally the correlation matrix, R, was non-invertible, which was remedied by diagonally augmenting the matrix with a positive constant as it was suggested in [15, 271, 272]. The addition of multipaths simply required the direction of arrival, and the strength of the multipath rays at the antenna array to be determined before adding these received signal vectors to the total received signal vector of the antenna array. In both the line-of-sight and the multipath scenarios, the transmit/receive channel was assumed to be frequency invariant, thus allowing the same antenna pattern to be used in both the uplink and the downlink.

4.3 Channel Allocation Techniques

P.J. Cherriman, L. Hanzo[1]

Channel assignment is the process of allocating a finite number of channels to the various base stations and mobile phones in the cellular network. In a system using fixed channel assignment, the channels are assigned to different cells during the network planning stage, and the assignment is rarely altered to reflect changes in traffic levels. A channel is assigned to a mobile at the commencement of the call and the mobile communicates with its base station on this channel until either the call terminates or the mobile leaves the current cell. Dynamic channel allocation, however, assigns a channel that best meets the channel selection criteria, which may be the channel experiencing the minimum interference level, depending upon the cost function used.

With the growth in the number of subscribers to mobile telecommunications systems worldwide and the expected introduction of multimedia services in handheld wireless terminals, a tremendous demand for bandwidth has arisen. Since bandwidth is scarce and becoming increasingly expensive, it must be utilized in an efficient manner.

The main limiting factor in radio spectrum reuse is co-channel interference. In reduced cell-size micro/picocellular architectures, the frequency reuse distance is reduced, thereby increasing the capacity and area spectral efficiency of the system. However, as the channel reuse distance is reduced, the co-channel interference increases. Co-channel interference caused by frequency reuse is the most severe limiting factor of the overall system capacity of mobile radio systems. The most important technique for reducing co-channel interference is power control, an issue, which will be discussed in detail in the context of adaptive modulation during our further discourse. Interference cancellation techniques [298] or adaptive antennas [299–301] can also be used to reduce co-channel interference. However, a simpler and more effective technique used in current systems is employing sectorized antennas [302].

Although handovers are necessary in mobile radio systems, they often cause several problems, and they constitute the major cause of calls being forcibly terminated. As the cell size is decreased, the average sojourn time or cell-crossing time for a user is reduced. This results

[1]This section is based on [151]

in an increased number of handovers, requiring more rapid handover completion. In practice a seamless handover is not always possible except when soft-handovers [303] are used in CDMA-based systems. Rapid and numerous handovers require a fast backbone network between the base stations and the mobile switching centers, or they necessitate an increased number of mobile switching centers. Clearly, the handover process is crucial with regard to the perceived Grade of Service (GOS), and a wide range of different complexity techniques have been proposed, for example, by Tekinay and Jabbari [304] and Pollini [305] for the forthcoming future systems. The related issue of time-slot reassignment was investigated by Bernhardt [306].

4.3.1 Overview of Channel Allocation

The purpose of channel allocation algorithms is to exploit the variability of the radio channel propagation characteristics in order to allow increased efficiency radio spectrum utilisation, while maintaining required signal quality. The most commonly used signal quality measure is the signal-to-interference ratio (SIR), also known as the carrier-to-interference ratio (CIR). The signal quality measure that we have used previously was the signal-to-interference+noise ratio (SINR). The SINR is approximately equal to the signal-to-noise ratio (SNR) in a noise-limited environment and approximately equal to the SIR in an interference-limited environment.

The radio spectrum is divided into sets of noninterfering physical radio channels, which can be achieved using orthogonal time or frequency slots, orthogonal user signature codes, and so on. The channel allocation algorithm attempts to assign these physical channels to mobiles requesting a channel, such that the required signal quality constraints are met. There are three main techniques for dividing the radio spectrum into radio channels. The first is frequency division (FD), in which the radio spectrum is divided into several nonoverlapping frequency bands. However, in practice the spectral spillage from one frequency band to another causes adjacent channel interference, which can be reduced by introducing frequency guard bands. However, these guard bands waste radio spectrum, and hence there is a compromise between adjacent channel interference and frequency band-packing efficiency. Tighter filtering can help reduce adjacent channel interference, allowing the guard bands to be reduced.

The second technique is time division (TD), in which the radio spectrum is divided into disjunct timeperiods, which are usually termed time-slots. However, using straight-forward rectangular windowing of the time-domain signal corresponds to convolving the signal spectrum with a frequency-domain sinc-function, resulting in Gibbs-oscillation. Hence, in practical systems a smooth time-domain ramp-up and ramp-down function associated with a time-domain guard period is employed. Therefore, there is a trade-off between complex synchronisation, time-domain guard periods, and adjacent channel interference.

The third technique for dividing the radio spectrum into channels is code division (CD). Code division multiple access (CDMA) [39–41, 307] has been used in military applications, in the IS-95 mobile radio system [308], and in the recently standardized Universal Mobile Telecommunications System (UMTS) [307, 309]. In code division, the physical channels are created by encoding different users with different user signature sequences.

In most systems a combination of these techniques is used. For example, the Pan-European GSM system [28] uses frequency division duplexing for up- and down-link trans-

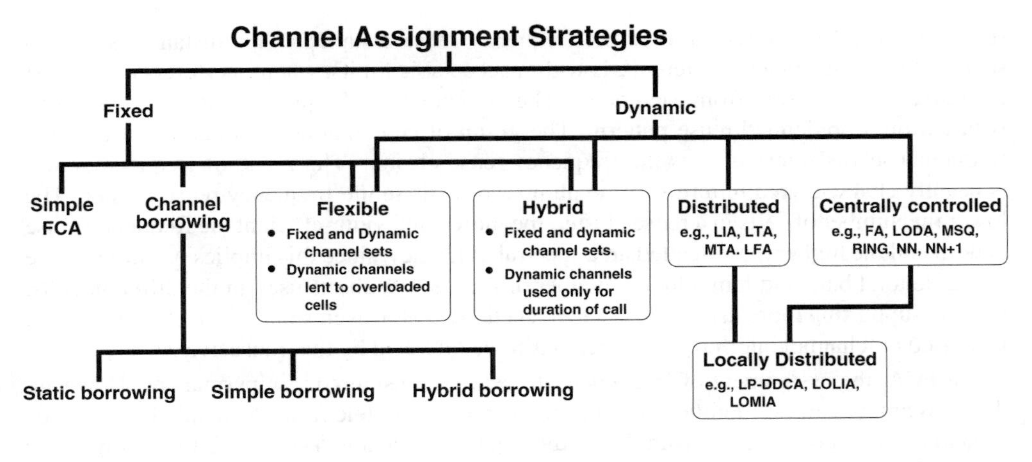

Simple FCA	Page 201	Static borrowing	Page 203
Simple borrowing	Page 203	Hybrid borrowing	Page 204
Flexible	Page 204	Hybrid	Page 209
Centrally controlled DCA	Page 206	Distributed DCA	Page 207
Locally distributed DCA	Page 208		

Figure 4.4: Family tree of channel allocation algorithms.

missions, while accommodating eight TDMA users per carrier. In this chapter, the term "channel" typically implies a physical channel, constituted by a time-slot of a given carrier frequency.

A wide variety of channel allocation algorithms have been suggested for mobile radio systems. The majority of these techniques can be classified into one of three main classes: fixed channel allocation (FCA), dynamic channel allocation (DCA), and hybrid channel allocation (HCA). Hybrid channel allocation is constituted by a combination of fixed and dynamic channel allocation, which is designed to amalgamate the best features of both, in order to achieve better performance or efficiency than DCA or FCA can provide. Several channel allocation schemes and the associated trade-offs in terms of performance and complexity are discussed in detail in the excellent overview papers of Katzela and Naghshineh [310] and those by Jabbari and Tekinay *et al.* [311, 312]. Figure 4.4 portrays the family tree for the main types of channel allocation algorithms, where the acronyms are introduced during our further discourse. Zander [313] investigated the requirements and limitations of radio resource management in general for future wireless networks. Everitt [314] compared various fixed and dynamic channel assignment techniques and investigated the effect of handovers in the context of CDMA-based systems.

4.3.1.1 Fixed Channel Allocation

In fixed channel allocation (FCA), the available radio spectrum is divided into sets of frequencies. One or more of these sets is then assigned to each base station on a semipermanent basis. The minimum distance between two base stations, they have been assigned the same set of frequencies is referred to as the frequency reuse distance. This distance is chosen such that the co-channel interference is within acceptable limits, when interferers are at least the reuse distance away from each other. The assignment of frequency sets to base stations is based on a predefined reuse pattern. The group of cells that contain one of each of the frequency sets is referred to as the frequency reuse cluster. The grade of frequency reuse is usually characterized in terms of the number of cells in the frequency reuse cluster. The lower the number of cells in a reuse cluster, the more bandwidth-efficient the frequency reuse pattern and the higher the so-called area spectral efficiency, since this implies partitioning the available total bandwidth in a lower number of frequency subsets used in the different cells, thereby supporting more users across a given cell area. However, small reuse clusters exhibit increased co-channel interference, which has to be tolerated by the transceiver.

In FCA, the assignment of frequencies to cells is considered semipermanent. However, the assignment can be modified in order to accommodate teletraffic demand changes. Although FCA schemes are very simple, modifying them to adapt to changing traffic conditions or user distributions can be problematic. Hence, FCA schemes have to be designed carefully, in order to remain adaptable and scalable, as the number of mobile subscribers increases. In this context, adaptability implies the ability to rearrange the network to provide increased capacity in a particular area on a long- or short-term basis, where scalability refers to the ability of easily increasing capacity across the whole network via tighter frequency reuse. For example, Dahlin *et al.* [315] suggested a reuse pattern structure for the GSM system that can be scaled to meet increased capacity requirements, as the number of subscribers increases. This is discussed in more detail in the overview paper by Madfors *et al.* [316]. Each measure invoked, in order to further increase the network capacity, increases the system's complexity and hence becomes expensive. Furthermore, such systems cannot be easily modified to provide increased capacity in the specific area of a traffic hot-spot on a short-term basis.

A commonly invoked reuse cluster/pattern is the seven-cell reuse cluster, providing coverage over regular hexagonal shaped cells, which is shown in Figure 4.5. Each cell in the seven-cell reuse cluster has six first-tier co-channel interfering cells at a distance D, the reuse distance. By exploiting the simple hexagonal geometry seen in Figure 4.5 it can be shown that for the seven-cell cluster the reuse distance, D, is 4.58 times the cell radius r [151]. This reuse pattern supports the same number of channels at each cell site, and hence the same system capacity. Therefore, the teletraffic capacity is distributed uniformly across all the cells. Since traffic distributions usually are not uniform in practice, such a system can lead to inefficiencies. For example, under nonuniform traffic loading, some cells may have no spare capacity; hence, new calls in these cells are blocked. However, nearby cells may have spare capacity.

Several studies have suggested techniques to find the optimal reuse pattern for particular traffic and users distributions, as exemplified by the work of Safak [317], on optimal frequency reuse with interference. While such contributions are useful, a practical system would need to modify the whole network configuration every time the traffic or user distributions changed significantly. Therefore, suboptimal but adaptable and scalable solutions are more desirable for practical implementations. When the traffic distribution changes, an alternative technique to modifying the reuse pattern is referred to as channel borrowing, which is the subject of the next section.

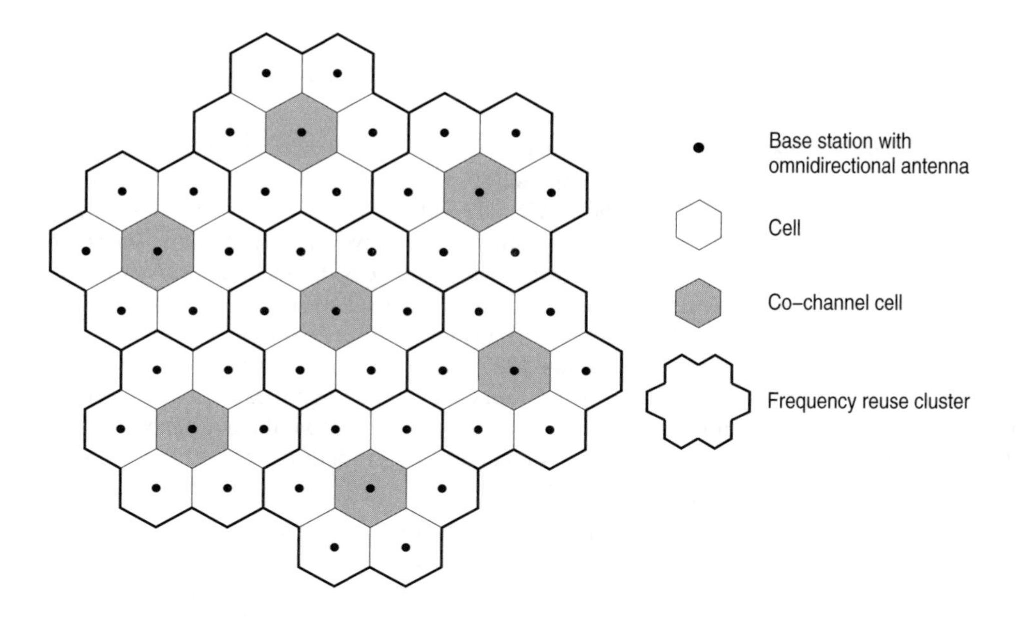

Figure 4.5: A commonly employed frequency reuse pattern for fixed channel assignment (FCA) algorithms. The frequency spectrum divided in seven frequency sets, one set assigned to each cell, yielding a seven-cell reuse cluster. Omnidirectional antennae were used, and the shaded cells represent cells assigned the same frequency set.

4.3.1.1.1 Channel Borrowing In channel-borrowing schemes, a cell that has a call setup request but no available channels (which is termed an acceptor cell), can borrow free channels from neighbouring cells referred to as donor cells in order to accommodate new calls, which would otherwise have been blocked. A channel can be borrowed only if its use will not interfere with existing ongoing calls. When a channel is borrowed, several cells are then prohibited from using the borrowed channel because it would cause interference. The process of prohibiting the use of borrowed channels is referred to as channel locking [318]. The various channel-borrowing algorithms differ in the way the free channel is chosen from a donor cell to be borrowed by an acceptor cell.

There are three main types of channel-borrowing algorithms: static, simple, and hybrid borrowing; a good overview of these algorithms can be found in [310–312]. Static borrowing could be described as a fixed channel re-allocation strategy rather than channel borrowing. In static borrowing, channels are reassigned from lightly loaded cells to heavily loaded cells, which are at distances in excess of the reuse distance. This reassignment is semipermanent and can be done based on measured or predicted changes in traffic. The other two types of channel borrowing (simple and hybrid) are different from static borrowing in that borrowed channels are returned when the call using the channels ends or is handed off to another base station. Therefore, the simple and hybrid channel borrowing schemes use short-term borrowing in order to cope with traffic excesses.

Simple channel- borrowing schemes allow any of the channels in a donor cell to be lent to an acceptor cell. Hybrid channel borrowing schemes split the channels assigned to each cell into two subsets. One subset of channels cannot be lent to other cells; hence, these are referred to as standard or local channels. The other subset can be lent to other cells, and so they are termed nonstandard or borrowable channels.

Simple borrowing [287,311,319] can reduce new call blocking, but it can cause increased interference in other cells; it can also prevent handovers of future calls in these cells. Experiments have shown that simple channel-borrowing algorithms perform better than static fixed channel allocation under light- and moderate traffic loads. However, at high traffic loads the borrowing of channels leads to channel locking, which reduces the channel utilisation and therefore results in an increase in new call blocking and in failed handovers. The various simple channel-borrowing algorithms differ in terms of flexibility, complexity and their reduction of channel locking. Some algorithms [287, 319] pick the channel to borrow, while taking into account the associated "cost" in terms of channel locking for each candidate channel. Other algorithms [319] invoke channel reassignment in order to reduce channel locking. The innovative technique used by Jiang and Rappaport [318] to reduce channel locking is to limit the transmission power of borrowed channels.

Hybrid channel borrowing [310, 311] is a hybrid of simple channel borrowing and static fixed channel allocation. By dividing the channels at each base station into two subsets, and only allowing channels of one of the subsets to be borrowed, the chance of channel locking or failed handovers can be mitigated under high traffic loads. A range of algorithms is discussed in the literature, each having different objectives in terms of improving performance in a particular area of operation. Some algorithms [320] have the ratio of channels in each subset assigned a priori, while others dynamically adapt the size of the subsets based on traffic measurements or predictions [321]. The algorithm may also check whether the candidate borrowed channel is free in the co-channel cells [322]. A common technique [319, 323] is to reassign calls using a borrowed channel to another borrowed channel in order to reduce channel locking. A better policy is to reassign a call currently using a borrowed channel to a local channel, thereby returning the borrowed channel to the donor cell. Another procedure [320, 322] to reduce channel locking is to estimate the direction of movement of the mobile in an attempt to reduce future channel locking and interference. A simple technique [324] is to subdivide cells into sectors and only allow borrowed channels to be used in particular sectors of the acceptor cell, thereby reducing channel locking.

4.3.1.1.2 Flexible Channel Allocation Flexible channel allocation schemes [310, 311, 325] are similar to hybrid channel allocation schemes (which are described in Section 4.3.1.3) in that they divide the available channels into fixed and dynamic allocation subsets. However, flexible channel allocation is similar to a fixed channel allocation strategy, such as that used in static channel borrowing. In flexible channel allocation, the fixed channel set is assigned to cells in the same way as in fixed and hybrid channel allocation. The dynamic or flexible channels can be assigned to cells depending on traffic measurements or predictions. The difference between so-called hybrid and flexible channel allocation schemes is that in hybrid channel allocation the dynamic channels are assigned to cells only for the duration of the call. In flexible channel allocation the dynamic channels are assigned to cells, when the blocking probability in these cells becomes intolerable. Flexible channel allocation requires much more centralized control than hybrid channel allocation.

4.3.1.2 Dynamic Channel Allocation

Although fixed channel allocation schemes are common in most existing cellular radio systems, the cost of increasing their teletraffic capacity can become high. In theory, the use of dynamic channel allocation allows the employment of all carrier frequencies in every cell, thereby ensuring much higher capacity, provided the transceiver-specific interference constraints can be met. Therefore, it is feasible to design a mobile radio system, which configures itself to meet the required capacity demands as and when they arise. However, in practice there are many complications, which make this simplistic view hard to implement in practice. Dynamic channel allocation is used, for example, in the Digital European Cordless Telephone (DECT) standard [257, 258, 326–328]. Law and Lopes [329] used the DECT system to compare the performance of two distributed DCA algorithms. However, DECT is a low-capacity system, where the time-slot utilisation is expected to be comparatively low. For low slot utilisation DCA is ideally suited. Dynamic channel allocation becomes more difficult to use in large-cell systems, which have higher channel utilisation. Salgado-Galicia *et al.* [330] discussed the practical problems that may be encountered in designing a DCA-based mobile radio system.

Even though much research has been carried out into channel allocation algorithms, particularly dynamic channel allocation, many unknowns remain. For example, the trade-offs and range of achievable capacity gains are not clearly understood. Furthermore, it is not known how to combine even two simple algorithms in order to produce a hybrid that has the best features of both. One reason that the issues of dynamic channel allocation are not well understood is the computational complexity encountered in investigating such algorithms. In addition, the algorithms have to be compared to others in a variety of scenarios. Furthermore, changing one algorithmic parameter in order to improve the performance in one respect usually has some effect on another aspect of the algorithm's performance, due to the parameters highly interrelated nature. This is particularly true, since experience showed that some handover algorithms are better suited for employment in certain dynamic channel allocation algorithms [304]. Therefore the various channel allocation algorithms have to be compared in conjunction with a variety of handover algorithms in order to ensure that the performance is not degraded significantly by a partially incompatible handover algorithm. The large number of parameters and the associated high computational complexity of implementing channel allocation algorithms complicate study of the trade-offs of the various algorithms.

Again, in dynamic channel allocation, typically all channels can be used at any base station as long as they satisfy the associated quality requirements. Channels are then allocated from this pool as and when they are required. This solution provides maximum flexibility and adaptability at the cost of higher system complexity. The various dynamic channel allocation algorithms have to balance allocating new channels to users against the potential co-channel interference they could inflict upon users already in the system. Dynamic channel allocation is better suited to microcellular systems [331] because it can handle the more nonuniform traffic distributions, the increased handover requests, and the more variable co-channel interference better than fixed channel allocation due to its higher flexibility. The physical implementation of DCA is more complex than that of FCA. However, with DCA the complex and labor-intensive task of frequency planning is no longer required.

The majority of DCA algorithms choose the channel to be used based on received signal quality measurements. This information is then used to decide which channel to allocate or whether to allocate a channel at all. It is sometimes better not to allocate a channel if it is likely to inflict severe interference on another user, forcibly terminating existing calls or preventing the setup of other new calls. Ideally, the channel quality measurements should be made at both the mobile and base station. If measurements are made only at the mobile or only at the base station, the channel allocation is partially blind [288]. Channel allocation decisions that are based on blind channel measurements can in some circumstances cause severe interference, leading to the possible termination of the new call as well as curtailing another user's call, who is using the same channel. If measurements are made at both the mobile and the base station, then the measurements need to be compared, requiring additional signaling, which increases the call setup time. The call setup time is longer in DCA algorithms than in FCA due to the time required to make measurements and to compare them. This can be a problem, when, for example, a handover is urgently required.

Probably the simplest dynamic channel allocation algorithm is to allocate the least interfered channel available to users requesting a channel. By measuring the received power within unused channels, effectively the noise plus interference on that channel can be measured. By allocating the least interfered channel, the new channel is not likely to encounter interference, and, due to semireciprocity, it is not likely to cause too much interference to channels already allocated. This works well for lightly loaded systems. However, this algorithm's performance is seriously impaired in high-load scenarios, where FCA would work better. However, the above is a very simple dynamic channel allocation algorithm. In Sections 4.3.4 and 4.4 we will demonstrate that it is possible to achieve a better performance and efficiency than that of FCA even at high traffic loads, when using certain channel allocation algorithms. For these reasons, some channel allocation algorithms use a combination of FCA and DCA to achieve better performance than simple DCA, and better reuse efficiency than FCA. These algorithms are classified as hybrid channel allocation (HCA) algorithms.

The difference between the various dynamic channel allocation algorithms is, essentially, how the allocated channel is chosen. All the algorithms assign a so-called cost to allocating each of the possible candidate channels, and the one with the lowest cost is allocated. The difference between the algorithms is how the "cost" is calculated using the cost function. The cost function can be calculated on the basis of one or more of the following aspects: future call blocking probability; usage frequency of the channel; distance to where the channel is already being used, that is, the actual reuse distance; channel occupancy distribution; radio signal quality measurements; and so on. Some algorithms may give better performance than others, but only in certain conditions. Most DCA algorithms' objectives can be classified into two types, where most of them attempt to reduce interference, while others try to maximizee channel utilisation in order to achieve spectral compactness.

There are three main types of DCA algorithms, namely:

- Centrally controlled algorithms
- Distributed algorithms
- Locally distributed algorithms (hybrid)

4.3.1.2.1 Centrally Controlled DCA Algorithms Centrally controlled DCA algorithms are also often referred to as centrally located or centralized DCA algorithms. These algorithms use interference measurements that are made by the mobiles and base stations that are then passed to a central controller, which in most cases would be a mobile switching center. The algorithm that determines the channel allocation is located at the central controller, and it decides on the allocation of channels based on the interference measurements provided by all the base stations and mobiles under its control. These algorithms provide very good performance even at high traffic loads. However, they are complex to implement and require a fast backbone network between the base stations and the central controller. The central controller can become a "bottleneck" and increase the call setup time, which may be critical for "emergency" handovers.

Centralized algorithms [320,322,332–334] have been researched actively for over twenty years. One of the simplest is referred to as the First Available (FA) [332, 335] algorithm, which allocates the first channel found that is not reused within a given preset reuse distance. The Locally Optimized Dynamic Assignment (LODA) [320, 322] algorithm bases its allocation decisions on the future blocking probability in the vicinity of the cell. Some algorithms exploit the amount of channel usage to make allocation decisions. The RING algorithm [310,334], for example, allocates the most often used channel within the cells, which are approximately at the reuse distance, and the terminology RING is justified by the fact that these cells effectively form a ring. There are also several algorithms, which attempt to optimize the reuse distance constraint. The Mean Square (MSQ) algorithm [335] attempts to minimize the mean square distance between cells using the same channel while maintaining the required signal quality. The Nearest Neighbour (NN) and Nearest Neighbour plus One (NN+1) algorithms [332, 335] pick a channel used by the nearest cell, which is at least at a protection distance amounting to the reuse distance (or reuse distance plus one cell radius for NN+1). Other algorithms [334] use channel reassignments to maintain the reuse distance constraint. Recall again that these algorithms were summarized in Figure 4.4.

4.3.1.2.2 Distributed DCA Algorithms In contrast to centrally controlled algorithms, distributed algorithms are the least complex DCA techniques, in which the same algorithm is used by each mobile or base station in order to determine the best channel for setting up a call. Each mobile and/or base station makes channel allocation decisions independently using the same algorithm — hence the name distributed algorithms. The algorithmic decisions are usually based on the interference measurements made by the mobile or the base station. These algorithms are easy to implement, and they perform well for low-slot occupancy systems. However, in high-load systems their performance is degraded. Distributed algorithms require less signaling than centralized algorithms. However, the allocation is generally suboptimal owing to their locally based decisions. One real advantage of distributed algorithms is that base stations can easily be added, moved, or removed because the system automatically reorganizes and reconfigures itself. However, the cost of this flexibility is that the local decision making generally leads to a suboptimal channel allocation solution and to a higher probability of interference in neighbouring cells. Furthermore, generally distributed algorithms are based on signal strength measurements and estimates of interference. However, these interference estimates can sometimes be poor, which can lead to bad channel allocation decisions. When a new allocation is made, the co-channel interference it inflicts may lead to an ongoing call to experience low-service quality, often termed a service interruption. If a service interruption leads to the ongoing call being terminated prematurely, this is referred to deadlock [310]. Successive service interruptions are termed as instability. A further problem with distributed algorithms is that the same channel can be allocated at the same time to two or more different users in adjacent cells. However, when the mobiles attempt to use the channel, they may find the quality unacceptably low. Therefore, distributed algorithms have to be able to check the quality of an allocation, before it is made permanent, which increases the call setup time further.

Chuang *et al.* [290] investigated the performance of several distributed DCA algorithms, arguing that under certain conditions these techniques can converge to a local minimum of the total interference averaged over the network. Grandhi *et al.* [336] and Chuang *et al.* [289] also evaluated the performance of combining dynamic channel allocation with transmission power control.

Examples of distributed algorithms are the Sequential Channel Search (SCS) and the least interference algorithm (LIA). The SCS algorithm [337] searches the available channels in a predetermined order, picking the first channel found, which meets the interference constraints. The LIA algorithm, alluded to earlier, picks the channel with the lowest measured interference that is available. One of the most complex distributed algorithms is the Channel segregation technique [338], which is a fully distributed, autonomous, self-organising assignment scheme. Each cell maintains a measure of the relative frequency of channel usage for each channel. This probability-based measure is modified every time an attempt to access a specific channel is made. The channel assigned to the new call is the one with the highest probability of being or having been idle. The algorithm has been shown to reduce blocking and adapt to traffic changes. Although the channel allocation may rapidly converge to a near-optimal solution, it may take a long time to reach a globally optimal solution. As before, for the family tree of these techniques, please refer to Figure 4.4.

4.3.1.2.3 Locally distributed DCA algorithms The third and final class of DCA algorithms are the locally distributed algorithms, which constitute a hybrid of distributed and centralized algorithms. These algorithms provide the greatest number of performance benefits of the centralized algorithms at a much lower complexity. Examples of locally distributed DCA algorithms are those proposed by Delli Priscoli *et al.* [293, 294] as an evolution of the Pan-European GSM system [28]. Locally distributed DCA algorithms use information from nearby base stations to augment their local channel quality information in order to make a more informed channel allocation decision. Most of the locally distributed algorithms maintain an Augmented Channel Occupancy (ACO) matrix [291]. This matrix contains the channel occupancy for the local and surrounding base stations from which information is received. After every channel allocation, the information to update the ACO matrices is sent to the nearby base stations. This signaling requires a fast backbone network, but it is far less complex than the signaling required for the centralized algorithms.

The Local Packing Dynamic Distributed Channel Assignment (LP-DDCA) algorithm, proposed in [291], maintains an ACO matrix for every base station for all surrounding cells within the co-channel interference distance or reuse distance from the base station. The LP-DDCA algorithm assigns the first channel available that is not used by the surrounding base stations, whose information is contained in the ACO matrix. There are several algorithms similar to this one, including those by Del Re *et al.* [339], and the Locally Optimized Least/Most Interference Algorithms (LOLIA/LOMIA) that we will use in Section 4.3.3.3 in the context of our performance comparisons.

An overview of the main differences between fixed and dynamic channel allocation is shown in Table 4.1; exploration of its detailed contents is left to the reader. However, this table

Fixed Channel Allocation (FCA)	Dynamic Channel Allocation (DCA)
• Better under heavy traffic loads	• Better under light/moderate traffic loads
• Low call setup delay	• Moderate to high call setup delay
• Suited to large-cell environment	• Suited to microcellular environment
• Low flexibility in channel assignment	• Highly flexible channel assignment
• Sensitive to time and spatial changes in traffic load	• Insensitive to time and spatial changes in traffic load
• Low computational complexity	• High computational complexity
• Labor-intensive and complex frequency planning	• No frequency planning required
• Radio equipment only covers channels assigned to cell	• Radio equipment may have to cover all possible channels available
• Low signaling load	• High signaling load
• Centralized control	• Control dependent on the specific scheme from centralized to fully distributed
• Low implementational complexity	• Medium to high implementational complexity
• Increasing system capacity is expensive and time-consuming	• Simple and quick to increase system capacity

Table 4.1: FCA and DCA Features

does not show the increase in spectral efficiency and channel utilisation that becomes possible with dynamic schemes, as will be demonstrated during our performance comparisons.

4.3.1.3 Hybrid Channel Allocation

Hybrid channel allocation schemes constitute a compromise between fixed and dynamic channel allocation schemes. They have been suggested in order to combine the benefits of DCA at low and medium traffic loads with the more stable performance of FCA at high traffic loads. Furthermore, hybrid schemes have been proposed as possible extensions to the fixed channel allocation used in second-generation mobile radio systems. In hybrid channel allocation schemes, the channels are divided into fixed and dynamic subsets. The fixed channels are assigned to the cells, as would be done for fixed channel allocation, and they are the preferred choice for channel allocation. When a cell exhausts all its fixed channels, it attempts to allocate a dynamically assigned channel from the central pool of channels. The algorithm used to pick the dynamically allocated channel depends on the hybrid scheme, but it can be any arbitrary DCA algorithm. The ratio of fixed and dynamic channels could be fixed [340] or varied dynamically, depending on the traffic load. At high loads, best performance is achieved, when the hybrid scheme behaves like FCA, by having none or a limited number of dynamically allocated channels [340, 341]. Some hybrid channel allocation algorithms reallocate fixed channels, which become free to calls using dynamic channels in order to free up the dynamic channels. This technique is known as channel reordering [334].

4.3.1.4 The Effect of Handovers

A handover or handoff event occurs when the quality of the channel being used degrades, and hence the call is switched to a newly allocated channel. If the new channel belongs to the same base station, then this is called an intra-cell handover. If the new channel belongs to a different base station, it is referred to as an inter-cell handover. Generally intra-cell handovers occur when the channel quality degrades due to interference or when the channel allocation algorithm decides that a channel reallocation will help increase the system's performance and capacity. Inter-cell handovers occur mainly because the mobile moves outside the cell area; hence, the signal strength degrades, requiring a handover to a nearer base station.

Handovers have a substantial effect on the performance of channel allocation algorithms. At high traffic loads, the majority of forced call terminations are due to the lack of channels available for handover rather than to interference. This can be a particular problem in microcellular systems, where the rate of handovers is significantly higher than that in normal cellular systems.

There are several known solutions to reduce the performance penalty caused by handovers. One of the simplest solutions is to reserve some channels exclusively for handovers, commonly referred to as cutoff priority [304,342,343] or guard channel [344] schemes. However, this solution reduces the maximum amount of carried traffic or system capacity and hence yields increased new call blocking. The guard or handover channels do not need to be permanently assigned to cells; they are invoked from an "emergency pool."

Algorithms that give higher priority to requests for handovers than to new calls are called Handover prioritisation schemes. Guard channel schemes are therefore a type of handover prioritisation arrangement. Another type of handover prioritisation is constituted by handover queuing schemes [310,311,342,343]. Normally, when an allocation request for handoff is rejected, the call is forcibly terminated. By allowing handover allocation requests to be queued temporarily, the forced termination probability can be reduced. The simplest handover queuing schemes use a First-In First-Out (FIFO) queuing regime [343]. Tekinay *et al.* [304] have suggested a nonpreemptive priority handover queuing scheme in which handover requests in the queue that are the most urgent ones are served first.

A further alternative to help reduce the probability of handover failure is to allow allocation requests for new calls to be queued [344]. New call allocation requests can be queued more readily than handovers because they are less sensitive to delay. Handover queuing reduces the forced termination probability owing to handover failures but increases the new call blocking probability. New call queuing reduces the new call blocking probability and also increases the carried teletraffic. This is because the new calls are not immediately blocked but queued, and in most cases they receive an allocation later.

4.3.1.5 The Effect of Transmission Power Control

Transmission power control is an effective way of reducing co-channel interference while also reducing the power consumption of the mobile handset. Jointly optimising transmission power control with the channel allocation decisions is promising in terms of increasing spectral efficiency. However, little research has been done into this area, apart from a contribution by Chuang and Sollenberger [289] showing the potential benefits. Transmission power control, like channel allocation, can be implemented in a centralized [345, 346] or distributed [347] manner.

An alternative fixed channel allocation strategy, referred to as Reuse partitioning [310], relies on transmission power control. In reuse partitioning, a cell is divided into two or more concentric subcells or zones. If a channel is used in the inner zone with transmission power control, the interference is reduced due to the reduced transmission power. Therefore, the interference from channels used in the inner zones is less than that by those channels, used in the outer zones. Channels used in the inner zones can thus be reused at much shorter distances than those utilized in the outer zones.

By combining transmission power control with dynamic channel allocation, the additional performance gains of reuse partitioning can be achieved. Using reuse partitioning with DCA is far simpler to implement than using FCA, since the system is self-configuring and does not require network reuse pattern planning.

4.3.2 Simulation of the Channel Allocation Algorithms

In this section, we highlight how we simulated the various channel allocation algorithms we investigated. Section 4.3.2.1 describes the simulation program, "Netsim," which was developed to simulate the performance of the channel allocation algorithms. The channel allocation algorithms that we simulated are described in detail in Section 4.3.3. In Section 4.3.3.4, we describe the performance metrics we have used to compare the performance of the channel allocation algorithms. Finally, in Section 4.3.3.5, we describe the model used to generate the nonuniform traffic distributions we used in our simulations.

4.3.2.1 The Mobile Radio Network Simulator, "Netsim"

In order to characterize the performance of the various channel allocation algorithms, we simulated a mobile radio network. The simulator program we developed is referred to as Netsim. The simulated base stations can be placed in a regular pattern or at arbitrary positions within the simulation area. Mobiles are distributed randomly across the simulation area. Each mobile can have different characteristics, such as a particular mobility model or velocity.

A screenshot from the simulator is shown in Figure 4.6. The figure shows a forty-nine-base station simulation, where the cell areas are represented by circles. The mobiles are shown as small squares, and when they become active, they change color on the video screen. The connection between an active mobile and a base station is represented by a line linking the base station and the mobile. The simulator has the following features:

New Call Queuing Channel allocation requests for new calls are queued if they cannot immediately be served [344]. The new call request is blocked if its request cannot be served within a preset timeout period, referred to as the Maximum new-call queue time.

Handover Prioritisation Channel allocation requests for handovers are given priority over new calls, supporting Handover Prioritisation [310].

Handover Urgency Prioritisation Channel allocation requests for handovers are processed by each base station, so that the more urgent handovers are served first [304].

Handover Hysteresis A call will not be handed over to another base station or channel unless the new channel has a signal quality better than the current channel by at least the

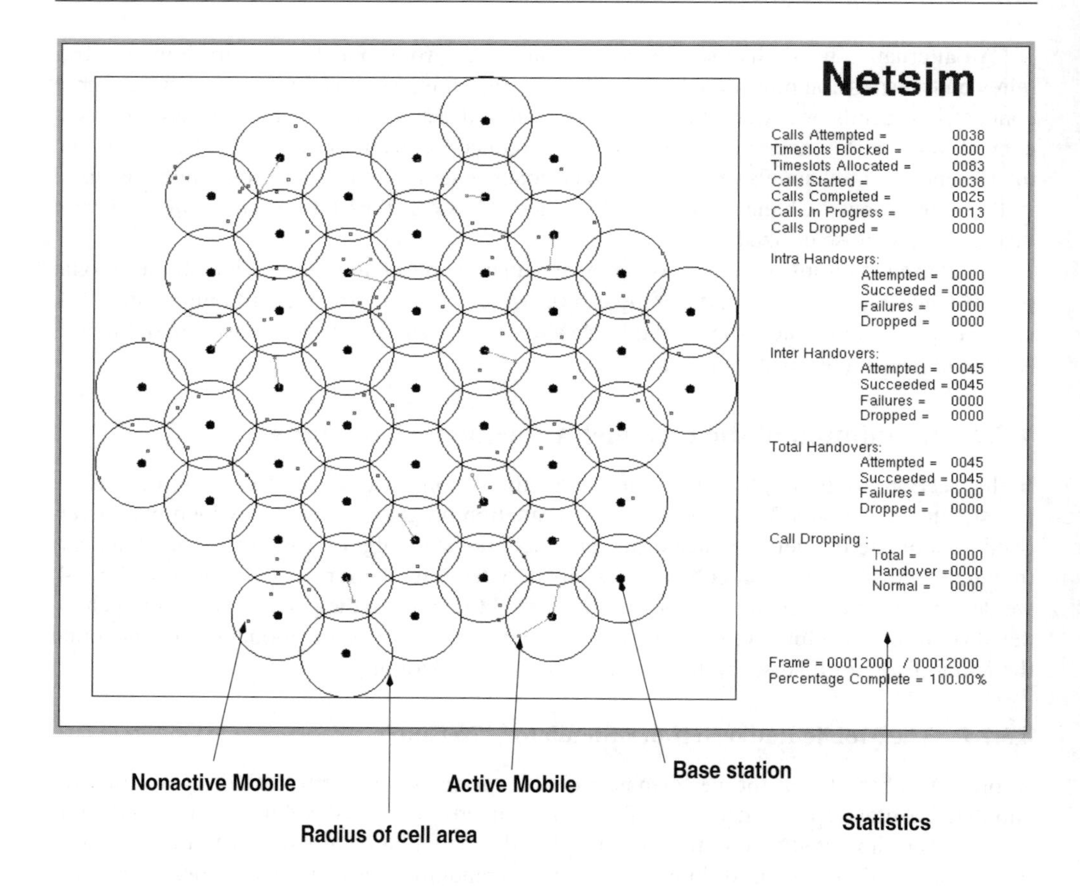

Figure 4.6: Screenshot of the Netsim program, showing 100 users in a 49-cell simulation. Each base station is located at the center of each cell, and the large circles represent the radius of the cell area. The connection between an active mobile and a base station is represented by a line.

preset handover hysteresis level. The only exception is when the current channel quality is below the signal quality level required to maintain the call and the new channel is above this quality level, but the difference between the quality of the new and current channel is less than the hysteresis threshold.

Channel Models The simulator models each propagation channel using one of several path-loss models and a shadow fading model. The shadow fading model can be turned off if necessary.

Call Generation Model Each mobile's activity is described by how much of the time the mobile is active (i.e., making a call). The activity of each mobile is controlled by two parameters, average call duration and average intercall time. The average call duration is the long-term mean of the length of all the calls made by the mobile. The duration of all the calls made by the mobile is Poisson-distributed [289,348]. The average intercall time is the long-term mean duration of time between calls being made. Similarly to the call durations, the time between calls is also Poisson distributed [289, 348].

Edge effects The cells at the edge of the simulation area behave differently from cells near the center of the simulation area. This is because the cells near the edge have fewer neighbouring cells and hence less interference. Therefore, in order to reduce the effect of these edge cells, the statistical results can be gathered only from the cells near the center of the grid (i.e., from the active cells). Furthermore, when a mobile reaches the edge of the simulation area, it is randomly repositioned somewhere else in the simulation area. In order that this does not cause handover problems, active mobiles reaching the edge of the simulation area finish their calls before they are repositioned.

Extensive Statistical data gathering The Netsim simulator stores a large range of statistics from each simulation. For example, the probability density function of the number of simultaneous calls at each base station is stored. Furthermore, the simulation area can be divided into a fine grid, the resolution of which depends on the required accuracy of the statistical evaluation aimed for. Statistics can be gathered separately for each grid square, allowing coverage maps of the simulation area to be generated.

Warmup period When the simulation is first begun, the number of active calls is far below the normal level. There is a latency, before the number of active calls is built up to the correct level, owing to the nature of the Poisson distributed call generation models [289, 348]. Therefore, in order not to bias the results, simulations are conducted for a sufficiently long period of time before the simulation statistics can be gathered. This period of time is referred to as a warmup period.

The Netsim simulator is a network layer-based framework employing a simple physical layer model in order to reduce the complexity of the simulations, which is described in the next section.

4.3.2.1.1 Physical Layer Model The physical layer, that is, the modulator and demodulator,are modeled using two parameters, Outage SINR and Reallocation SINR. The Reallocation SINR threshold is always set above the Outage SINR threshold. When the signal quality measured in terms of the signal-to-interference+noise ratio (SINR) (defined in Equation 4.14 drops below the reallocation SINR level, the mobile requests a new channel to hand over to. This handover request can be asking for another channel from the same base station to which the mobile is currently connected and is called an intra-cell handover. Alternatively, the handover can be initiated to a channel from a different base station and is called an inter-cell handover.

If, while waiting for a reallocation handover, the signal quality drops further, below the so-called Outage SINR threshold, the signal is deemed to be lost for that time period. This is referred to as an outage. If a channel is in outage for several consecutive time periods, then the call is forcibly terminated. The parameter termed the Maximum Consecutive Outage reflects the number of consecutive outages that need to occur to cause a call to be forcibly terminated.

The Reallocation SINR threshold should be set at the average SINR required to maintain marginal signal quality. The Outage SINR threshold should be set as the SINR, below which the demodulated signal cannot be decoded error free. This twin-threshold physical layer model is similar to those described by Tekinay *et al.* [311] and by Katzela *et al.* [310]. The difference is that our model is based on SINR thresholds instead of received power thresholds used in these references. Since the computational complexity would be too high to simulate fast Rayleigh fading in a network-layer simulation, the SINR threshold of the physical layer model should include a margin to emulate the effects of fast fading, thereby increasing the required outage level.

The simulator calculates the probability of outage as the proportion of time in which a channel was below the Outage SINR threshold (i.e., in outage). The simulator can also calculate the low signal quality probability, as the proportion of time a channel is below the Reallocation SINR threshold.

The next section describes the model used to simulate shadow fading of the radio channels.

4.3.2.1.2 Shadow Fading Model The channel model used by the Netsim simulator is fairly simple in order to reduce the computational complexity of the simulations. The channel can be modeled using a variety of path-loss models and an optional shadow fading model. This section is concerned with the shadow fading model. Network simulations are particularly complex, since all the possible interfering channels may need to be modeled, that is, from each transmitter to every receiver tuned to the same carrier frequency at the same time.

Shadow fading can be modeled using a correlated signal, which is log-normally distributed [52]. In our previous chapters, shadow fading was modeled by using precalculated shadow fading signal envelopes. However, because of the high number of interfering channels, where the channels should be uncorrelated, a large number of precalculated shadow fading envelopes would be needed. This is impractical because of the associated high storage requirements, and the increased simulation time resulting from storage access delays.

We decided to invoke a method originally used to generate Rayleigh fading rather than shadow fading in order to produce the correlated log-normally distributed shadow fading envelope required. Jakes' method [47] was originally proposed to produce Rayleigh-distributed correlated signal envelope and phase. Jakes' technique is also often called the sum of sinusoids method, which uses the summation of several low-frequency sinusoids with regularly spaced phase differences in order to produce the desired signal. A signal, $r(t)$, exhibiting Rayleigh-distributed envelope or magnitude fluctuations can be produced from the complex summation of two independent Gaussian random variables, which is formulated as:

$$r(t) = X_1 + jX_2. \tag{4.6}$$

Jakes' method produces the required pair of correlated independent Gaussian distributed random variables, X_1, X_2, which are approximated by $x_1(t)$ and $x_2(t)$, given by:

$$x_1(t) = 2\left[\sum_{n=1}^{N_o} \cos(\beta_n)\cos(\omega_n t)\right] + \sqrt{2}\cos(a)\cos(\omega_m t) \tag{4.7}$$

$$x_2(t) = 2\left[\sum_{n=1}^{N_o} \sin(\beta_n)\cos(\omega_n t)\right] + \sqrt{2}\sin(a)\cos(\omega_m t) \tag{4.8}$$

$$\beta_n = \frac{n\pi}{(N_o+1)} \tag{4.9}$$

$$N = 2(2N_o+1) \tag{4.10}$$

$$\omega_n = \omega_m \cos\left(\frac{2\pi n}{N}\right) \tag{4.11}$$

$$\omega_m = 2\pi f_d, \tag{4.12}$$

where the functions $x_1(t)$ and $x_2(t)$ produce the in-phase and quadrature components of the Rayleigh-fading signal, $r(t)$. Both the in-phase and quadrature components are the sum of $(N_o + 1)$ oscillators, yielding the sum of sinusoids. The maximum Doppler frequency (f_d) sets the highest oscillator's frequency (ω_m), the phase of which is set by a. The remaining N_o oscillators have frequencies of less than ω_m set by ω_n, the phase of which is set by β_n. Therefore, $x_1(t)$ and $x_2(t)$ are functions of t, with parameters f_d and N_o.

Either one of the variables $x_1(t)$ or $x_2(t)$ can be used to produce the log-normally distributed shadow fading envelope $s(t)$, given by:

$$s(t) = 10^{[x_1(t)/10]} \quad \text{or} \quad s(t) = 10^{[x_2(t)/10]}. \tag{4.13}$$

In the next sections, we describe the investigated algorithms in detail.

4.3.3 Overview of Channel Allocation Algorithms

In this section, we describe the channel allocation algorithms that we have investigated in order to identify the most attractive performance trade-offs. Our simulations have concentrated on dynamic channel allocation (DCA) algorithms (Section 4.3.1.2). However, we have also performed experiments using a basic fixed channel allocation (FCA) algorithm (Section 4.3.1.1) as a benchmarker.

We investigated two classes of dynamic channel allocation (DCA) algorithms, namely, distributed and locally distributed algorithms, described previously in Sections 4.3.1.2.2 and 4.3.1.2.3. We studied four distributed DCA algorithms, which are characterized in Section 4.3.3.2, while Section 4.3.3.3 portrays the two locally distributed DCA algorithms that we investigated. In the next section, we introduce the fixed channel allocation algorithm employed.

4.3.3.1 Fixed Channel Allocation Algorithm

In order to benchmark our dynamic channel assignment (DCA) algorithms, a fixed channel allocation (FCA) scheme was required. We decided to employ a basic fixed channel assignment algorithm, which uses omnidirectional antennas and a reuse cluster size of seven cells. This structure is commonly used to provide coverage over a grid of regular hexagonally shaped cells. The frequency spectrum was divided into seven frequency sets, and one set was assigned to each cell.

Figure 4.5 shows such a reuse structure, where the shaded cells represent cells assigned the same set of carrier frequencies. The figure shows the center cell and its six first-tier interfering cells. This fixed channel allocation reuse structure provides uniform capacity across all cells, since each cell site has the same number of carrier frequencies. In the next section we describe the distributed DCA algorithms investigated.

4.3.3.2 Distributed Dynamic Channel Allocation Algorithms

In this section we highlight four well-known distributed DCA algorithms that we have studied comparatively. The most plausible technique is the Least Interference Algorithm (LIA) [290], which allocates the channel suffering from the least received instantaneous interference power; hence, it attempts to minimize the total interference within the system. More specifically, this algorithm minimizes the interference at low traffic loads but increases it at high loads. This is because at high loads the LIA algorithm will still attempt to allocate a channel to a new call, even when all the slots have a high level of interference. Again, this increases the total interference load of the system.

The second distributed DCA algorithm we studied is a refinement of the LIA algorithm, which is referred to as the Least interference below Threshold Algorithm (LTA) [290]. This algorithm attempts to reduce the interference caused by the LIA algorithm at high loads by blocking calls from using those channels, where the interference measured is deemed excessive for the transceiver to sustain adequate communications quality. The algorithm allocates the least interfered channel, whose interference is below a preset maximum tolerable interference threshold. Therefore, the LTA algorithm attempts to minimize the overall interference in the system, while maintaining the quality of each call above the minimum acceptable level.

The third algorithm we investigated attempts to utilize the frequency spectrum more efficiently while maintaining acceptable call quality. This algorithm works in a similar way to the LTA algorithm, and it is termed the Highest (or Most) interference below Threshold Algorithm (HTA or MTA) [290]. Since its goal is not to reduce the interference, but to maximize the spectral efficiency, it allocates the most interfered channel, whose interference is below the maximum tolerable interference threshold. The interference threshold is determined by the transceiver's interference resilience.

The final distributed DCA algorithm can be characterized as the Lowest Frequency below Threshold Algorithm (LFA) [290]. This algorithm is a derivative of the LTA algorithm, the difference being that the LFA algorithm attempts to reduce the number of carrier frequencies being used concurrently. This has the advantage that, statistically speaking, fewer transceivers may then be required at each base station. The algorithm allocates the least interfered channel below the maximum tolerable interference threshold, while also attempting to reduce the number of carrier frequencies used. Therefore, no new carrier frequency is invoked from the set of carriers, unless all the available time-slots on the currently used carrier frequencies are considered too interfered. In the next section, we describe the two locally distributed DCA algorithms, whose performance we have compared to the above algorithms using simulations.

4.3.3.3 Locally Distributed Dynamic Channel Allocation Algorithms

We have investigated the performance of two locally distributed dynamic channel allocation algorithms, both of which are quite similar. The Locally Optimized Least Interference Algorithm (LOLIA) attempts to reduce the overall interference in a system, like the LIA and LTA algorithms, while the Locally Optimized Most Interference Algorithm (LOMIA) attempts to increase the spectral efficiency in a similar way to the HTA algorithm.

Specifically, the locally distributed DCA algorithms constitute a hybrid of distributed and centralized channel allocation decisions. They exploit the information provided by neighbouring base stations in order to improve the channel allocation decisions, which constitute the centrally controlled part of the distributed/centralized hybrid solution. Their complexity is therefore somewhere between that required for centralized and distributed algorithms.

The LOLIA algorithm carries out its channel allocation decisions in the same way as the distributed LIA algorithm. However, it will not allocate a channel, if it is used in the nearest "n," neighbouring cells by another subscriber. Therefore, the nearby base stations exchange information concerning the channels that are currently being used. This requires a fast backbone network but does not rely on central control. The overall level of interference in the system can be reduced by increasing the number of cells, which are classed as neighbouring cells. However, the larger "n." the more calls are blocked, since there will be fewer available channels, which are not being used in the nearest "n" base stations. Figure 4.7 shows the arrangement of neighbouring cells for $n = 7$ and $n = 19$. The "n" parameter of the algorithm effectively imposes a minimum reuse distance constraint on the algorithm.

The second locally distributed DCA algorithm we consider is similar to LOLIA, but it is based on the HTA and not the LIA distributed algorithm. The LOMIA algorithm picks the most interfered channel, provided that this channel is not used in the nearest "n" neighbouring cells. The LOLIA and LOMIA algorithms are similar to those proposed by De Re *et al.* [339] and ChihLin *et al.* [291].

Having described the algorithms that we have simulated in order to identify the performance trade-offs of the various channel allocation algorithms, in the next section we describe the metrics used to compare the performance of the various algorithms.

4.3.3.4 Performance Metrics

Several performance metrics can be used to quantify the performance or quality of service provided by a particular channel allocation algorithm. The five performance metrics defined below have been widely used in the literature [290], and we also opted for their employment:

- New call blocking probability, P_B
- Call dropping or forced termination probability, P_D or P_{FT}
- Probability of low-quality connection, P_{low}
- Probability of outage, P_{out}
- Grade of Service, GOS

The new call blocking probability, P_B, is defined as the probability that a new call is denied access to the network. This may be the case because there are no available channels

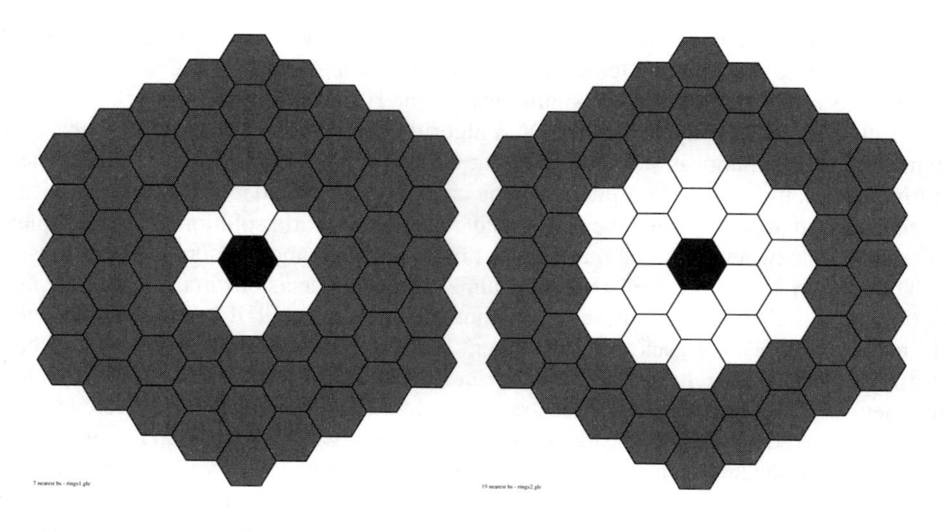

(a) 7 nearest base stations monitored

(b) 19 nearest base stations monitored

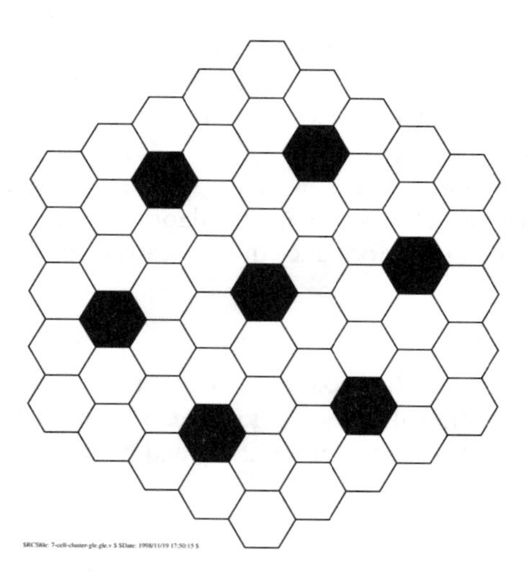

(c) 7-cell cluster FCA

Figure 4.7: The nearest neighbour constraint for $n = 7$ and $n = 19$ for the locally optimized algorithms, LOLIA and LOMIA, compared to a seven-cell reuse cluster for FCA.

or the channel allocation algorithm decided that to allow the new call to access any of the available channels would cause increased interference, which might lead to loss of the new call or calls in progress. Ideally, a low call blocking probability is desired. However, it is even more undesirable when calls in progress are lost, and this is where the second performance metric, namely, P_{FT} is useful.

The call dropping probability, P_D, also widely known as the forced termination probability, P_{FT}, is the probability that a call is forced to terminate prematurely. This can be caused by excessive interference. However, generally when a channel becomes excessively interfered with, the mobile or base station will request a new channel. If no channels are available and the quality of the call degrades significantly because of interference or low signal strength, then the call may be forcibly terminated. Calls can also be forcibly terminated when a mobile moves across a cell boundary into a heavily loaded cell. If there are no available channels in the new cell to hand over to, then the call may be lost prematurely. Since premature call termination is annoying to mobile subscribers, the channel allocation algorithm should attempt to keep the call dropping probability low.

The third performance metric we have used is the probability of a low-quality connection or access, P_{low}. This is the probability that either the up-link or down-link signal quality is below the level required by the specific transceiver to maintain a good-quality connection. A low-quality access could be due to low signal strength or high interference, which is defined as:

$$\begin{aligned} P_{low} &= P\{SINR_{up-link} < SINR_{req} \; or \; SINR_{down-link} < SINR_{req}\} \\ &= P\{min(SINR_{up-link}, SINR_{down-link}) < SINR_{req}\}. \end{aligned} \tag{4.14}$$

This metric allows different channel allocation algorithms, which may have similar call dropping and blocking probability to be compared, in order to identify which is better, when calls are in progress. The quantity $SINR_{req}$ is the required reallocation SINR threshold described in Section 4.3.2.1.1. The probability of outage is similar to the probability of low communications quality metric (P_{low}), which was defined in Equation 4.14, except in this case the quantity $SINR_{req}$ is the required SINR value, below which the call is deemed to be in outage, as described in Section 4.3.2.1.1.

The final metric we have used to evaluate the performance of various channel allocation algorithms is the grade of service (GOS). The definition we have used is that proposed by Cheng and Chuang [290] which is stated as follows:

$$\begin{aligned} GOS &= P\{\text{unsuccessful or low-quality call accesses}\} \\ &= P\{\text{call is blocked}\} + P\{\text{call is admitted}\} \times \\ &\quad\; P\{\text{low signal quality and call is admitted}\} \\ &= P_B + (1 - P_B)P_{low}. \end{aligned} \tag{4.15}$$

The grade of service is the probability of unsuccessful network access (blocking, P_B) or low-quality access, when a call is admitted into the system (P_{low}). This performance metric is a hybrid of the new call blocking probability (P_B) and the low-quality access probability (P_{low}), when calls are not blocked and it is therefore an important performance metric. Now that we have described the algorithms and the metrics used to compare their performance, the next section describes the model used to generate nonuniform traffic distributions.

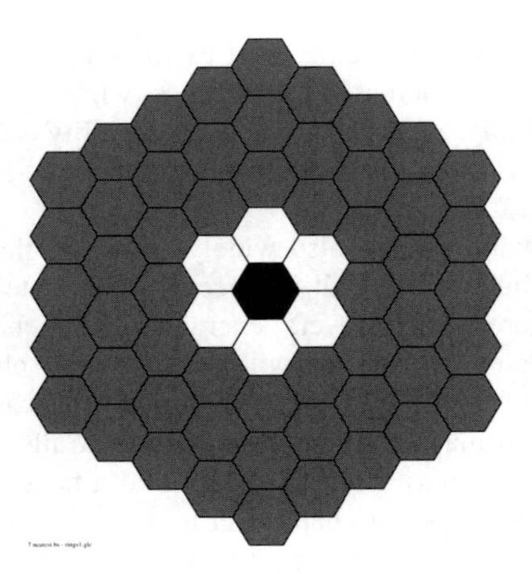

Figure 4.8: Nonuniform traffic conditions exhibiting a traffic "hot spot" in the central cell (black), and a "warm spot" (white) surrounding it. Mobiles in the gray cells move at the standard speed of 13.4 m/s (30 mph). Mobiles in the white ("warm-spot cells") can move at a speed of 9 m/s (20 mph). Mobiles in the black "hot-spot cell" are limited to a speed of 4 m/s (9 mph).

4.3.3.5 Nonuniform Traffic Model

Generally, investigations using fixed channel allocation assume a uniform traffic distribution and therefore a uniform carrier frequency allocation per base station. In practice some base stations have more channels, where demand is expected to be increased, for example, at airports and railway stations. However, fixed channel allocation cannot cope with unexpected traffic demand peaks [349], which are sometimes referred to as traffic "hot spots" [324]. Dynamic channel allocation algorithms are better equipped to cope with these unexpected traffic demands, since a DCA system is effectively self-adapting. Furthermore, DCA schemes typically have more potential channels available at each base station. This is an area in which DCA algorithms have a clear advantage over FCA.

Therefore we defined a model to generate a sudden unexpected traffic "hot spot" in order to measure the performance benefits that DCA algorithms provide over FCA. The model we developed is very simple and causes an increase in teletraffic in the cells affected. The model simply limits the maximum velocity of mobile terminals within a particular geographical area. Mobile users can still enter and leave a "hot spot" cell. However, since the users slow down as they enter the cell, the average cell crossing time is increased. This leads to a higher mobile terminal density in the cell, which in turn leads to increased generated teletraffic.

As an example, we refer to Figure 4.8, presented later in this chapter, in which the speed of mobiles in the gray cells is not limited by the model. For our simulations, however, the mobiles all travel at 30 mph. Upon roaming and entering the white cells, these mobiles reduced their speed to 20 mph. The white cells could represent the outskirts of a city. Upon entering the black cell, which could represent a city center, the speed of mobiles is again reduced to 9 mph.

In order to compare our network performance results attained by fixed and various dynamic channel allocation algorithms, with and without adaptive antenna arrays at the base station, it was necessary to consider more than one performance metric. For example, an algorithm may perform very well in one respect, yet have poor performance when measured using an alternative metric. Therefore, it was decided to invoke two different scenarios:

- A *conservative scenario*, where the maximum acceptable value for the call blocking probability, P_B, is 3%, for the call dropping probability, P_{FT}, is 1%, for P_{low} is 1%, and for the GOS is 4%.
- A *lenient scenario*, where the maximum acceptable value for the call blocking probability, P_B, is 5%, for the call dropping probability, P_{FT}, is 1%, for P_{low} is 2%, and for the GOS is 6%.

It must be noted that the maximum allowable GOS does not have to obey Equation 4.15 for the given values of P_b and P_{low}, since they may be traded off against each other. Hence the GOS may be interpreted as a form of 'user satisfaction'. As a consequence, for example in the lenient scenario the GOS is 6%, rather than the expected 7%, since it may be unacceptable for the user to simultaneously tolerate both a P_b of 5% and at the same time a low-quality link probability of P_{low}=2%. Therefore the required 'user satisfaction' may be maintained with the proviso of satisfying any acceptable combination of P_b and P_{low} values, as long as their sum remains below the required GOS level.

The next section presents a summary of the results obtained for the previously described channel allocation algorithms.

4.3.4 DCA Performance without Adaptive Arrays

In our previous work [23, 151, 295, 350] a comparative study of a range of DCA algorithms was conducted and it was found that the algorithm which provided the best overall compromise in terms of the desired performance measures was the Locally Optimised Least Interference Algorithm (LOLIA). The results in Table 4.2 indicate the achievable network capacities, without AAAs and without shadow fading, for various DCA algorithms and for the FCA algorithm. Hence, our further investigations presented here we focus our attention on the LOLIA by combining it with adaptive beamforming and other network capacity enhancement techniques.

4.4 Employing Adaptive Antenna Arrays

Here, a study into the usage of an adaptive antenna array in a cellular network is conducted. A theoretical analysis of such a system is performed and the results are presented for later comparison with simulated results. To simplify this process the following assumptions were made:

- There is a uniform distribution of users in each cell.
- There is a blocking probability of P_B in all cells.
- The omni-directional base station antenna has an ideal beam pattern, giving a uniform circular coverage.

Algorithm	Number of users supported by network Conservative P_{FT}=1%, P_{low}=1% GOS=4%, P_B=3%	Lenient P_{FT}=1%, P_{low}=2% GOS=6%, P_B=5%
FCA	820	1120
HTA	1435	1520
LFA	1555	1705
LOMIA (n=19)	1505	2040
LTA	1815	1830
LIA	1820	1820
LOLIA (n=7)	1860	2115
LOLIA (n=19)	1935	2005

Table 4.2: Maximum number of mobile users that can be supported by the various DCA algorithms [151, 295].

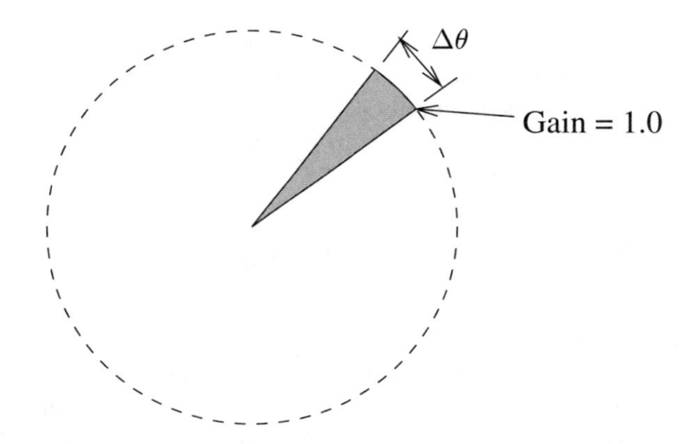

Figure 4.9: Beam pattern of an ideal beamformer with beamwidth $\Delta\theta$.

- The adaptive base station antenna array can generate m ideal beams, each with a gain of 1.0 over a beamwidth of $\Delta\theta = 2\pi/m$ radians, and a gain of 0.0 over the remaining angular sector, as shown in Figure 4.9

The blocking probability, P_B, is the fraction of attempted calls that cannot be allocated a channel. If the traffic intensity offered is a Erlangs, then the actual traffic carried is $a(1-P_B)$ Erlangs. The Erlang is a measure of offered tele-traffic, which indicates the quantity of traffic on a channel or group of channels per unit time. This gives a channel usage efficiency of [2]:

$$\eta = \frac{a(1-P_B)}{N}, \tag{4.16}$$

where N is the total number of channels allocated per cell.

It was also assumed that the main beam formed by the adaptive antenna was centred about the angle of arrival of the desired mobile's signal and that the mobile was tracked with

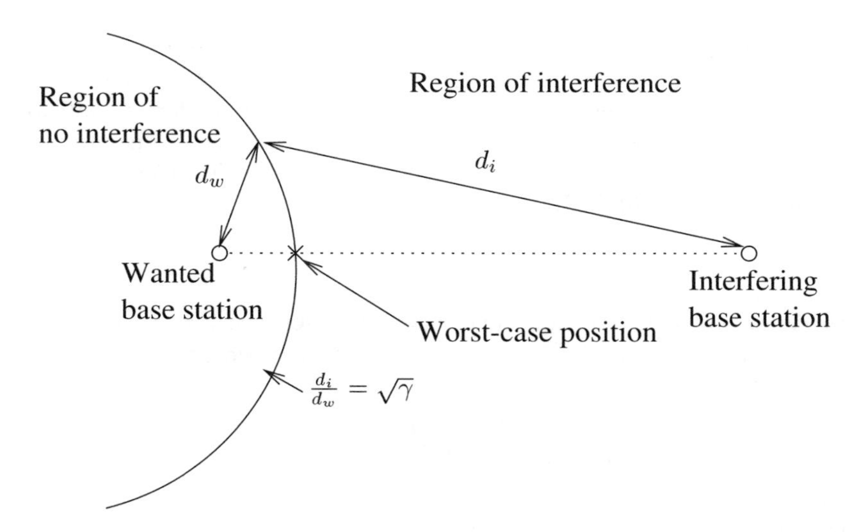

Figure 4.10: Contour defining interference regions in a downlink scenario using omnidirectional antennas.

no error. Additionally, all interfering sources outside the main beam were assumed to be nulled successfully. The ideal beamformer model used has a single mainlobe with a unity-gain beamwidth of $\Delta\theta$ and sidelobes of zero gain, as shown in Figure 4.9. When the desired signal's power, S, does not exceed the co-channel interference power, I, by the required protection ratio, γ. In this situation an 'outage' will occur, i.e. we fail to achieve satisfactory reception at the mobile in the presence of interference with the probability of [2, 351–353]:

$$P(\text{outage}) = P(S \leq \gamma I) = P(S/I \leq \gamma) = P(SIR \leq \gamma), \tag{4.17}$$

where SIR is the signal-to-interference ratio. In other words, $P(\text{outage})$ is the probability of the power of the signal being insufficient to provide reliable communications due to the interference in the channel. Considering only the propagation path loss, but no fast- and shadow-fading, we have $SIR = S/I = d_i^2/d_w^2 \leq \gamma$, hence for a given interference protection ratio, a locus defined by $d_i/d_w = \sqrt{\gamma}$ can be drawn, as in Figure 4.10. This defines a region, where the signal-to-interference ratio necessary for reliable downlink (DL) communications is maintained, and a region where interference occurs.

In a cellular network employing base station (BS) adaptive antenna arrays, the occurrence of co-channel interference is a statistical phenomenon dependent upon the number of co-channel interferers and on the positions of these interferers in the co-channel cells. In general the uplink (UL) and downlink (DL) interference calculations are different and hence they have to be considered separately. The total probability of co-channel interference-induced outage can be evaluated by [2, 283, 353]:

$$P(\text{outage}) = P(SIR \leq \gamma) = \sum_{n=1}^{N} P(SIR \leq \gamma|n)P(n), \tag{4.18}$$

where N is the total number of co-channel interferers, usually restricted to the first tier of interferers, shown in white in Figure 4.7(a), i.e. to six, $P(SIR \leq \gamma|n)$ is the conditional probability of co-channel interference, $P(SIR \leq \gamma)$ given n interferers. Furthermore, $P(n)$ is the probability that there are n active interfering co-channel cells. Therefore, if the activation of channels is assumed to be independent and identically distributed, $P(n)$ has the form of a binomial PDF [2, 265, 353]:

$$P(n) = \binom{6}{n} p^n (1-p)^{6-n}, \tag{4.19}$$

where p is the probability of finding one interfering co-channel active. The probability p that a single co-channel BS has an active DL co-channel interferer, given that the wanted mobile has been assigned that DL channel already, is [2]:

$$p = \frac{\text{number of active channels}}{\text{total number of channels}} = \frac{a(1-P_B)}{N} = \eta. \tag{4.20}$$

Therefore, the probability that n co-channel interfering BSs are using the same DL channel as the wanted mobile for its reception becomes:

$$P(n) = \binom{6}{n} \eta^n (1-\eta)^{6-n}. \tag{4.21}$$

Hence, from Equations (4.18) and (4.21) we have:

$$P(\text{outage}) = P(SIR \leq \gamma) = \sum_{n=1}^{N} P(SIR \leq \gamma|n) \binom{6}{n} \eta^n (1-\eta)^{6-n}. \tag{4.22}$$

In conjunction with an omnidirectional BS antenna, the probability of an active DL co-channel interferer was given by η, the channel usage efficiency. For an adaptive BS antenna, forming m beams per cell, there will always be six DL beams targeted at the wanted mobiles from the six co-channel base stations. Therefore, for an adaptive base station antenna [2, 265] we have:

$$\begin{aligned} p &= \begin{pmatrix} \text{probability that the beam pointing at the desired} \\ \text{mobile also contains an interfering mobile} \end{pmatrix} \\ &= \frac{\text{number of active channels in beam}}{\text{total number of channels}} \\ &= \frac{a(1-P_B)/m}{N} = \frac{\eta}{m}. \end{aligned} \tag{4.23}$$

Hence, for an adaptive BS antenna array with m beams per BS we have:

$$P(n) = \binom{6}{n} \left(\frac{\eta}{m}\right)^n \left(1 - \frac{\eta}{m}\right)^{6-n}, \tag{4.24}$$

leading to the overall outage probability for a BS adaptive antenna array in the form of:

$$P(\text{outage}) = P(SIR \leq \gamma) = \sum_{n=1}^{N} P(SIR \leq \gamma|n) \binom{6}{n} \left(\frac{\eta}{m}\right)^n \left(1 - \frac{\eta}{m}\right)^{6-n}. \tag{4.25}$$

where $P(SIR \leq \gamma|n)$ is the conditional outage probability, which is dependent on the mean received signal power and the mean received interference power.

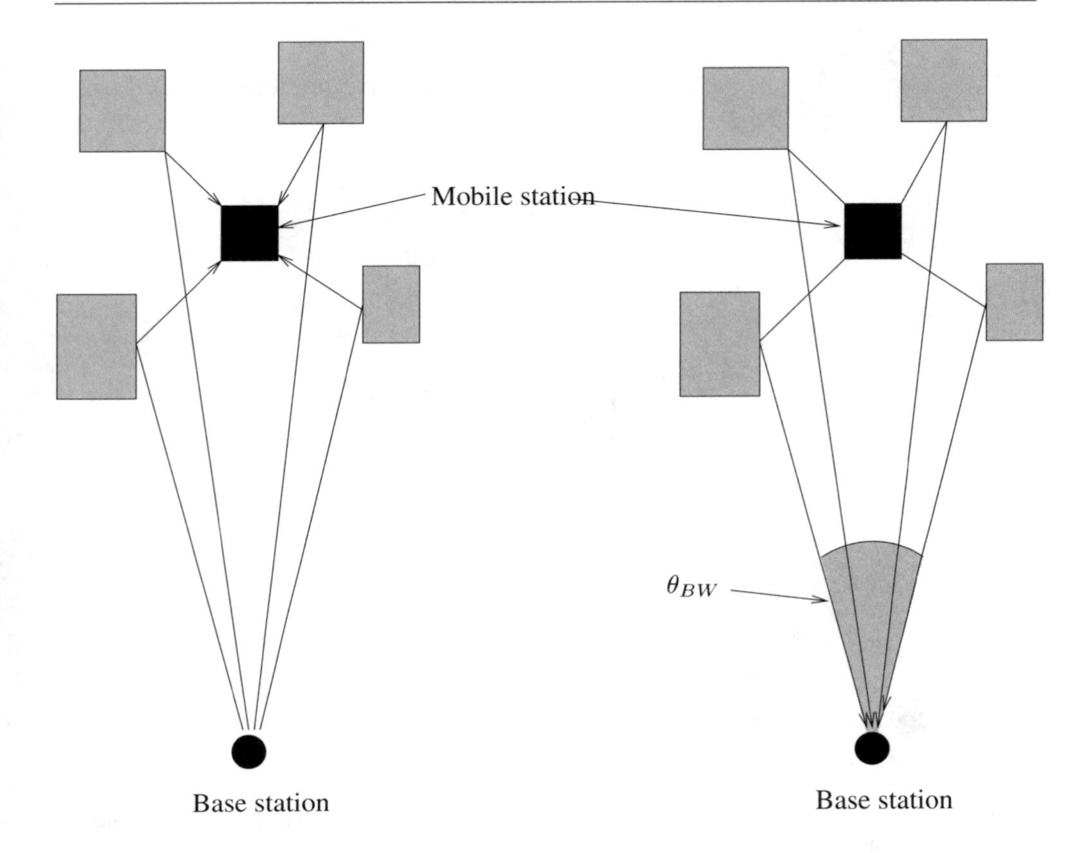

Figure 4.11: Macrocellular uplink and downlink multipath scattering scenarios.

4.5 Multipath Propagation Environments

In Section 4.2 various situations were investigated where only a direct Line-Of-Sight (LOS) link existed between the base station and the mobile handset. However, in a real environment, a phenomenon known as multipath scattering takes place, which results in the presence of numerous signal components, or multipath components, at the receiver. This is due to reflections, diffractions and signal scattering, caused by objects in the path between the transmitter and the receiver. A simple figure showing an example of the multipath propagation channel is shown in Figure 4.11. Each signal component experiences a different path attenuation and phase rotation, which determines the received signal's amplitude, carrier phase shift, time delay, angle of arrival and Doppler shift [21]. In general, each of these components will be time-varying. We note here that the various uplink and downlink scenarios will be considered in more depth in Figure 4.22 during our further discourse.

Figure 4.11 shows the multipath environment that may be found on the uplink and downlink in a macrocellular environment. It is usually assumed that the scatterers surrounding the

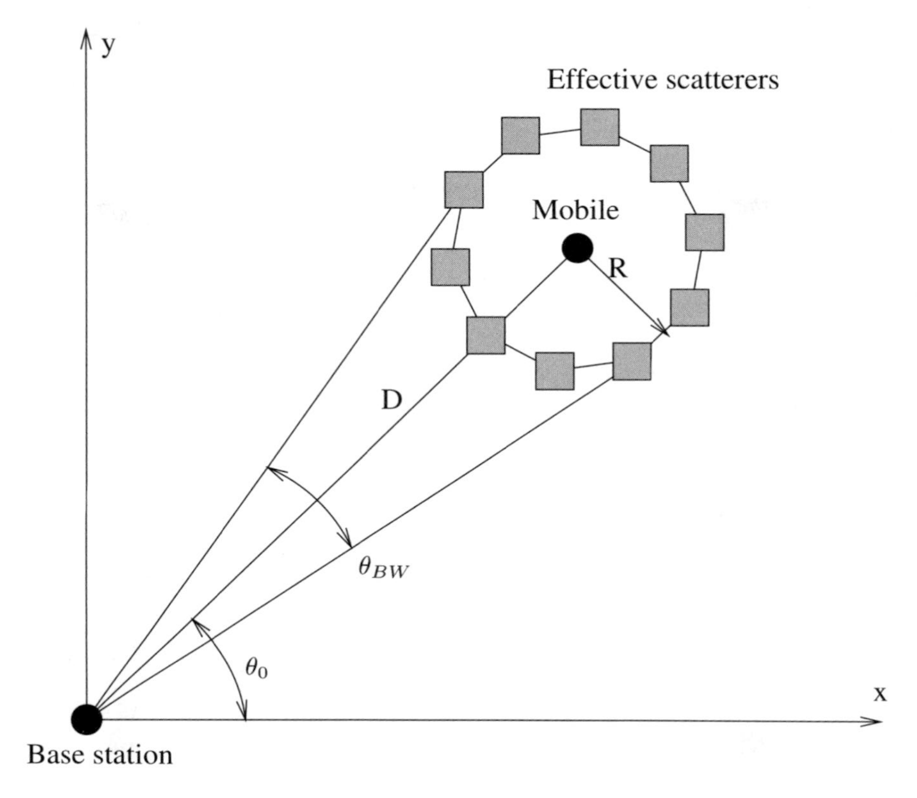

Figure 4.12: Lee's model for multipath scattering using N scatterers in a circle of radius R around the mobile station.

mobile station are at about the same height as or are higher than the mobile. This implies that the received signal at the mobile antenna arrives from all directions after bouncing from the surrounding scatterers, as illustrated in Figure 4.11. Under these conditions it is assumed that the downlink Direction-Of-Arrival (DOA) at the mobile is uniformly distributed over $[0, 2\pi]$ [21, 296]. However, the uplink DOA of the received signal at the base station is quite different. In a macrocellular environment, the base station is typically positioned higher than the surrounding scatterers. Hence, the received signals at the base station result predominantly from the scattering process in the vicinity of the mobile station, as it may be seen in Figure 4.11. The uplink multipath components are restricted to a smaller angular region, θ_{BW}, and hence the distribution of the uplink DOA is no longer uniform over $[0, 2\pi]$. Many different models have been developed for use in different applications. Below a brief description of some of the models follows, but for a more detailed exposition the reader is referred for example, to Ertel *et al.* [296]. The macrocellular models are all based around the same principle of placing a number of scatterers near the mobile station in a given pattern, obeying a geographic probability distribution. In Lee's model, the scatterers are evenly spaced on a circular ring about the mobile, as shown in Figure 4.12. Assuming that the N scatterers are uniformly spaced on the circle having a radius R and orientated such that a scatterer is located

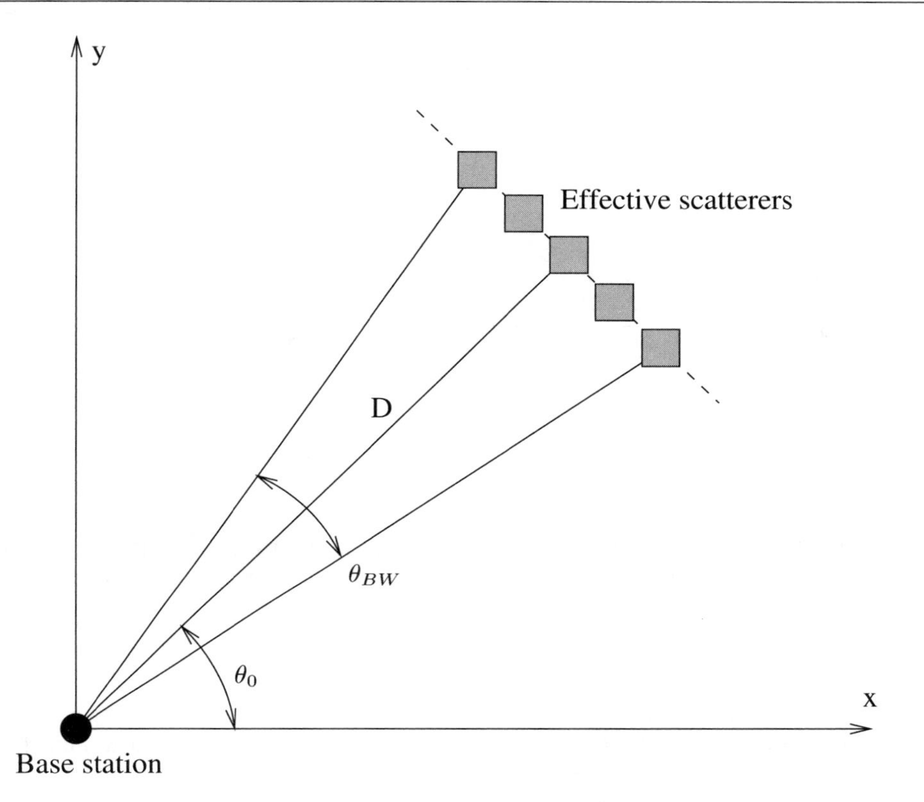

Figure 4.13: The Discrete Uniform Distribution model for multipath scattering using a line of N scatterers centred about the line of sight to the mobile.

along the LOS path, the discrete DOAs are [296]:

$$\theta_i \approx \frac{R}{D} \sin\left(\frac{2\pi}{N} i\right), \qquad i = 0, 1, \ldots, N-1. \tag{4.26}$$

However, the model was originally designed simply for providing information regarding the signal correlations of the multipath components and when used to provide DOA and Time-Of-Arrival (TOA) information, the simulated results are not consistent with measurements [296].

A model similar to Lee's, known as the discrete uniform distribution, evenly spaces N scatterers within a narrow beamwidth centred about the LOS to the mobile, as shown in Figure 4.13. According to [296], the discrete possible DOAs, assuming that N is odd, are given by:

$$\theta_i = \frac{1}{N-1} \theta_{BW} i, \qquad i = -\frac{N-1}{2}, \ldots, \frac{N-1}{2}. \tag{4.27}$$

The Geometrically Based Single-Bounce (GBSB) Statistical Channel Models are defined by a spatial scatterer density function. This model involves randomly placing scatterers in

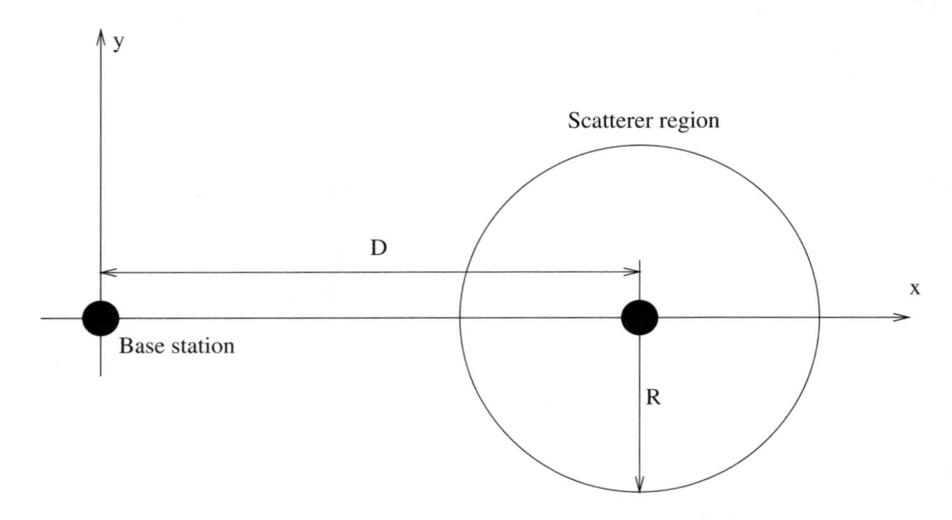

Figure 4.14: The Geometrically Based Single-Bounce Circular Model (GBSBCM), which is suitable for use as a macrocellular model, showing the region in which the scatterers are located.

the scatterer region according to the spatial scatterer density function. From the location of each of the scatterers, the DOA, TOA, and signal amplitude can be determined. The Geometrically Based Single-Bounce Circular Model (GBSBCM) is shown in Figure 4.14, which was found to be suitable for macrocellular modelling, since it assumes that all the scatterers lie within the radius R about the mobile and $R < D$ [296]. An alternative spatial distribution of the scatterers, known as the Geometrically Based Single-Bounce Elliptical Model (GBSBEM) [296, 297], assumes that the scatterers are uniformly distributed within an ellipse, as shown in Figure 4.15, where the base station and the mobile station are the foci of the ellipse, and the parameters a_m and b_m are the semi-major and semi-minor axis values, which may be calculated as [296, 297]:

$$a_m = \frac{c\tau_m}{2} \qquad b_m = \frac{1}{2}\sqrt{c^2\tau_m^2 - D^2}, \tag{4.28}$$

where τ_m is the maximum time of arrival to be considered, D is the distance between the transmitter and the receiver and c is the velocity of light in free space. This model was proposed for microcellular environments [297], where the antenna heights are relatively low, and therefore, multipath scattering near the base station is equally likely, as scattering near the mobile station [297].

The GBSBEM may be used to generate the path time delay, τ_i, the angle of arrival, ϕ_i, the direction of departure, Φ_i, the power of the multipath component, P_i, and the phase angle, α_i. However, here we are only concerned with the angle of arrival information at the base station. The Cumulative Density Function (CDF) of the angle of arrival, ϕ_i, conditioned on

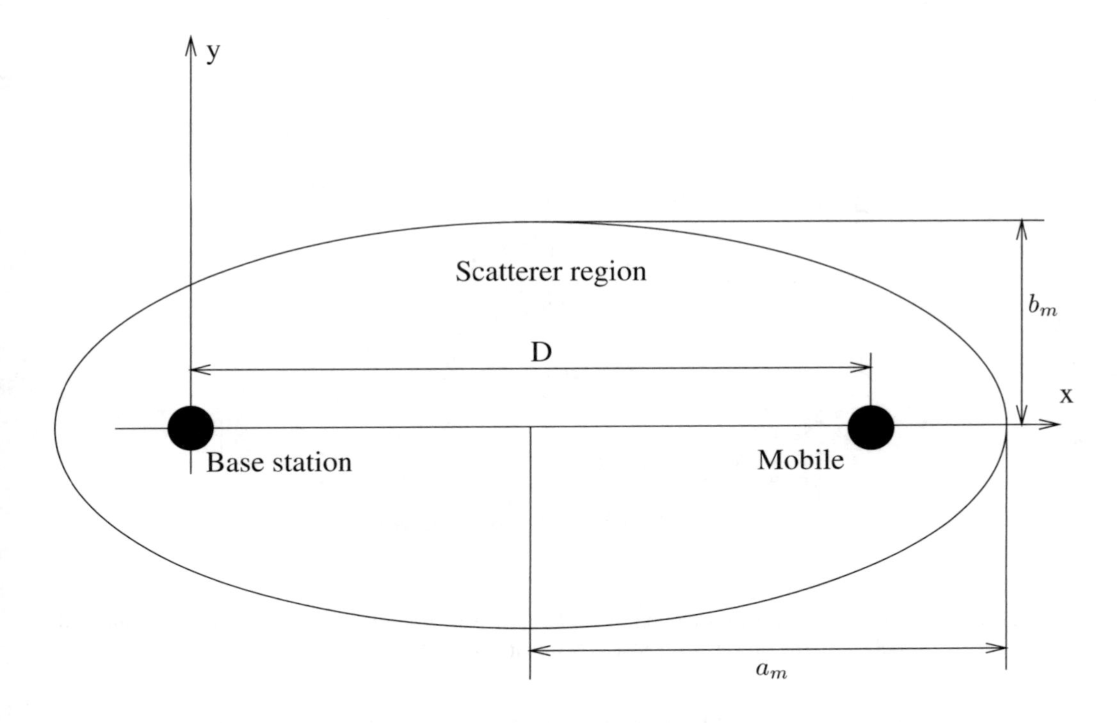

Figure 4.15: The Geometrically Based Single-Bounce Elliptical Model (GBSBEM), which is suitable for use as a microcellular model, showing the region in which the scatterers are located.

the normalised multipath delay, $r_i = c\tau_i/D = \tau_i/\tau_0$, is given as [297]:

$$F_{\phi|r}(\phi_i|r_i) = \begin{cases} \frac{1}{2\pi}\cos^{-1}\left(\frac{1-r_i\cos\phi_i}{r_i-\cos\phi_i}\right) - \frac{\sqrt{r_i^2-1}\sin(-\phi_i)(1-r_i\cos\phi_i)}{2\pi(2r_i^2-1)(r_i-\cos\phi_i)^2} & -\pi \leq \phi_i \leq 0 \\ 1 - \frac{1}{2\pi}\cos^{-1}\left(\frac{1-r_i\cos\phi_i}{r_i-\cos\phi_i}\right) + \frac{\sqrt{r_i^2-1}\sin(\phi_i)(1-r_i\cos\phi_i)}{2\pi(2r_i^2-1)(r_i-\cos\phi_i)^2} & 0 \leq \phi_i \leq \pi. \end{cases} \quad (4.29)$$

The conditional probability density function of ϕ_i, may be found by differentiating Equation (4.29) with respect to Φ leading to:

$$f_{\phi|r}(\phi|r_i) = \frac{(r_i^2-1)^{3/2}(r_i^2-2r_i\cos\phi+1)}{\pi(2r_i^2-1)(r_i-\cos\phi)^3} \qquad -\pi \leq \phi \leq \pi, \quad (4.30)$$

which is plotted in Figure 4.16 for various values of the normalised multipath delay, r_i. From this figure it can be seen that as the normalised multipath delay increases, the distribution of the angles-of-arrival tends to the uniform distribution, since the longer the delays, the greater the distance travelled, which results in a wider range of angles-of-arrival. In contrast, a small value of r_i concentrates the multipath components around the angle-of-arrival of the direct path component.

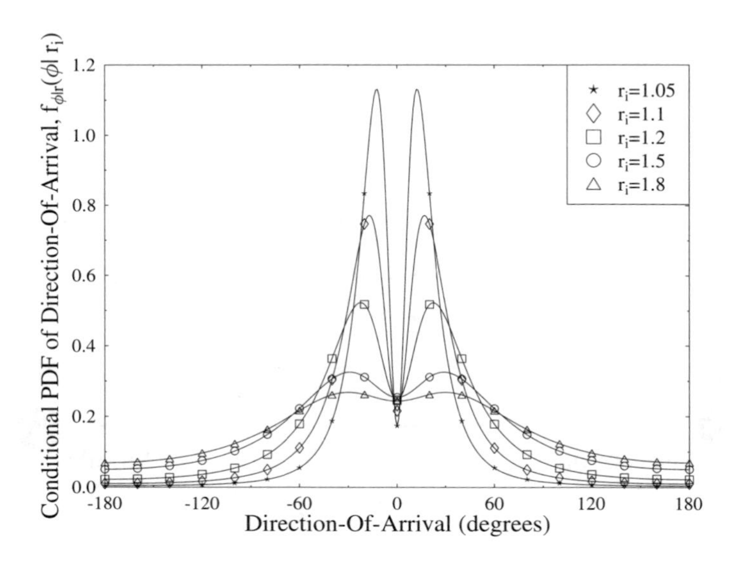

Figure 4.16: Probability density function of angle-of-arrival conditioned on the normalised multipath delay, r_i, for various values of r_i, evaluated from Equation 4.30.

In simulating multipath component parameters, it is necessary to generate samples of random variables from specified distributions. The normalised path delay, r_i, of the ith multipath component, may be calculated thus as [297]:

$$r_i = \sqrt{\frac{1}{2} + \frac{1}{2}\sqrt{1 + 4\beta^2 x_i^2}}, \tag{4.31}$$

where x_i is a uniformly distributed random variable, denoted by $U(0, 1)$, ranging from 0 to 1 and $\beta = r_m\sqrt{r_m^2 - 1}$ depends on the maximum value of the normalised path delay, r_m. The maximum normalised path delay, r_m, may be determined by the four different selection criteria summarised in Table 4.3 [297].

Criteria	Expression
Maximum path delay, τ_m	$r_m = \tau_m/\tau_0$
Fixed threshold, T (in dB), with path loss exponent n	$r_m = 10^{(T-L_r)/10n}$
Fixed delay spread, σ_τ	$r_m = 3.24(\sigma_\tau/\tau_0) + 1$
Maximum excess delay, σ_e	$r_m = (\tau_0 + \tau_e)/\tau_0$

Table 4.3: Selection criteria for choosing r_m, the maximum normalised path delay [296, 297].

Again, using large values of r_m results in a near-uniform distribution of the angles of arrival, whereas small values of r_m gives low-delay multipath components clustered in angle of arrival about the direct LOS path component.

From normalised path delay r_i and y_i, a uniformly distributed random variable, again formulated as $U(0, 1)$, over 0 to 1, it is now possible to determine the angle-of-arrival of the

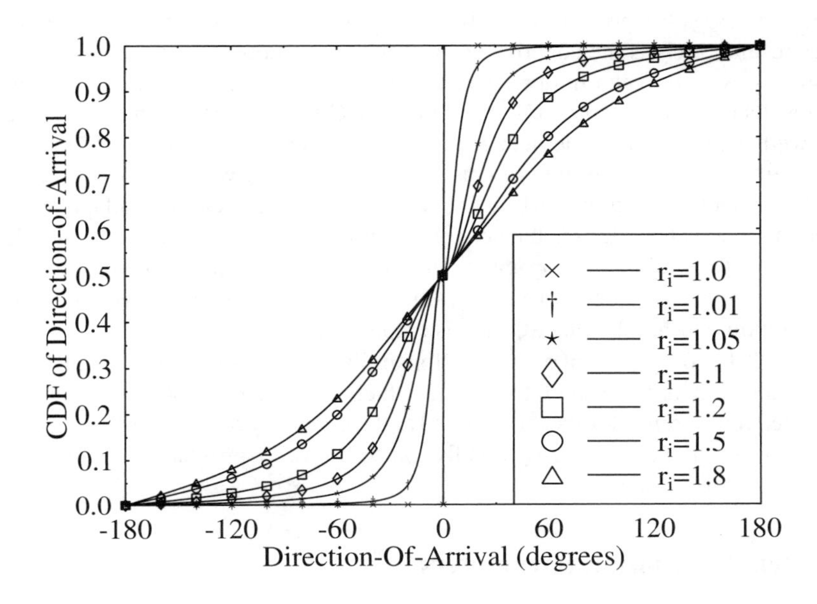

Figure 4.17: Cumulative density function of the angle-of-arrival conditioned on the normalised multipath delay, r_i, for various values of r_i.

ith multipath component by solving $y_i = F_{\phi|r}(\phi_i|r_i)$ for ϕ_i, where $F_{\phi|r}(\phi_i|r_i)$ is defined in Equation (4.29).

The corresponding Cumulative Density Function (CDF) is a smooth and monotonic function of the angle-of-arrival, as illustrated in Figure 4.17. The figure shows that, if the normalised path delay, $r_i = 1$, then the angle-of-arrival is $0°$, and that as r_i increases, so does the spread of values of the angle-of-arrival.

Therefore, to summarise, the process of generating the angles-of-arrival obeying the required distribution the following sequence of operations must be performed:

- Determine r_m for the scenario under consideration.
- Calculate $\beta = r_m\sqrt{r_m^2 - 1}$.
- Generate $x_i = U(0, 1)$.
- Calculate $r_i = \sqrt{\frac{1}{2} + \frac{1}{2}\sqrt{1 + 4\beta^2 x_i^2}}$.
- Generate $y_i = U(0, 1)$.
- Solve Equation (4.29) for ϕ_i, given y_i using numerical methods.

4.6 Network Performance Results

Section 4.6.1 describes the processes involved in the simulator used to obtain the network performance results, such as the adaptive beamforming techniques, new call generation and handover queues as well as the multipath propagation model. Section 4.6.2.1 presents our simulation results obtained for the FCA and LOLIA DCA algorithms with a single element antenna, as well as two and four element adaptive antenna arrays, assuming a LOS propagation environment. Further results are presented in Section 4.6.2.2 which were obtained using the multipath channel of Section 4.6.1, using two, four and eight element adaptive antenna arrays. Section 4.6.2.3 characterises the network performance of using two and four element antenna arrays, in the multipath propagation environment, in conjunction with power control. This is further expanded upon in Section 4.6.2.4, where power control assisted Adaptive Quadrature Amplitude Modulation (AQAM) is employed.

Sections 4.6.3.1-4.6.3.4 present our results for similar scenarios generated using the 'wrap-around' rather than the 'desert-island' technique, which eliminates the edge effects associated with the reduced interference levels encountered at the boundary of the simulation area. This process is described in Section 4.6.1. Finally, Section 4.6.2.7 provides a summary of the results obtained in this section.

4.6.1 System Simulation Parameters

The performance of the various channel allocation algorithms was investigated in a GSM-like [28] microcellular system, the parameters of which are defined in Table 4.4. The propagation environment was modelled using the power pathloss model having a pathloss exponent of -3.5. The mobile and base station transmit powers were fixed at 10 dBm (10 mW) for the simulations using no power control. The mobile and base station transmit powers were restricted to the range of -20 dBm to +10 dBm for the power control assisted and adaptive modulation based simulations. The number of carrier frequencies in the whole system was limited to seven, each supporting eight timeslots, in order to maintain an acceptable computational load. This implied that the DCA system employing seven carrier frequencies in conjunction with eight time slots, as in GSM for example, was potentially capable of handling a maximum of $7 \times 8 = 56$ (or $12 \times 8 = 96$) instantaneous calls at one base station, provided that their quality was adequate. If a channel allocation request for a new call could not be satisfied immediately, it was queued for a duration of up to 5 s, after which time, if not satisfied, it was classed as blocked. It was assumed that the network was synchronous from cell to cell, thus channels on different time slots of the same frequency were orthogonal in the time-domain and hence did not interfere with each other. The GSM-like system used a channel bandwidth of 200 kHz, but instead of the Gaussian Minimum Shift Keying (GMSK) [11] based modulation scheme, 4-QAM was employed for the sake of increasing the achievable bandwidth efficiency from 1.35 bps/Hz to 1.64 bps/Hz. Hence, the achievable bit rate was 200 kHz $\times$ 1.64 bps/Hz = 328 kbps. When dividing this bit rate amongst the eight users supported by the eight timeslots, the channel rate of the users - when for the sake of a simple argument neglecting transmission overheads, such as the equaliser training sequences, tailing sequences, guard periods and channel coding - became 328/8 = 41 kbps. The call arrivals were Poisson distributed, and hence the call duration and inter-call periods were exponentially distributed [289, 348] with the mean values shown in Table 4.4.

Parameter	Value	Parameter	Value
Noisefloor	-104 dBm	Multiple Access	F/TDMA
Frame duration	0.4615 ms	Cell radius	218 m
BS transmit power	10 dBm	MS transmit power	10 dBm
BS power control	No	MS power control	No
Number of base stations	49	Handover hysteresis	2 dB
Outage SINR threshold	17 dB	Re-alloc. SINR threshold	21 dB
Modulation scheme	4-QAM	Pathloss exponent	-3.5
Number of timeslots	8	Number of carriers	7
Average inter-call-time	300 s	Max new-call queue-time	5 s
Average call duration	60 s	Ref. signal modulation	BPSK
Beamforming algorithm	SMI	Reference signal length	8 bits
MS speed	30 mph	No. of antenna elements	2, 4 & 8
Pathloss at 1 m ref. point	0 dB	Shadow fading	No
Geometry of antenna array	Linear	Array element spacing	$\lambda/2$
Channel/carrier bandwidth	200 kHz		

Table 4.4: Network simulation parameters.

The physical layer was modelled using two parameters, namely the 'Outage SINR' and 'Reallocation SINR', defined as the average Signal-to-Interference+Noise Ratio (SINR) required by a transceiver in order to satisfy the FER requirements over a narrowband Rayleigh fading channel. More specifically, Pilot Symbol Assisted (PSA) 4-QAM transmitting 2 bits per symbol was assumed, which had an outage SINR of 17 dB and a reallocation SINR of 21 dB [12, 13]. When the signal quality, expressed in terms of the SINR, drops below the 'Reallocation SINR', a low quality access is encountered, and the mobile requests a new physical channel to handover to, thus initiating an intra- or inter-cell handover. If, while waiting for a reallocation handover, the signal quality drops further, below the so-called 'Outage SINR', defined as the SINR required to maintain a 10% FER, then an outage is encountered. A prolonged outage leads to the call being dropped or forcibly terminated. Since a user typically views a dropped call as less desirable than a blocked call, a Handover Queueing System (HQS) was employed. By forming a queue of the handover requests, which have a higher priority during contention for network resources than new calls, it is possible to reduce the number of dropped calls at the expense of a higher blocked call probability. A further advantage of the HQS is that a time window is formed, during which the handover may take place, enabling the user to wait, if necessary, for a slot to become free, thus increasing its chances of a successful handover. This twin-threshold physical layer model is similar to those described by Tekinay and Jabbari [311] and Katzela and Naghshineh [310]. However, the model described here is based on SINR thresholds, rather than on the received power thresholds of Tekinay and Naghshineh [311] and Katzela and Naghshineh [310]. A further metric, namely the low signal quality probability, is calculated as the proportion of time that the SINR is below the 'Reallocation SINR' threshold.

Again, the 'Outage SINR' and 'Reallocation SINR' threshold were determined, with the aid of independent bit-level simulations, for BPSK, QPSK/4-QAM and 16-QAM [12, 13], conducted in a Rayleigh fading environment using approximately half-rate Bose-Chaudhuri-

Modulation Scheme	Reallocation SINR threshold (dB) for 5% FER	Outage SINR threshold (dB) for 10% FER
BPSK	17	13
4-QAM	21	17
16-QAM	27	24

Table 4.5: The 'Reallocation SINRs' and 'Outage SINRs' used in the handover process, found by bit-level simulations for BPSK, QPSK/4-QAM, and 16-QAM modems. The 'Reallocation SINR' is the SINR, below which a channel reallocation will be requested, while the 'Outage SINR' is the SINR, below which a service outage is declared. Successive service outages render the call to be forcibly terminated.

Hocquenghem (BCH) codes, which employed bit interleaving over the different number of bits per transmission frame conveyed by the different modem modes [354]. Thus, the 'Reallocation SINR' threshold was determined to be the average SINR required by the specific transceiver employed for maintaining a 5% transmission FER. This SINR value is transceiver dependent and in general can be reduced at the cost of increased transceiver complexity and power consumption. The loss of a maximum of 5% of the speech or video frames can be considered a worst-case scenario for modern 'wireless-oriented', i.e. error-resilient source codecs. Therefore, by setting the reallocation threshold at this level, the system requested handovers to new channels, before the speech or video quality degradation due to excessive FERs became objectionable.

The 'Outage SINR' threshold defines the SINR, below which the system declares that the radio channel has degraded to such a level as to cause a service outage. If the radio channel continues to be in outage, then the call is forcibly terminated. The 'Outage SINR' threshold was determined by bit-level or physical layer simulations to be the average SINR required for maintaining a 10% FER. Therefore, if the radio channel degrades such that at least 10% of the speech or video frames were lost for some period of time, then the call would be forcibly terminated. The corresponding SINR thresholds based on bit-level simulations of BPSK, QPSK/4-QAM and 16-QAM modems are shown in Table 4.5.

The mobiles were capable of moving freely, at a speed of 30 mph, in a fixed random direction, selected at the start of the simulation from a uniform distribution, within the simulation area of 49 traffic cells, each having a radius of 218 m. Two different types of simulation area were invoked, the classical 'desert island' type and the 'wraparound' type. The 'desert-island' or 'urban, sub-urban, rural' environment neglects the interference emanating from the cells surrounding the outside of the simulation area. In other words, the traffic cells at the centre of the simulation area are surrounded by interfering cells and thus are subjected to the highest levels of interference. However, the cells at the edges of the simulation area are not surrounded by interfering cells and hence are subjected to a lower level of interference. This can be likened to an 'urban, sub-urban, rural' environment, where the centre cells represent the urban environment and the outer cells are considered to be low traffic-density rural cells in nature. However, this can lead to optimistic results, and hence often a 'wraparound' simulation area is used [355, 356]. In order to facilitate the employment of an infinite plane of simulation area, a tessellating rhombic simulation area was used. Hence, the simulation area was replicated around itself, or tiled to form a larger, or effectively infinite, simulation

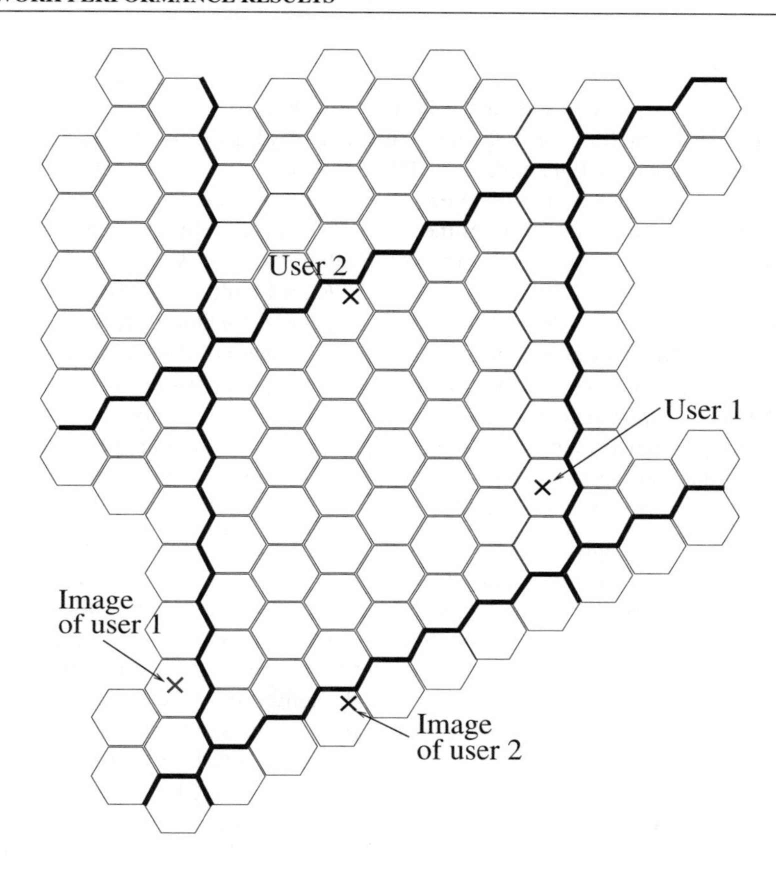

Figure 4.18: The 7x7 rhombic simulation area showing a user and its 'wrapped' image.

area. More explicitly, mobile stations and their signals were 'wrapped around' from one side of the network to the other [355, 356]. Hence, for example, a mobile station in call, which leaves the network at its edge, re-enters the network at the opposite side, whilst inflicting Co-Channel Interference (CCI) to **all** users, which may be positioned at any location in the network. Figure 4.18 depicts this scenario graphically.

The receiver antenna array weights were calculated using the Sample Matrix Inversion (SMI) algorithm [242, 244, 249], which determines the value of the AAA weights, such that they are optimised with respect to the received SINR [249]. In order to calculate the receiver antenna weights using the SMI algorithm, an eight-symbol long BPSK reference signal was assigned to the desired mobile. The remaining seven orthogonal eight-symbol duration BPSK reference symbols were then assigned to the interfering mobiles. However, any of these seven codes were allocated to more than one mobile, if the number of interferers was higher than seven. Thus, the desired mobile was uniquely identifiable, with the aid of its reference signal and the receiver antenna weights were optimised for obtaining the maximum received SINR, as detailed in Section 3.3.2.3. The calculation of the receiver antenna array weights was performed on a transmission frame-by-frame basis, leading to updated uplink and downlink

SINRs every transmission frame.

The base station's receiver antenna weights calculated for uplink reception may not be suitable for the downlink transmission due to the generally uncorrelated uplink and downlink channels of Frequency Division Duplexed (FDD) systems. However, forming a feedback loop from the mobile to the base stations for conveying the mobile's received reference signal and thus effectively conveying the quality of the mobile's received reference signals for use in an iterative adaptive beamforming algorithm, would allow the base station to use the downlink weights as proposed in [278, 279]. In a Time Division Duplexed (TDD) system having a sufficiently short dwell time, the AAA weights calculated for uplink reception can also be used for downlink transmission, since the propagation channel does not vary significantly between the uplink and downlink timeslots [6]. However, the system considered here is an FDD based network, and hence the assumption of channel predictability should therefore give an upper limit to the performance gains that may be achieved using an adaptive array. From now on we assume that the base station's receive and transmit, in other words the uplink and downlink beam patterns are identical.

An example of the adaptive antenna array beam patterns generated by two element adaptive antenna arrays is shown in Figure 4.19. In this figure the mobiles are denoted by the use of small squares, while the base stations are represented by black filled circles. The solid black lines from the base stations to the users show the direction that the antenna array is steered in, and the gain in that direction. The half-tone grey lines pointing towards the mobiles represent the interfering signals, where the length of these lines is proportional to the gain of the antenna array in that direction. As it can be seen from the figure, the main beamwidth is large and, although there is some beneficial interference nulling, its extent is limited. For the four element adaptive antenna array, as in Figure 4.20, the beams in the direction of the desired users are significantly narrower, and hence the interference sources are nulled much more strongly, as indicated by the shortened half-tone lines. We can observe in both Figures 4.23 and 4.24 that the antenna array beam patterns formed are symmetrical in the y-axis, as a direct consequence of the linear array geometry with the antenna array elements located on the y-axis. Using an alternative array geometry, such as a square or circle shaped one, would prevent this beam pattern symmetry from occuring and thus could potentially improve the achievable performance.

Both a purely Line-Of-Sight (LOS) propagation environment and a multipath propagation environment were considered. This multipath environment consisted of the Line-Of-Sight (LOS) ray and two additional rays, each having a third of the power of the LOS ray. The angles-of-arrival at the base station were determined using the Geometrically Based Single-Bounce Elliptical Model (GBSBEM) of Section 4.5 [296, 297], with its parameters chosen such that the multipath rays had one-third of the received power of the direct ray. The multipath received power criteria of Table 4.3 was used to determine the value of r_m to be used in the GBSBEM. Specifically, we opted for $r_m = 10^{(T-L_r)/10n}$, where T is the received power value in dB, L_r is the reflection loss and n is the pathloss exponent. Furthermore, T was set to 4.8 dB with L_r equal to zero in conjunction with a pathloss exponent of 3.5, in order to achieve the desired received signal power of one-third that of the LOS ray. Hence, using the formulae of Section 4.5, $r_m = 10^{(4.8-0.0)/35} = 1.36874$, leading to, $\beta = r_m\sqrt{r_m^2 - 1} = 1.2792$. Since $r_i = \sqrt{\frac{1}{2} + \frac{1}{2}\sqrt{1 + 4\beta^2 x_i^2}}$ where x_i is a uniformly distributed random variable over $[0, 1]$, r_i varies from 1.0 to 1.36874. The PDF of the angle-

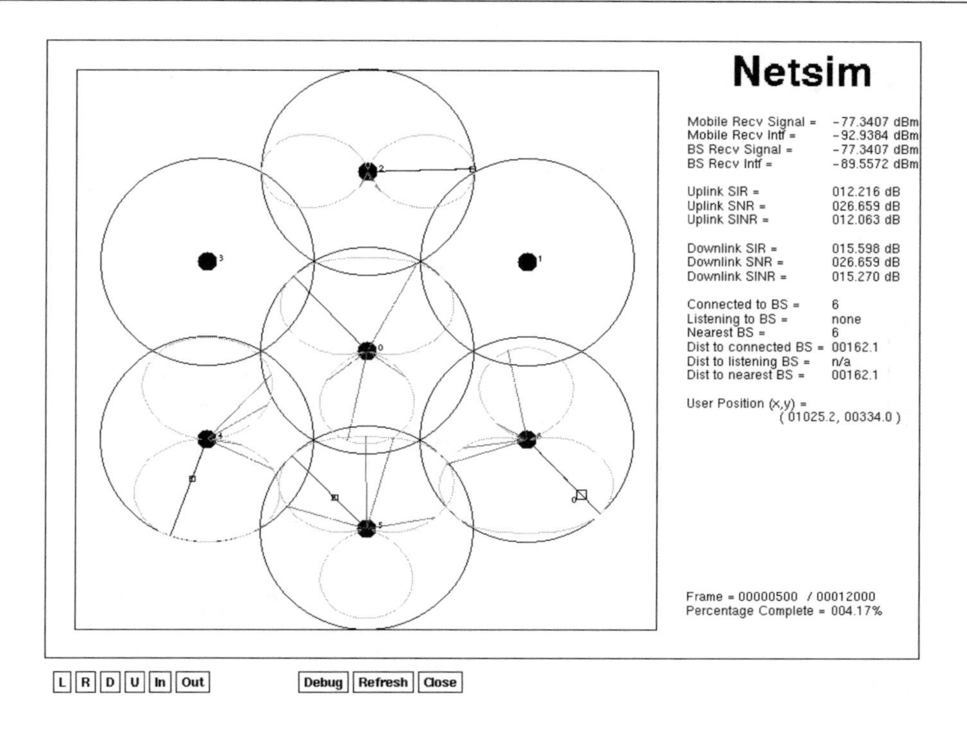

Figure 4.19: Screenshot of the simulation software, 'Netsim', for a 7-cell, 5-user simulation, showing the identical uplink and downlink, or receive and transmit, beam patterns generated by the adaptive antenna arrays using 2 elements. The squares represent the mobiles, with the large black circles denoting the base stations. The black lines from the base stations, passing through the squares, show the array gain in the desired direction. While the half-tone grey lines point in the direction of interfering sources, where the length of the lines indicates the antenna gain in that direction.

of-arrival for r_m=1.36874 is shown in Figure 4.21, which was generated using the GBSBEM algorithm of Section 4.5 for 100 000 trials. It was assumed that all of these multipath rays arrived with zero time delay relative to the LOS path, or that a space-time equaliser [18, 38] was employed, thus making full use of the additional received signal energy. However, the numerous extra desired and interfering signals incident upon the antenna array rapidly consume the finite degrees of freedom of the antenna array, limiting its ability to fully cancel each source of interference.

The addition of multipath rays, for both the desired signal and the interference sources, results in many more received uplink signals impinging upon the antenna array at the base station. A result of the increased number of received uplink signals is that the limited degrees of freedom of the base station's adaptive antenna array are exhausted, resulting in reduced nulling of the interference sources. A solution to this limitation is to increase the number of antenna elements in the base station's adaptive array, although this has the side effect of raising the cost and complexity of the array. In a macro-cellular system it may be possible to

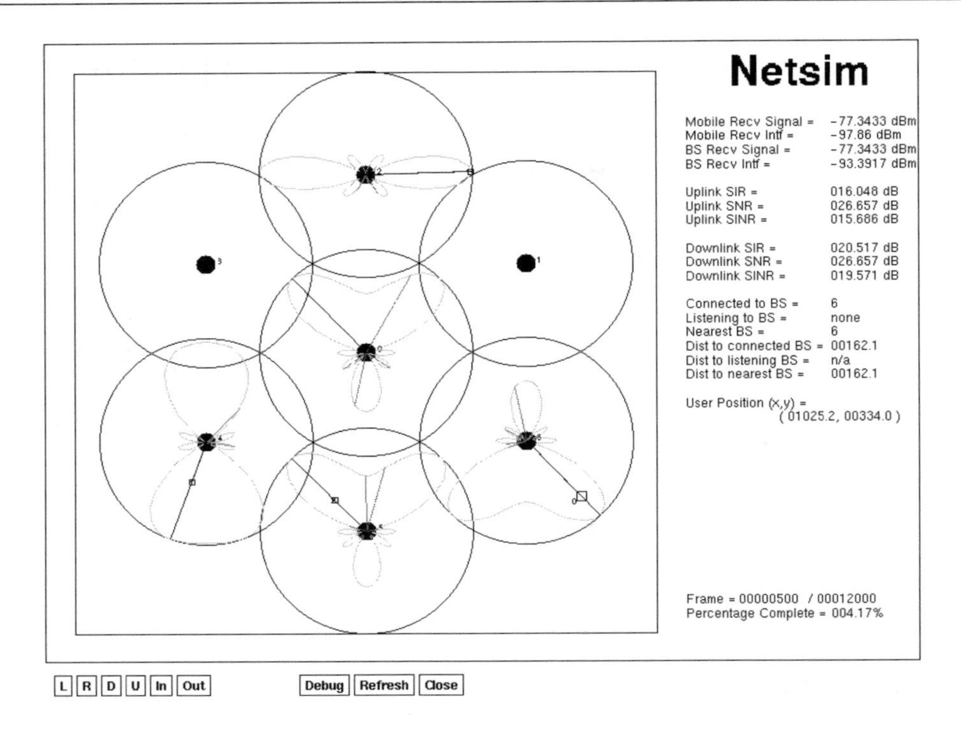

Figure 4.20: Screenshot of the simulation software, 'Netsim', for a 7-cell, 5-user simulation, showing the identical uplink and downlink, or receive and transmit, beam patterns generated by the adaptive antenna arrays using 4 elements. The squares represent the mobiles, with the large black circles denoting the base stations. The black lines from the base stations, passing through the squares, show the array gain in the desired direction. While the half-tone grey lines point in the direction of interfering sources, where the length of the lines indicates the antenna gain in that direction.

neglect multipath rays arriving at the base station from interfering sources since the majority of the scatterers are located close to the mobile station [21]. In contrast, in a micro-cellular system the scatterers are located in both the region of the reduced-elevation base station and that of the mobile, and hence multipath propagation must be considered. Figure 4.22 shows the simulated environment for both the uplink and the downlink, with the multipath components of the desired signal and interference signals clearly illustrated, where the uplink and downlink are assumed to be reciprocal. When the DCA algorithm is 'listening', in order to determine the best channel to be selected, only the LOS signals are considered, while the multipath signals are neglected. However, at all other times the multipath signals are used in the calculation of the received signal and interference levels.

Figures 4.23, 4.24 and 4.25 show examples of the beam patterns obtained for two, four and eight element adaptive antenna arrays in the presence of multipath propagation. For the two element antenna array, as illustrated in Figure 4.23, the beamwidth of the antenna array is large, thus limiting its efficiency in nulling the sources of interference. Nonetheless,

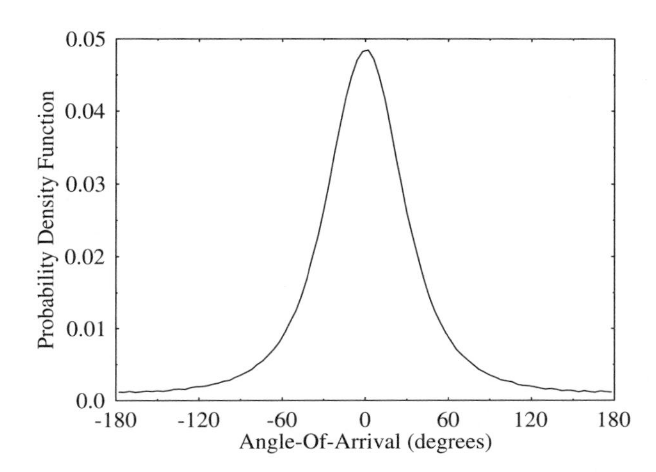

Figure 4.21: Probability density function of the angle-of-arrival of the uplink multipath rays, centred about the angle-of-arrival of the line-of-sight path. Furthermore, $r_m = 1.36874$ and 1 000 000 trials were used.

it can be seen that the arrays are attempting to steer towards the desired signals, and away from the sources of interference. Here the desired signals are represented by the three black lines, where the black line passing through a square is the direct ray, while the remaining two black lines indicate the multipath rays arriving from the desired user. Observe in the figures that most of these lines end on the unity-gain circles, implying that they are received with a unity gain by the base station. Furthermore, the dark grey lines indicate the LOS paths from the interference sources, while the corresponding two multipath rays of the interferers are denoted by the light grey lines. The beam pattern of base station '1' is a good example of how the array is steering towards the desired signal paths, and away from the interference. For base station '5', at the bottom of Figure 4.23, the small angular separation between the arriving signals, and the end-fire location of these sources, makes rejection of the interference harder to accomplish. The use of a four element antenna array, depicted in Figure 4.24, results in more successful nulling of the interference sources, but again, for base station '5' at the bottom, the similar angular location of the desired and interfering sources results in poor interference cancellation performance. Figure 4.25 shows that an eight element adaptive antenna array performs well in most cases, nulling the sources of interference strongly, whilst efficiently steering towards the desired signals. Using an alternative layout of the antenna elements, rather than the uniform linear array, should minimise the possibility of a situation, similar to that of base station '5', where all the sources are located at end-fire, which is the area of poorest performance of the array.

Having described the simulation parameters, in the next section we present our simulation results, quantifying the amount of traffic that can be carried by each of the simulated networks, whilst maintaining the required network quality.

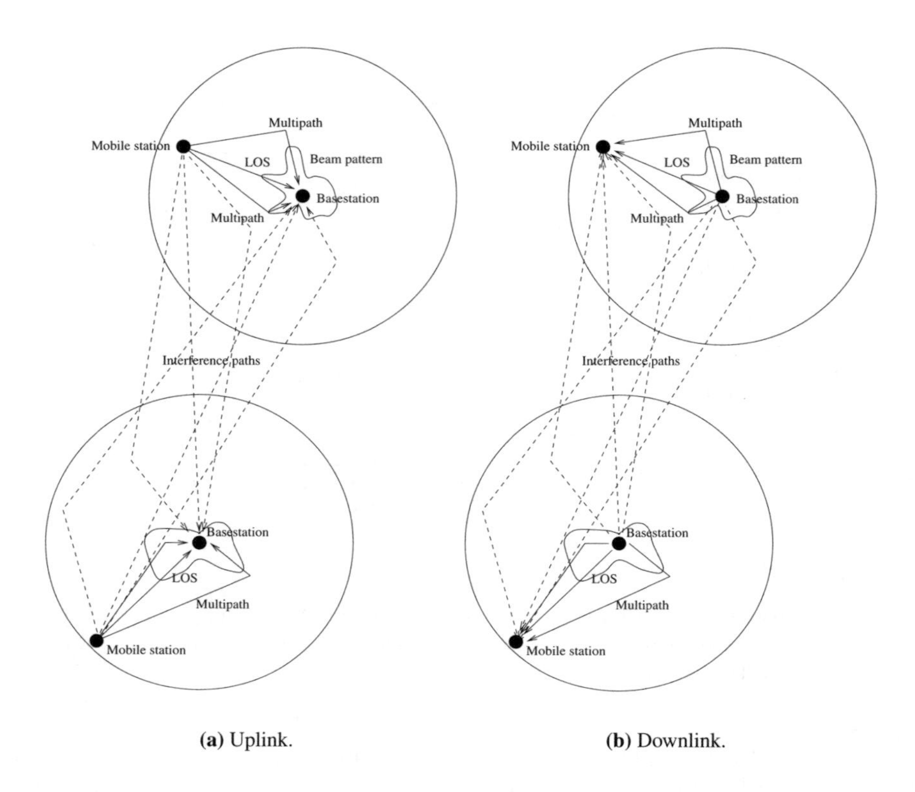

(a) Uplink. (b) Downlink.

Figure 4.22: The multipath environments of both the uplink and downlink, showing the multipath components of the desired signals, the line-of-sight interference and the associated base station antenna array beam patterns.

4.6.2 Non-Wraparound Network Performance Results

The results presented in this section were obtained for the 'desert-island' or 'urban, suburban, rural' scenario, i.e. with the highest levels of interference present at the centre of the simulation area. Results were obtained for single, two and four element antenna arrays over an LOS channel for both the FCA algorithm and the LOLIA with exclusion zones of 7 and 19 cells. This work was then extended to provide network capacity estimates for non-LOS or multipath channels using adaptive antenna arrays comprising two, four and eight elements. Power control and adaptive modulation techniques were also employed for increasing the network capacity further.

4.6.2.1 Performance Results over a LOS Channel

Figure 4.26 shows the new call blocking probability for a variety of uniform traffic loads, measured in terms of the mean normalised carried traffic, with units of Erlangs/km^2/MHz. The figure shows that for a given traffic load, both FCA and the LOLIA, using an exclusion zone of $n = 19$ maintained a fairly similar probability of new call blocking, regardless of

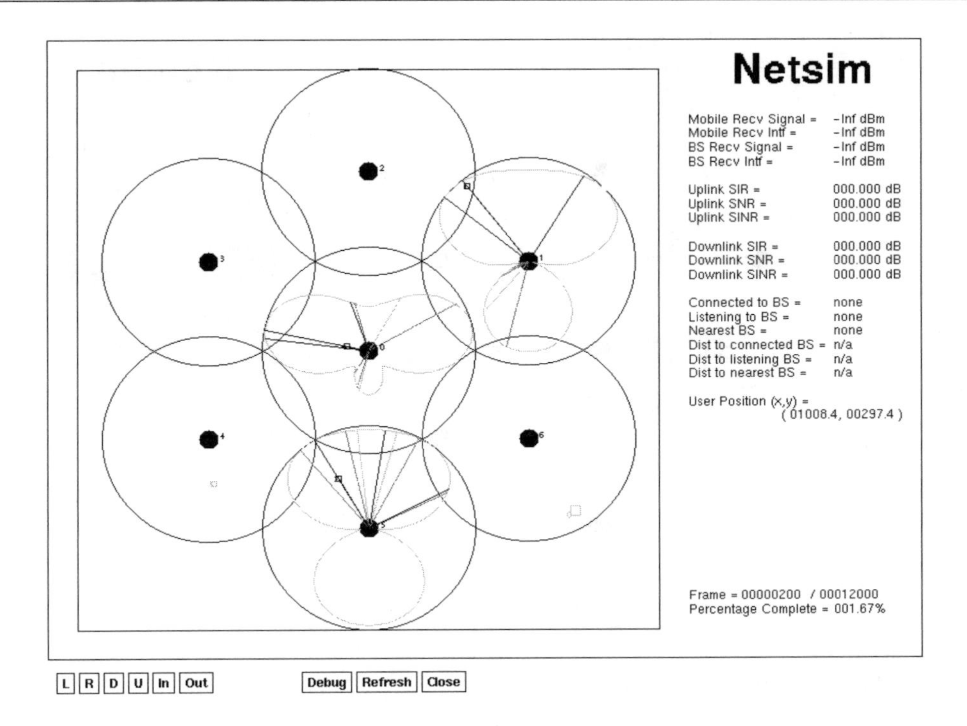

Figure 4.23: Screenshot of the simulation software, 'Netsim', for a 7-cell, 5-user scenario, showing the beam patterns generated by the adaptive antenna arrays using 2 elements. The squares represent the users, with the large black circles denoting the base stations. The black lines from the base stations, passing through the squares, show the array gain in the desired direction, the black lines not passing through the squares are the desired user's multipath rays. The dark grey lines are the LOS interference paths, while the interferer's multipath components are illustrated by the light grey lines, where the length of the lines is proportional to the corresponding antenna gains in their directions.

the number of elements in the antenna array. In the case of the FCA algorithm, this was due to the limited number of frequency/timeslot combinations available as a direct result of the fixed nature of the network. However, for the LOLIA having an exclusion zone of 19 cells, the lack of frequency/timeslot combinations was due to the large exclusion zone. Thus, using the smaller exclusion zone of 7 cells led to a significantly reduced new call blocking probability. The figure also shows that, since the new call blocking probability of the LOLIA using $n = 7$ was reduced, thanks to the adaptive antenna arrays, the new call blocking performance was interference limited. This contrasts with the FCA algorithm and the LOLIA using $n = 19$, whose new call blocking performance was limited by the availability of frequency/timeslot combinations. It is interesting to note that, in terms of its new call blocking probability, the FCA algorithm performed better using only one antenna element as a result of its significantly increased call dropping probability, which freed up network resources, thus enabling more new calls to start.

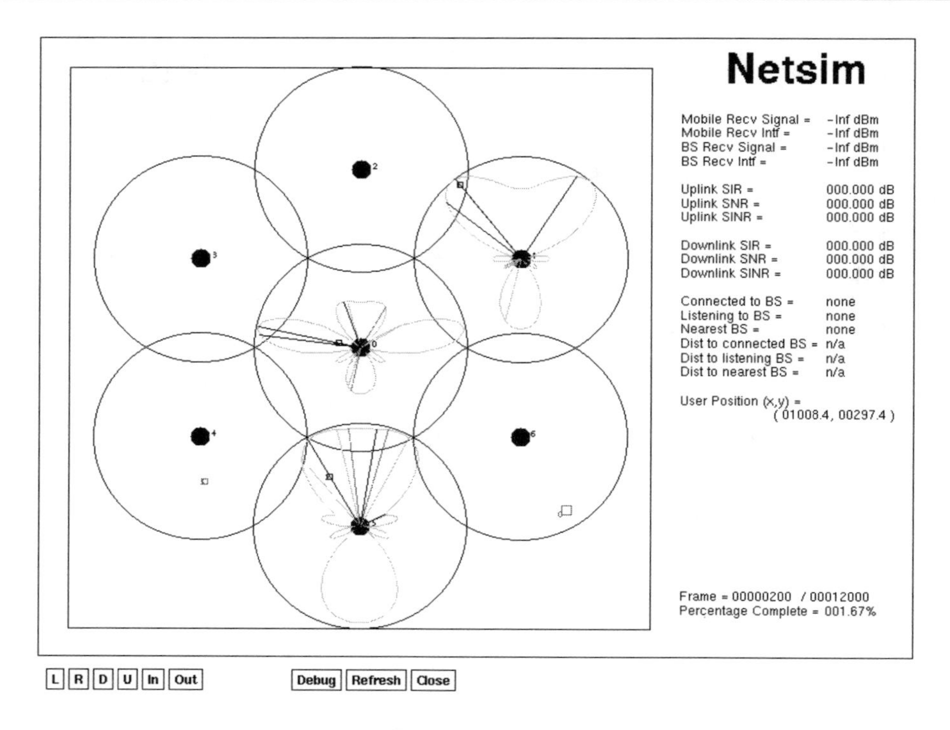

Figure 4.24: Screenshot of the simulation software, 'Netsim', for a 7-cell, 5-user scenario, showing the beam patterns generated by the adaptive antenna arrays using 4 elements. The squares represent the users, with the large black circles denoting the base stations. The black lines from the base stations, passing through the squares, show the array gain in the desired direction, the black lines not passing through the squares are the desired user's multipath rays. The dark grey lines are the LOS interference paths, while the interferer's multipath components are illustrated by the light grey lines, where the length of the lines is proportional to the corresponding antenna gains in their directions.

The call dropping probability of the FCA algorithm, and that of the LOLIAs is depicted in Figure 4.27 for one, two and four element antenna arrays, when subjected to varying uniform traffic loads. The FCA algorithm suffered from the highest call dropping probability of the three channel allocation schemes. In conjunction with a four element adaptive antenna array it is similar to the LOLIA using $n = 7$ and a single antenna element for teletraffic loads higher than 10 Erlang/km^2/MHz. For teletraffic levels below this point, the FCA algorithm offered superior performance due to the call dropping probability 'floor' experienced by the LOLIA using $n = 7$. The large exclusion zone of the LOLIA using $n = 19$ resulted in a very low probability of forced termination until the system approached its maximum capacity of around 12 Erlang/km^2/MHz, where the dropping probability increased rapidly. However, the performance of the LOLIA with $n = 19$ still exceeded that of both the FCA algorithm and the LOLIA with $n = 7$ due to the low levels of co-channel interference resulting from the high frequency re-use distance associated with the large exclusion zone.

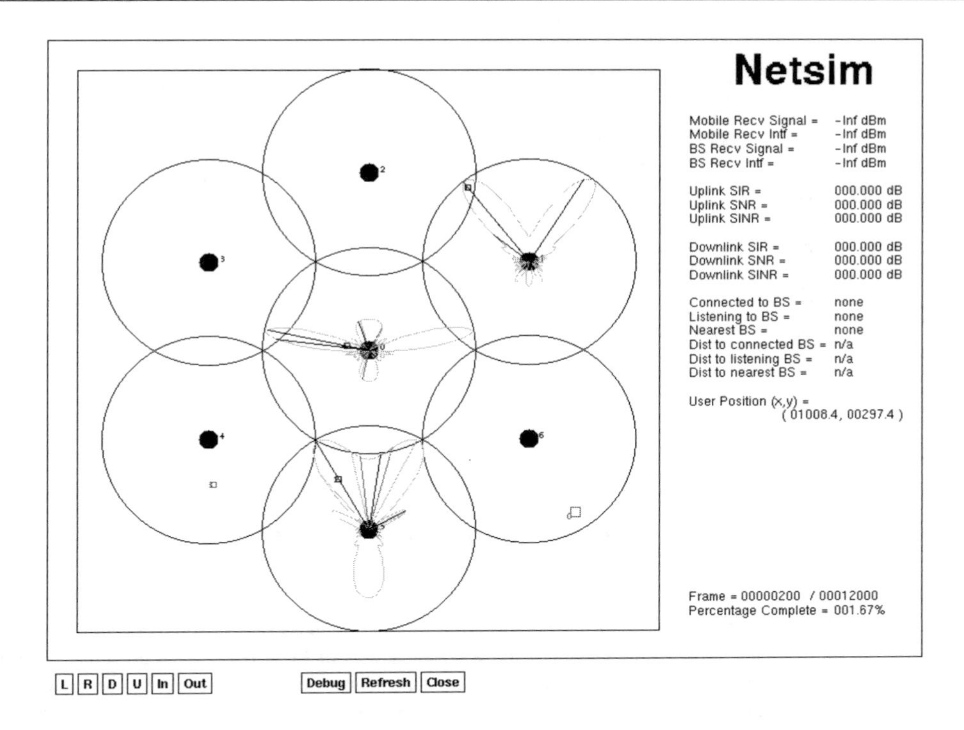

Figure 4.25: Screenshot of the simulation software, 'Netsim', for a 7-cell, 5-user scenario, showing the beam patterns generated by the adaptive antenna arrays using 8 elements. The squares represent the users, with the large black circles denoting the base stations. The black lines from the base stations, passing through the squares, show the array gain in the desired direction, the black lines not passing through the squares are the desired user's multipath rays. The dark grey lines are the LOS interference paths, while the interferer's multipath components are illustrated by the light grey lines, where the length of the lines is proportional to the corresponding antenna gains in their directions.

Figure 4.28 shows the probability of low quality access versus various uniform traffic loads. The figure shows our results for the FCA algorithm and the LOLIA for nearest base station constraints of 7 and 19 cells. Again, the LOLIA with $n = 19$ offered the best performance at the lower traffic levels, but the low-quality access probability increased the most rapidly as the traffic load increased. For a given traffic load the LOLIA using $n = 19$ provided the lowest probability of a low quality access. This resulted from the low level of co-channel interference of the network and the interference rejection capabilities of the adaptive antenna arrays. The figure shows that all of the channel allocation schemes benefited from the use of the adaptive antenna arrays.

Figure 4.29 shows the Grade-Of-Service (GOS) for a range of uniform teletraffic loads. The figure shows results for the FCA algorithm and the LOLIAs with nearest base station constraints of 7 and 19 cells, for cases of a single antenna element as well as for two and four element adaptive antenna arrays. The grade of service is better, i.e. lower, for larger

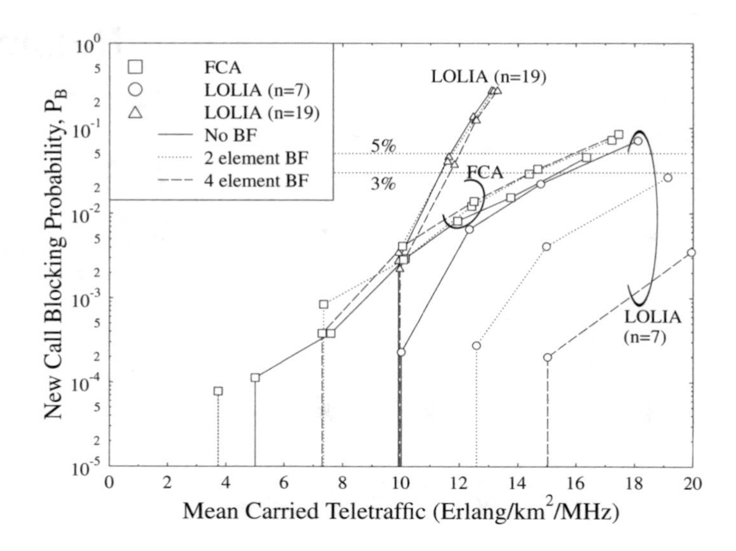

Figure 4.26: New call blocking probability performance versus mean carried traffic, for comparison of the LOLIA, with 7 and 19 'local' base stations, and of FCA using a 7-cell reuse cluster, under a uniform geographic traffic distribution, for a single antenna element as well as for two and four element antenna arrays with beamforming in an **LOS environment**. See Figure 4.32 for the corresponding multipath results.

exclusion zone size when the traffic load is low, which is reversed for high traffic loads. This is mainly attributable to the higher call blocking probability of the larger exclusion zone of 19 cells, particularly in the region of the highest traffic loads. The GOS for the FCA scheme follows the probability of a blocked call and the dropping probability trends by increasing smoothly and monotonically with the traffic load.

The effect of beamforming on the number of handovers performed can be seen in Figure 4.30. The performance of the LOLIAs was barely altered by the use of beamforming, with both performing the lowest number of handovers per call. At the highest teletraffic loads it can be seen that the LOLIA using an exclusion zone of 7 base stations benefited slightly from the use of the adaptive antenna arrays. In contrast, the number of handovers performed by the FCA algorithm was reduced significantly as a benefit of using adaptive antennas with a maximum reduction in the mean number of handovers performed per call of 69% for two elements, and of 86% for four elements. This translates into a significantly reduced load for the network, since it has to manage far less handovers, therefore reducing the complexity of the network infrastructure. As the network load exceeded about 12 Erlangs/km^2/MHz, the mean number of handovers performed per call dropped due to the excessive call dropping probability, since calls were being dropped before they could handover, thus reducing the number of handovers.

Figure 4.31 portrays the mean carried teletraffic versus the number of mobiles in the simulated system. The figure shows that at low traffic loads both FCA and the LOLIA carry virtually identical amounts of traffic. However, as the mobile density, and hence the traffic

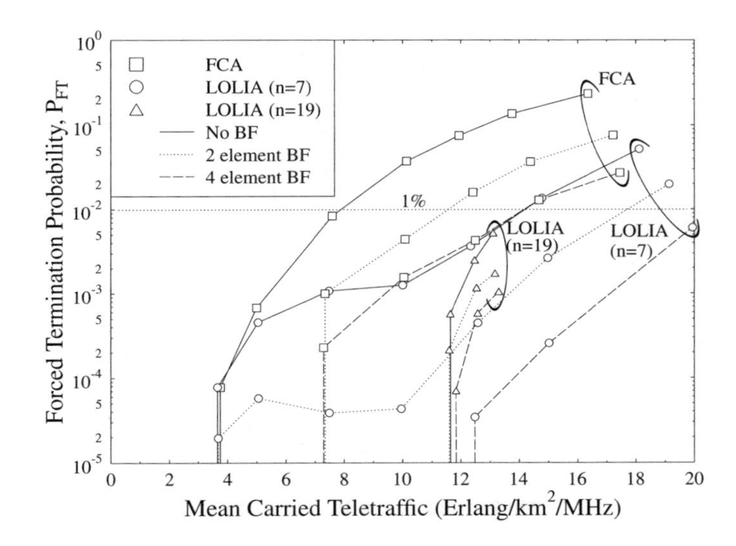

Figure 4.27: Call dropping probability performance versus mean carried traffic, for comparison of the LOLIA, with 7 and 19 'local' base stations, and of FCA using a 7-cell reuse cluster, under a uniform geographic traffic distribution, for a single antenna element as well as for two and four element antenna arrays with beamforming in an **LOS environment**. See Figure 4.33 for the corresponding multipath results.

load, is increased, the LOLIA with the nearest base station limit of 19 reaches its maximum traffic load and cannot carry further traffic. In other words, the employment of adaptive antennas does not enable the network to carry more traffic, since the performance of the network is resource limited, not interference limited. The limiting factor is effectively the high frequency reuse distance of the LOLIA in conjunction with n=19, since the associated low level of interference cannot be substantially further reduced by the adaptive arrays. Hence in Figure 4.31 increasing the number of antenna elements in the adaptive array does not support a susbtantially increased teletraffic capacity in terms of the number of users supported, since the number of available frequency/timeslot combinations is limited, as indicated by the flattening performance curves.

By contrast, for FCA and the LOLIA in conjunction with $n = 7$, the advantage of using adaptive antennas can be explicitly seen from the figure. Specifically, the FCA and the LOLIA in conjunction with $n = 7$, enable a higher level of traffic to be carried, at a higher quality than a system without adaptive antenna arrays. The performance gain attained by the LOLIA, over the FCA algorithm, is also shown in Figure 4.31, which illustrates the increase in carried traffic as a result of the dynamic configurability of DCA schemes.

It can be seen from Table 4.6 that for all of the channel allocation schemes, the use of adaptive antenna arrays at the receiver resulted in increased carried teletraffic, hence supporting a higher number of simultaneous users. The FCA algorithm benefited most from the use of adaptive antennas with a 67% increase in the number of users supported when using two antenna elements and a 144% rise in the carried traffic, when using an adaptive array

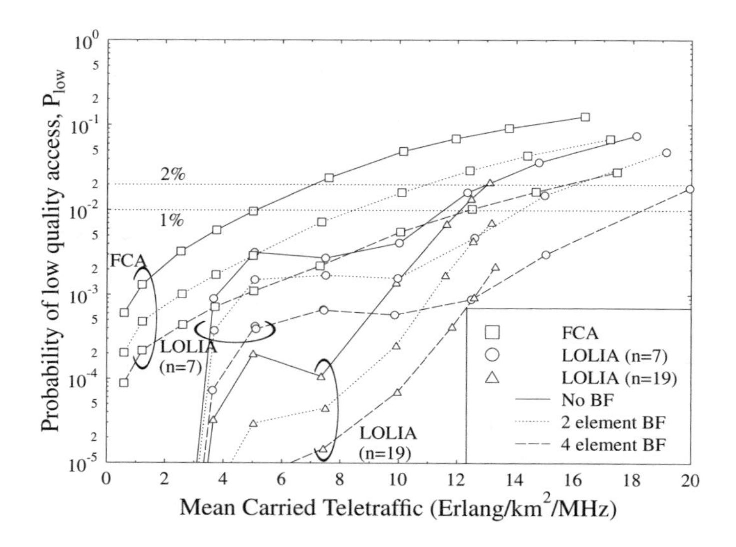

Figure 4.28: Probability of low quality access versus mean carried traffic, for comparison of the LOLIA, with 7 and 19 'local' base stations, and of FCA using a 7-cell reuse cluster, under a uniform geographic traffic distribution, for a single antenna element as well as for two and four element antenna arrays with beamforming in an **LOS environment**. See Figure 4.34 for the corresponding multipath results.

with four elements. The LOLIA associated with $n = 7$, supported a higher number of users than FCA although the capacity increases obtained through the use of adaptive antenna arrays were more limited. Specifically, a two element array carried an extra 22% of users and with the aid of four elements it supported 58% more users. Using a channel exclusion zone of 19 base stations gave a slight performance advantage over the 7-cell variant for the conservative scenario of Section 4.3.3.4, but only without adaptive antennas. Employing adaptive antennas had little effect on the number of users supported by the network using the LOLIA with $n = 19$, increasing the traffic carried by only a small margin. The corresponding multipath results are summarised in Table 4.7 with network configurations common between the two highlighted in bold.

4.6.2.2 Performance Results over a Multipath Channel

Following our previous simulations, where a purely LOS environment existed between the mobiles and their base stations, this section presents our performance results for the multipath environment described in Section 4.6.1, using two, four and eight element adaptive antenna arrays.

Comparing the blocking probabilities of the multipath environment, in Figure 4.32, with those of the LOS environment, which were portrayed in Figure 4.26, reveals that the FCA algorithm and both the LOLIAs behaved similarly in both propagation environments. Again, only the LOLIA with an exclusion zone of 7 base stations benefited from the use of the

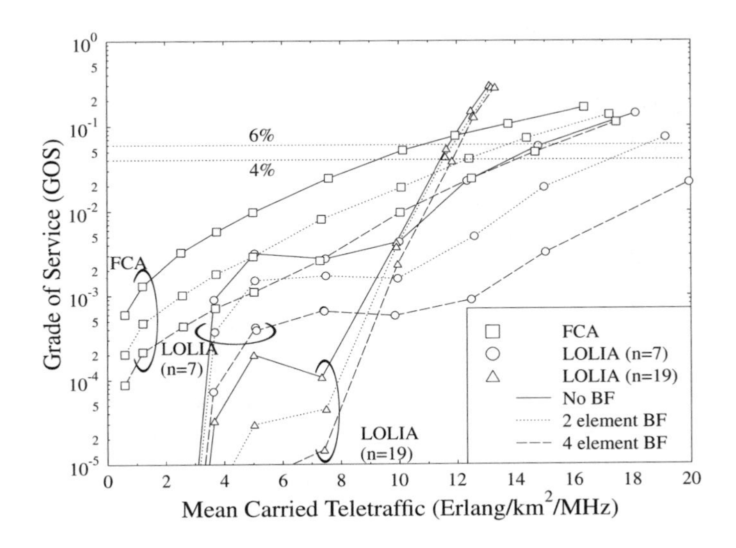

Figure 4.29: Grade-Of-Service (GOS) performance versus mean carried traffic, for comparison of the LOLIA, with 7 and 19 'local' base stations, and of FCA using a 7-cell reuse cluster, under a uniform geographic traffic distribution, for a single antenna element as well as for two and four element antenna arrays with beamforming in an **LOS environment**. See Figure 4.35 for the corresponding multipath results.

Algorithm	Conservative $P_{FT}=1\%, P_{low}=1\%$ $GOS=4\%, P_B=3\%$			Lenient $P_{FT}=1\%, P_{low}=2\%$ $GOS=6\%, P_B=5\%$		
	Users	**Traffic**	**Limiting Factor**	**Users**	**Traffic**	**Limiting Factor**
FCA, 1 element (el.)	815	5.10	P_{low}	1115	7.05	P_{low}
FCA, 2 elements	**1360**	8.45	P_{low}	**1755**	11.00	P_{low}
FCA, 4 elements	**1985**	12.40	P_{low}	**2710**	15.75	P_{low}
LOLIA (n=7), 1 el.	1855	11.50	P_{low}	2110	13.00	P_{low}
LOLIA (n=7), 2 el.	**2260**	14.15	P_{low}	**2600**	16.00	P_{low}
LOLIA (n=7), 4 el.	**2935**	18.30	P_{low}	**>3200**	>20.00	P_{low}
LOLIA (n=19), 1 el.	1935	11.35	P_B	2010	11.65	P_B
LOLIA (n=19), 2 el.	**1940**	11.35	P_B	**2045**	11.70	P_B
LOLIA (n=19), 4 el.	**1960**	11.65	P_B	**2090**	12.00	P_B

Table 4.6: Maximum mean carried traffic, and the maximum number of mobile users that can be supported by each configuration whilst meeting the preset quality constraints of Section 4.3.3.4. The carried traffic is expressed in terms of normalised Erlangs (Erlang/km^2/MHz), for the network described in Table 4.4 in an **LOS environment**.

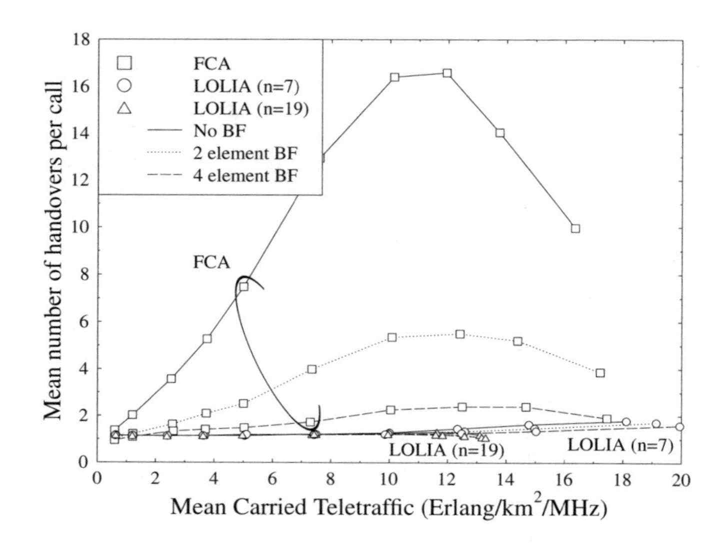

Figure 4.30: The mean number of handovers per call versus mean carried traffic, for comparison of the LOLIA, with 7 and 19 'local' base stations, and of FCA using a 7-cell reuse cluster, under a uniform geographic traffic distribution, for a single antenna element as well as for two and four element antenna arrays with beamforming in an **LOS environment**. See Figure 4.36 for the corresponding multipath results.

adaptive antenna arrays in terms of the new call blocking probability.

In Figure 4.33 the probability of a dropped call in a multipath environment is presented which, for the FCA algorithm, was similar under the multipath propagation conditions to that of the LOS scenario in Figure 4.27. The LOLIA using an exclusion zone of 7 base stations also exhibited call dropping probabilities close to those observed in the LOS scenario, when using a two element adaptive antenna array. In conjunction with a four element antenna array the performance was slightly degraded in the multipath scenario, but using the eight element antenna array resulted in superior performance to that of the four element array in the LOS environment. There was a slight call dropping performance improvement for the LOLIA using $n = 19$.

The probability of low quality access is depicted in Figure 4.34. The FCA algorithm did not perform as well, with respect to the probability of a low quality access, in the multipath propagation environment, when compared to the LOS case of Figure 4.28. The same was true of the LOLIA using $n = 7$ at higher traffic levels, although, at lower levels of traffic the performance in the multipath case was superior. At low levels of traffic the average level of interference was relatively low, and hence the extra signal power received in the multipath environment resulted in a reduced chance of a low quality access occurring. However, at higher levels of teletraffic, the background interference level was higher than in the LOS scenario of Figure 4.28, and hence the extra received power had a less beneficial impact, in fact the multipath components created additional interference. The LOLIA using an exclusion zone of 19 base stations and an adaptive antenna array of two elements performed better in the

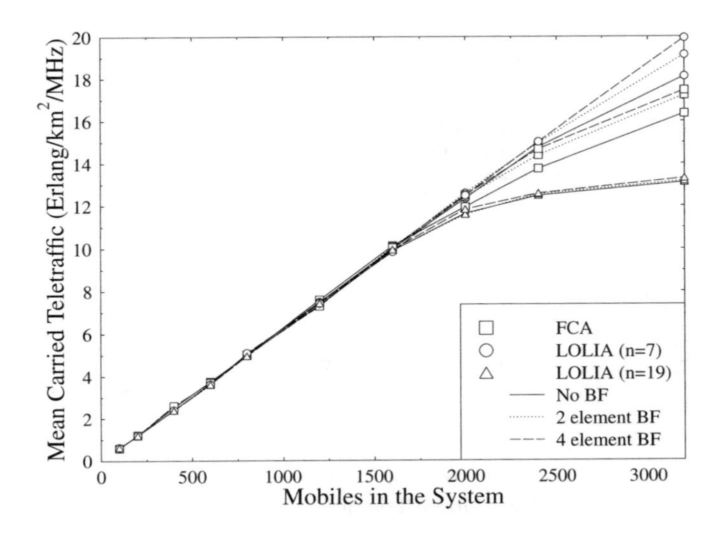

Figure 4.31: Mean traffic carried versus the number of mobiles in the system, for comparison of the LOLIA, with 7 and 19 'local' base stations, and of FCA using a 7-cell reuse cluster, under a uniform geographic traffic distribution, for a single antenna element as well as for two and four element antenna arrays with beamforming in an **LOS environment**. See Figure 4.37 for the corresponding multipath results.

multipath case. However, in conjunction with four elements it offered a superior performance in the LOS scenario of Figure 4.28. Overall, the improvement in the probability of low quality access through increasing the number of adaptive antenna array elements, was reduced in the multipath propagation environment, since the added interference power outweighed the increased received signal power. This ultimately reduced the prevalent SINR even when using adaptive antenna arrays.

As expected on the basis of Equation 4.15, the FCA algorithm and the LOLIA with $n =$ 19, offered a similar GOS performance for both the LOS scenario of Figure 4.29 and for the multipath environment. Figure 4.35 also shows that the GOS of the FCA algorithm using a given number of antenna elements is inferior to the GOS of the LOS propagation environment characterised in Figure 4.29, as for the probability of low quality access seen in Figures 4.28 and 4.34. At network loads of less than about 13 Erlang/km^2/MHz, the GOS of the LOLIA with $n = 7$ was superior to that of the LOS environment in Figure 4.29, however, above this carried traffic value the performance was worse.

Figure 4.36 demonstrates the significant impact that adaptive antennas have on the mean number of handovers per call for the FCA algorithm in a multipath environment. As in the LOS propagation environment characterised in Figure 4.30, more handovers per call were initiated when using FCA system employing two or four element antenna arrays, than for either of the LOLIAs using a single antenna element. Furthermore, a higher number of handovers was required in the multipath environment than in the LOS scenario of Figure 4.30, for a given antenna array configuration. The LOLIA schemes performed much fewer han-

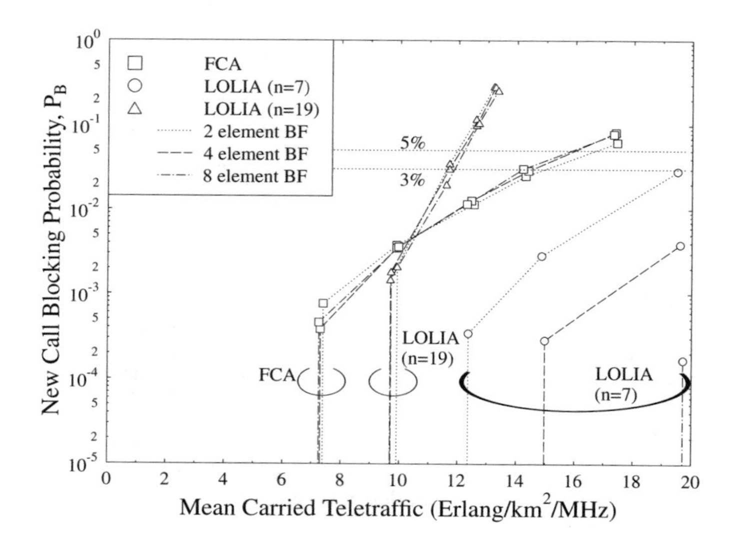

Figure 4.32: New call blocking probability performance versus mean carried traffic, for comparison of the LOLIA, with 7 and 19 'local' base stations, and of FCA using a 7-cell reuse cluster, under a uniform geographic traffic distribution, for two, four and eight element antenna arrays with beamforming in a **multipath environment**. See Figure 4.26 for the corresponding LOS results.

dovers than FCA, irrespective of the propagation environment, and generally did not appear to benefit from the employment of adaptive antennas in terms of the required handovers per call.

As it can be seen in Figure 4.37 for the adaptive array, the mean levels of carried teletraffic against the number of mobiles in the system followed a near-linear trend, with the capacity of the LOLIA 19 system rolling off above 2000 users, as for the LOS scenario in Figure 4.31. Above this number of users, very little extra teletraffic was carried, with corresponding several orders of magnitude increases of the blocking, dropping and low quality access probabilities as well as that of the GOS measure. For the channel allocation algorithms operating in a multipath rather than LOS environment, increasing the number of antenna elements did not significantly increase the levels of traffic carried, although the network performance improved in other respects, such as for example the call dropping probability.

Table 4.7 presents similar results to Table 4.6, but for a multipath environment. From this table it can be seen that LOLIA 19 actually performed slightly better in the multipath scenario, than in a LOS situation. This was due to the large reuse distance of the system, resulting in the sum of the three desired multipath signals versus the sum of the interfering signals being higher than the ratio of the LOS desired signal power to the LOS interference power. The LOLIA 7 algorithm, however, did not generally benefit from the multipath environment, since the smaller reuse distance resulted in numerous sources of relatively strong interference, all requiring cancellation. Therefore, as the number of antenna elements increased, so should the number of users supported by the network, as a result of the increased number of degrees

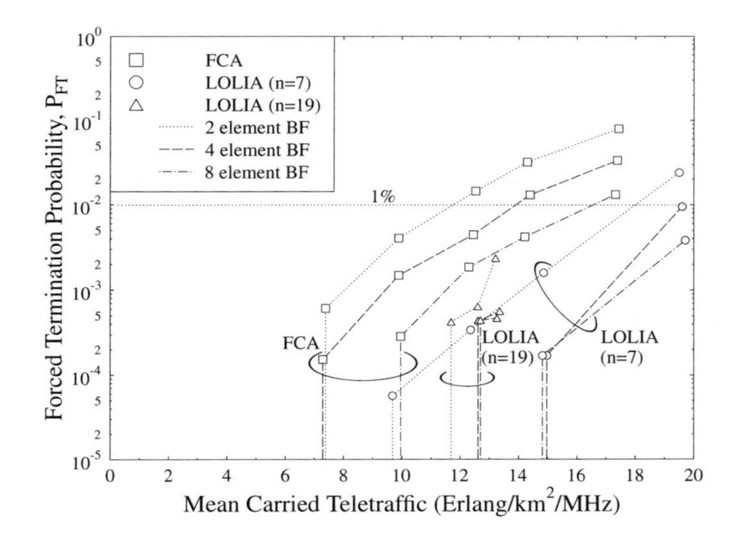

Figure 4.33: Call dropping probability performance versus mean carried traffic, for comparison of the LOLIA, with 7 and 19 'local' base stations, and of FCA using a 7-cell reuse cluster, under a uniform geographic traffic distribution, for two, four and eight element antenna arrays with beamforming in a **multipath environment**. See Figure 4.27 for the corresponding LOS results.

of freedom, and therefore, the increased number of sources that may be nulled. The results support this expectation with a 17% gain in the number of users, when upgrading the system from two element to four element arrays, and a further 15% improvement in the number of supported users with the aid of eight element antenna arrays instead of the four element arrays. As for the LOS results, the FCA algorithm, again, benefited the most in terms of the number of users supported by the network from the employment of adaptive antenna arrays. The number of users increased by 35%, when doubling the number of antenna elements from two to four, and on doubling from four to eight delivered a further 29% user capacity improvement.

4.6.2.3 Performance over a Multipath Channel using Power Control

This section builds on the results obtained in the previous section for a multipath propagation environment. Simulations were conducted for a standard 7-cell FCA scheme and a LOLIA-assisted system using $n = 7$, both invoking power control. The power control algorithm implemented attempted to independently adjust the mobile and base station transmit powers, such that the uplink and downlink SINRs were within a given target SINR window. The use of a target window avoided constantly increasing and decreasing the transmission powers, which could lead to potential power control instabilities within the network. Furthermore, using a range of possible transmission powers is analogous to accounting for an inherent power control error plus slow fading phenomenon. The 'Target SINR' given in Table 4.8 is

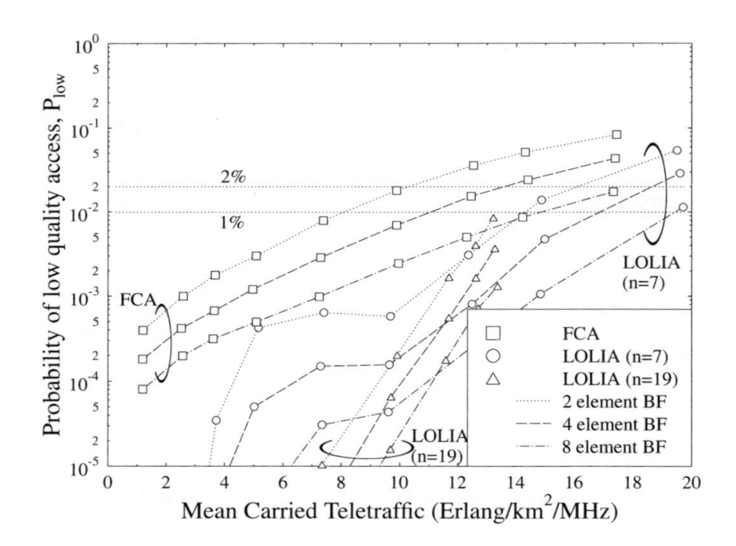

Figure 4.34: Probability of low quality access versus mean carried traffic, for comparison of the LOLIA, with 7 and 19 'local' base stations, and of FCA using a 7-cell reuse cluster, under a uniform geographic traffic distribution, for two, four and eight element antenna arrays with beamforming in a **multipath environment**. See Figure 4.28 for the corresponding LOS results.

Algorithm	Conservative $P_{FT} = 1\%, P_{low} = 1\%$ $GOS = 4\%, P_B = 3\%$			Lenient $P_{FT} = 1\%, P_{low} = 2\%$ $GOS = 6\%, P_B = 5\%$		
	Users	**Traffic**	**Limiting Factor**	**Users**	**Traffic**	**Limiting Factor**
FCA, 2 elements (el.)	**1315**	**8.10**	P_{low}	**1660**	**10.30**	P_{low}
FCA, 4 elements	**1790**	**11.10**	P_{low}	**2240**	**13.60**	P_{low}
FCA, 8 elements	2400	14.20	P_B	2780	15.70	GOS
LOLIA (n=7), 2 el.	**2310**	**14.30**	P_{low}	**2610**	**16.10**	P_{low}
LOLIA (n=7), 4 el.	**2735**	**16.90**	P_{low}	**3035**	**18.65**	P_{low}
LOLIA (n=7), 8 el.	3155	19.45	P_{low}	>3200	>20.00	P_{low}
LOLIA (n=19), 2 el.	**1970**	**11.55**	P_B	**2110**	**11.95**	P_B
LOLIA (n=19), 4 el.	**1990**	**11.65**	P_B	**2155**	**12.05**	P_B
LOLIA (n=19), 8 el.	2095	11.85	P_B	2220	12.20	P_B

Table 4.7: Maximum mean carried traffic, and maximum number of mobile users that can be supported by each configuration, whilst meeting the preset quality constraints of Section 4.3.3.4. The carried traffic is expressed in terms of normalised Erlangs (Erlang/km^2/MHz), for the network described in Table 4.4 in a **multipath environment**. The corresponding LOS results are summarised in Table 4.6 with network configurations common between the two highlighted in bold.

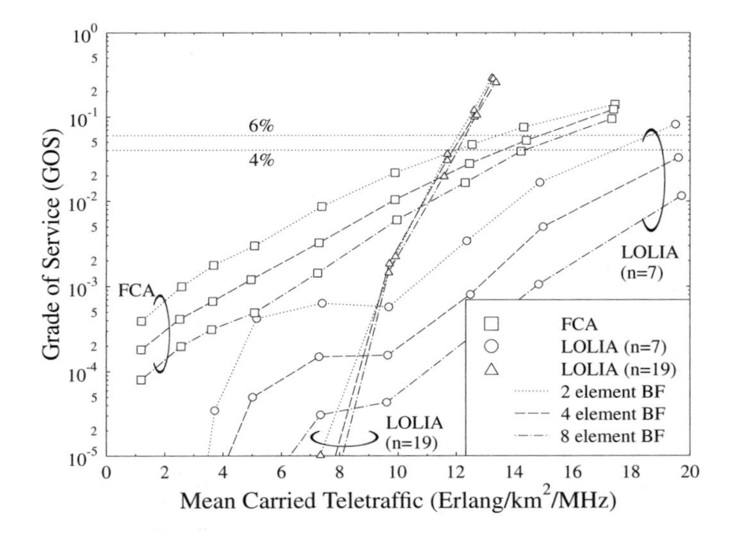

Figure 4.35: Grade-Of-Service (GOS) performance versus mean carried traffic, for comparison of the LOLIA, with 7 and 19 'local' base stations, and of FCA using a 7-cell reuse cluster, under a uniform geographic traffic distribution, for two, four and eight element antenna arrays with beamforming in a **multipath environment**. See Figure 4.29 for the corresponding LOS results.

the SINR to be maintained by the power control algorithm. The immediate effect of power control on the SINR versus the mobile's distance from the base station can be seen in Figure 4.38. This figure shows that power control attempts to maintain a constant SINR, sufficiently high for reliable communications across the network, rather than allowing for unnecessarily high SINRs near the base station and providing insufficient levels of SINR far from the base stations, evident for a cordless telephone type network using no power control. It was found that in conjunction with 4-QAM using a target SINR of 27 dB was most suitable, when using the FCA algorithm. However, the LOLIA required a higher target SINR of 31 dB in order to obtain satisfactory call dropping performance, as a result of its dynamic nature causing the interference levels to vary more rapidly than for the FCA algorithm. In other words, the LOLIA required a higher SINR 'headroom' above the re-allocation SINR threshold.

Figure 4.39 shows the new call blocking probability versus the mean normalised carried traffic, expressed in terms of Erlangs/km^2/MHz. The figure shows that the blocking performance of the FCA algorithm is limited by the availability of frequency/timeslot combinations, and hence the addition of power control does not improve the new call blocking performance. However, the blocking performance of the LOLIA is not dominated by the availability of frequency/timeslot combinations and hence it can be seen to benefit significantly from using power control.

From Figure 4.40 it can be seen that the Power Control (PC) algorithm substantially improved the call dropping probability of the FCA algorithm in comparison to the scenario without PC in Figure 4.32. Specifically, at the highest traffic loads, the PC-assisted perfor-

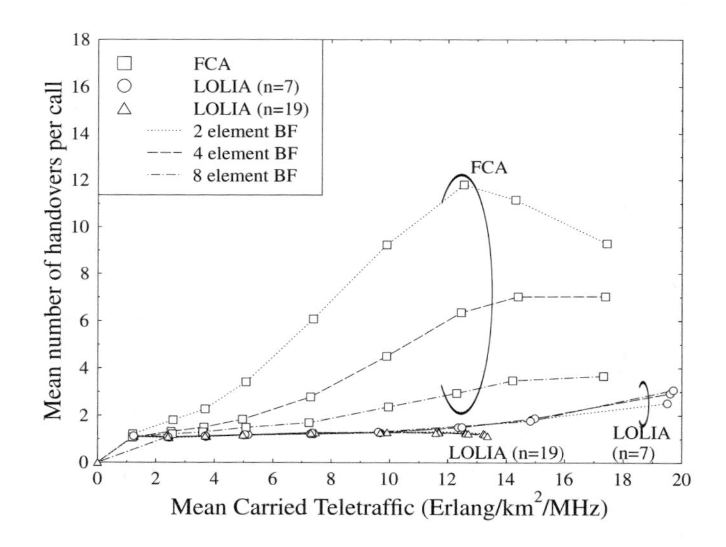

Figure 4.36: Mean number of handovers per call versus mean carried traffic, for comparison of the LOLIA, with 7 and 19 'local' base stations, and of FCA using a 7-cell reuse cluster, under a uniform geographic traffic distribution, for two, four and eight element antenna arrays with beamforming in a **multipath environment**. See Figure 4.30 for the corresponding LOS results.

Parameter	Value	Parameter	Value
Noisefloor	-104 dBm	Multiple Access	F/TDMA
Frame duration	0.4615 ms	Cell radius	218 m
Max. BS transmit power	10 dBm	Maximum MS transmit power	10 dBm
Min. BS transmit power	-20 dBm	Minimum MS transmit power	-20 dBm
Power control stepsize	1 dB	Power control hysteresis	3 dB
Number of base stations	49	Handover hysteresis	2 dB
Outage SINR threshold	17 dB	Power control FCA target SINR	27 dB
Re-alloc. SINR threshold	21 dB	Pow. cont. LOLIA7 target SINR	31 dB
Number of timeslots	8	Number of carriers	7
Average inter-call-time	300 s	Max new-call queue-time	5 s
Average call length	60 s	Reference signal modulation	BPSK
Beamforming algorithm	SMI	Reference signal length	8 bits
MS speed	13.4 m/s	Number of antenna elements	2 &4
Pathloss at 1 m ref. point	0 dB	Pathloss exponent	-3.5
Geometry of antenna array	Linear	Array element spacing	$\lambda/2$
Modulation scheme	4-QAM	Channel/carrier bandwidth	200 kHz

Table 4.8: Simulation parameters for the FCA, and DCA-assisted networks using power control.

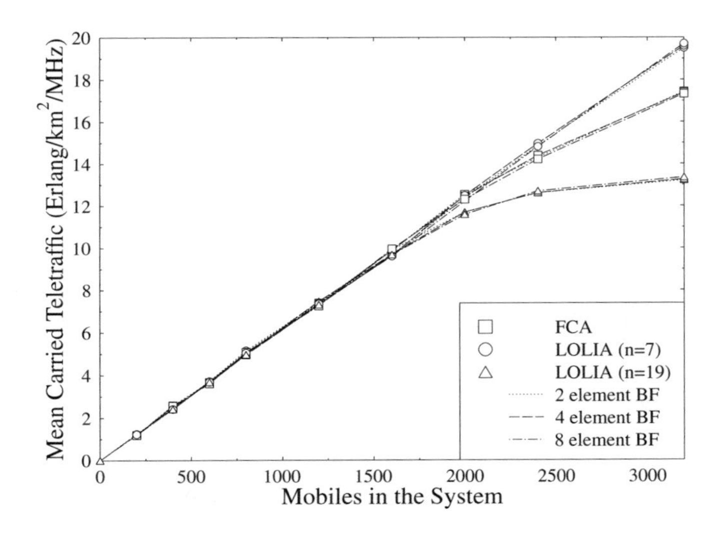

Figure 4.37: Mean traffic carried versus the number of mobiles in the system, for comparison of the LOLIA, with 7 and 19 'local' base stations, and of FCA using a 7-cell reuse cluster, under a uniform geographic traffic distribution, for two, four and eight element adaptive antenna arrays in a **multipath environment**. See Figure 4.31 for the corresponding LOS results.

mance matched that without power control but using antenna arrays with twice the number of antenna elements. At lower levels of traffic, the performance improvement obtained with the aid of power control was even higher, with the two element array results approaching those of the eight element array without power control. However, below approximately 10 Erlang/km^2/MHz a forced termination probability performance plateau was reached as a result of the power control algorithm limiting the maximum SINR. In contrast, when no power control is used and there are few users, the average SINR is very high and consequently fewer calls are dropped.

The performance gain of the LOLIA using power control is lower than that of the FCA algorithm, but still significant, since its performance is about halfway between that of the LOLIA without power control and using the same number of antenna elements, and that with twice the number of antenna elements.

The probability of low quality access of the PC-assisted scenario is shown in Figure 4.41. The corresponding curves for using no PC were plotted in Figure 4.33. The power controlled variant of the FCA algorithm offered a significantly reduced probability of low quality access for a given number of antenna elements. In fact, the probability of low quality access, when using power control and a two element adaptive antenna array, was lower than that when using a four element array without power control. The LOLIA also benefited to the same extent, with the probability of low quality access when using the power control algorithm equalling that obtained with the aid of twice the number of antenna elements and no power control.

The GOS illustrated in Figure 4.42 is related to the probability of low quality access by

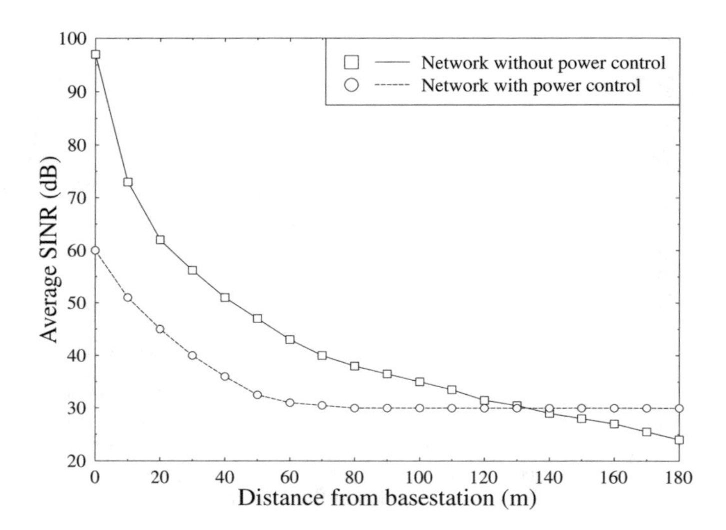

Figure 4.38: Signal-to-Interference plus Noise Ratio (SINR) versus mobile station distance measured from the base station, for networks with and without power control. The unnecessarily high SINR near the base station was a consequence of the base station's inability to power down below the minimum transmit power of -20 dBm, when the mobile station was within a distance of about 60 m from the base station.

Equation 4.15, hence the close resemblance to Figure 4.41. However, it can be seen that the performance difference of the FCA algorithm using two and four element antenna arrays diminished as a result of their similar new call blocking performances, which dominate the GOS metric of Equation 4.15.

Figure 4.43 shows the mean number of handovers performed per call versus the mean carried teletraffic. From this figure it can be seen that the performance of the FCA algorithm was improved significantly as a result of using the power control algorithm. However, the FCA algorithm still required significantly more handovers per call for maintaining the desired call quality than the equivalent LOLIA based network without power control.

From the mean transmission power results of Figure 4.44 it can be seen that, as expected, the mean transmission power increased as the amount of teletraffic carried increased due to the higher levels of interference to be overcome. At high traffic loads the difference between the mean transmission powers of the mobile stations and the base stations, became more significant for the FCA algorithm. This resulted from the downlink interfering base stations being, on average, farther away from the served mobile, than the interfering mobiles were from the serving base station on the uplink. This was further exacerbated by the omnidirectional nature of the mobiles' antennas and the directional nature of the antennas at the base stations. The LOLIA using a 7-cell exclusion zone required a higher mean transmission power than the FCA algorithm, which was attributed to the higher target SINR required by the LOLIA for maintaining an acceptable call dropping performance. When compared to the fixed transmission power of 10 dBm for an identical network operating without power

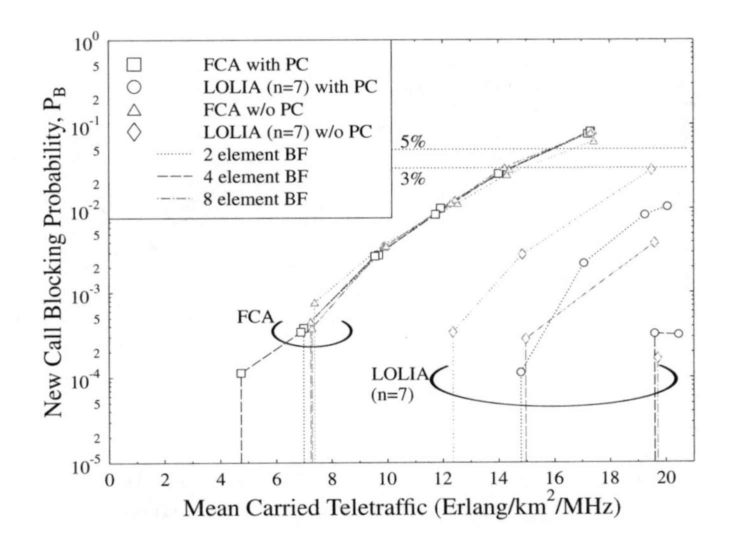

Figure 4.39: New call blocking performance versus mean carried traffic, for comparison of the LOLIA, with 7 'local' base stations, and of FCA using a 7-cell reuse cluster, under a uniform geographic traffic distribution, **with and without power control**, for two and four element antenna arrays with beamforming in a multipath environment.

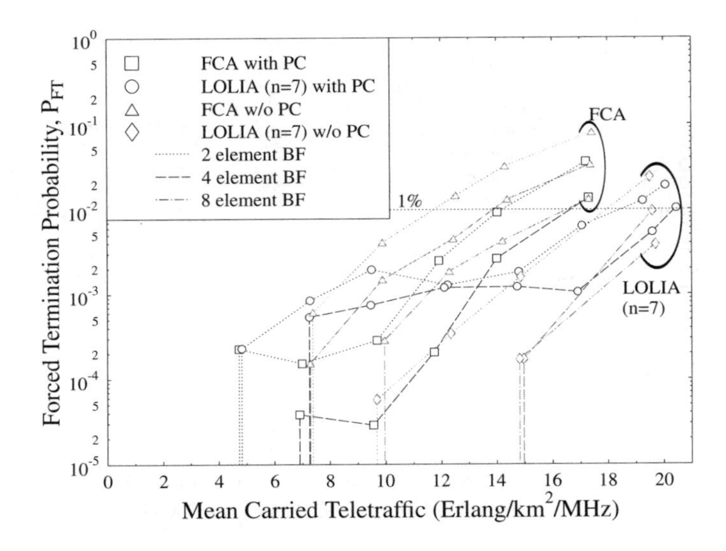

Figure 4.40: Call dropping performance versus mean carried traffic, for comparison of the LOLIA, with 7 'local' base stations, and of FCA using a 7-cell reuse cluster, under a uniform geographic traffic distribution, **with and without power control**, for two and four element antenna arrays with beamforming in a multipath environment.

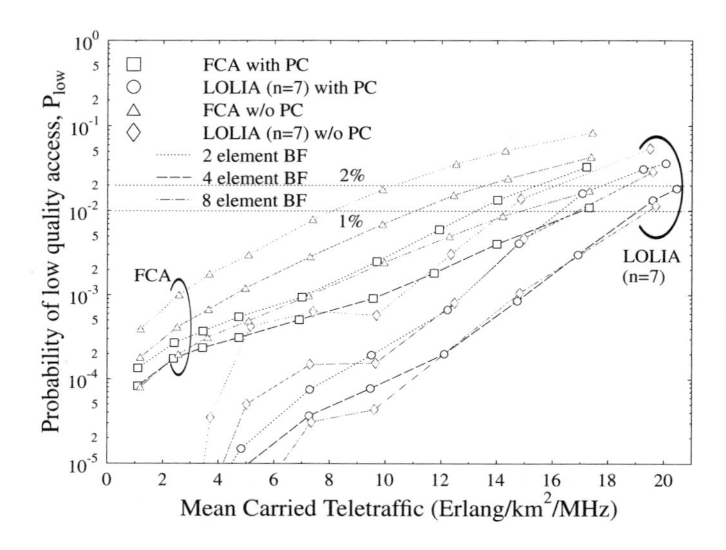

Figure 4.41: Probability of low quality access per call versus mean carried traffic, for comparison of the LOLIA, with 7 'local' base stations, and of FCA using a 7-cell reuse cluster, under a uniform geographic traffic distribution, **with and without power control**, for 2 and 4 element antenna arrays with beamforming in a multipath environment.

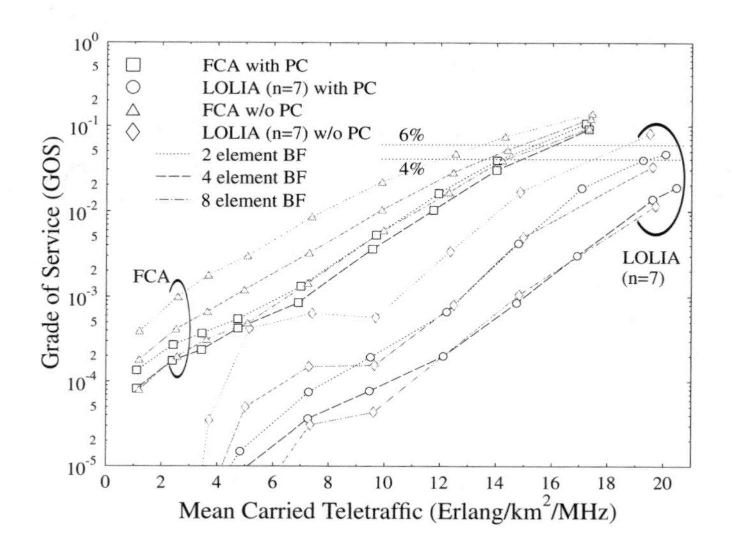

Figure 4.42: Grade-Of-Service (GOS) performance versus mean carried traffic, for comparison of the LOLIA, with 7 'local' base stations, and of FCA using a 7-cell reuse cluster, under a uniform geographic traffic distribution, **with and without power control**, for two and four element antenna arrays with beamforming in a multipath environment.

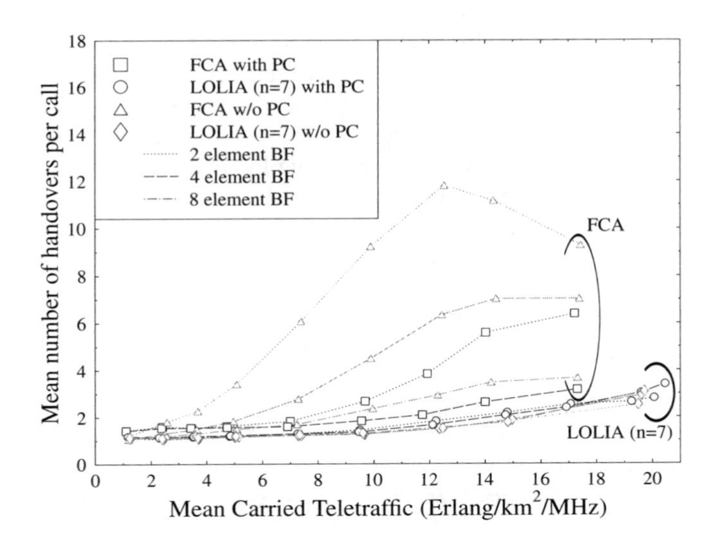

Figure 4.43: Mean number of handovers per call versus mean carried traffic, for comparison of the LOLIA, with 7 'local' base stations, and of FCA using a 7-cell reuse cluster, under a uniform geographic traffic distribution, **with and without power control**, for two and four element antenna arrays with beamforming in a multipath environment.

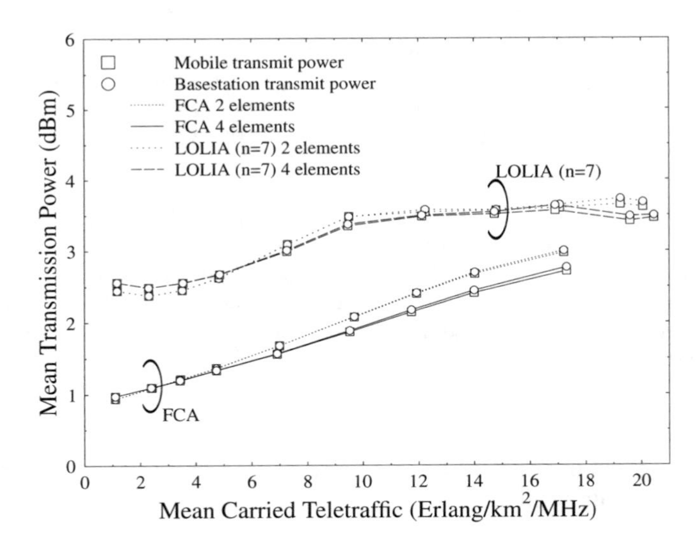

Figure 4.44: Mean transmission power versus mean carried traffic, of the LOLIA, with 7 'local' base stations, under a uniform geographic traffic distribution, **with power control**, for two and four element antenna arrays with beamforming in a multipath environment.

control, the reductions in transmitted power are significant, with a minimum average transmit power reduction of 6 dB, which substantially extends the mobile stations' battery lives.

Table 4.9 presents the summary of our results obtained for a network using power control in a multipath environment. The table shows that the use of power control has increased the number of users that may be serviced according to the required network performance criteria. The number of users supported by the network using the FCA algorithm increased by 28% to 70%, with a mean of 54% over the conservative and lenient scenarios. The capacity gains obtained with the aid of power control in a network using the LOLIA 7, however, were lower, namely between 9% and 15%, with a mean of almost 13% for both the conservative and lenient scenarios. Whilst the LOLIA 7 capacity gains are fairly modest, the overall call quality of the channel allocation techniques has improved for a given level of traffic, when compared to an identical network without power control.

Algorithm	**Conservative** $P_{FT}=1\%, P_{low}=1\%$ $GOS=4\%, P_B=3\%$			**Lenient** $P_{FT}=1\%, P_{low}=2\%$ $GOS=6\%, P_B=5\%$		
	Users	**Traffic**	**Limiting Factor**	**Users**	**Traffic**	**Limiting Factor**
Without power cont.						
FCA, 2 elements (el.)	**1315**	**8.10**	P_{low}	**1660**	**10.30**	P_{low}
FCA, 4 elements	**1790**	**11.10**	P_{low}	**2240**	**13.60**	P_{low}
FCA, 8 elements	2400	14.20	P_B	2780	15.70	GOS
LOLIA (n=7), 2 el.	**2310**	**14.30**	P_{low}	**2610**	**16.10**	P_{low}
LOLIA (n=7), 4 el.	**2735**	**16.90**	P_{low}	**3035**	**18.65**	P_{low}
LOLIA (n=7), 8 el.	3155	19.45	P_{low}	>3200	>20.00	P_{low}
With power cont.						
FCA, 2 elements	**2260**	**13.30**	P_{low}	**2455**	**14.25**	P_{FT}
FCA, 4 elements	**2510**	**14.45**	P_B	**2870**	**15.95**	P_B
LOLIA (n=7), 2 el.	**2665**	**16.30**	P_{low}	**2935**	**17.80**	P_{low}
LOLIA (n=7), 4 el.	**3125**	**19.08**	P_{low}	**3295**	**20.42**	P_{FT}

Table 4.9: Maximum mean carried traffic, and maximum number of mobile users that can be supported by each configuration whilst meeting the preset quality constraints of Section 4.3.3.4. The carried traffic is expressed in terms of normalised Erlangs (Erlang/km^2/MHz) for the network described in Table 4.8 both **with and without power control** in a **multipath environment**. The figures in bold indicate common network configurations to both the results without power and those with.

4.6.2.4 Transmission over a Multipath Channel using Power Control and Adaptive Modulation

The idea behind adaptive modulation is to select a modulation mode according to the instantaneous radio channel quality [12, 13]. Thus, if the channel quality exhibits a high instantaneous SINR, then a high order modulation mode may be employed, enabling the exploitation of the temporarily high channel capacity. In contrast, if the channel has a low instantaneous SINR, using a high-order modulation mode would result in an unacceptable Frame Error

Ratio (FER), and hence a more robust, but lower throughput modulation mode would be invoked. Hence, adaptive modulation not only combats the effects of a poor quality channel, but also attempts to maximise the throughput, whilst maintaining a given target FER. Thus, there is a trade-off between the mean FER and the data throughput, which is governed by the modem mode switching thresholds. These switching thresholds define the SINRs, at which the channel is considered unsuitable for a given modulation mode, where an alternative AQAM mode must be invoked.

The power control algorithm invoked attempted to independently adjust the mobile and base station powers, such that the uplink and downlink SINRs were within a given target SINR window. The employment of a target window avoided constantly increasing and decreasing the transmission powers, which could lead to potential power control instabilities within the network. Furthermore, the affect of a range of different possible transmission powers is analogous to an inherent power control error plus slow fading envelope.

The combination of power control with adaptive modulation leads to several performance trade-offs, which must be considered when designing the power control and modulation mode switching algorithm. For example, the transmitted power could be minimised, which would result in either a high FER and a high throughput, or a low BER and a low throughput. Alternatively, the FER could be lowered even while maintaining a high throughput, when tolerating high transmission powers.

The power control and modulation mode switching algorithm invoked in our simulations attempted to minimise the transmitted power, whilst maintaining a high throughput with a less than 5% target FER. The pseudo-code of the proposed algorithm is described in the next section.

4.6.2.5 Power Control and Adaptive Modulation Algorithm

```
determine lowest SINR out of uplink and downlink SINRs
if in 16-QAM mode
  if lowest SINR < 16-QAM drop SINR
    drop to 4-QAM mode
  else if lowest SINR < 16-QAM reallocation SINR
    if at maximum transmit power
      revert to 4-QAM
    else
      increase transmit power
  else if lowest SINR < 16-QAM target SINR
    if not at maximum power
      increase transmit power
  else if lowest SINR > 16-QAM target SINR+hysteresis
    decrease transmit power
else if in 4-QAM mode
  if lowest SINR < 4-QAM drop SINR
    drop to BPSK mode
  else if lowest SINR < 4-QAM reallocation SINR
    if at maximum transmit power
      revert to BPSK
```

```
    else
      increase transmit power
  else if lowest SINR < 4-QAM target SINR
    if not at maximum power
      increase transmit power
  else if lowest SINR > 16-QAM target SINR+hysteresis
    change to 16-QAM mode
  else if lowest SINR > 4-QAM target SINR+hysteresis
    if at maximum transmit power
      reduce transmit power
    else
      if lowest SINR > 16-QAM drop SINR
        change to 16-QAM
      else
        decrease transmit power
else if in BPSK
  if lowest SINR < BPSK drop SINR
    outage occurs
  else if lowest SINR < BPSK reallocation SINR
    if not at maximum transmit power
      increase transmit power
  else if lowest SINR > 4-QAM target SINR+hysteresis
    change to 4-QAM
  else if lowest SINR > BPSK target_hysteresis
    if at maximum transmit power
      reduce transmit power
    else
      change to 4-QAM
```

Figure 4.45 shows the flowchart of the AQAM and power control decision tree, when in the 4-QAM mode. The first step in the process is to determine the lower of the uplink and the downlink SINRs. The next step is to determine whether the BPSK modulation mode should be selected. When in the BPSK mode, outages may occur due to an insufficiently high SINR level and after a given number of BPSK outages the call is dropped. The conditions for this to occur are that either the lower SINR is below the 4-QAM call dropping threshold or that it is below the 4-QAM call reallocation threshold and currently the maximum possible transmission power is used. If the lower SINR is below the 4-QAM call reallocation threshold, or the SINR is below the 4-QAM target SINR, and the maximum transmission power has not been reached, then the transmit power is increased. However, if the SINR is below the 4-QAM target SINR and the maximum possible transmit power is currently used, then the modem remains in the 4-QAM mode. The 16-QAM mode is chosen, if the SINR is higher than the 16-QAM target SINR, plus the associated hysteresis. Alternatively, the 16-QAM mode is invoked if the SINR is higher than the 4-QAM target SINR plus the hysteresis, furthermore the transmission power required to obtain this SINR is lower than the maximum transmit power, and the SINR is higher than the 16-QAM call dropping SINR. However, if the SINR is below the 16-QAM call dropping SINR or the maximum transmission power is

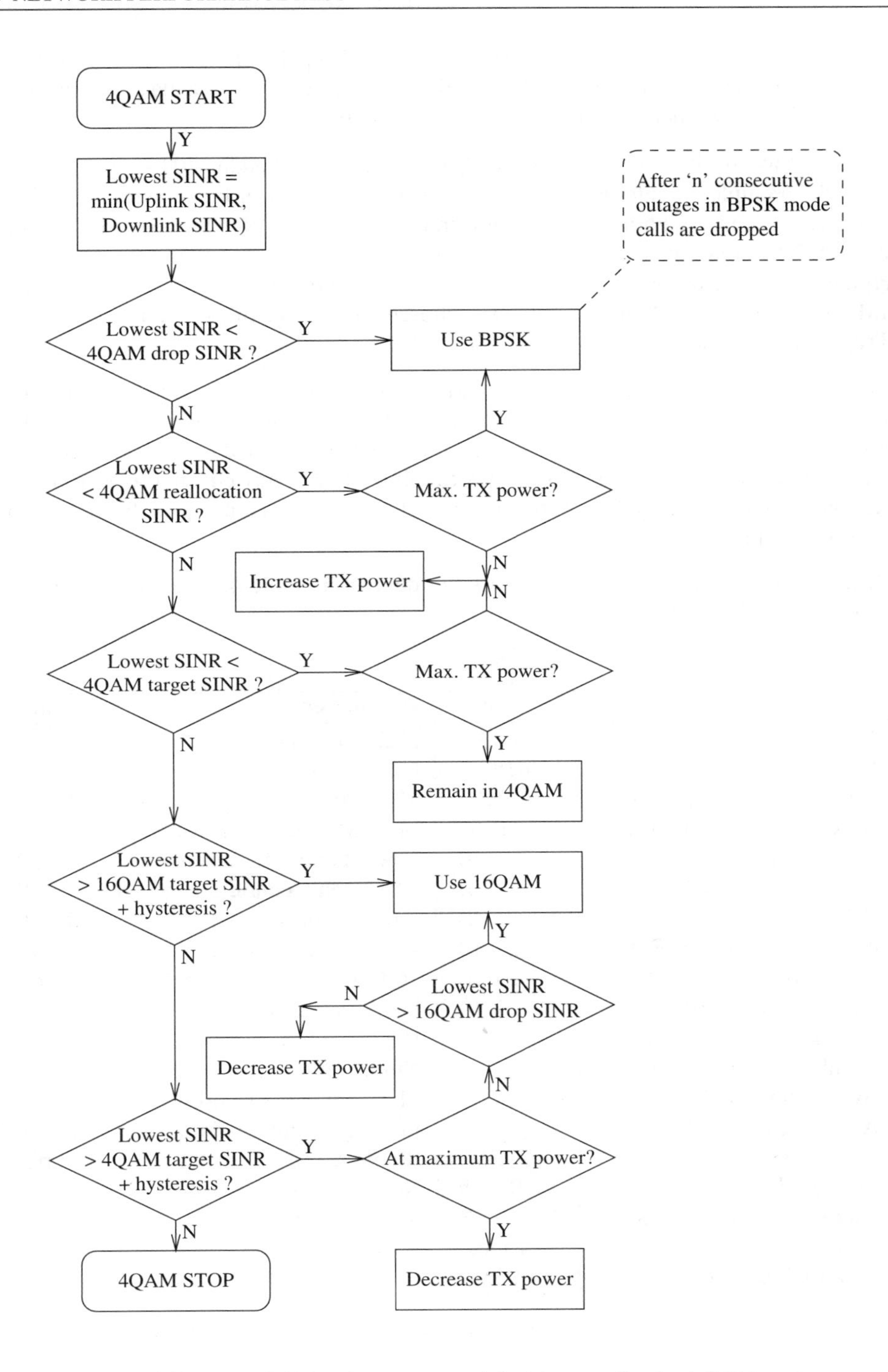

Figure 4.45: The AQAM and power control decision tree for the 4-QAM mode.

in use, then the transmit power is reduced in an effort to keep the SINR in the 4-QAM mode's target SINR window. The improved SINR achieved using adaptive antenna arrays at the base station facilitates a higher mean network data throughput.

The FER was evaluated for approximately half-rate Bose-Chaudhuri-Hocquenghem (BCH) codes, which employed interleaving over the different number of bits conveyed by the different modem modes within a transmission frame [106]. The 'Reallocation SINR' and the 'Outage SINR' are defined as the average SINRs necessary for satisfying the 5% and 10% maximum FER constraints, respectively, using a given modulation mode such as BPSK, 4-QAM, or 16-QAM. The 'Target SINR' was chosen so as to maximise the network capacity and represents an FER of approximately 2%.

The calculation of the receive antenna arrays weights was performed on a transmission frame-by-frame basis, leading to updated uplink and downlink SINRs every transmission frame. These SINR values were then used for selecting the modulation mode and transmission power to be employed, and for determining whether any channel re-allocation was necessary. Hence, frame-by-frame adaptive modulation, power control and dynamic channel allocation was jointly performed.

The system parameters for the network are defined in Table 4.10 and our performance results are provided in the next section.

Parameter	**Value**	**Parameter**	**Value**
Noisefloor	-104 dBm	Multiple Access	TDMA
Frame length	0.4615 ms	Cell radius	218 m
Min. BS transmit power	-20 dBm	Min. MS transmit power	-20 dBm
Max. BS transmit power	10 dBm	Max. MS transmit power	10 dBm
Power control stepsize	1 dB	Power control hysteresis	3 dB
BPSK outage SINR	13 dB	BPSK reallocation SINR	17 dB
BPSK target SINR	21 dB	4-QAM outage SINR	17 dB
4-QAM reallocation SINR	21 dB	4-QAM target SINR	27 dB
16-QAM outage SINR	24 dB	16-QAM reallocation SINR	27 dB
16-QAM target SINR	32 dB	Pathloss exponent	-3.5
Number of base stations	49	Handover hysteresis	2 dB
Number of timeslots/carrier	8	Number of carriers	7
Average inter-call-time	300 s	Max new-call queue-time	5 s
Average call length	60 s	Ref. signal modulation	BPSK
Beamforming algorithm	SMI	Reference signal length	8 bits
MS speed	30 mph	No. of antenna elements	2 & 4
Pathloss at 1 m ref. point	0 dB	Shadow fading	No
Geometry of ant. array	Linear	Array element spacing	$\lambda/2$
Channel/carrier bandwidth	200 kHz		

Table 4.10: Simulation parameters for the AQAM based network using power control.

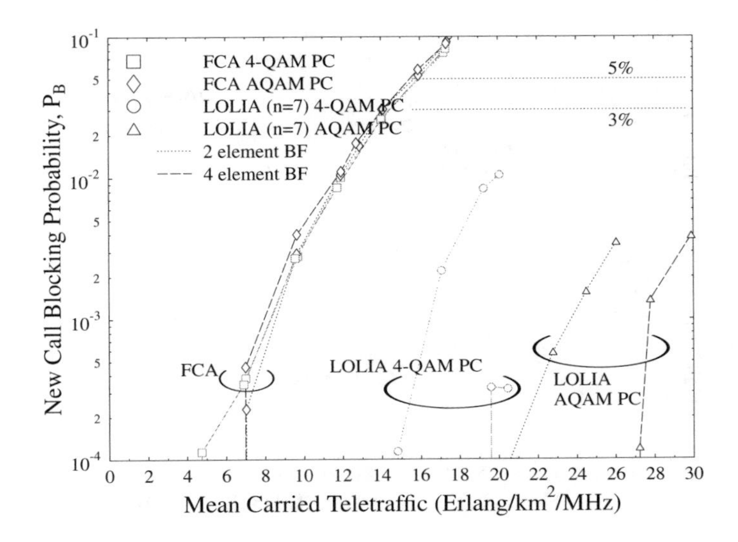

Figure 4.46: New call blocking probability versus mean carried traffic of the LOLIA, with 7 'local' base stations, and of FCA employing a 7-cell reuse cluster, for 2 and 4 element antenna arrays, **with and without AQAM**.

4.6.2.6 Performance of PC-assisted, AQAM-aided Dynamic Channel Allocation

This section presents the simulation results obtained for a network using burst-by-burst adaptive modulation in order to improve the network's performance. Simulations were conducted for both a standard 7-cell FCA scheme and for the LOLIA using $n = 7$. The benchmark results obtained for a 4-QAM based network using power control were included for comparison purposes. Due to the enhanced network performance resulting from the employment of AQAM, a further constraint of a minimum throughput of 2 bits/symbol was invoked. This ensured a fair comparison with the fixed 4-QAM based network.

Figure 4.46 shows the new call blocking probability versus the mean normalised carried traffic. From this figure it can be seen that in conjunction with the LOLIA there are no blocked calls, except for the highest levels of traffic. In contrast, the performance of the FCA algorithm was degraded by using AQAM. This was the result of the limited availability of frequency/timeslot combinations restricting the achievable performance gain, since the reduced call dropping probability encouraged the prolonged utilisation of the limited resources. This however, prevented new call setups.

The corresponding call dropping probability is depicted in Figure 4.47, which shows that when invoking adaptive modulation, the FCA algorithm performs better than the LOLIA below a traffic load of about 14 Erlangs/km^2/MHz. Both channel allocation algorithms consistently offered a lower call dropping probability, when employing AQAM compared to when using the fixed-mode 4-QAM modulation scheme. This reduction in the call dropping rate using adaptive modulation was brought about by the inherent ability of the AQAM scheme to be reconfigured to a lower-order, and hence more interference resistant modulation mode,

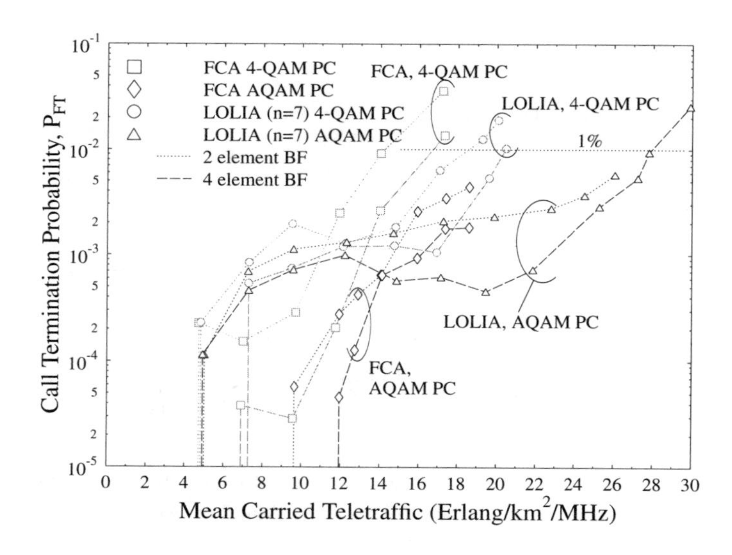

Figure 4.47: Call dropping or forced termination performance versus mean carried traffic of the LOLIA, with 7 'local' base stations, and of FCA employing a 7-cell reuse cluster, for two and four element antenna arrays, **with and without AQAM.**

in order to prevent calls from being dropped.

Figure 4.48 shows that the probability of a low quality access was substantially reduced by AQAM for both the FCA scheme and the LOLIA. At lower traffic loads the probability of low quality outage was higher than when using the fixed 4-QAM modulation mode for both of the channel allocation schemes. This was due to the frequent use of the highest order modulation mode, 16-QAM, which was more susceptible to low quality outages. The more frequent usage of the 16-QAM mode by the four element adaptive antenna arrays also explains their greater probability of low quality outage at the lower traffic levels. However, as the traffic levels increased, the lower order modulation modes were invoked more frequently, and hence when combined with the four element arrays, the system guaranteed a lower probability of low quality outage than the two element arrays.

From Figure 4.49 it can be seen that the GOS of the FCA algorithm did not benefit from employing AQAM to the same extent as the LOLIA, except at the lower traffic levels when the new call blocking probability does not dominate the overall GOS performance. The LOLIA, however, benefited substantially, as we have also seen for the probability of low quality outages, since its performance was not constrained by its new call blocking probability observed in Figure 4.46 for both 4-QAM and AQAM.

The employment of AQAM, in Figure 4.50, reduced the mean number of handovers per call of the LOLIA at all traffic loads, and of the FCA for the highest traffic loads, although an increased number of handovers were performed by the FCA at lower traffic loads. At these lower traffic loads, more intra-cell handovers were performed by the FCA algorithm, due to the employment of the 16-QAM modulation mode, which required more frequent intra-cell handovers in order to maintain a sufficiently high SINR. However, as the traffic

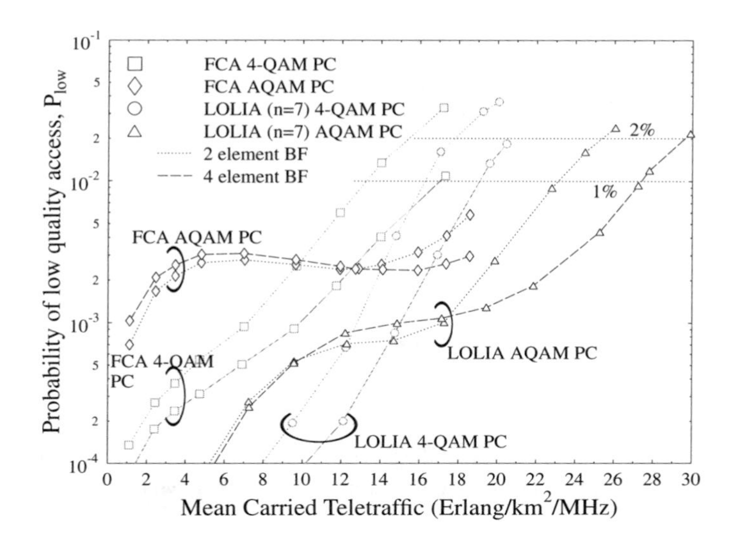

Figure 4.48: Probability of low quality access versus mean carried traffic of the LOLIA, with 7 'local' base stations, and of FCA employing a 7-cell reuse cluster, for two and four element antenna arrays, **with and without AQAM**.

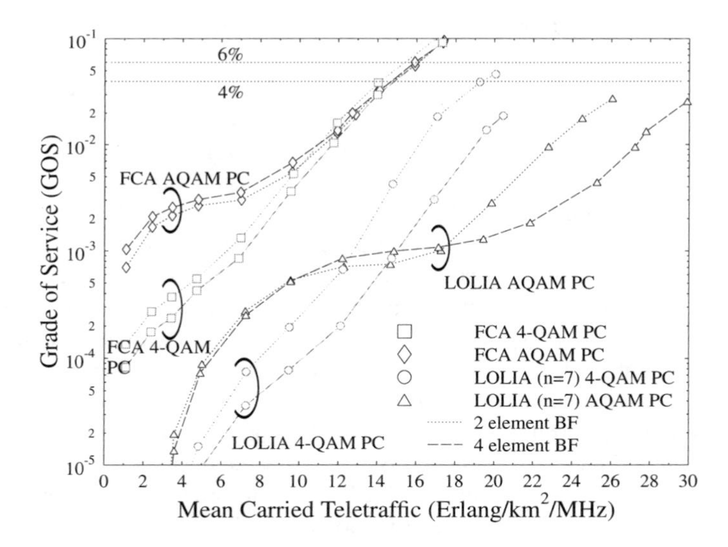

Figure 4.49: GOS performance versus mean carried traffic of the LOLIA, with 7 'local' base stations, and of FCA employing a 7-cell reuse cluster, for two and four element antenna arrays, **with and without AQAM**.

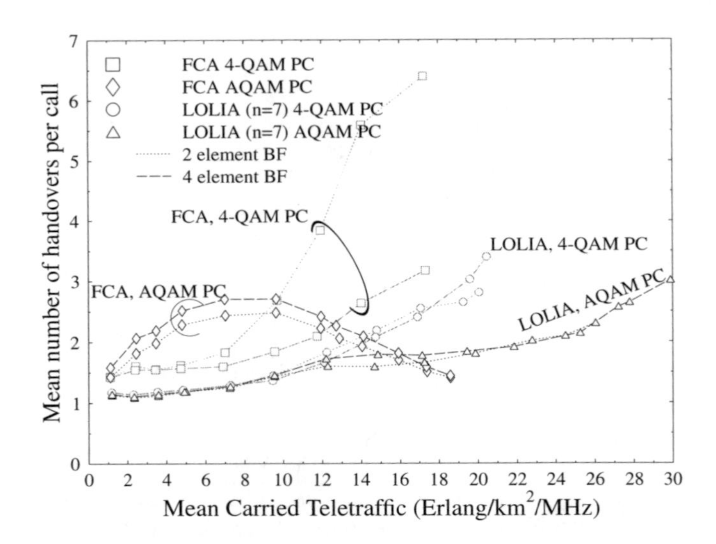

Figure 4.50: Mean number of handovers per call versus mean carried traffic of the LOLIA, with 7 'local' base stations, and of FCA employing a 7-cell reuse cluster, for two and four element antenna arrays, **with and without AQAM**.

load increased, the lower-order modulation modes were used more frequently, and hence less intra-cell handovers were required, leading to a reduction in the number of handovers performed.

The mean transmission power results of Figure 4.51 demonstrate how the employment of AQAM can reduce the power transmitted both for the uplink and the downlink. At low traffic load levels the FCA algorithm performed slightly worse in transmitted power terms, than the LOLIA. However, as the traffic loads increased, the gap became negligible when using two element antenna arrays. By contrast, when using four element antenna arrays, the LOLIA required a higher transmission power at these higher teletraffic loads. When compared to the fixed transmission power of 10 dBm for a network using no power control, the employment of AQAM resulted in a significant reduction of the mean transmission power. Specifically, the minimum reduction of the transmitted power was more than 4 dB and a maximum reduction of more than 7 dB was attained in addition to achieveing a superior call quality and an increased mean modem throughput.

The average modem throughput expressed in bits per symbol versus the mean carried teletraffic is shown in Figure 4.52. The figure shows how the mean number of bits per symbol decreased as the network traffic increased. The FCA algorithm offered the lowest throughput with its performance degrading near-linearly upon increasing the network's traffic load. The LOLIA, especially for the lower levels of traffic, offered a greater modem throughput for a given level of teletraffic carried, with the achievable performance gracefully decreasing, as the carried teletraffic continued to increase. Table 4.11 shows the mean modem throughput in bits per symbol, for the maximum mean carried traffic levels, whilst meeting the predefined quality constraints of Section 4.3.3.4.

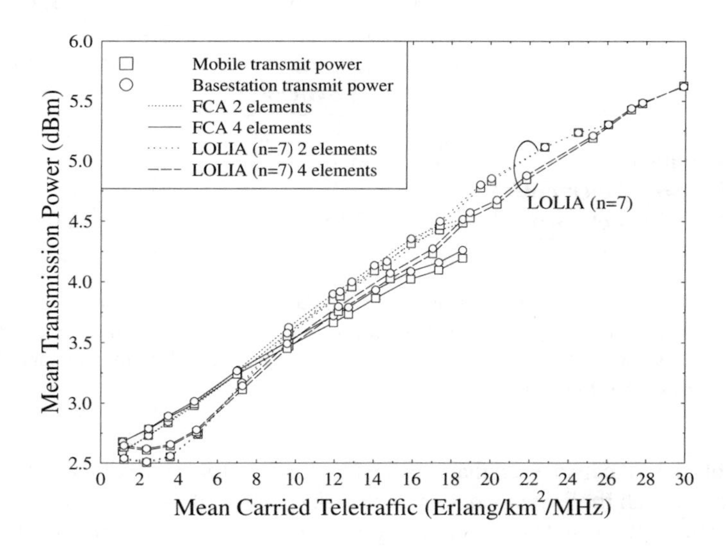

Figure 4.51: Mean transmit power versus mean carried traffic of the LOLIA, with 7 'local' base stations, and of FCA employing a 7-cell reuse cluster, for two and four element antenna arrays, **with and without AQAM**.

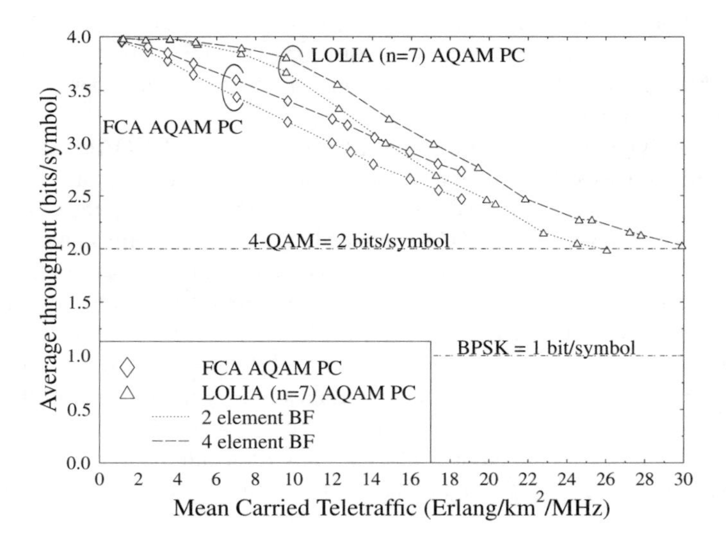

Figure 4.52: Mean throughput in terms of bits per symbol versus mean carried traffic of the LOLIA, with 7 'local' base stations, and of FCA employing a 7-cell reuse cluster, for two and four element antenna arrays, in conjunction **with AQAM**.

	Conservative $P_{FT} = 1\%$, $P_{low} = 1\%$ $GOS = 4\%$, $P_B = 3\%$ Bits per Symbol	Lenient $P_{FT} = 1\%$, $P_{low} = 2\%$ $GOS = 6\%$, $P_B = 5\%$ Bits per Symbol
FCA, 2 elements	2.8	2.7
FCA, 4 elements	3.1	2.9
LOLIA (n=7), 2 elements	2.1	$\approx$2.0
LOLIA (n=7), 4 elements	2.15	2.05

Table 4.11: Mean modem throughput, when supporting the maximum mean carried traffic, whilst meeting the preset quality constraints of Section 4.3.3.4. The carried traffic is expressed in terms of normalised Erlangs (Erlang/km^2/MHz) for the network described in Table 4.10 in a **multipath environment with AQAM**.

From Table 4.12 it can be seen that it is the blocking performance of the network using the FCA algorithm which limits its associated network capacity, thus leading to a relatively high mean modem throughput at its user capacity limits. The increase in the modem throughput for the FCA algorithm varied from 35% to 55%, with corresponding user capacity improvements of 6% and -4%, when comparing the AQAM network to 4-QAM. The table also shows that the number of users supported by the FCA network using two element adaptive antenna arrays increased when using AQAM, which was restricted by the probability of low quality access when using 4-QAM. In contrast, when using four element antenna arrays, the network capacity was limited by the network's new call blocking performance. Hence, using AQAM techniques did not increase the number of users supported. In fact, due to the superior call dropping performance of AQAM, the new call blocking probability increased as a result of the lack of available frequency/timeslot combinations, and hence the number of users supported by the network decreased.

However, the dynamic nature of the LOLIA limited its fixed 4-QAM based network capacity due to its excessive low quality access probability, and thus in all cases, AQAM increased the number of users supported, by 38% to 50%, whilst meeting the required call quality criteria of Section 4.3.3.4. The AQAM-induced improvement in mean modem throughput varied from 0% to 7.5% as a result of the particular AQAM implementation used in the simulations. This can be further verified with the aid of Figure 4.51, which shows that the mean transmission powers were not at their maxima and hence both the modem throughput and the probability of low quality access were sub-optimal. In other words, had the AQAM algorithm been more aggressive in terms of its transmitted power usage, a reduced probability of low quality access and an increased mean modem throughput would have occurred. However, a trade-off existed where both the number of users supported and the mean modem throughput were increased, whilst achieving a significant reduction in the mean transmission powers.

4.6.2.7 Summary of Non-Wraparound Network Performance

The performance results summarised in this section can be gleaned from Tables 4.6-4.12. Specifically, in this section simulation results were obtained for a LOS scenario, for both the FCA algorithm and for the LOLIA, which showed that the FCA algorithm benefited

Algorithm	**Conservative** $P_{FT} = 1\%, P_{low} = 1\%$ $GOS = 4\%, P_B = 3\%$			**Lenient** $P_{FT} = 1\%, P_{low} = 2\%$ $GOS = 6\%, P_B = 5\%$		
	Users	**Traffic**	**Limiting Factor**	**Users**	**Traffic**	**Limiting Factor**
4-QAM with PC						
FCA, 2 elements (el.)	2260	13.30	P_{low}	2455	14.25	P_{FT}
FCA, 4 elements	2510	14.45	P_B	2870	15.95	P_B
LOLIA (n=7), 2 el.	2665	16.30	P_{low}	2935	17.80	P_{low}
LOLIA (n=7), 4 el.	3125	19.08	P_{low}	3295	20.42	P_{FT}
AQAM with PC						
FCA, 2 elements	2400	14.00	P_B	2760	15.75	P_B
FCA, 4 elements	2400	14.10	P_B	2710	15.50	P_B
LOLIA (n=7), 2 el.	3675	23.10	P_{low}	4115	25.4	P_{low}
LOLIA (n=7), 4 el.	4460	27.40	P_{low}	4940	29.6	P_{low}

Table 4.12: Maximum mean carried traffic, and maximum number of mobile users that can be supported by each configuration, whilst meeting the preset quality constraints of Section 4.3.3.4. The carried traffic is expressed in terms of normalised Erlangs (Erlang/km^2/MHz) for the network described in Table 4.10 in a **multipath environment**.

the most from the employment of adaptive antenna arrays, with an increase of 144% in the number of users supported by four element antenna arrays. The corresponding figure was 67% with the aid of two element arrays. The performance of the LOLIA with a 19 base station constraint improved least using adaptive antenna arrays due to the inherently low interference levels present. However, for the LOLIA with a base station constraint of 7, using two element adaptive antenna arrays, an extra 22% additional users were supported with the desired performance metric limits of Section 4.3.3.4 observed. Using four element adaptive antenna arrays at the base stations led to an increase of 58% in the number of users supported.

Identical simulations with the addition of two multipath rays were then performed. These simulations demonstrated that the LOLIA 19 actually performed better in a multipath scenario, than in a LOS situation. This was due to the large reuse distance of the system, resulting in the sum of the powers of the three desired multipath signals versus the sum of the interfering signal powers being higher than the ratio of the LOS desired signal power to the interference power. The FCA algorithm, which offered the lowest network capacity in the LOS simulations, also suffered from the greatest capacity reduction in the multipath scenarios. The corresponding network capacities, expressed in terms of the number of users supported, decreased by between 3% and 17%. The number of users supported by the network using the LOLIA 7, was not significantly affected by the multipath propagation environment, with the highest reduction of almost 7% occurring using a four element antenna array employed in the conservative network scenario of Section 4.3.3.4. The FCA algorithm benefited the most from increasing the number of elements comprising the adaptive antenna arrays, with a minimum increase of 25% in the number of users supported upon doubling the number of antenna elements. The LOLIA employing a reuse cluster size of seven also performed well, with a user capacity increase of at least 15% for each doubling of the number of antenna elements.

Simulations were then performed in the multipath environment, where the network used the power control algorithm to maintain a fairly constant received SINR across the cell area. It was found that the power control algorithm increased the number of users carried in all the scenarios considered. The FCA algorithm exhibited the greatest gains in terms of the number of users supported by the network. When compared to an identical network without power control, the user capacity increased by 28%-72%, with an average increase of 47%. The LOLIA 7 using power control carried more traffic than the equivalent power control assisted FCA networks, and the LOLIA 7 system using no power control. When compared to the LOLIA 7 network using no power control, 9% to 15% more users were carried with a satisfactory performance. With respect to an FCA based network using power control, the increase in the number of supported users varied from 9% to almost 25%.

Further experiments were conducted in order to investigate the potential of AQAM techniques to increase network capacity. The gains achievable by the FCA algorithm were restricted by the number of available frequency/timeslot combinations, for both new calls and handovers, and hence the capacity increases were constrained by the new call blocking probability to 6%. However, this limitation to the number of supported users resulted in an increased mean modem throughput of between 2.7 and 3.1 bits per symbol, a reduced mean transmission power, and an overall improvement in call quality. The LOLIA, however, was not constrained by its new call blocking probability and was able to fully exploit the advantages of adaptive modulation. Thus, the LOLIA achieved a minimum network capacity increase of 38% over an identical scenario not using adaptive modulation.

The next section presents similar results but obtained using the 'wraparound' technique in an effort to provide an effectively infinite simulation plane with, on average, constant interference levels present over the entire simulation area.

4.6.3 Wrap-around Network Performance Results

This section presents a range of performance results similar to those obtained in the previous section. However, in this section the 'wrap-around' technique of Section 4.6.1 was used to generate results not subjected to the edge effects present at the perimeter of the simulation area. This process was described in Section 4.6.1. Results were obtained for the LOS propagation environment in Section 4.6.3.1 and for the multipath propagation environment of Section 4.6.3.2. Section 4.6.3.3 portrays the results obtained for the multipath propagation environment using power control, and Section 4.6.3.4 presents the network performance using adaptive modulation techniques.

4.6.3.1 Performance Results over a LOS Channel

Firstly we compared the FCA and the LOLIA under uniform geographic traffic distribution conditions using both a single antenna element and adaptive antenna arrays consisting of two and four elements in a LOS propagation environment. The FCA scheme employed a seven-cell reuse cluster, corresponding to one carrier frequency per base station. The LOLIA was used in conjunction with the constraints of seven and nineteen nearest base stations, i.e., $n = 7$ or 19.

As seen in Figure 4.53 the LOLIA using $n = 19$ offered the worst call blocking performance of the three channel allocation schemes, with the AAAs having little beneficial effect.

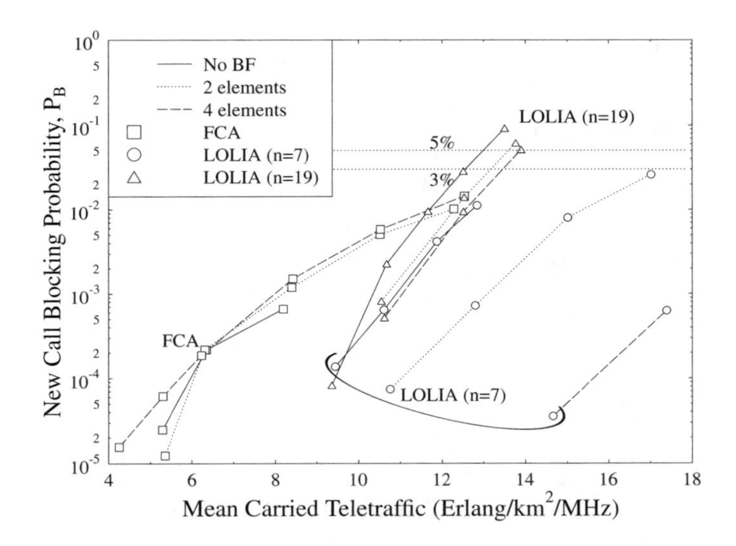

Figure 4.53: New call blocking probability performance versus mean carried traffic, for the LOLIA using 7 and 19 'local' base stations, and for FCA employing a 7-cell reuse cluster, under uniform geographic traffic distribution, for a single antenna element, as well as for two and four element antenna arrays with beamforming in a **LOS environment** using wrap-around. See Figure 4.26 for the corresponding 'desert-island' scenario.

This demonstrated that the limiting factor was not inadequate signal quality for a call to be setup, but the lack of available frequency/timeslot combinations due to the large exclusion zone. The FCA algorithm benefited only to a limited extent from the employment of the AAAs, suggesting that the majority of the blocked calls were as a result of the limited availability of frequency/timeslot combinations. Inadequate signal quality caused the remainder of the blocked calls. The call blocking performance of the LOLIA using $n = 7$ appeared mainly to be interference limited, hence the AAAs guaranteed a significant reduction of the number of blocked calls, particularly for mean carried traffic levels in excess of 9 Erlang/km^2/MHz.

Figure 4.54 shows that - as expected - the FCA algorithm performed the least satisfactorily of the three channel allocation schemes investigated with respect to its call dropping performance. Even in conjunction with a four-element adaptive antenna array, it exhibited a higher call dropping rate than that of either of the LOLIAs ($n = 19$ and $n = 7$). The large exclusion zone of the LOLIA using $n = 19$ led to a low dropping probability of less than 1×10^{-3} for teletraffic loads below approximately 12 Erlang/km^2/MHz. However, the rapid rise in the call dropping probability upon increasing the teletraffic became unacceptable for teletraffic loads in excess of about 13 Erlang/km^2/MHz. The large exclusion zone of the algorithm prevented from handovers occurring, since there were no free channels available in the vicinity, hence resulting in a high number of dropped calls. Thus, for $n = 19$ the employment of adaptive antenna arrays at the base stations did not improve the performance significantly, unlike for the FCA and LOLIA using $n = 7$, which were predominantly interference limited. The call dropping performance of the LOLIA using $n = 7$ benefited the most from the assis-

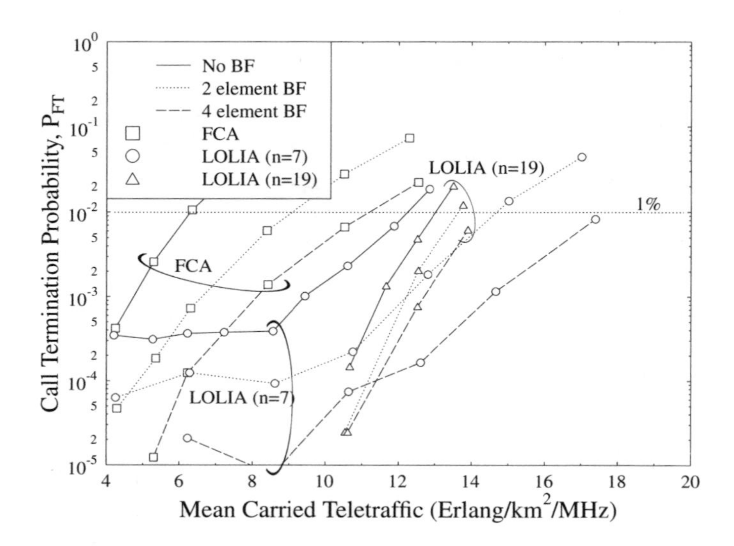

Figure 4.54: Dropping probability performance versus mean carried traffic, for the LOLIA using 7 and 19 'local' base stations, and for FCA employing a 7-cell reuse cluster, under a uniform geographic traffic distribution, for a single antenna element, as well as for two and four element antenna arrays with beamforming in a **LOS environment** using wrap-around. See Figure 4.27 for the corresponding 'desert-island' scenario.

tance of adaptive antenna arrays, with the most dramatic gains in call dropping performance at the higher teletraffic levels.

Figures 4.55 and 4.56 show the probability of low quality access and the GOS, which are similar in terms of their trends and are closely related to each other by Equation 4.15. The GOS of the FCA algorithm was dominated by the probability of low quality access, since it had a higher value than the blocking probability. However, the rapid rise of the new call blocking probability of the LOLIA with $n = 19$ caused a steep increase in its GOS, especially when coupled with its rapidly degrading probability of low quality access. All of the algorithms benefited substantially from the employment of adaptive antenna arrays.

The effect of beamforming on the number of handovers performed can be seen in Figure 4.57. The LOLIAs required the least frequent handovers, with beamforming barely altering the results. In contrast, the number of handovers performed when using the FCA algorithm was reduced significantly due to employing AAAs with a maximum reduction of 72% for two elements, and of 89% for four elements. This translates into a significantly reduced signalling load for the network, since it has to manage far less handovers, therefore reducing the complexity of the network infrastructure.

It can be seen from Table 4.13 that in a LOS environment all of the channel allocation schemes benefit from the use of base station AAAs in terms of an increased level of teletraffic carried, hence supporting an increased number of users. The FCA algorithm benefited most from the employment of AAAs, with a 160% increase in terms of the number of users supported, when using a four-element antenna array. The performance improvements of the

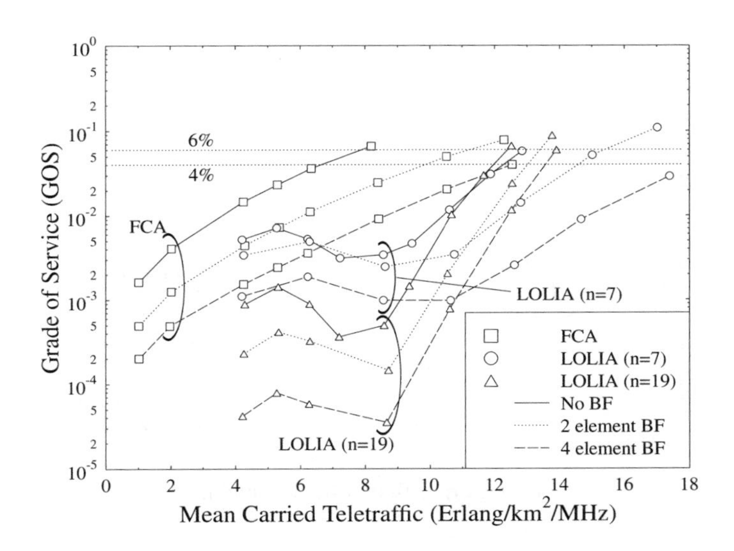

Figure 4.55: GOS performance versus mean carried traffic, for the LOLIA using 7 and 19 'local' base stations, and for FCA employing a 7-cell reuse cluster, under a uniform geographic traffic distribution, for a single antenna element, as well as for two and four element antenna arrays with beamforming in a **LOS environment** using wrap-around. See Figure 4.29 for the corresponding 'desert-island' scenario.

Algorithm	Conservative $P_{FT}=1\%, P_{low}=1\%$ $P_B=3\%, GOS=4\%$			Lenient $P_{FT}=1\%, P_{low}=2\%$ $P_B=5\%, GOS=6\%$		
	Users	**Traffic**	**Limiting Factor**	**Users**	**Traffic**	**Limiting Factor**
FCA, 1 element (el.)	340	3.6	P_{low}	465	4.9	P_{low}
FCA, 2 elements	**575**	**6.1**	P_{low}	**755**	**7.9**	P_{low}
FCA, 4 elements	**885**	**9.3**	P_{low}	**1105**	**11.2**	P_{FT}
LOLIA (n=7), 1 el.	990	10.5	P_{low}	1065	11.45	P_{low}
LOLIA (n=7), 2 el.	**1155**	**12.35**	P_{low}	**1260**	**13.5**	P_{low}
LOLIA (n=7), 4 el.	**1420**	**14.9**	P_{low}	**1535**	**16.5**	P_{low}
LOLIA (n=19), 1 el.	1020	10.9	P_{low}	1090	11.6	P_{low}
LOLIA (n=19), 2 el.	**1200**	**12.5**	P_{low}	**1330**	**13.35**	P_{low}
LOLIA (n=19), 4 el.	**1335**	**13.45**	P_B	**1400**	**13.9**	P_B

Table 4.13: Maximum mean carried traffic, and maximum number of mobile users that can be supported by each configuration, whilst meeting the preset quality constraints defined in Section 4.3.3.4. The carried traffic is expressed in terms of normalised Erlangs (Erlang/km^2/MHz), for the network described in Table 4.4 in a **LOS environment** using wrap-around. See Table 4.6 for the corresponding 'desert-island' results.

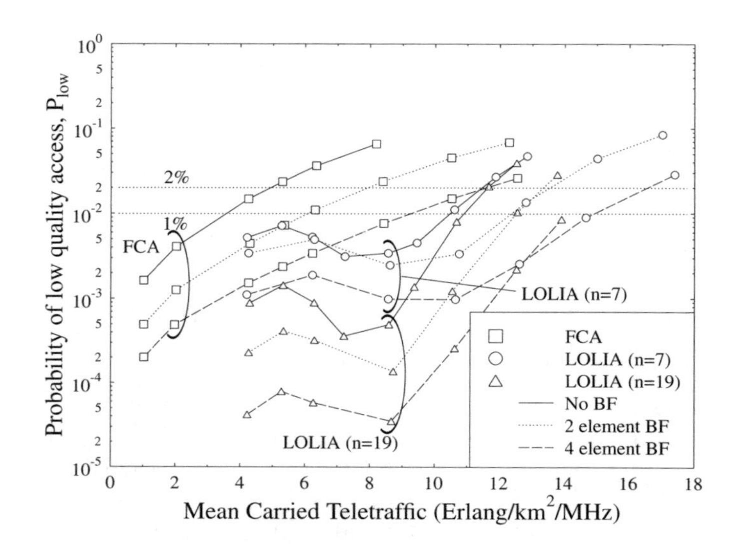

Figure 4.56: Probability of low quality access performance versus mean carried traffic, for the LOLIA using 7 and 19 'local' base stations, and for FCA employing a 7-cell reuse cluster, under a uniform geographic traffic distribution, for a single antenna element, as well as for two and four element antenna arrays with beamforming in a **LOS environment** using wrap-around. See Figure 4.28 for the corresponding 'desert-island' scenario.

LOLIA in conjunction with $n = 7$ due to using AAAs were more modest than for the FCA system. Specifically, 44% more users were supported by the four element AAA-assisted LOLIA using $n = 7$, when compared to the single antenna element based results. The network capacity of the LOLIA along with a 19-cell exclusion zone was higher than that of the LOLIA using $n = 7$, until the limited number of channels available in conjunction with such a large exclusion zone became significant. Up to this point, the AAAs reduced the levels of interference, thus improving the network capacity. However, when using a four-element AAA, the new call blocking probability became the dominant network performance limiting factor.

4.6.3.2 Performance Results over a Multipath Channel

Following our previous experiments, where a purely LOS environment existed between the mobiles and their base stations, this section presents results for a multipath environment using two-, four- and eight-element AAAs.

Comparing the call blocking probabilities of the multipath environment, shown in Figure 4.58, with those of the LOS environment, shown in Figure 4.53, reveals that all of the channel allocation algorithms behave similarly for both radio environments. The FCA scheme actually behaved more unfavourably in terms of its new call blocking probability, as the number of AAA elements was increased. However, this is a consequence of the additional antenna elements improving the other performance measures, such as the call dropping rate. This

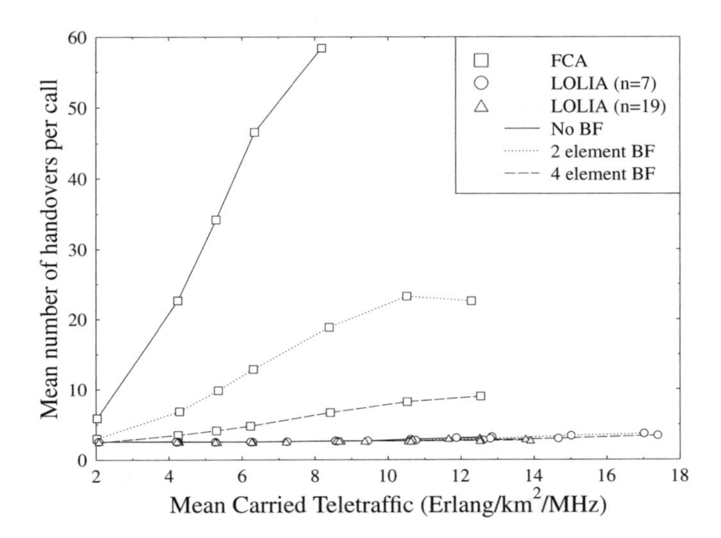

Figure 4.57: Mean number of handovers per call versus mean carried traffic, for comparison of the LOLIA using 7 and 19 'local' base stations, and for FCA employing a 7-cell reuse cluster, under a uniform geographic traffic distribution, for a single antenna element, as well as for two and four element antenna arrays with beamforming in a **LOS environment** using wrap-around. See Figure 4.30 for the corresponding 'desert-island' scenario.

enabled additional calls to be sustained at a given time, leading to a higher call blocking rate. In conjunction with an exclusion zone of 19 cells we found that the LOLIA's blocking performance was barely affected by the adaptive antenna arrays, whilst for $n = 7$ the blocked call rate was improved by a factor of 10 at a traffic load of 14-17 Erlang/km^2/MHz.

Figure 4.59 shows the probability of a dropped call in a multipath propagation environment, which was slightly higher than for the LOS scenario of Figure 4.54, when considered in the context of a given channel allocation algorithm and for a given antenna array size. The call dropping rate was improved with the aid of adaptive antenna arrays for all of the channel allocation algorithms, though the LOLIA using $n = 19$ did not benefit to the same extent as the other algorithms.

Again, as expected, the GOS curves in Figure 4.60 and the probability of low quality access curves of Figure 4.61 are similar in shape, with the differences resulting from the blocked call probability according to Equation 4.15. Hence, the GOS of the LOLIA having an exclusion zone of 19 base stations increases more rapidly than its probability of low quality access. In addition, the gain in its low quality of access performance achieved by using the adaptive antenna arrays is reduced, in terms of the GOS, due to the limited blocking probability improvement offered by the adaptive antenna arrays. All three algorithms benefit significantly in terms of their low quality access performance from the employment of the adaptive antenna arrays. However, the significant blocking performance limitations of the LOLIA using $n = 19$ restricts its GOS performance gains.

Figure 4.62 demonstrates the significant impact that adaptive antenna arrays have on the

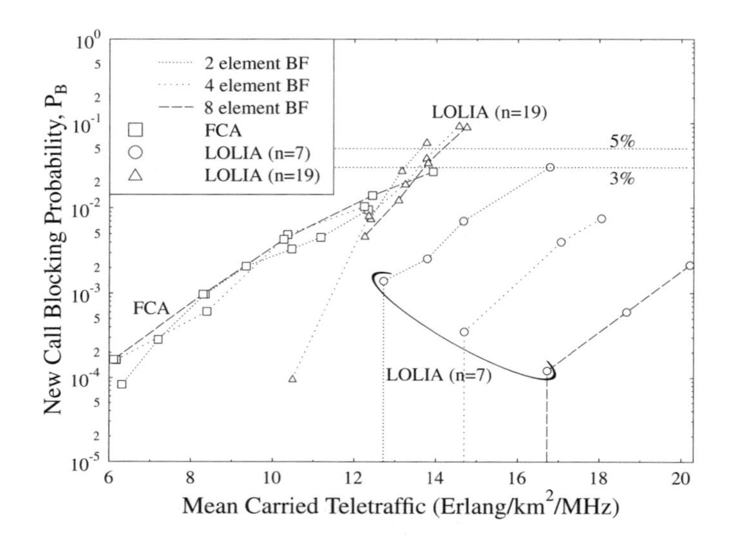

Figure 4.58: New call blocking probability performance versus mean carried traffic, for comparison of the LOLIA using 7 and 19 'local' base stations, and for FCA using a 7-cell reuse cluster, under a uniform geographic traffic distribution, for two, four and eight element antenna arrays with beamforming in a **multipath environment** using wrap-around. See Figure 4.32 for the corresponding 'desert-island' scenario.

mean number of handovers per call for the FCA algorithm in a multipath environment. Even in conjunction with adaptive antenna arrays more handovers per call were invoked, when using the FCA system, than for either of the LOLIAs using a single antenna element. Furthermore, a higher number of handovers was required in the multipath environment, than in the LOS scenario, for a given size of adaptive antenna array. The LOLIAs required significantly fewer handovers than the FCA, irrespective of the propagation environment, and did not benefit from the employment of adaptive antenna arrays in terms of the required handovers per call.

Table 4.14 presents results similar to those in Table 4.13, but for a multipath environment, with the bold values highlighting the adaptive antenna array sizes common to both sets of investigations. From this table it can be seen that the LOLIA using $n = 19$ carries approximately the same amount of traffic in the multipath scenario, which translates into a similar network capacity to that of the LOS scenario of Table 4.13. Again, the number of users supported by the network is limited by the probability of a low quality access and by the new call blocking probability. The performance of the LOLIA using $n = 7$ was interference limited, where the smaller reuse distance or exclusion zone led to numerous sources of relatively strong interference, all requiring interference cancellation. Hence, as the number of adaptive antenna array elements increased, so did the number of users supported, with an average improvement of about 15% for each doubling of the number of array elements.

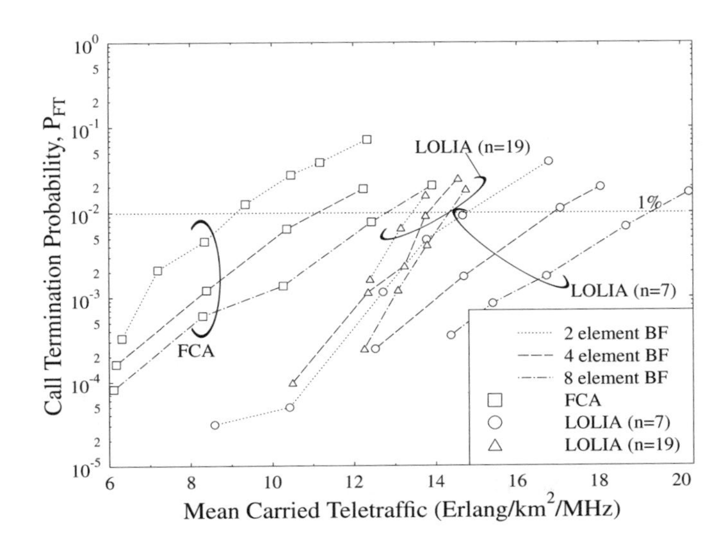

Figure 4.59: Call dropping probability performance versus mean carried traffic, for comparison of the LOLIA using 7 and 19 'local' base stations, and for FCA using a 7-cell reuse cluster, under a uniform geographic traffic distribution, for two, four and eight element antenna arrays with beamforming in a **multipath environment** using wrap-around. See Figure 4.33 for the corresponding 'desert-island' scenario.

Algorithm	Conservative $P_{FT}=1\%, P_{low}=1\%$ $P_B=3\%, GOS=4\%$			Lenient $P_{FT}=1\%, P_{low}=2\%$ $P_B=5\%, GOS=6\%$		
	Users	**Traffic**	**Limiting Factor**	**Users**	**Traffic**	**Limiting Factor**
FCA, 2 element (el.)	**600**	**6.0**	P_{low}	**740**	**7.65**	P_{low}
FCA, 4 elements	**790**	**8.3**	P_{low}	**995**	**10.3**	P_{low}
FCA, 8 elements	1085	11.2	P_{low}	1250	12.8	P_{FT}
LOLIA (n=7), 2 el.	**1195**	**12.65**	P_{low}	**1290**	**13.7**	P_{low}
LOLIA (n=7), 4 el.	**1370**	**14.35**	P_{low}	**1475**	**15.6**	P_{low}
LOLIA (n=7), 8 el.	1555	16.15	P_{low}	1700	17.7	P_{low}
LOLIA (n=19), 2 el.	**1235**	**12.65**	P_{low}	**1325**	**13.3**	P_{low}
LOLIA (n=19), 4 el.	**1360**	**13.55**	P_B	**1410**	**13.8**	P_{FT}
LOLIA (n=19), 8 el.	1385	13.7	P_B	1475	14.15	P_B

Table 4.14: Maximum mean carried traffic, and maximum number of mobile users that can be supported by each configuration, whilst meeting the preset quality constraints defined in Section 4.3.3.4. The carried traffic is expressed in terms of normalised Erlangs (Erlang/km^2/MHz), for the network described in Table 4.4 in a **multipath environment** using wrap-around. See Table 4.7 for the corresponding 'desert-island' results.

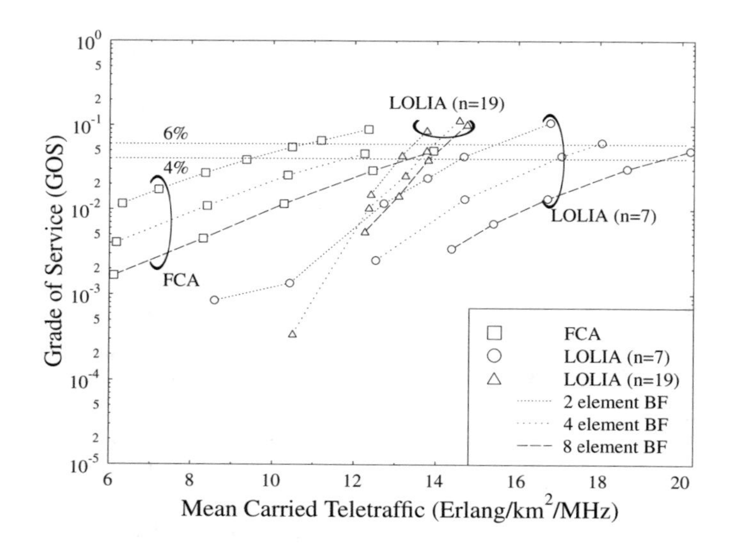

Figure 4.60: GOS performance versus mean carried traffic, for the comparison of the LOLIA with 7 and 19 'local' base stations, and for FCA using a 7-cell reuse cluster, under a uniform geographic traffic distribution, for two, four and eight element antenna arrays with beamforming in a **multipath environment** using wrap-around. See Figure 4.35 for the corresponding 'desert-island' scenario.

4.6.3.3 Performance over a Multipath Channel using Power Control

This section presents results obtained using the same wrap-around scenario of Section 4.6.1 over a multipath channel using power control. The power control algorithm was the same as that described in Section 4.6.2.3. The power control algorithm implemented attempted to independently adjust the mobile and base station transmit powers, such that the uplink and downlink SINRs were within a given target SINR window. The use of a target SINR window allowed us to avoid constantly increasing and decreasing the transmission powers, which could lead to potential power control instabilities within the network. Furthermore, the effect of employing a range of possible transmission powers is analogous to an inherent power control error plus slow fading phenomenon.

Figure 4.63 portrays the new call blocking probability versus the mean normalised carried traffic, expressed in terms of Erlangs/km^2/MHz. The figure shows that using power control in conjunction with the FCA algorithm resulted in a slight increase in the new call blocking probability as a direct consequence of the improved call dropping probability shown in Figure 4.64. In contrast, the blocking probability of the LOLIA improved significantly due to using power control, achieving a reduction by a factor of 4 to 34.

The new call blocking performance of the LOLIA was superior to that of the FCA algorithm both with and without power control, as seen in Figure 4.63, which is a result of the dynamic nature of the LOLIA. This enabled the LOLIA to allocate any of the available channels not used within the 7-cell exclusion zone (maximum of 7×8=56 channels in this

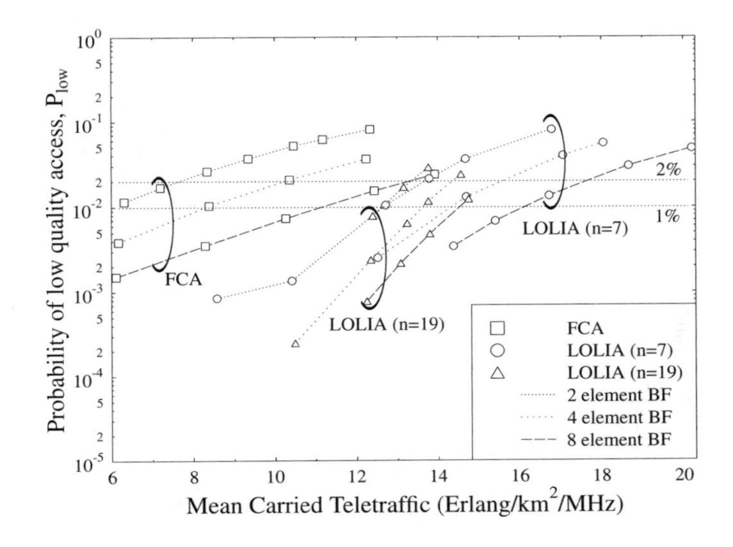

Figure 4.61: Probability of low quality access performance versus mean carried traffic, for the comparison of the LOLIA using 7 and 19 'local' base stations, and for FCA using a 7-cell reuse cluster, under a uniform geographic traffic distribution, for two, four and eight element antenna arrays with beamforming in a **multipath environment** using wrap-around. See Figure 4.34 for the corresponding 'desert-island' scenario.

scenario) to a new call request. However, the FCA algorithm only had one carrier frequency per base station, and therefore was less likely to be able to satisfy a new call request. The addition of power control to the LOLIA in conjunction with $n = 7$ led to a reduced new call blocking probability. Specifically, the new call blocking probability with power control was reduced to near that achieved using twice the number of antenna elements without power control. The higher new call blocking probability of the network using no power control can be attributed to the lower average SINR values, which prevent new call initiation, whereas the higher average SINR level of the network observed in Figure 4.38 in conjunction with power control enables additional calls to commence.

Figure 4.64 shows that the call dropping probability was significantly reduced for both the FCA algorithm and the LOLIA using $n = 7$, in conjunction with power control. The FCA algorithm in conjunction with power control offered a call dropping probability close to that of a similar network without power control, and using twice the number of adaptive antenna elements. However, at traffic loads of below approximately 7 Erlangs/km^2/MHz the call dropping probability began to level off for the FCA algorithm. This phenomenon was also noticeable in the context of the LOLIA and resulted from the power control algorithm limiting the maximum SINR, leading to a flatter call dropping profile than that of the network without power control. Thus, at lower traffic loads the network without power control had a higher average SINR as was evidenced by Figure 4.38, leading to less dropped calls. However, at higher levels of teletraffic the power control algorithm offered a lower call dropping rate, as a consequence of the lower levels of interference present when using the power control scheme.

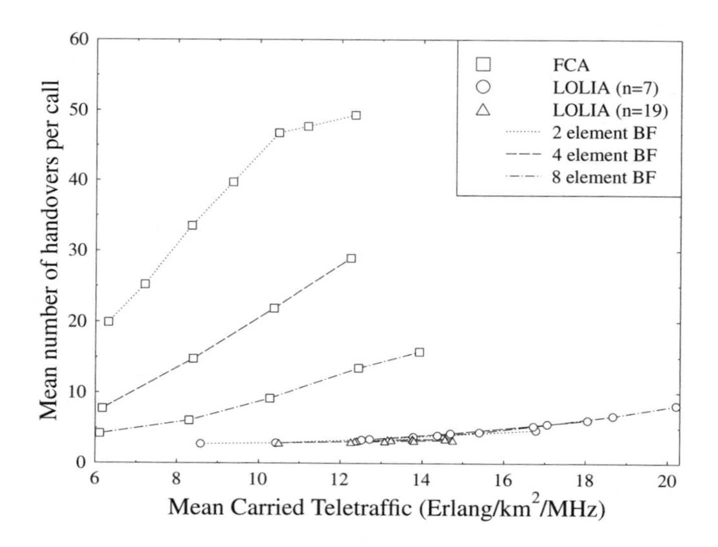

Figure 4.62: Mean number of handovers per call versus mean carried traffic, for comparison of the LOLIA using 7 and 19 'local' base stations, and for FCA using a 7-cell reuse cluster, under a uniform geographic traffic distribution, for two, four and eight element antenna arrays with beamforming in a **multipath environment** using wrap-around. See Figure 4.36 for the corresponding 'desert-island' scenario.

The FCA algorithm exhibited the greatest improvement in the probability of a low quality access due to the implementation of power control, as shown in Figure 4.65. Using a two element adaptive antenna array in conjunction with the power control algorithm resulted in a probability of low quality access approximately equal to that obtained using an eight element adaptive array without power control. The LOLIA also benefited from invoking the power control algorithm, but to a lesser extent, offering a performance close to that of an array with twice the number of elements without power control.

The GOS performance gains of the FCA algorithm using power control seen in Figure 4.66, were somewhat reduced compared to those of the probability of a low quality access in Figure 4.65, due to the similar blocking performances of the power-controlled and non-power-controlled scenarios seen in Figure 4.63. Nonetheless, the GOS gains remained quite high in Figure 4.66. The GOS gains of the LOLIA due to power control were also quite substantial, as seen in Figure 4.66.

The effect of power control on the mean number of handovers performed per call becomes explicit in Figure 4.67. From this figure it can be seen for the FCA algorithm that with respect to the number of handovers per call, the performance of the network employing power control significantly exceeded that of the network without power control using an adaptive antenna array of twice the number of antenna elements. The employment of power control in conjunction with the FCA algorithm led to a mean reduction by a factor of 4.4 in the number of handovers. The inherently good performance of the LOLIA was also slightly improved on average.

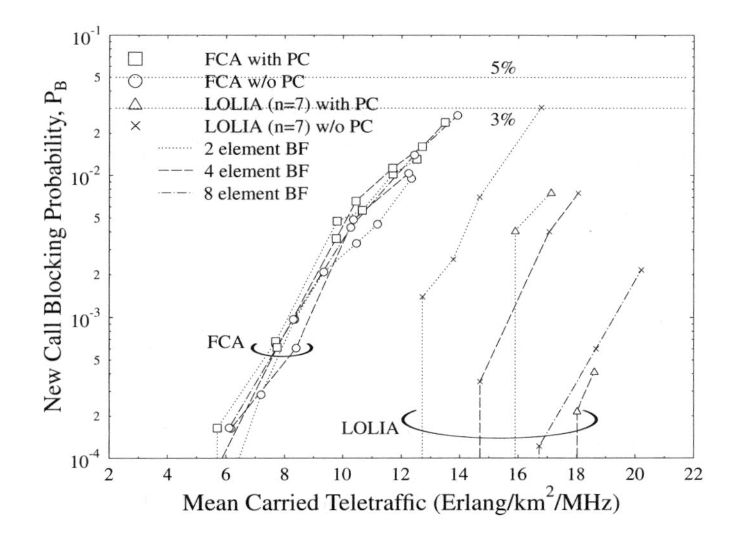

Figure 4.63: New call blocking probability versus mean carried traffic of the LOLIA using 7 'local' base stations, and for FCA employing a 7-cell reuse cluster, for two and four element antenna arrays, **with and without power control** using wrap-around. See Figure 4.39 for the corresponding 'desert-island' scenario.

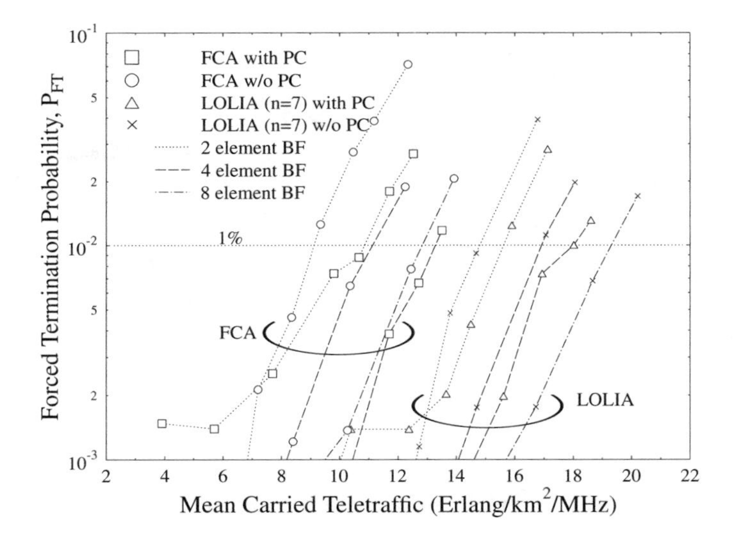

Figure 4.64: Call dropping probability versus mean carried traffic of the LOLIA using 7 'local' base stations, and for FCA employing a 7-cell reuse cluster, for two and four element antenna arrays, **with and without power control** using wrap-around. See Figure 4.40 for the corresponding 'desert-island' scenario.

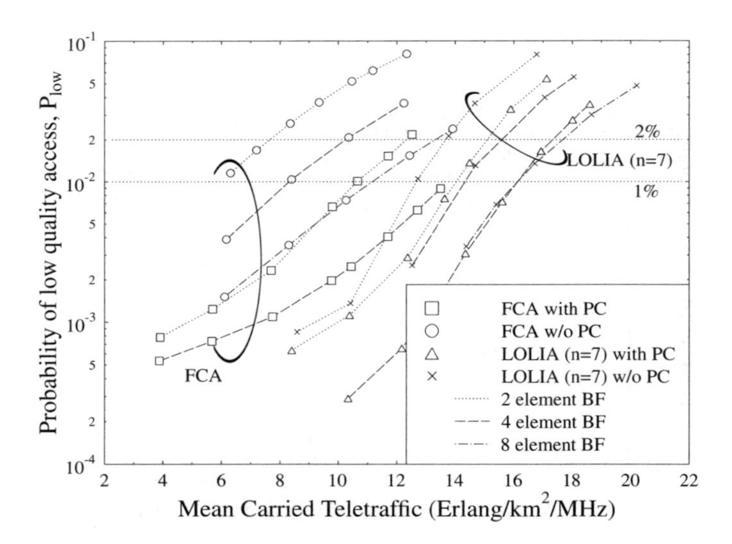

Figure 4.65: Probability of low quality outage versus mean carried traffic of the LOLIA using 7 'local' base stations, and for FCA employing a 7-cell reuse cluster, for two and four element antenna arrays, **with and without power control** using wrap-around. See Figure 4.41 for the corresponding 'desert-island' scenario.

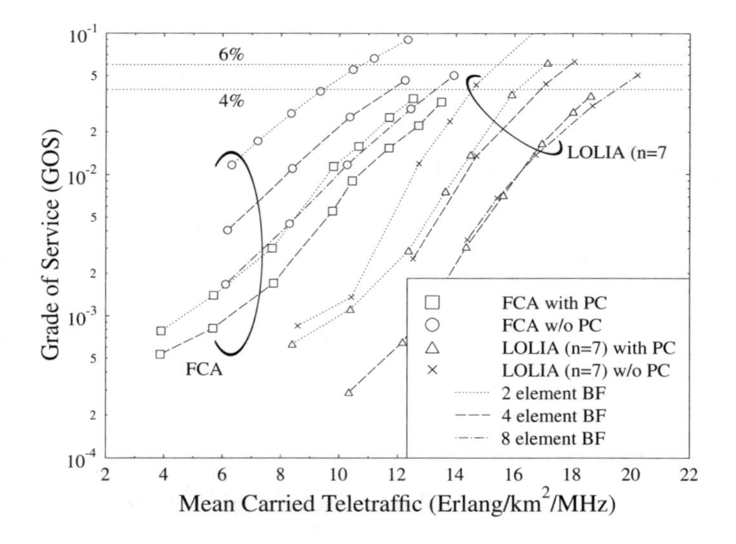

Figure 4.66: GOS performance versus mean carried traffic of the LOLIA using 7 'local' base stations, and for FCA employing a 7-cell reuse cluster, for two and four element antenna arrays, **with and without power control** using wrap-around. See Figure 4.42 for the corresponding 'desert-island' scenario.

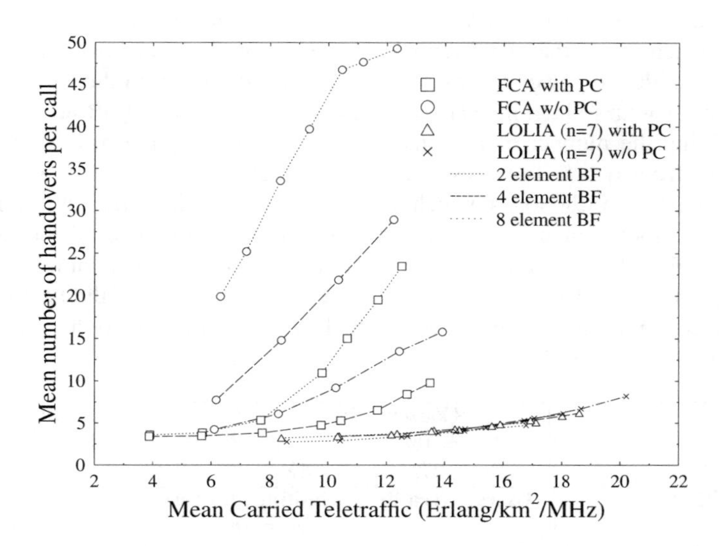

Figure 4.67: Mean number of handovers per call versus mean carried traffic of the LOLIA using 7 'local' base stations, and for FCA employing a 7-cell reuse cluster, for two and four element antenna arrays, **with and without power control** using wrap-around. See Figure 4.43 for the corresponding 'desert-island' scenario.

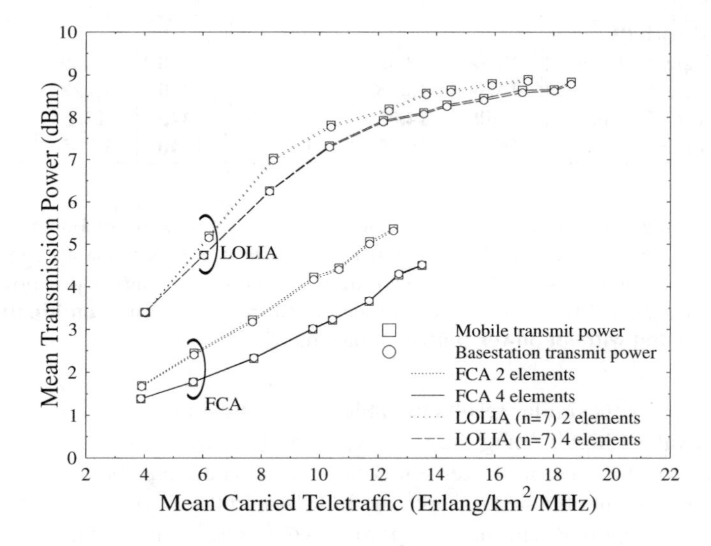

Figure 4.68: Mean transmit power versus mean carried traffic of the LOLIA using 7 'local' base stations, and for FCA employing a 7-cell reuse cluster, for two and four element antenna arrays, **with and without power control** using wrap-around. See Figure 4.44 for the corresponding 'desert-island' scenario.

A further advantage of using power control in a cellular mobile network is portrayed in Figure 4.68, which shows that the mean transmit power was reduced from the fixed transmit power of 10 dBm due to power control. The mean transmit power of the FCA algorithm was reduced the most with reductions varying from 4.5 dB to almost 9 dB at the lowest traffic levels. Doubling the number of antenna elements comprising the base stations' adaptive antenna arrays from two to four, resulted in additional mean transmission power gains of almost 1 dB at higher traffic loads, which is a consequence of the extra interference rejection capability of the four element array. The mean transmission powers of the LOLIAs were significantly higher due to the higher target SINRs required for maintaining an acceptable call dropping rate. This was a consequence of the dynamic nature of the LOLIA, leading to more rapidly changing interference levels, which required a relatively high target SINR of 31 dB as seen in Table 4.8.

Algorithm	**Conservative** $P_{FT}=1\%, P_{low}=1\%$ $P_B=3\%, GOS=4\%$			**Lenient** $P_{FT}=1\%, P_{low}=2\%$ $P_B=5\%, GOS=6\%$		
	Users	**Traffic**	**Limiting Factor**	**Users**	**Traffic**	**Limiting Factor**
4-QAM without PC						
FCA, 2 element (el.)	**600**	**6.0**	P_{low}	**740**	**7.65**	P_{low}
FCA, 4 elements	**790**	**8.3**	P_{low}	**995**	**10.3**	P_{low}
FCA, 8 elements	1085	11.2	P_{low}	1250	12.8	P_{FT}
LOLIA (n=7), 2 el.	**1195**	**12.65**	P_{low}	**1290**	**13.7**	P_{low}
LOLIA (n=7), 4 el.	**1370**	**14.35**	P_{low}	**1475**	**15.6**	P_{low}
LOLIA (n=7), 8 el.	1555	16.15	P_{low}	1700	17.7	P_{low}
4-QAM with PC						
FCA, 2 elements (el.)	**1090**	**10.6**	P_{low}	**1120**	**10.85**	P_{FT}
FCA, 4 elements	**1370**	**13.28**	P_{FT}	**1370**	**13.28**	P_{FT}
LOLIA (n=7), 2 el.	**1350**	**14.05**	P_{low}	**1445**	**15.1**	P_{low}
LOLIA (n=7), 4 el.	**1540**	**16.15**	P_{low}	**1640**	**17.35**	P_{low}

Table 4.15: Maximum mean carried traffic, and maximum number of mobile users that can be supported by each configuration, whilst meeting the preset quality constraints defined in Section 4.3.3.4. The carried traffic is expressed in terms of normalised Erlangs (Erlang/km^2/MHz), for the network described in Table 4.4 in a **multipath environment with and without power control** using wrap-around.

Table 4.15 presents similar results to Table 4.14, but using our power control algorithm, with the bold values highlighting the adaptive antenna array sizes common to both sets of investigations, for the sake of convenient comparison. The table shows the significant performance improvement obtained for both the LOLIA and the FCA algorithm in terms of the number of users supported with the advent of power control, whilst maintaining the desired network quality. In the conservative scenario, for example, the FCA algorithm using a two element adaptive antenna array and power control supported the same number of users as the network using an eight element adaptive antenna array without power control. The LOLIA-based network, however, did not benefit from the employment of the power control algorithm to the same extent, although it still offered similar performance to that of a network with-

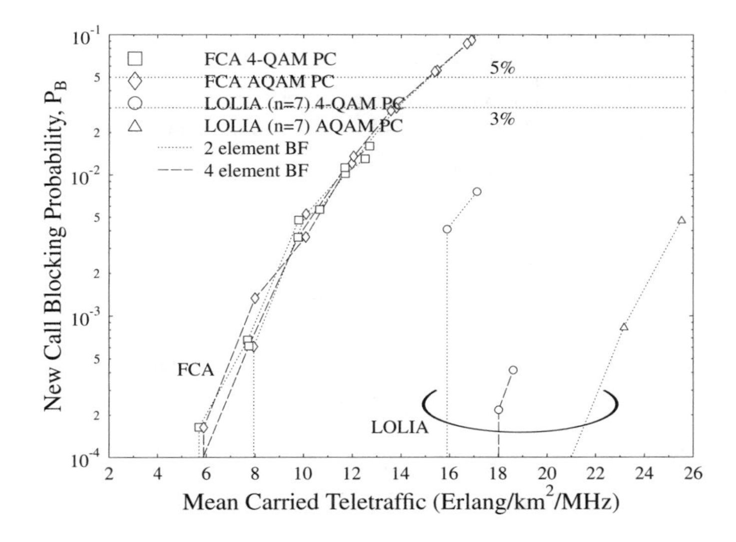

Figure 4.69: New call blocking probability versus mean carried traffic of the LOLIA using 7 'local' base stations, and for FCA employing a 7-cell reuse cluster, for two and four element antenna arrays, **with and without AQAM** using wrap-around. See Figure 4.46 for the corresponding 'desert-island' scenario.

out power control and using adaptive antenna arrays having twice the number of antenna elements.

4.6.3.4 Performance of an AQAM based Network using Power Control

This section presents our simulation results obtained for a network using burst-by-burst adaptive modulation [13, 168, 357, 358] invoked in order to improve the network's performance. Simulations were conducted for both a standard 7-cell FCA scheme and a 7-cell LOLIA assisted system. The results obtained for a 4-QAM based network using power control were included for comparison purposes.

The new call blocking probability depicted in Figure 4.69 was essentially unchanged for the FCA algorithm using power control in conjunction with 4-QAM or AQAM, suggesting that the new call blocking performance of the FCA algorithm was limited by the lack of available frequency/timeslot combinations, rather than by inadequate signal quality. This hypothesis was confirmed by the improvement in the new call blocking performance of the LOLIA resulting from the superior signal quality of AQAM.

The corresponding call dropping probability is depicted in Figure 4.70. The AQAM LOLIA using $n = 7$ in conjunction with a two element adaptive antenna had, in general, a higher call dropping probability compared to that of power control assisted 4-QAM. However, the power control algorithm, when used in conjunction with AQAM, maintained the call dropping probability below the given threshold for a significantly higher traffic load. Similar performance trends were observed for both the two element and the four element adaptive

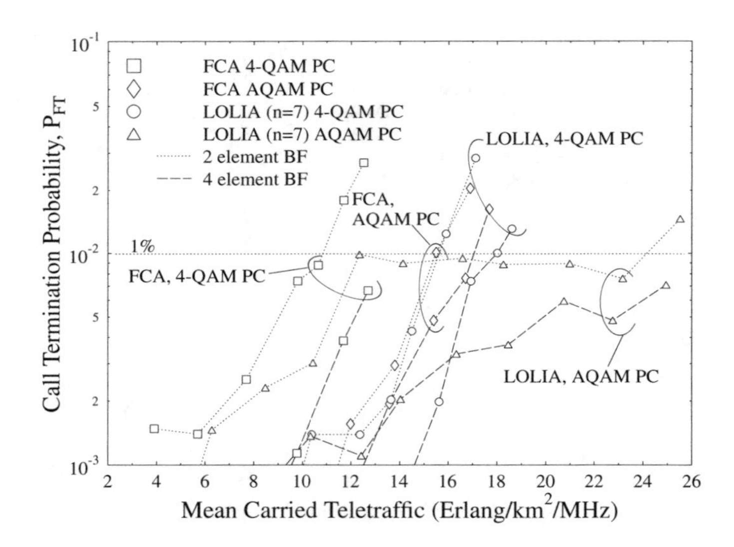

Figure 4.70: Call dropping, or forced termination, performance versus mean carried traffic of the LOLIA using 7 'local' base stations, and for FCA employing a 7-cell reuse cluster, for two and four element antenna arrays, **with and without AQAM** using wrap-around. See Figure 4.47 for the corresponding 'desert-island' scenario.

array, although the higher interference rejection capability offered by the four element array resulted in a substantially reduced call dropping probability. The dropped calls were caused almost exclusively by insufficient signal quality during the intra-cell handover process, thus increasing the number of adaptive antenna elements from two to four improved the call dropping performance. The high call dropping probability observed for traffic loads between 12 and 20 Erlangs/km^2/MHz when using the two element adaptive antenna array was due to the power control and AQAM attempting to trade-off modem throughput and transmit power against each other, whilst attempting to minimise the number of dropped calls. The extra interference suppression capability of the four-element adaptive antenna array led to a reduced call dropping probability. Hence, altering the AQAM mode selection algorithm of Figure 4.45, may improve its performance at these traffic loads, when used in conjunction with a two element antenna array.

The FCA algorithm dropped all of its calls during the inter-cell handover process due to the lack of available slots to handover to. However, since inter-cell handovers could be performed, if necessary, in order to improve the signal quality, the number of dropped calls was reduced when using the four element adaptive array, due to its better interference rejection capability. All the calls were dropped during the inter-cell handover process, which means that no calls were dropped due to insufficient SINR or through the intra-cell handover process. This can be attributed to the AQAM scheme, which enabled users to drop to lower order modulation modes of the AQAM scheme, when the SINR became poor.

Figure 4.71 characterises the mean number of handovers per call for 4-QAM and AQAM, both using power control. The LOLIA using $n = 7$ performed a lower total number of

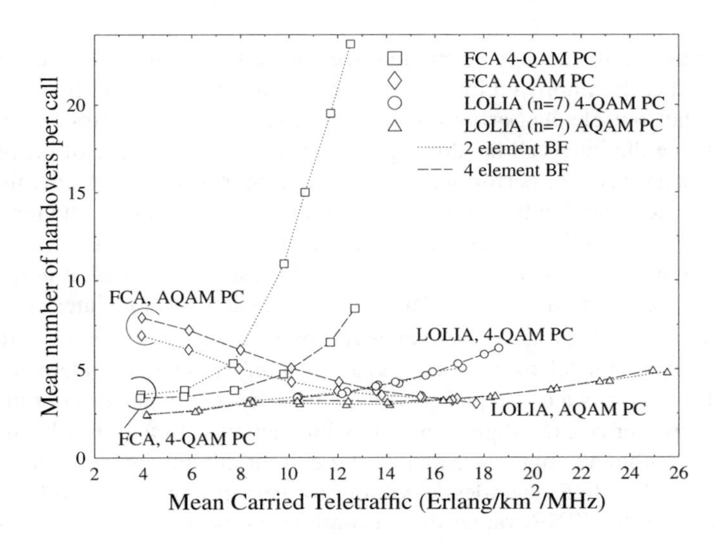

Figure 4.71: Mean number of handovers per call versus mean carried traffic of the LOLIA using 7 'local' base stations, and for FCA employing a 7-cell reuse cluster, for two and four element antenna arrays, **with and without AQAM** using wrap-around. See Figure 4.50 for the corresponding 'desert-island' scenario.

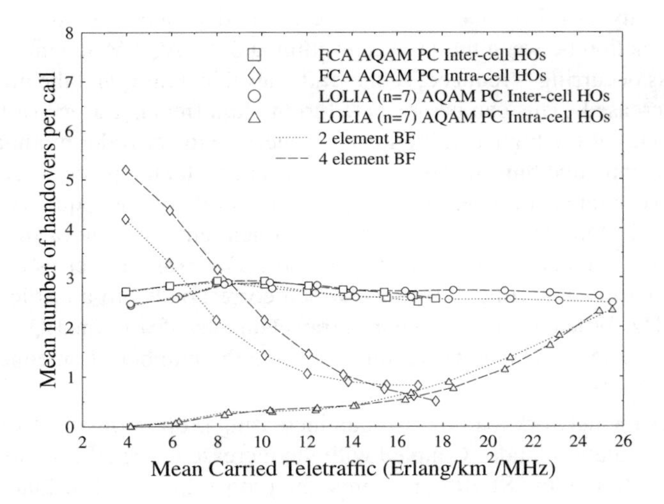

Figure 4.72: Mean number of inter-cell and intra-cell handovers per call versus mean carried traffic of the LOLIA using 7 'local' base stations, and for FCA employing a 7-cell reuse cluster, for two and four element antenna arrays, in conjunction **with AQAM** using wrap-around.

handovers per call, when using AQAM, due its inherent resilience to poor signal quality conditions.

The breakdown of the handovers into inter-cell and intra-cell handovers is given in Figure 4.72. Observe that the improved interference rejection capability, and the associated superior SINR of the four-element array results in a lower number of intra-cell handovers for the LOLIA. Since the intra-cell handover process is the primary cause of dropped calls and less intra-cell handovers are performed when using a four-element antenna, more inter-cell handovers are necessitated in the network using four-element adaptive antenna arrays, as the users roam from cell to cell. In other words, since the LOLIA using a four element array drops less calls than when using a two element array, more users are in call at a given time, and hence these users cross more cell boundaries, thus necessitating more inter-cell handovers.

In contrast, the number of intra-cell handovers performed in conjunction with the FCA algorithm decreases, as the teletraffic rises, and as the number of antenna elements is increased from two to four. This is a consequence of the particular implementation of the modulation mode selection/power control algorithm and its interaction with the FCA handover process. The AQAM algorithm attempts to remain in the current modulation mode as long as possible, and hence as the SINR degrades, it will opt for performing an intra-cell handover in an attempt to maintain the SINR, rather than reconfiguring itself in order to use a lower-order modulation mode suitable for the reduced SINR level. Thus, when using a four-element adaptive antenna array, the average (and instantaneous) SINR is typically higher than that of a two-element array, leading to a more frequent employment of the less resilient higher-order modulation modes, which potentially requires additional intra-cell handovers. However, as the mean teletraffic increases, so does the level of interference in the network and a greater proportion of transmission time is spent in the lower-order modulation modes, thus requiring less intra-cell handovers, as illustrated in Figure 4.72.

The probability of a Low Quality (LQ) access is depicted in Figure 4.73, showing an interesting interaction between the FCA algorithm and the AQAM scheme. The probability of an LQ access occurring is reduced, as the traffic level increases and the number of antenna elements is decreased. This can be attributed to the less frequent usage of the higher-order modulation modes at the higher traffic loads. Hence the lower-order modulation modes are used more frequently and thus the chance of an LQ access taking place is reduced. The four-element adaptive antenna array leads to a higher probability of a low quality access, since its higher associated SINR levels activate a more frequent employment of the less robust, but higher-throughput, higher-order modulation modes. For example, let us consider the FCA AQAM PC scenario supporting 400 users, which corresponded to a traffic load of about 4 Erlang/km^2/MHz. When using two antenna array elements, 85% of the LQ accesses occurred whilst in the 16-QAM mode, however, on increasing the number of antenna array elements to four this rose to 93%.

However, as the network loading rises, an increasing proportion of the LQ outages occur in the BPSK modulation mode. Coupled with the increase in the BPSK modulation mode's employment due to the low SINR constraints, the probability of a low quality outage is expected to increase at a certain traffic load. This can be seen in Figure 4.73, where the LQ outage probability is starting to rise for FCA in conjunction with both two and four elements, though the extra interference suppression capability of the four element array allows extra traffic to be carried, before this phenomenon commences. More specifically, although not explicit in Figure 4.73, we found that for a network supporting 1200 users, corresponding

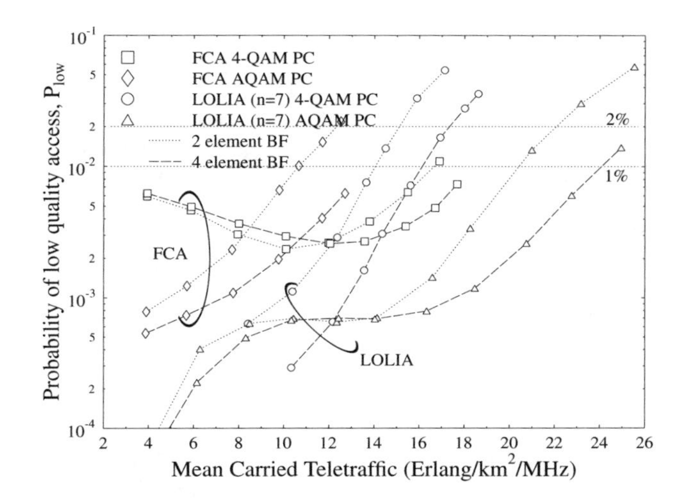

Figure 4.73: Probability of low quality access versus mean carried traffic of the LOLIA using 7 'local' base stations, and for FCA employing a 7-cell reuse cluster, for two and four element antenna arrays, **with and without AQAM** using wrap-around. See Figure 4.48 for the corresponding 'desert-island' scenario.

to a traffic load of about 12 Erlang/km^2/MHz, and employing two element adaptive antenna arrays, 43% of the LQ accesses occurred, whilst in the 16-QAM mode, versus 72% with four-element antenna arrays. Again, not explicitly shown in the figure, but increasing the number of users to 1400, or a traffic load of just less than 14 Erlang/km^2/MHz, reduced the number of 16-QAM LQ accesses, but increased the BPSK LQ outages to 69% and 31% for the two- and four-element arrays respectively, with reductions to 21% and 53% of the LQ outages in the 16-QAM mode.

From Figure 4.74 it can be seen that the GOS, as defined in Section 4.3.3.4, of the FCA algorithm did not benefit from invoking AQAM to the same extent as the LOLIA. This resulted from the fairly similar probability of low quality access performance of the two and four element antenna array assisted systems in Figure 4.73, and the limiting blocking performance observed in Figure 4.69. However, since the LOLIA did not suffer from these limiting factors, its GOS improved due to the employment of both adaptive antenna arrays and AQAM techniques.

The average modem throughput expressed in bits per symbol versus the mean carried teletraffic is shown in Figure 4.75, demonstrating that the mean number of bits per symbol throughput of the users decreased, as the number of users supported increased. The FCA algorithm offered the lowest throughput and its performance degraded near-linearly upon increasing the number of users supported. At the user capacity limits of 1400 and 1565 users, the mean modem throughput was 2.45 BPS and 2.35 BPS for the conservative and lenient scenarios, respectively, using two element adaptive antenna arrays. Using four element adaptive antenna arrays the corresponding throughputs were 2.7 BPS and 2.6 BPS. The LOLIA, especially for lower levels of traffic, offered a higher modem throughput for a given level of

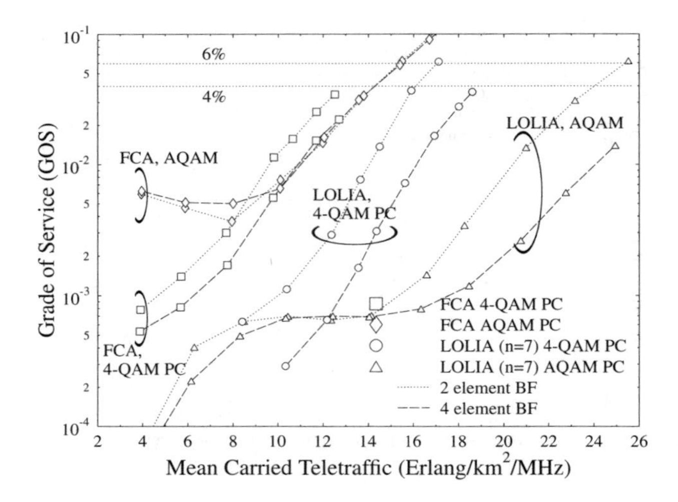

Figure 4.74: GOS performance versus mean carried traffic of the LOLIA using 7 'local' base stations, and for FCA employing a 7-cell reuse cluster, for two and four element antenna arrays, **with and without AQAM** using wrap-around. See Figure 4.49 for the corresponding 'desert-island' scenario.

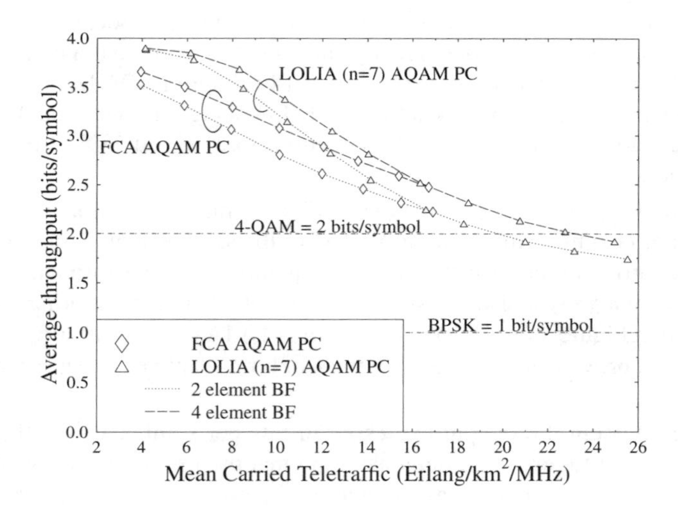

Figure 4.75: Mean throughput of users in terms of bits per symbol versus mean carried traffic of the LOLIA using 7 'local' base stations, and for FCA employing a 7-cell reuse cluster, for two and four element antenna arrays, **using AQAM** using wrap-around. See Figure 4.52 for the corresponding 'desert-island' scenario.

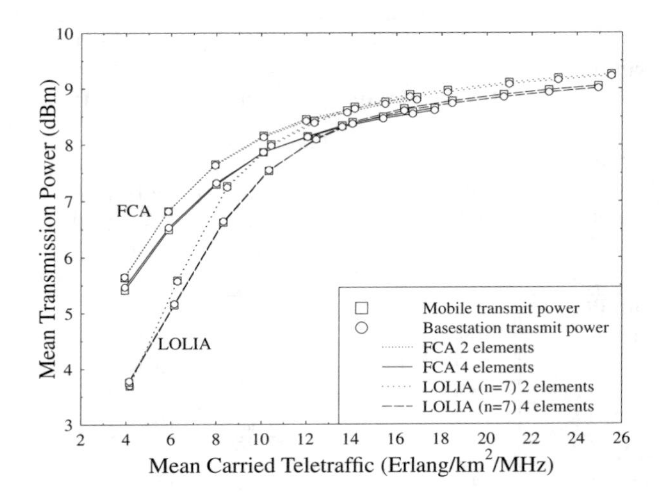

Figure 4.76: Mean transmit power versus mean carried traffic of the LOLIA using 7 'local' base stations, and for FCA employing a 7-cell reuse cluster, for two and four element antenna arrays, **using AQAM** using wrap-around. See Figure 4.48 for the corresponding 'desert-island' scenario.

teletraffic carried, with the BPS throughput performance gracefully decreasing, as the carried teletraffic continued to increase. The capacity limiting factor of the LOLIA was the throughput restriction of 2.0 BPS.

The mean transmission power results of Figure 4.76 demonstrate that the employment of AQAM is capable of reducing the power transmitted, both for the uplink and the downlink. At low traffic levels the FCA algorithm performed noticeably worse in transmitted power terms, than the LOLIA. However, as the traffic load increased, the difference became negligible. The mean power reduction, when compared to a fixed transmission power of 10 dBm, varied from approximately 1 dB to more than 6 dB. A 1 dB reduction in transmission power is not particularly significant for the mobile user, especially since at this network load a throughput of just 2 bits/symbol is possible. The difference between the network using AQAM and that without, though, is the overall improved call quality that can be achieved in the context of our performance metrics, and the significantly increased number of users that can be supported by the network Again, the constraint of a minimum throughput of 2 bits/symbol was invoked in order to ensure a fair comparison with the fixed 4-QAM based network.

Table 4.16 shows the performance of the various networks using AQAM with power control, as well as 4-QAM with and without power control, in terms of the number of users supported. A mean increase of 61% was achieved in terms of the number of users by the addition of power control to the FCA algorithm based 4-QAM network. Invoking AQAM and power control led to a further average user capacity increase of almost 22%, with any further gains limited by the lack of free frequency/timeslot combinations available for new calls to start. Therefore, since the network capacity of the FCA algorithm when using adaptive modulation was not limited by co-channel interference, it would be possible to reduce the

Algorithm	Conservative $P_{FT} = 1\%, P_{low} = 1\%$ $GOS = 4\%, P_B = 3\%$			Lenient $P_{FT} = 1\%, P_{low} = 2\%$ $GOS = 6\%, P_B = 5\%$		
	Users	Traffic	Limiting Factor	Users	Traffic	Limiting Factor
4-QAM without PC						
FCA, 2 elements (el.)	600	6.0	P_{low}	740	7.65	P_{low}
FCA, 4 elements	790	8.3	P_{low}	995	10.3	P_{low}
LOLIA (n=7), 2 el.	1195	12.65	P_{low}	1290	13.7	P_{low}
LOLIA (n=7), 4 el.	1370	14.35	P_{low}	1475	15.6	P_{low}
4-QAM with PC						
FCA, 2 elements	1090	10.6	P_{low}	1120	10.85	P_{FT}
FCA, 4 elements	1370	13.275	P_{FT}	1370	13.275	P_{FT}
LOLIA (n=7), 2 el.	1350	14.05	P_{low}	1445	15.1	P_{low}
LOLIA (n=7), 4 el.	1540	16.15	P_{low}	1640	17.35	P_{low}
AQAM with PC						
FCA, 2 elements	1400	13.8	P_B	1565	15.20	P_B
FCA, 4 elements	1415	13.7	P_B	1575	15.15	P_B
LOLIA (n=7), 2 el.	1910	19.75	BPS	1910	19.75	BPS
LOLIA (n=7), 4 el.	2245	23.25	BPS	2245	23.25	BPS

Table 4.16: Maximum mean carried traffic, and maximum number of mobile users that can be supported by each configuration, whilst meeting the preset quality constraints defined in Section 4.3.3.4. The carried traffic is expressed in terms of normalised Erlangs (Erlang/km^2/MHz), for the network described in Table 4.4 in a **multipath environment with and without power control and AQAM** using wrap-around.

frequency re-use distance to increase the network capacity.

The performance of the LOLIA was not limited in this sense, however, and the addition of power control to the 4-QAM network provided an mean increase of 12% extra users supported. In conjunction with AQAM techniques this user capacity was further extended by an average of 39%, thus supporting an additional 56% more users, when compared to the 4-QAM network using no power control.

4.7 Summary and Conclusions

In this chapter we have examined the network capacity and performance of the FCA algorithm and the LOLIA using an exclusion zone of seven or 19 base stations, in the context of LOS and multipath propagation environments. We have shown that the addition of power control results in a substantially increased number of supported users, additionally benefiting from a superior call quality, and reduced transmission power for a given number of adaptive antenna array elements located at the base stations. The advantages of using AQAM within a mobile cellular network have also been illustrated, resulting in performance improvements in terms of the mean modem throughout, call quality, mean transmission power and the number of supported users. The next chapter involves the investigation of network capacity in the context of a CDMA-based UMTS-type FDD mode network.

Chapter 5

UTRA Network Performance Using Adaptive Arrays and Adaptive Modulation

5.1 Introduction

In January 1998, the European standardisation body for third generation mobile radio systems, the European Telecommunications Standards Institute - Special Mobile Group (ETSI SMG), agreed upon a radio access scheme for third generation mobile radio systems, referred to as the Universal Mobile Telecommunication System (UMTS) [11,32]. Although this chapter was detailed in Chapter 1, here we provide a rudimentary introduction to the system, in order to allow readers to consult this chapter directly, without having to read Chapter 1 first. Specifically, the UMTS Terrestrial Radio Access (UTRA) supports two modes of duplexing, namely Frequency Division Duplexing (FDD) , where the uplink and downlink are transmitted on different frequencies, and Time Division Duplexing (TDD) , where the uplink and the downlink are transmitted on the same carrier frequency, but multiplexed in time. The agreement recommends the employment of Wideband Code Division Multiple Access (W-CDMA) for UTRA FDD and Time Division - Code Division Multiple Access (TD-CDMA) for UTRA TDD. TD-CDMA is based on a combination of Time Division Multiple Access (TDMA) and CDMA, whereas W-CDMA is a pure CDMA-based system. The UTRA scheme can be used for operation within a minimum spectrum of 2 x 5 MHz for UTRA FDD and 5 MHz for UTRA TDD. Both duplex or paired and simplex or unpaired frequency bands have been identified in the region of 2 GHz to be used for the UTRA third generation mobile radio system. Both modes of UTRA have been harmonised with respect to the basic system parameters, such as carrier spacing, chip rate and frame length. Thereby, FDD/TDD dual mode operation is facilitated, which provides a basis for the development of low cost terminals. Furthermore, the interworking of UTRA with GSM [11] is ensured.

In UTRA, the different service needs are supported in a spectrally efficient way by a combination of FDD and TDD. The FDD mode is intended for applications in both macro- and micro-cellular environments, supporting data rates of up to 384 kbps and high mobility. The TDD mode, on the other hand, is more suited to micro and pico-cellular environments, as well as for licensed and unlicensed cordless and wireless local loop applications. It makes efficient use of the unpaired spectrum - for example in wireless Internet applications, where much of the teletraffic is in the downlink - and supports data rates of up to 2 Mbps. Therefore, the TDD mode is particularly well suited for environments generating a high traffic density (e.g. in city centres, business areas, airports etc.) and for indoor coverage, where the applications require high data rates and tend to have highly asymmetric traffic again, as in Internet access.

In parallel to the European activities, extensive work has been carried out also in Japan and the USA on third generation mobile radio systems. The Japanese standardisation body known as the Association of Radio Industry and Business (ARIB) also opted for using W-CDMA, and the Japanese as well as European proposals for FDD bear strong similarities. Similar concepts have also been developed by the North-American T1 standardisation body for the pan-American third generation (3G) system known as cdma2000, which was also described in Chapter 1 [11].

In order to work towards a truly global third generation mobile radio standard, the Third Generation Partnership Project (3GPP) was formed in December 1998. 3GPP consists of members of the standardisation bodies in Europe (ETSI), the US (T1), Japan (ARIB), Korea (TTA - Telecommunications Technologies Association), and China (CWTS - China Wireless Telecommunications Standard). 3GPP merged the already well harmonised proposals by the regional standardisation bodies and now works towards a single common third generation mobile radio standard under the terminology UTRA, retaining its two modes, and aiming to operate on the basis of the evolved GSM core network. The Third Generation Partnership Project 2 (3GPP2), on the other hand, works towards a third generation mobile radio standard, which is based on an evolved IS-95 type system which was originally referred to as cdma2000 [11]. In June 1999, major international operators in the Operator Harmonisation Group (OHG) proposed a harmonised G3G (Global Third Generation) concept, which has been accepted by 3GPP and 3GPP2. The harmonised G3G concept is a single standard with the following three modes of operation:

- CDMA direct spread (CDMA-DS), based on UTRA FDD as specified by 3GPP.
- CDMA multi-carrier (CDMA-MC), based on cdma2000 using FDD as specified by 3GPP2.
- TDD (CDMA TDD) based on UTRA TDD as specified by 3GPP.

5.2 Direct Sequence Code Division Multiple Access

A rudimentary introduction to CDMA was provided in Chapter 1 in the context of single-user receivers, while in Chapter 2 the basic concepts of multi-user detection have been introduced. However, as noted earlier, our aim is to allow reader to consult this chapter directly, without having to refer back to the previous chapters. Hence here a brief overview of the undrlying CDMA basics is provided.

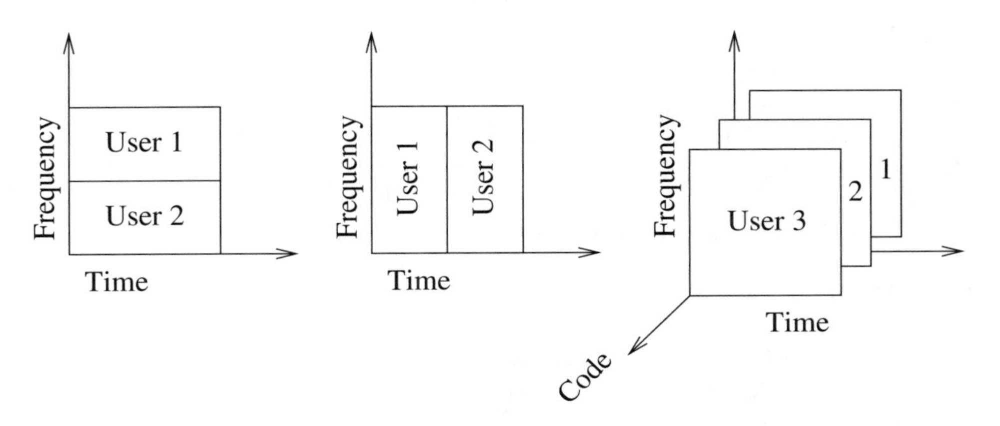

Figure 5.1: Multiple access schemes: FDMA (left), TDMA (middle) and CDMA (right).

Traditional ways of separating signals in time using TDMA and in frequency ensure that the signals are transmitted orthogonal in either time or frequency and hence they are non-interfering. In CDMA different users are separated employing a set of waveforms exhibiting good correlation properties, which are known as spreading codes. Figure 5.1 illustrates the principles of FDMA, TDMA and CDMA. More explicitly, FDMA uses a fraction of the total FDMA frequency band for each communications link for the whole duration of a conversation, while TDMA uses the entire bandwidth of a TDMA channel for a fraction of the TDMA frame, namely for the duration of a time slot. Finally, CDMA uses the entire available frequency band all the time and separates the users with the aid of unique, orthogonal user signature sequences.

In a CDMA digital communications system, such as that shown in Figure 5.2, the data stream is multiplied by the spreading code, which replaces each data bit with a sequence of code chips. A chip is defined as the basic element of the spreading code, which typically assumes binary values. Hence, the spreading process consists of replacing each bit in the original user's data sequence with the complete spreading code. The chip rate is significantly higher than the data rate, hence causing the bandwidth of the user's data to be spread, as shown in Figure 5.2.

At the receiver, the composite signal containing the spread data of multiple users is multiplied by a synchronised version of the spreading code of the wanted user. The specific auto-correlation properties of the codes allow the receiver to identify and recover each delayed, attenuated and phase-rotated replica of the transmitted signal, provided that the signals are separated by more than one chip period and the receiver has the capability of tracking each significant path. This is achieved using a Rake receiver [5] that can process multiple delayed received signals. Coherent combination of these transmitted signal replicas allows the original signal to be recovered. The unwanted signals of the other simultaneous users remain wideband, having a bandwidth equal to that of the noise, and appear as additional noise with respect to the wanted signal. Since the bandwidth of the despread wanted signal is reduced relative to this noise, the signal-to-noise ratio of the wanted signal is enhanced by the despreading process in proportion to the ratio of the spread and despread bandwidths, since

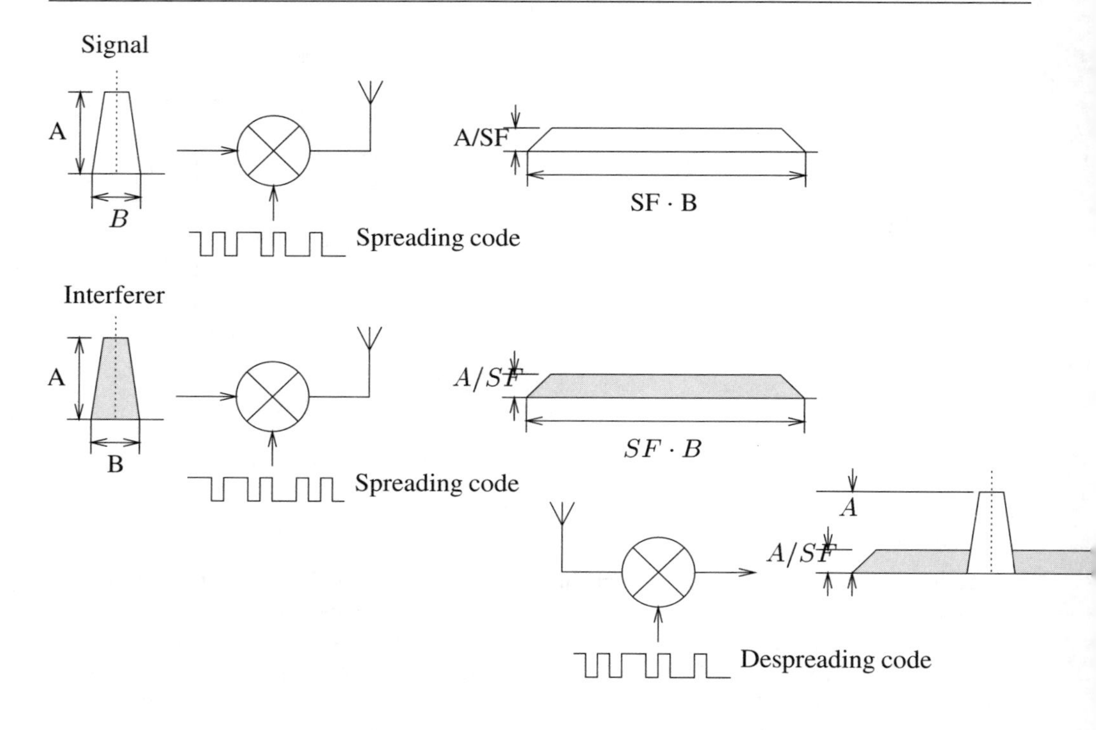

Figure 5.2: CDMA Spreading and Despreading Processes

the noise power outside the useful despread signal's bandwidth can be removed by a low-pass filter. This bandwidth ratio is equal to the ratio of the chip rate to the data rate, which is known as the Processing Gain (PG). For this process to work efficiently, the signals of all of the users should be received at or near the same power at the receiver. This is achieved with the aid of power control, which is one of the critical elements of a CDMA system. The performance of the power control scheme directly affects the capacity of the CDMA network.

5.3 UMTS Terrestrial Radio Access

A bandwidth of 155 MHz has been allocated for UMTS services in Europe in the frequency region of 2.0 GHz. The paired bands of 1920-1980 MHz (uplink) and 2110-2170 MHz (downlink) have been set aside for FDD W-CDMA systems, and the unpaired frequency bands of 1900-1920 MHz and 2010-2025 MHz for TDD CDMA systems.

A UTRA Network (UTRAN) consists of one or several Radio Network Sub-systems (RNSs), which in turn consist of base stations (referred to as Node Bs) and Radio Network Controllers (RNCs). A Node B may serve one or multiple cells. Mobile stations are known as User Equipment (UE), which are expected to support multi-mode operation in order to enable handovers between the FDD and TDD modes and, prior to complete UTRAN coverage, also to GSM. The key parameters of UTRA have been defined as in Table 5.1.

Duplex scheme	FDD	TDD
Multiple access scheme	W-CDMA	TD-CDMA
Chip rate	3.84 Mchip/s	3.84 Mchip/s
Spreading factor range	4-512	1-16
Frequency bands	1920-1980 MHz (UL)	1900-1920 MHz
	2110-2170 MHz (DL)	2010-2025 MHz
Modulation mode	4-QAM/QPSK	4-QAM/QPSK
Bandwidth	5 MHz	5 MHz
Nyquist pulse shaping	0.22	0.22
Frame length	10 ms	10 ms
Number of timeslots per frame	15	15

Table 5.1: Key UTRA Parameters.

5.3.1 Spreading and Modulation

As usual, the uplink is defined as the transmission path from the mobile station to the base station, which receives the unsynchronised channel impaired signals from the network's mobiles. The base station has the task of extracting the wanted signal from the received signal contaminated by both intra- and inter-cell interference. However, as described in Section 5.2, some degree of isolation between interfering users is achieved due to employing unique orthogonal spreading codes, although their orthogonality is destroyed by the hostile mobile channel.

The spreading process consists of two operations. The first one is the channelisation operation, which transforms every data symbol into a number of chips, thus increasing the bandwidth of the signal, as seen in Figure 5.2 of Section 5.2. The channelisation codes in UTRA are Orthogonal Variable Spreading Factor (OVSF) codes [11] that preserve the orthogonality between a given user's different physical channels, which are also capable of supporting multirate operation. These codes will be further discussed in the context of Figure 5.4. The second operation related to the spreading, namely the 'scrambling' process then multiplies the resultant signals separately on the I- and Q-branches by a complex-valued scrambling code, as shown in Figure 5.3. The scrambling codes may be one of either 2^{24} different 'long' codes or 2^{24} 'short' uplink scrambling codes.

The Dedicated Physical Control CHannel (DPCCH) [11, 359] is spread to the chip rate by the channelisation code C_c, while the nth Dedicated Physical Data CHannel (DPDCH), namely DPDCH$_n$, is spread to the chip rate by the channelisation code $C_{d,n}$. One DPCCH and up to six parallel DPDCHs can be transmitted simultaneously, i.e. $1 \leq n \leq 6$ as seen in Figure 5.3). However, it is beneficial to transmit with the aid of a single DPDCH, if the required bit-rate can be provided by a single DPDCH for reasons of terminal amplifier efficiency. This is because multi-code transmissions increase the peak-to-average ratio of the transmission, which reduces the efficiency of the terminal's power amplifier [32]. The maximum user data rate achievable with the aid of a single code is derived from the maximum channel bit rate, which is 960 kbps using a spreading factor of four without channel coding in the 1999 version of the UTRA standard. However, at the time of writing a spreading factor of one is being considered by the standardisation body. With channel coding the maximum

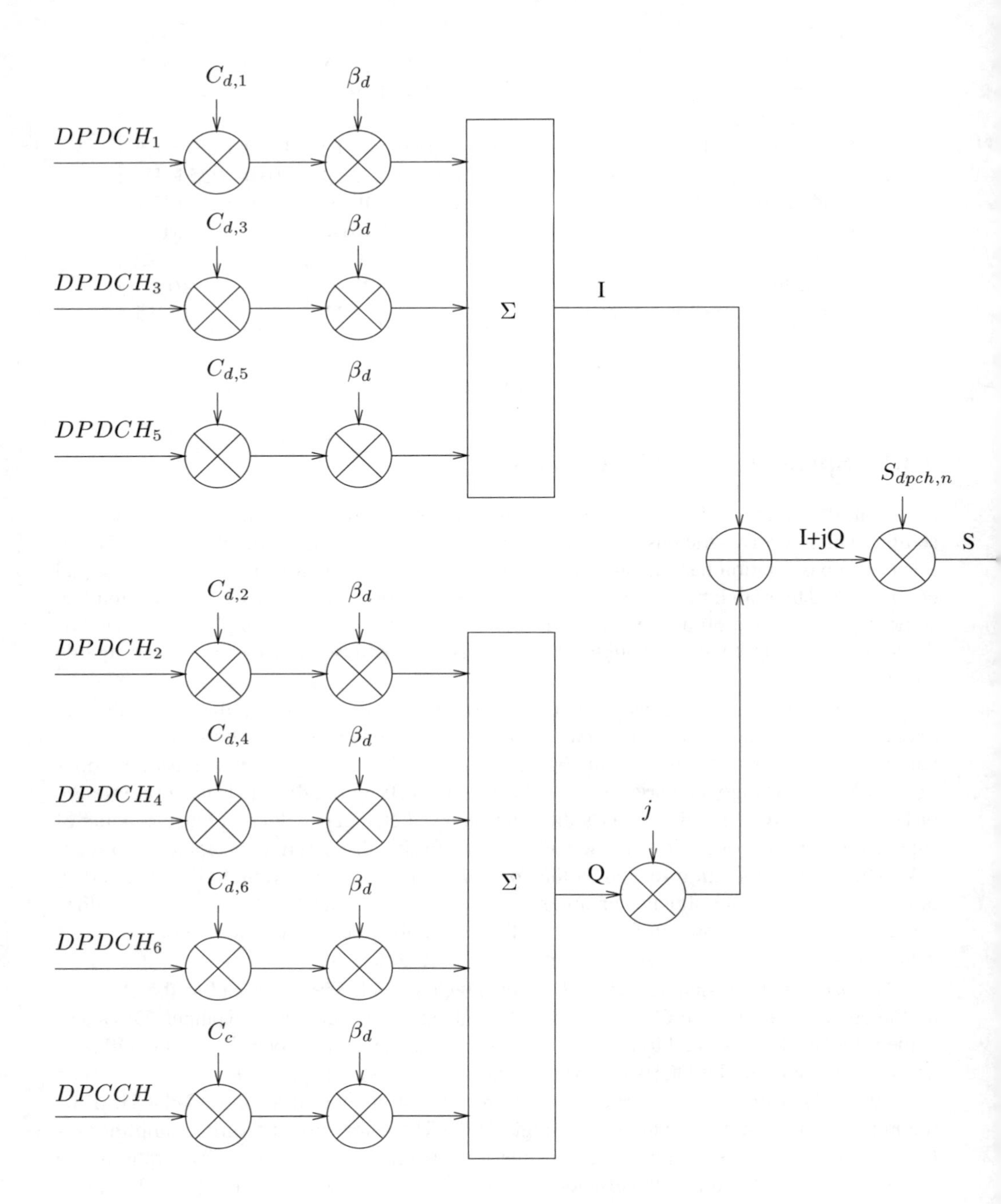

Figure 5.3: Spreading for uplink DPCCH and DPDCHs

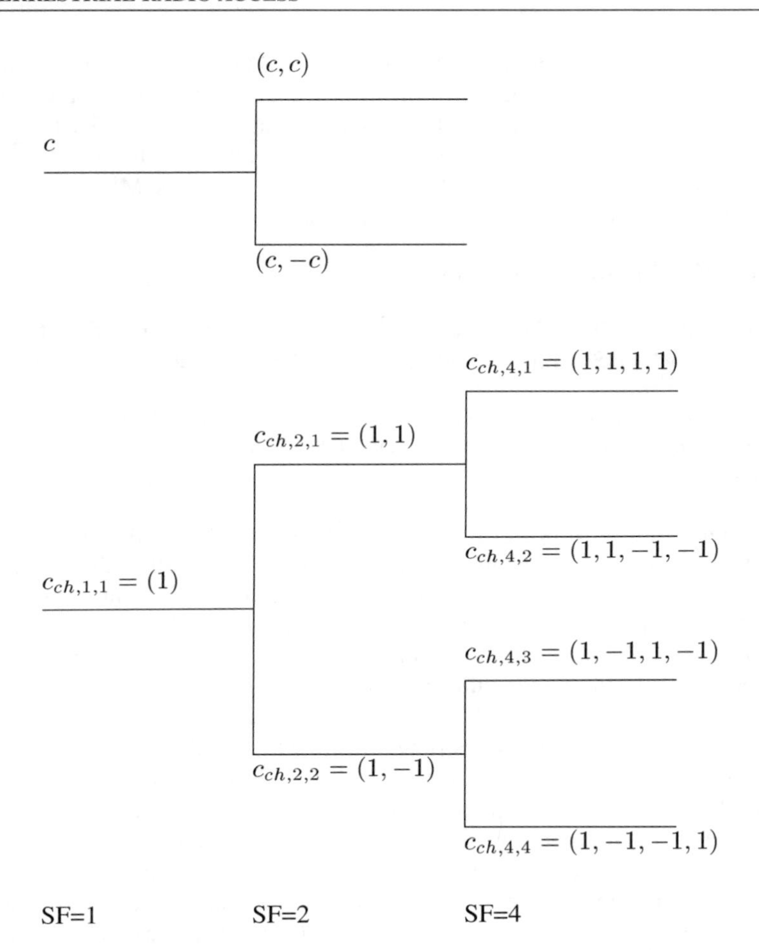

Figure 5.4: Code tree for the generation of Orthogonal Variable Spreading Factor (OVSF) codes

practical user data rate for single code transmission is of the order of 400-500 kbps. For achieving higher data rates parallel multi-code channels are used. This allows up to six parallel codes to be used, increasing the achievable channel bit rate up to 5740 kbps, which can accommodate a 2 Mbps user data rate or even higher data rates, when the channel coding rate is $1/2$.

The OVSF codes [103] can be defined using the code tree of Figure 5.4. In Figure 5.4, the channelisation codes are uniquely described by $C_{ch,SF,k}$, where SF is the spreading factor of the codes, and k is the code index where $0 \leq k \leq SF - 1$. Each level in the code tree defines spreading codes of length SF, corresponding to a particular spreading factor of SF. The number of codes available for a particular spreading factor is equal to the spreading factor itself. All the codes of the same level in the code tree constitute a set and they are orthogonal to each other. Any two codes of different levels are also orthogonal to each other, as long as one of them is not the mother of the other code. For example, the codes $c_{15}(2)$,

$c_7(1)$ and $c_3(1)$ are all the mother codes of $c_{31}(3)$ and hence are not orthogonal to $c_{31}(3)$, where the number in the round bracket indicates the code index. Thus not all the codes within the code tree can be used simultaneously by a mobile station. Specifically, a code can be used by an MS if and only if no other code on the path from the specific code to the root of the tree, or in the sub-tree below the specific node is used by the same MS.

For the DPCCH and DPDCHs the following applies:

- The PDCCH is always spread by code $C_c = C_{ch,256,0}$.
- When only one DPDCH is to be transmitted, DPDCH_1 is spread by the code $C_{d,1} = C_{ch,SF,k}$, where SF is the spreading factor of DPDCH_1 and $k = SF/4$.
- When more than one DPDCHs have to be transmitted, all DPDCHs have spreading factors equal to four. Furthermore, DPDCH_n is spread by the code $C_{d,n} = C_{ch,4,k}$, where $k = 1$ if $n \subset \{1, 2\}$, $k = 3$ if $n \subset \{3, 4\}$, and $k = 2$ if $n \subset \{5, 6\}$.

A fundamental difference between the uplink and the downlink is that in the downlink synchronisation is common to all users and channels of a given cell. This enables us to exploit the cross-correlation properties of the OVSF codes, which were originally proposed in [103]. These codes offer perfect cross-correlation in an ideal channel, but there is only a limited number of these codes available. The employment of OVSF codes allows the spreading factor to be changed and orthogonality between the spreading codes of different lengths to be maintained. The codes are selected from the code tree, which is illustrated in Figure 5.4. As illustrated above, there are certain restrictions as to which of the channelisation codes can be used for transmission from a single source. Another physical channel may invoke a certain code from the tree, if no other physical channel to be transmitted employing the same code tree is using a code on an underlying branch, since this would be equivalent to using a higher spreading factor code generated from the spreading code to be used, which are not orthogonal to each other on the same branch of the code tree. Neither can a smaller spreading factor code on the path to the root of the tree be used. Hence, the number of available codes depends on the required transmission rate and spreading factor of each physical channel.

In the UTRA downlink a part of the multi-user interference can be orthogonal - apart from the channel effects. The users within the same cell share the same scrambling code, but use different channelisation/OVSF codes. In a non-dispersive downlink channel, all intra-cell users are synchronised and therefore they are perfectly orthogonal. Unfortunately, in most cases the channel will be dispersive, implying that non-synchronised interference will be suppressed only by a factor corresponding to the processing gain, and thus they will interfere with the desired signal. The interference from other cells which is referred to as inter-cell interference, is non-orthogonal, due to employing different scrambling but possibly the same channelisation codes. Therefore inter-cell interference is also suppressed by a factor corresponding to the processing gain.

The channelisation code used for the Primary Common PIlot CHannel (CPICH) is fixed to $C_{ch,256,0}$, while the channelisation code for the Primary Common Control Physical CHannel (CCPCH) is fixed to $C_{ch,256,1}$ [359]. The channelisation codes for all other physical channels are assigned by the UTRAN [359].

A total of $2^{18} - 1 = 262143$ scrambling codes, numbered as $0 \ldots 262142$ can be generated. However, not all of the scrambling codes are used. The scrambling codes are divided into 512 sets, each consisting of a primary scrambling code and 15 secondary scrambling codes [359].

More specifically, the primary scrambling codes consist of scrambling codes $n = 16 * i$, where $i = 0 \ldots 511$. The ith set of secondary scrambling codes consists of scrambling codes $16 * i + k$ where $k = 1 \ldots 15$. There is a one-to-one mapping between each primary scrambling code and the associated 15 secondary scrambling codes in a set, such that the ith primary scrambling code uniquely identifies the ith set of secondary scrambling codes. Hence, according to the above statement, scrambling codes $k = 0 \ldots 8191$ are used. Each of these codes is associated with a left alternative scrambling code and a right alternative scrambling code, that may be used for the so-called compressed frames. Specifically, compressed frames are shortened duration frames transmitted right before a handover, in order to create an inactive period during which no useful data is transmitted. This allows the transceivers to carry out operations necessary for the handover to be successful. The left alternative scrambling code associated with scrambling code k is the scrambling code $k + 8192$, while the corresponding right alternative scrambling code is scrambling code $k + 16384$. In compressed frames, the left alternative scrambling code is used, if $n < SF/2$ and the right alternative scrambling code is used, if $n \geq SF/2$, where $C_{ch,SF,n}$ is the channelisation code used for non-compressed frames.

The set of 512 primary scrambling codes is further divided into 64 scrambling code groups, each consisting of 8 primary scrambling codes. The jth scrambling code group consists of primary scrambling codes $16 * 8 * j + 16 * k$, where $j = 0 \ldots 63$ and $k = 0 \ldots 7$.

Each cell is allocated one and only one primary scrambling code. The primary CCPCH and primary CPICH are always transmitted using this primary scrambling code. The other downlink physical channels can be spread and transmitted with the aid of either the primary scrambling code or a secondary scrambling code from the set associated with the primary scrambling code of the cell.

5.3.2 Common Pilot Channel

The Common PIlot CHannel (CPICH) is an unmodulated downlink code channel, which is scrambled with the aid of the cell-specific primary scrambling code. The function of the downlink CPICH is to aid the Channel Impulse Response (CIR) estimation necessary for the detection of the dedicated channel at the mobile station and to provide the CIR estimation reference for the demodulation of the common channels, which are not associated with the dedicated channels.

UTRA has two types of common pilot channels, namely the primary and secondary CPICHs. Their difference is that the primary CPICH is always spread by the primary scrambling code defined in Section 5.3.1. More explicitly, the primary CPICH is associated with a fixed channelisation code allocation and there is only one such channel and channelisation code for a cell or sector. The secondary CPICH may use any channelisation code of length 256 and may use a secondary scrambling code as well. A typical application of secondary CPICHs usage would be in conjunction with narrow antenna beams intended for service provision at specific teletraffic ‘hot spots’ or places exhibiting a high traffic density [32].

An important application of the primary common pilot channel is during the collection of channel quality measurements for assisting during the handover and cell selection process. The measured CPICH reception level at the terminal can be used for handover decisions.

Furthermore, by adjusting the CPICH power level the cell load can be balanced between different cells, since reducing the CPICH power level encourages some of the terminals to handover to other cells, while increasing it invites more terminals to handover to the cell, as well as to make their initial access to the network in that cell.

5.3.3 Power Control

Agile and accurate power control is perhaps the most important aspect in W-CDMA, in particular on the uplink, since a single high-powered rogue mobile can cause serious performance degradation to other users in the cell. The problem is referred to as the 'near-far effect' and occurs when, for example, one mobile is near the cell edge, and another is near the cell centre. In this situation, the mobile at the cell edge is exposed to a significantly higher pathloss, say 70 dB higher, than that of the mobile near the cell centre. If there were no power control mechanisms in place, the mobile near the base station could easily 'overpower' the mobile at the cell edge, and thus may block a large part of the cell. The optimum strategy in the sense of maximising the system's capacity is to equalise the received power per bit of all mobile stations at all times.

A so-called open-loop power control mechanism [32] attempts to make a rough estimate of the expected pathloss by means of a downlink beacon signal, but this method can be highly inaccurate. The prime reason for this is that the fast fading is essentially uncorrelated between the uplink and downlink, due to the large frequency separation of the uplink and downlink band of the W-CDMA FDD mode. Open-loop power control is however, used in W-CDMA, but only to provide a coarse initial power setting of the mobile station at the beginning of a connection.

A better solution is to employ fast closed-loop power control [32]. In closed-loop power control in the uplink, the base station performs frequent estimates of the received SIR and compares it to the target SIR. If the measured SIR is higher than the target SIR, the base station commands the mobile station to reduce the power, while if it is too low it will instruct the MS to increase its power. Since each 10 ms UTRA frame consists of 15 time slots, each corresponding to one power control power adjustment period, this procedure takes place at a rate of 1500 Hz. This is far faster than any significant change of pathloss, including street corner effects, and indeed faster than the speed of Rayleigh fading for low to moderate mobile speeds. The street corner effect occurs when a mobile turns the street corner and hence the received signal power drops markedly. Therefore the mobile responds by rapidly increasing its transmit power, which may inflict sever interference upon other closely located base stations. In response, the mobiles using these base stations increase their transmit powers in an effort to maintain their communications quality. This is undesirable, since it results in a high level of co-channel interference, leading to excessive transmission powers and to a reduction of the battery recharge period.

The same closed-loop power control technique is used on the downlink, although the rationale is different. More specifically, there is no near-far problem due to the one-to-many distributive scenario, i.e. all the signals originate from the single base station to all mobiles. It is, however, desirable to provide a marginal amount of additional power to mobile stations near the cell edge, since they suffer from increased inter-cell interference. Hence, the closed loop power control in CDMA systems ensures that each mobile transmits just sufficient power to satisfy the outer-loop power control scheme's SIR target. The SIR target is controlled by an outer-loop power control process that adjusts the required SIR in order to meet the Bit Error Ratio (BER) requirements of a particular service. At higher mobile speeds typically a higher SIR is necessary for attaining a given BER/FER.

5.3.3.1 Uplink Power Control

The uplink's inner-loop power control adjusts the mobile's transmit power in order to maintain the received uplink SIR at the given SIR target, namely at SIR_{target}. The base stations that are communicating with the mobile generate Transit Power Control (TPC) commands and transmit them, once per slot, to the mobile. The mobile then derives from the TPC commands of the various base stations, a single TPC command, TPC_cmd, for each slot, combining multiple received TPC commands if necessary. In [360] two algorithms were defined for the processing of TPC commands and hence for deriving TPC_cmd.

Algorithm 1: [360]

When not in soft-handover, i.e. when the mobile communicates with a single base station, only one TPC command will be received in each slot. Hence, for each slot, if the TPC command is equal to 0 ($SIR > SIR_{target}$) then $TPC\_cmd = -1$, otherwise, if the TPC command is 1 ($SIR < SIR_{target}$) then $TPC\_cmd = 1$, which implies powering down or up, respectively.

When in soft handover, multiple TPC commands are received in each slot from the different base stations in the active base station set. If all of the base station's TPC commands are identical, then they are combined to form a single TPC command, namely TPC_cmd. However, if the TPC commands of the different base stations differ, then a soft decision W_i is generated for each of the TPC commands, TPC_i, where $i = 1, 2, \ldots, N$, and N is the number of TPC commands. These N soft decisions are then used to form a combined TPC command TPC_cmd according to:

$$TPC\_cmd = \gamma(W_1, W_2, \ldots, W_N) \tag{5.1}$$

where TPC_cmd is either -1 or +1 and $\gamma()$ is the decision function combining the soft values, $W_1, \ldots, W_N$.

If the N TPC commands appear to be uncorrelated, and have a similar probability of being 0 or 1, then function $\gamma()$ should be defined such that the probability that the output of the function $\gamma()$ is equal to 1, is greater than or equal to $1/2^N$, and the probability that the output of $\gamma()$ is equal to -1, shall be greater than or equal to 0.5 [360]. Alternatively, the function $\gamma()$ should be defined such that $P(\gamma() = 1) \geq 1/2^N$ and $P(\gamma() = -1) \geq 0.5$.

Algorithm 2: [360]

When not in soft handover, only one TPC command will be received in each slot, and the mobile will process the maximum 15 TPC commands in a five-slot cycle, where the sets of five slots are aligned with the frame boundaries and the sets do not overlap. Therefore, when not in soft handover, for the first four slots of a five-slot set $TPC\_cmd = 0$ is used for indicating that no power control adjustments are made. For the fifth slot of a set the mobile performs hard decisions on all five of the received TPC commands. If all five hard decisions result in a binary 1, then we set $TPC\_cmd = 1$. In contrast, if all five hard decisions yield a binary 0, then $TPC\_cmd = -1$ is set, else $TPC\_cmd = 0$.

When the mobile is in soft handover, multiple TPC commands will be received in each slot from each of the base stations in the set of active base stations. When the TPC commands of the active base stations are identical, then they can be combined into a single TPC command. However, when the received TPC commands are different, the mobile makes a hard decision concerning the value of each TPC command for three consecutive slots, resulting in N hard decisions for each of the three slots, where N is the number of base stations within the active set. The sets of three slots are aligned to the frame boundaries and do not overlap. Then $TPC\_cmd = 0$ is set for the first two slots of the three-slot set, and then TPC_cmd is determined for the third slot as follows.

The temporary command TPC_temp_i is determined for each of the N sets of three TPC commands of the consecutive slots by setting $TPC\_temp_i = 1$ if all three TPC hard decisions are binary 1. In contrast, if all three TPC hard decisions are binary 0, $TPC\_temp_i = -1$ is set, otherwise we set $TPC\_temp_i = 0$. These temporary TPC commands are then used to determine the combined TPC command for the third slot invoking the decision function $\gamma(TPC\_temp_1, TPC\_temp_2, \ldots, TPC\_temp_N)$ defined as:

$$\begin{aligned} TPC\_cmd &= 1 && \text{if } \frac{1}{N}\sum_{i=1}^{N} TPC\_temp_i > 0.5 \\ TPC\_cmd &= -1 && \text{if } \frac{1}{N}\sum_{i=1}^{N} TPC\_temp_i < -0.5 \\ TPC\_cmd &= 0 && \text{otherwise.} \end{aligned} \tag{5.2}$$

5.3.3.2 Downlink Power Control

The downlink transmit power control procedure simultaneously controls the power of both the DPCCH and its corresponding DPDCHs, both of which are adjusted by the same amount, and hence the relative power difference between the DPCCH and DPDCHs remains constant.

The mobile generates TPC commands for controlling the base station's transmit power and sends them in the TPC field of the uplink DPCCH. When the mobile is not in soft handover, the TPC command generated is transmitted in the first available TPC field using the uplink DPCCH. In contrast, when the mobile is in soft handover, it checks the downlink power control mode (DPC_MODE) before generating the TPC command. If $DPC\_MODE = 0$, the mobile sends a unique TPC command in the first available TPC field in the uplink DPCCH. If however, $DPC\_MODE = 1$, the mobile repeats the same TPC command over three consecutive slots of the same frame and the new TPC command is transmitted to the base station in an effort the control its power at the beginning of the next frame. The minimum required transmit power step size is 1 dB, with a smaller step size of 0.5 dB being optional. The power control step size can be increased from 1 dB to 2 dB, thus allowing a 30 dB correction range during the 15 slots of a 10 ms frame. The maximum transmit powers are +21 dBm and +24 dBm, although it is likely that in the first phase of network deployment most terminals will belong to the 21 dBm power class [32].

5.3.4 Soft Handover

Theoretically, the ability of CDMA to despread the interfering signals, and thus adequately operate at low signal-to-noise ratios, allows a CDMA network to have a frequency reuse factor of one [32]. Traditionally, non-CDMA based networks have required adjacent cells to have different carrier frequencies, in order to reduce the co-channel interference to acceptable levels. Therefore, when a mobile hands over from one cell to another, it has to re-tune its synthesiser to the new carrier frequency, i.e. it performs an inter-frequency handover. This process is a 'break-before-make' procedure, known as a hard handover, and hence call disruption or interruption is possible. However, in a CDMA based network, having a frequency reuse factor of one, so-called soft handovers may be performed, which is a 'make-before-break' process, potentially allowing for a smoother handover between cells. During a soft handover a mobile is connected to two or more base stations simultaneously, thus utilising more network resources and transmitting more signals, which interfere with other users. Therefore, it is in the network operator's interests to minimise the number of users in soft handover, whilst maintaining a satisfactory quality of service. In soft handover, each connected base station receives and demodulates the user's data, and selection diversity is performed between the base stations, i.e. the best version of the uplink frame is selected. In the downlink, the mobile station performs maximal ratio combining [5] of the signal received from the multiple base stations. This diversity combining improves the coverage in regions of previously low-quality service provision, but at the expense of increased backhaul connections.

The set of base stations engaged in soft handover is known as the *active set*. The mobile station continuously monitors the received power level of the PIlot CHannels (PICHs) transmitted by its neighbouring base stations. The received pilot power levels of these base stations are then compared to two thresholds, the acceptance threshold, T_{acc} and the dropping threshold T_{drop}. Therefore, as a mobile moves away from base station 1, and towards base station 2, the pilot signal strength received from base station 2 increases. When the pilot strength exceeds the *acceptance threshold*, T_{acc}, the mobile station enters the soft handover state, as shown in Figure 5.5. As the mobile continues to move away from base station 1, its pilot strength decreases, until it falls below the *drop threshold*. After a given time interval, T_{drop}, during which the signal strength from base station 1 has not exceeded the drop threshold, base station 1 is removed from the active set.

5.3.5 Signal-to-Interference plus Noise Ratio Calculations

5.3.5.1 Downlink

The interference received at the mobile can be divided into interference due to the signals transmitted to other mobiles from the same base station, which is known as intra-cell interference, and that received due to the signals transmitted to other mobiles from other base stations, which is termed inter-cell interference. In an ideal case, the intra-cell interference would be zero, since all the signals from the base station are subjected to the same channel conditions, and orthogonal channelisation codes are used for separating the users. However, after propagation through a dispersive multipath channel, this orthogonality is eroded. The intra-cell and inter-cell interference values are always non-zero when in a single-user scenario due to the inevitable interference inflicted by the common pilot channels.

The instantaneous SINR is obtained by dividing the received signal powers by the total interference plus thermal noise power, and then by multiplying this ratio by the spreading factor, SF, yielding

$$SINR_{DL} = \frac{SF \cdot S}{(1-\alpha)I_{Intra} + I_{Inter} + N_0}, \tag{5.3}$$

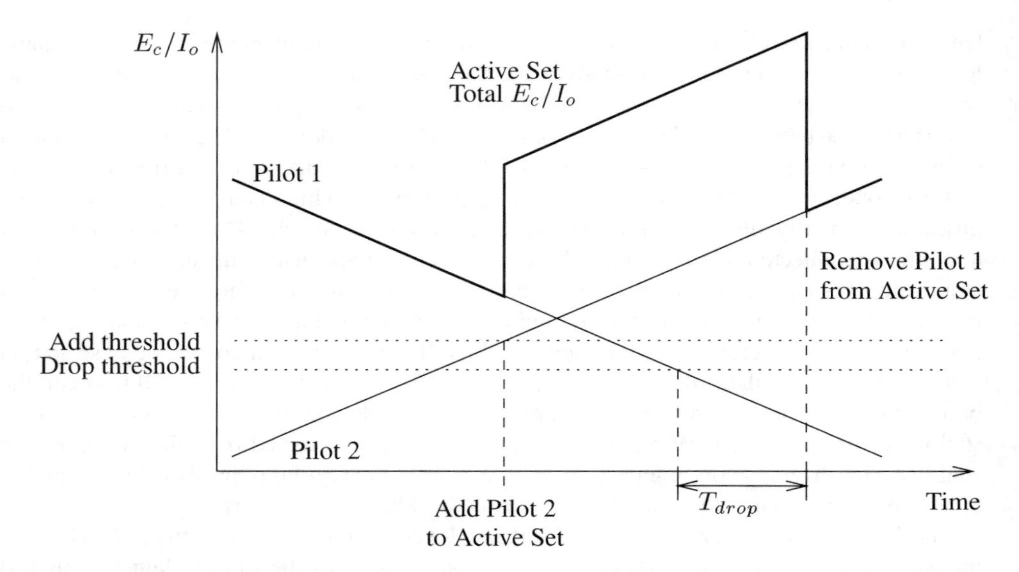

Figure 5.5: The soft handover process showing the process of adding and dropping base stations from the active set.

where $\alpha = 1$ corresponds to the ideal case of perfectly orthogonal intra-cell interference, and $\alpha = 0$ is for completely asynchronous intra-cell interference. Furthermore, N_0 is the thermal noise's power spectral density, S is the received signal power, I_{Intra} is the intra-cell interference and I_{Inter} is the inter-cell interference. Again, the interference plus noise power is scaled by the spreading factor, SF, since after the low-pass filtering the noise bandwidth is reduced by a factor of SF during the despreading process.

When in soft handover, the maximum ratio combining is performed on the N received signals of the N active base stations. Therefore, provided that the active base stations' received signals are independent, the SINR in this situation is:

$$SINR_{DL} = SINR_{DL_1} + SINR_{DL_2} + \ldots + SINR_{DL_N}. \tag{5.4}$$

5.3.5.2 Uplink

The uplink differs from the downlink in that the multiple access interference is asynchronous in the uplink due to the un-coordinated transmissions of the mobile stations, whereas it may remain quasi-synchronous in the downlink. Therefore, the intra-cell uplink interference is not orthogonal. A possible solution for mitigating this problem is employing Multi-User Detectors (MUDs) [66] at the base stations.

Thus, we define β as the MUD's efficiency, which effectively gives the percentage of the intra-cell interference that is removed by the MUD. Setting $\beta = 0.0$ implies 0% efficiency, when the intra-cell interference is not reduced by the MUD, whereas $\beta = 1.0$ results in the perfect suppression of all the intra-cell interference. Therefore, the expression for the uplink

SINR is:

$$SINR_{UL} = \frac{SF \cdot S}{(1-\beta)I_{Intra} + I_{Inter} + N_0}. \tag{5.5}$$

When in soft handover, selection diversity is performed on the N received signals at each of the active base stations. Therefore, the SINR in this situation becomes:

$$SINR_{UL} = \max(SINR_{UL_1}, SINR_{UL_2}, \cdots, SINR_{UL_N}). \tag{5.6}$$

5.3.6 Multi-User Detection

Multiple access communications using DS-CDMA is interference limited due to the Multiple Access Interference (MAI) generated by the users transmitting simultaneously within the same bandwidth. The signals received from the users are separated with the aid of the despreader using spreading sequences that are unique to each user. Again, these spreading sequences are usually non-orthogonal. Even if they are orthogonal, the asynchronous uplink transmissions of the users or the time-varying nature of the mobile radio channel may partially destroy this orthogonality. The non-orthogonal nature of the codes results in residual MAI, which degrades the performance of the system. The frequency selective mobile radio channel also gives rise to Inter-Symbol Interference (ISI) due to dispersive multipath propagation. This is exacerbated by the fact that the mobile radio channel is time-varying.

Conventional CDMA detectors - such as the matched filter [5, 361] and the RAKE combiner [362] - are optimised for detecting the signal of a single desired user. RAKE combiners exploit the inherent multi-path diversity in CDMA, since they essentially consist of matched filters combining each resolvable path of the multipath channel. The outputs of these matched filters are then coherently combined according to a diversity combining technique, such as maximal ratio combining [282], equal gain combining or selective diversity combining . These conventional single-user detectors are inefficient, because the interference is treated as noise, and our knowledge concerning the CIR of the mobile channel, or that of the spreading sequences of the interferers is not exploited. The efficiency of these detectors is dependent on the cross-correlation (CCL) between the spreading codes of all the users. The higher the cross-correlation, the higher the MAI. This CCL-induced MAI is exacerbated by the effects of the dispersive multi-path channel and asynchronous transmissions. The utilisation of these conventional receivers results in an interference-limited system. Another weakness of the above-mentioned conventional CDMA detectors is the phenomenon known as the 'near-far effect' [363, 364]. For conventional detectors to operate efficiently, the signals received from all the users have to arrive at the receiver with approximately the same power. A signal that has a significantly weaker signal strength compared to the other signals will be 'swamped' by the relatively higher powers of the other signals and the quality of the weaker signal at the output of the conventional receiver will be severely degraded. Therefore, stringent power control algorithms are needed to ensure that the signals arrive at similar powers at the receiver, in order to achieve a similar quality of service for different users [364, 365]. Using conventional detectors to detect a signal corrupted by MAI, while encountering a hostile channel results in an irreducible BER, even if the E_s/N_0 ratio is increased. This is because at high E_s/N_0 values the probability of errors due to thermal noise is insignificant compared to the errors caused by the MAI and the channel. Therefore, detectors that can reduce or remove the effects of MAI and ISI are needed in order to achieve user capacity gains. These detectors also have to be 'near-far resistant', in order to avoid the need for stringent power control requirements. In order to mitigate the problem of MAI, Verdú [66] proposed the optimum multi-user detector for asynchronous Gaussian multiple access channels. This optimum detector significantly outperforms the conventional detector and it is near-far resistant, but unfortunately its complexity increases exponentially according to the order of $O(2^{NK})$, where N is the number of overlapping asynchronous bits considered in the detector's window, and K is the number of interfering users. In order to reduce the complexity of the receiver and yet to provide an acceptable BER performance, significant research efforts have been invested in the field of sub-optimal CDMA multiuser receivers [66, 366].

In summary, multi-user detectors reduce the error floor due to MAI and this translates into user capacity gains for the system. These multi-user detectors are also near-far resistant to a certain extent and this results in less stringent power control requirements. However, multi-user detectors are more complex than conventional detectors. Coherent detectors require the explicit knowledge of the channel impulse response estimates, which implies that a channel estimator is needed in the receiver, and hence training sequences have to be included in the transmission frames. Training sequences are specified in the TDD mode of the UTRA standard and enable the channel impulse response of each simultaneously communicating user to be derived, which is necessary for the multi-user detectors to be able to separate the signals received from each user. These multi-user detectors also exhibit an inherent latency, which results in delayed reception. Multi-user detection is more suitable for the uplink receiver since the base station has to detect all users' signals anyway and it can tolerate a higher complexity. In contrast, a hand-held mobile receiver is required to be compact and lightweight, imposing restrictions on the available processing power. Recent research into blind MUDs has shown that data detection is possible for the desired user without invoking the knowledge of the spreading sequences and channel estimates of other users. Hence using these detectors for downlink receivers is becoming feasible.

5.4 Simulation Results

This section presents simulation results obtained for an FDD mode UMTS type CDMA cellular network, investigating the applicability of various soft handover metrics when subjected to different propagation conditions. This is followed by performance curves obtained using adaptive antenna arrays, when subjected to both non-shadowed as well as shadowed propagation conditions. The performance of adaptive modulation techniques used in conjunction with adaptive antenna arrays in a shadow faded environment is then characterised.

5.4.1 Simulation Parameters

Simulations of an FDD mode UMTS type CDMA based cellular network were conducted for various scenarios and algorithms in order to study the interactions of the processes involved in such a network. As in the standard, the frame length was set to 10 ms, containing 15 power control timeslots. The power control target SINR was chosen to give a Bit Error Ratio (BER) of 1×10^{-3}, with a low quality outage occurring at a BER of 5×10^{-3} and an outage taking place at a BER of 1×10^{-2}. The received SINRs at both the mobile and the base stations were required for each of the power control timeslots, and hence the outage and low quality outage statistics were gathered. If the received SINR was found to be below the outage SINR for 75 consecutive power control timeslots, corresponding to 5 consecutive transmission frames or 50 ms, the call was dropped. The post despreading SINRs necessary for obtaining the target BERs were determined with the aid of physical-layer simulations using a 4-QAM modulation scheme, in conjunction with $1/2$ rate turbo coding and joint detection over a COST 207 seven-path Bad Urban channel [367]. For a spreading factor of 16, the post-de-spreading SINR required to give a BER of 1×10^{-3} was 8.0 dB, for a BER of 5×10^{-3} it was 7.0 dB, and for a BER of 1×10^{-2} was about 6.6 dB. These values can be seen along with the other system parameters in Table 5.2. The-pre de-spreading SINR is related to E_b/N_o and to the spreading factor by :

$$SINR = (E_b/N_o)/SF, \tag{5.7}$$

where the spreading factor $SF = W/R$, with W being the chip rate and R the data rate. A receiver noise figure of 7 dB was assumed for both the mobile and the base stations [32]. Thus, in conjunction with a thermal noise density of -174 dBm/Hz and a noise bandwidth of 3.84 MHz, this resulted in a receiver noise power of -100 dBm. The power control algorithm used was relatively simple, and unrelated to the previously introduced schemes of Section 5.3.3. Furthermore, since it allowed a full transmission power change of 15 dB within a 15-slot UTRA data frame, the power control scheme advocated is unlikely to limit the network's capacity.

Specifically, for each of the 15 timeslots per transmitted frame, both the mobile and base station transmit powers were adjusted such that the received SINR was greater than the target SINR, but less than the target SINR plus 1 dB of hysteresis. When in soft handover, a mobile's transmission power was only increased if all of the base stations in the Active Base station Set (ABS) requested a power increase, but was it decreased if any of the base stations in the ABS had an excessive received SINR. In the downlink, if the received SINR at the mobile was insufficiently high then all of the active base stations were commanded to increase their transmission powers. Similarly, if the received SINR was unnecessarily high, then the active base stations would reduce their transmit powers. The downlink intra-cell interference orthogonality factor, α, as described in Section 5.3.5, was set to 0.5 [368–370]. Due to the frequency reuse factor of one, with its associated low frequency reuse distance, it was necessary for both the mobiles and the base stations, when initiating a new call or entering soft handover, to increase their transmitted power gradually. This was required to prevent sudden increases in the level of interference, particularly on links using the same base station. Hence, by gradually increasing the transmit power to the desired level, the other users of the network were capable of compensating for the increased interference by increasing their transmit powers, without encountering undesirable outages. In an FDMA/TDMA network this effect is less noticeable due to the significantly higher frequency reuse distance.

Since a dropped call is less desirable from a user's viewpoint than a blocked call, two resource allocation queues were invoked, one for new calls and the other - higher priority - queue, for handovers. By forming a queue of the handover requests, which have a higher priority during contention for network resources than new calls, it is possible to reduce the number of dropped calls at the expense of an increased blocked call probability. A further advantage of the Handover Queueing System (HQS) is that during the time a handover is in the queue, previously allocated resources may become available, hence increasing

Parameter	Value	Parameter	Value
Frame length	10 ms	Timeslots per frame	15
Target E_b/N_o	8.0 dB	Outage E_b/N_o	6.6 dB
Low Quality (LQ) Outage E_b/N_o	7.0 dB	BS Pilot Power	-5 dBm
BS/MS Minimum TX Power	-44 dBm	BS Antenna Gain	11 dBi
BS/MS Maximum TX Power	+21 dBm	MS Antenna Gain	0 dBi
Attenuation at 1 m reference point	39 dB	Pathloss exponent	-3.5
Power control SINR hysteresis	1 dB	Cell radius	150 m
Downlink scrambling codes per BS	1	Modulation scheme	4-QAM
Downlink OVSF codes per BS	Variable	Max new-call queue-time	5 s
Uplink scrambling codes per BS	Variable	Average inter-call time	300 s
Uplink OVSF codes per BS	Variable	Average call length	60 s
Spreading factor	Variable	Data/voice bit rate	Variable
Remove BS from ABS threshold	Variable	Add BS to ABS threshold	Variable
User speed	1.34 m/s	Noisefloor	-100 dBm
	(3 mph)	Size of ABS	2

Table 5.2: Simulation parameters of the UTRA-type CDMA based cellular network.

the probability of a successful handover. However, in a CDMA based network the capacity is not hard-limited by the number of frequency/timeslot combinations available, like in an FDMA/TDMA based network, such as GSM. The main limiting factors are the number of available spreading and OVSF codes, where the number of the available OVSF codes is restricted to the spreading factor minus one, since an OVSF code is reserved for the pilot channel. This is because, although the pilot channel has a spreading factor of 256, it removes an entire branch of the OVSF code generation tree. Other limiting factors are the interference levels in conjunction with the restricted maximum transmit power, resulting in excessive call dropping rates. New call allocation requests were queued for up to 5 s, if they could not be immediately satisfied, and were blocked if the request had not been completed successfully within the 5 s.

Similarly to our TDMA-based investigations portrayed in Chapter 4, several network performance metrics were used in order to quantify the quality of service provided by the cellular network, namely the:

- New Call Blocking probability, P_B,
- Call Dropping or Forced Termination probability, P_{FT},
- Probability of low quality connection, P_{low},
- Probability of Outage, P_{out},
- Grade Of Service, GOS.

The new call blocking probability, P_B, is defined as the probability that a new call is denied access to the network. In an FDMA/TDMA based network, such as GSM, this may occur because there are no available physical channels at the desired base station or the available channels are subject to excessive interference. However, in a CDMA based network this does not occur, provided that no interference level based admission control is performed and hence the new call blocking probability is typically low.

The call dropping probability, P_{FT}, is the probability that a call is forced to terminate prematurely. In a GSM type network, an insufficiently high SINR, which inevitably leads to dropped calls, may be remedied by an intra- or inter-cell handover. However, in CDMA either the transmit power must be increased, or a soft handover must be performed in order to exploit the available diversity gain.

Again, the probability of a low quality connection is defined as:

$$\begin{aligned} P_{low} &= P\{SINR_{uplink} < SINR_{req} \text{ or } SINR_{downlink} < SINR_{req}\} \\ &= P\{min(SINR_{uplink}, SINR_{downlink}) < SINR_{req}\}. \end{aligned} \quad (5.8)$$

The GOS was defined in [290] as:

$$\begin{aligned} GOS &= P\{\text{unsuccessful or low-quality call access}\} \\ &= P\{\text{call is blocked}\} + P\{\text{call is admitted}\} \times \\ &\quad P\{\text{low signal quality and call is admitted}\} \\ &= P_B + (1 - P_B)P_{low}, \end{aligned} \quad (5.9)$$

and is interpreted as the probability of unsuccessful network access (blocking), or low quality access, when a call is admitted to the system.

In our forthcoming investigations, in order to compare the network capacities of different networks, similarly to our TDMA-based investigations in Chapter 4, it was decided to use two scenarios defined as :

- A *conservative scenario*, where the maximum acceptable value for the new call blocking probability, P_B, is 3%, the maximum call dropping probability, P_{FT}, is 1%, and P_{low} is 1%.
- A *lenient scenario*, where the maximum acceptable value for the new call blocking probability, P_B, is 5%, the maximum call dropping probability, P_{FT}, is 1%, and P_{low} is 2%.

In the next section we consider the network's performance considering both fixed and normalised soft handover thresholds using both received pilot power and received pilot power versus interference threshold metrics. A spreading factor of 16 was used, corresponding to a channel data rate of 3.84 Mbps/16 = 240 kbps with no channel coding, or 120 kbps when using $1/2$ rate channel coding. It must be noted at this stage that the results presented in the following sections are network capacities obtained using a spreading factor of 16. The network capacity results presented in the previous chapter, which were obtained for an FDMA/TDMA GSM-like system, were achieved for speech-rate users. Here we assumed that the channel coded speech-rate was 15 kbps, which is the lowest possible Dedicated Physical Data CHannel (DPDCH) rate. Speech users having a channel coded rate of 15 kbps may be supported by invoking a spreading factor of 256. Hence, subjecting the channel data rate of 15 kbps to $1/2$ rate channel coding gives a speech-rate of 7.5 kbps, or if protected by a $2/3$ rate code the speech-rate becomes 10 kbps, which are sufficiently high for employing the so-called Advanced MultiRate (AMR) speech codec [371–373] capable of operating at rates between 4.7 kbps and 12.2 kbps. Therefore, by multiplying the resultant network capacities according to a factor of 256/16=16, it is possible to estimate the number of speech users supported by a speech-rate network. However, with the aid of our exploratory simulations, conducted using a spreading factor of 256, which are not presented here, we achieved network capacities higher than 30 times the network capacity supported in conjunction with a spreading factor of 16. Therefore, it would appear that the system is likely to support more than 16 times the number of 240 kbps data users, when communicating at the approximately 16 times lower speech-rate, employing a high spreading factor of 256. Hence, using the above-mentioned scaling factor of 16 we arrive at the lower bound of network capacity. A mobile speed of 3 mph was used in conjunction with a cell size of 150 m radius, which was necessarily small in order to be able to support the previously assumed 240 kbps high target data rate. The performance advantages of using both adaptive beamforming and adaptive modulation assisted networks are also investigated.

5.4.2 The Effect of Pilot Power on Soft Handover Results

In this section we consider the settings of the soft handover thresholds, for an IS-95 type handover algorithm [31], where the handover decisions are based on downlink pilot power measurements. Selecting inappropriate values for the soft handover thresholds, namely for the *acceptance threshold* and the *drop threshold*, may result in an excessive number of blocked and dropped calls in certain parts of the simulation area. For example, if the *acceptance threshold* that has to be exceeded by the signal level for a base station to be added to the active set is too high (Threshold B in Figure 5.6), then a user may be located within a cell, but it would be unable to add any base stations to its active base station set. Hence this user is unable to initiate a call. Figure 5.6 illustrates this phenomenon and shows that the *acceptance thresholds* must be set sufficiently low for ensuring that at least one base station covers every part of the network.

Another consequence of setting the *acceptance threshold* to an excessively high value, is that soft handovers may not be completed. This may occur when a user leaving the coverage area of a cell, since the pilot signal from that cell drops below the *drop threshold*, before the signal from the adjacent cell becomes sufficiently strong for it to be added to the active base station set. However, if the *acceptance threshold*, in conjunction with the *drop threshold*, is set correctly, then new calls and soft handovers should take place as required, so long as the availability of network resources allows it. Care must be taken however, not to set the soft handover threshold too low, otherwise the mobiles occupy additional network resources and create extra interference, due to initiating unnecessary soft-handovers.

5.4.2.1 Fixed Received Pilot Power Thresholds without Shadowing

Figure 5.7 shows the new call blocking probability of a network using a spreading factor of 16, in conjunction with fixed received pilot signal strength based soft handover thresholds without imposing any shadowing effects. The figure illustrates that reducing both the acceptance and the dropping soft handover thresholds results in an improved new call blocking performance. Reducing the threshold at which further base stations may be added to the Active Base station Set (ABS) increases the probability that base stations exist within the ABS, when a new call request is made. Hence, as expected, the new call blocking probability is reduced, when the acceptance threshold is reduced. Similarly, dropping the threshold at which base stations are removed from the ABS also results in an improved new call blocking

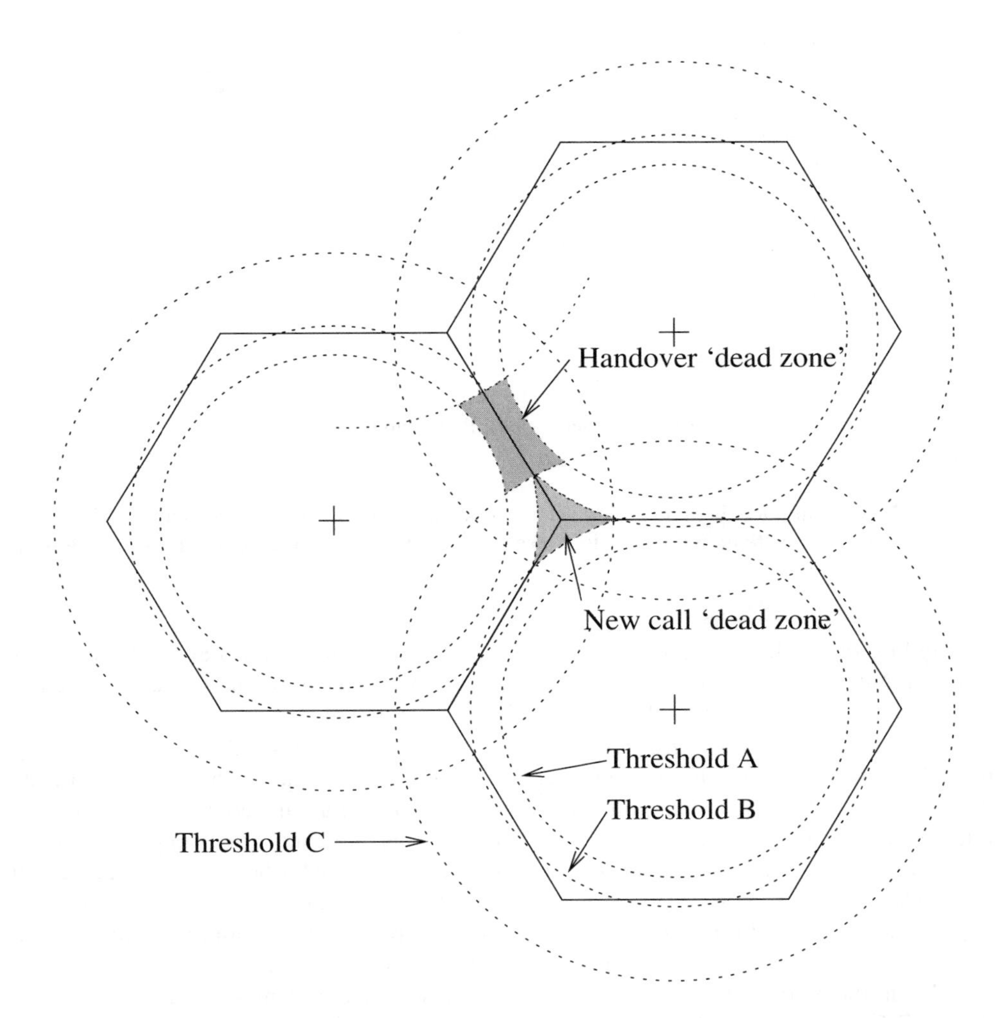

Figure 5.6: This figure indicates that using inappropriate soft handover thresholds may lead to blocked and dropped calls due to insufficient pilot coverage of the simulation area. Threshold A is the drop threshold, which when combined with the acceptance threshold C can fail to cover the simulation area sufficiently well, thus leading to soft handover failure. When combining threshold A with the acceptance threshold B, users located in the 'new call dead zone' may become unable to initiate calls.

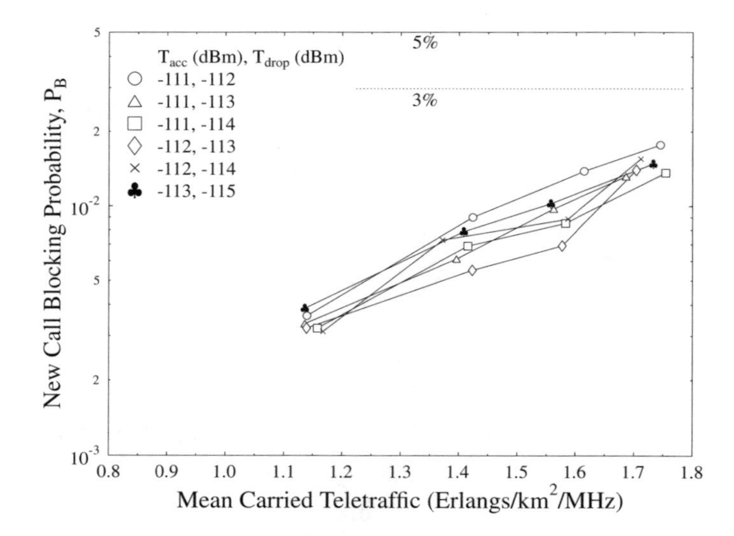

Figure 5.7: New call blocking probability versus mean carried traffic of a CDMA based cellular network using **fixed received pilot power** based soft handover thresholds **without shadowing** for SF=16.

probability, since a base station is more likely to be retained in the ABS as a mobile moves away from it. Therefore, should a mobile attempt to initiate a call in this situation, there is a greater chance that the ABS will contain a suitable base station.

The associated call dropping probability is depicted in Figure 5.8, indicating that reducing the soft handover thresholds, and thus increasing the time spent in soft handover, improved the performance up to a certain point. However, above this point the additional interference inflicted by the soft handover process led to a degraded performance. For example, in this figure the performance associated with $T_{acc} = -111$ dBm improved, when T_{drop} was decreased from -112 dBm to -113 dBm. However, at high traffic levels the performance degraded when T_{drop} was decreased further, to -114 dBm. The call dropping probability obtained using T_{acc}=-113 dBm and T_{drop}=-115 dBm was markedly lower for the lesser levels of traffic carried due to the extra diversity gain provided by the soft handover process. However, since these soft handover thresholds resulted in a greater proportion of time spent in soft handover, the levels of interference were increased, and thus at the higher traffic levels the performance degraded rapidly, as can be seen in Figure 5.8. Hence, the call dropping performance is based on a trade-off between the diversity gain provided by the soft handover process and the associated additional interference.

The probability of low quality access (not explicitly shown) was similar in terms of its character to the call dropping probability, since reducing T_{drop} improved the performance to a certain point, after which it degraded.

The mean number of base stations in the ABS is shown in Figure 5.9, illustrating that reducing the soft handover thresholds leads, on average, to a higher number of base stations in the ABS. Therefore, a greater proportion of call time is spent in soft handover. The asso-

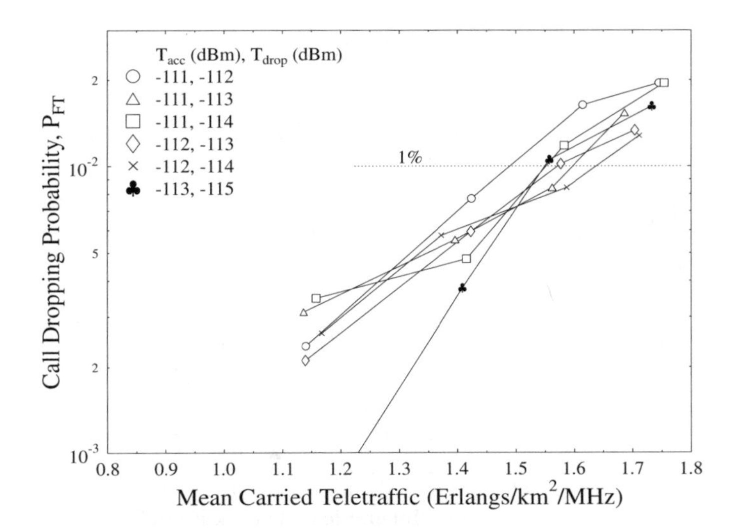

Figure 5.8: Call dropping probability versus mean carried traffic of a CDMA based cellular network using **fixed received pilot power** based soft handover thresholds **without shadowing** for SF=16.

ciated diversity gain improves the link quality of the reference user but additional co-channel interference is generated by the diversity links, thus ultimately reducing the call quality, as shown in Figure 5.8. Additionally, this extra co-channel interference required more transmission power for maintaining the target SINR as depicted in Figure 5.10. This figure shows that when lower soft handover thresholds are used, and thus a greater proportion of time is spent in soft handover, greater levels of co-channel interference are present, and thus the required mean transmission powers became higher. It is interesting to note that for the highest soft handover thresholds employed in Figure 5.10, the downlink transmission power required for maintaining the target SINR is lower than the uplink transmission power, whereas for the lower soft handover thresholds, the required mean uplink transmission power is lower than the downlink transmission power. The required downlink transmission power was, in general, lower than the uplink transmission power due to the mobile stations' ability to perform maximal ratio combining when in soft handover. This was observed despite the absence of the pilot interference in the uplink, and despite the base stations' ability to perform selective diversity which offers less diversity gain when compared to maximal ratio combining. However, reducing the soft handover thresholds to the lowest levels shown in Figure 5.10, led to increased co-channel interference on the downlink, thus requiring higher base station transmission powers, as clearly seen in the figure.

In summary, as seen by comparing Figures 5.7-5.10 the maximum capacity of the network using fixed received pilot power based soft handover thresholds was limited by the call dropping probability. The new call blocking probability remained below the 3% limit, thanks to the appropriate choice of thresholds used, whilst the probability of low quality access was constantly below the 1% mark. Therefore, the maximum normalised teletraffic load was

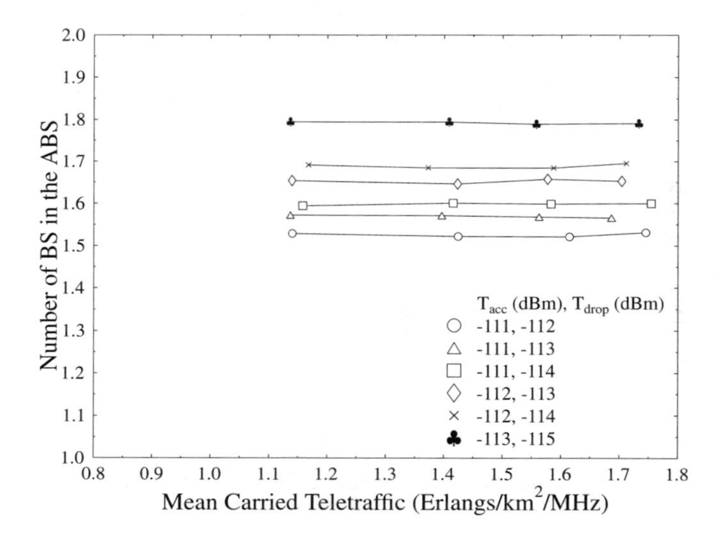

Figure 5.9: Mean number of base stations in the active base station set versus mean carried traffic of a CDMA based cellular network using **fixed received pilot power** based soft handover thresholds **without shadowing** for SF=16.

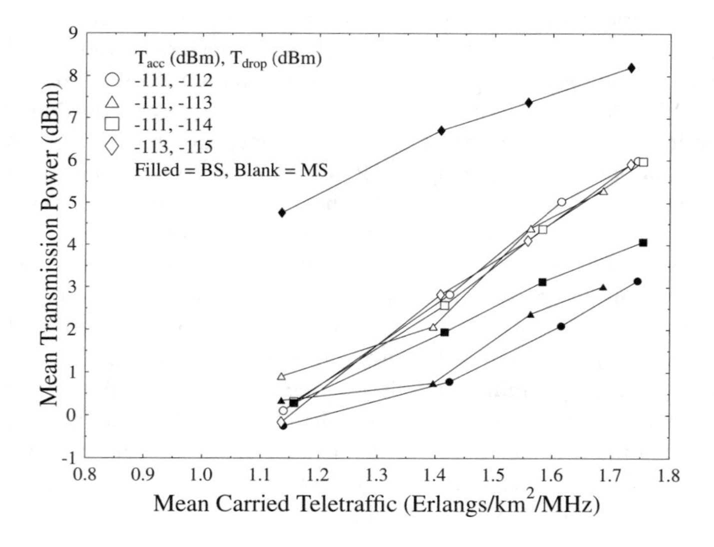

Figure 5.10: Mean transmission power versus mean carried traffic of a CDMA based cellular network using **fixed received pilot power** based soft handover thresholds **without shadowing** for SF=16.

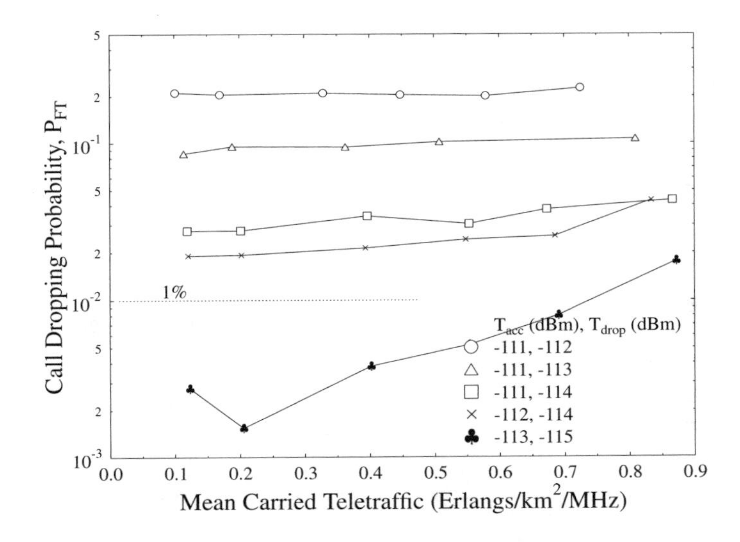

Figure 5.11: Call dropping probability versus mean carried traffic of a CDMA based cellular network using **fixed received pilot power** based soft handover thresholds in conjunction **with 0.5 Hz shadowing having a standard deviation of 3 dB** for SF=16.

1.64 Erlangs/km^2/MHz, corresponding to a total network capacity of 290 users, while satisfying both quality of service constraints, was achieved with the aid of an acceptance threshold of -112 dBm and a dropping threshold of -114 dBm. A mean ABS size of 1.7 base stations was registered at this traffic level, and both the mobile and base stations exhibited a mean transmission power of 5.1 dBm.

5.4.2.2 Fixed Received Pilot Power Thresholds with 0.5 Hz Shadowing

In this section we examine the achievable performance, upon using fixed received pilot power based soft handover thresholds when subjected to log-normal shadow fading having a standard deviation of 3 dB and a maximum frequency of 0.5 Hz.

The call dropping results of Figure 5.11 suggested that the network's performance was poor when using fixed received pilot power soft handover thresholds in the above mentioned shadow fading environment. The root cause of the problem is that the fixed thresholds must be set such that the received pilot signals, even when subjected to shadow fading, are retained in the active set. Therefore, setting the thresholds too high results in the base stations being removed from the active set, thus leading to an excessive number of dropped calls. However, if the thresholds are set too low, in order to counteract this phenomenon, then the base stations can be in soft handover for too high a proportion of time, and thus an unacceptable level of low quality accesses is generated due to the additional co-channel interference inflicted by the high number of active base stations. Figure 5.11 shows that reducing the soft handover thresholds improved the network's call dropping probability, but Figure 5.12 illustrates that reducing the soft handover thresholds engendered an increase in the probability of a low

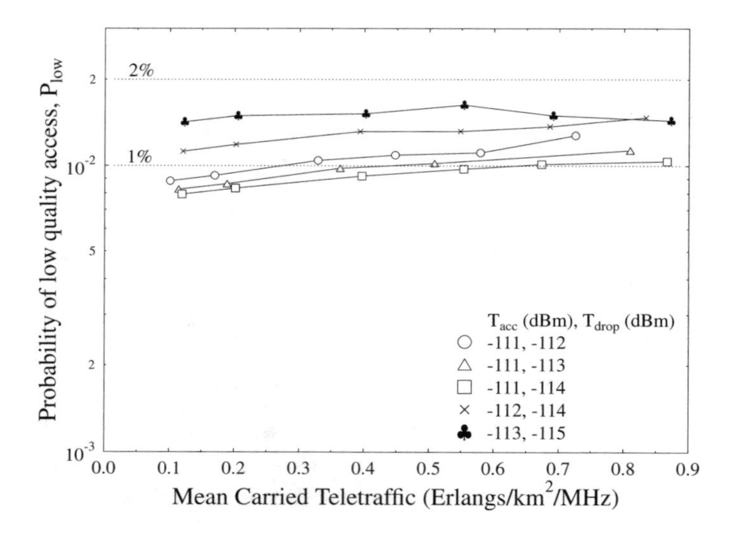

Figure 5.12: Probability of low quality access versus mean carried traffic of a CDMA based cellular network using **fixed received pilot power** based soft handover thresholds in conjunction **with 0.5 Hz shadowing having a standard deviation of 3 dB** for SF=16.

quality access.

The network cannot satisfy the quality requirements of the conservative scenario, namely that of maintaining a call dropping probability of 1% combined with a maximum probability of low quality access below 1%. However, the entire network supported 127 users, whilst meeting the lenient scenario's set of criteria, which consists of a maximum call dropping probability of 1% and a probability of low quality access of below 2%, using the thresholds of T_{acc}=-113 dBm and T_{drop}=-115 dBm.

5.4.2.3 Fixed Received Pilot Power Thresholds with 1.0 Hz Shadowing

This section presents results obtained using fixed receiver pilot power based soft handover thresholds in conjunction with log-normal shadow fading having a standard deviation of 3 dB and a maximum fading frequency of 1.0 Hz.

The corresponding call dropping probability is depicted in Figure 5.13, showing that using fixed thresholds in a propagation environment exposed to shadow fading resulted in a very poor performance. This was due to the shadow fading induced fluctuations of the received pilot signal power, which resulted in removing base stations from the ABS mid-call, which ultimately engendered dropped calls. Hence, lowering the fixed thresholds significantly reduced the call dropping probability. However, this led to a deterioration of the low quality access probability, as shown in Figure 5.14. The probability of low quality access was also very poor due to the rapidly fluctuating interference-limited environment. This was shown particularly explicitly in conjunction with T_{acc}=-113 dBm and T_{drop}=-115 dBm, where reducing the number of users resulted in a degradation of the low quality access performance

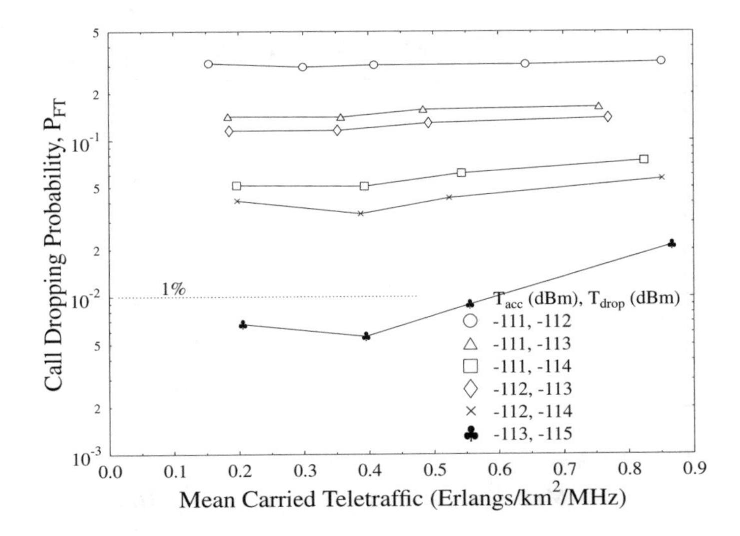

Figure 5.13: Call dropping probability versus mean carried traffic of a CDMA based cellular network using **fixed received pilot power** based soft handover thresholds in conjunction **with 1 Hz shadowing having a standard deviation of 3 dB** for SF=16.

due to the higher deviation of the reduced number of combined sources of interference. In contrast, adding more users led to a near-constant level of interference that varied less dramatically.

It was found that the network was unable to support any users at the required service quality, since using the thresholds that allowed the maximum 1% call dropping probability restriction to be met, led to a greater than 2% probability of a low quality outage occurring.

5.4.2.4 Summary

In summary of our findings in the context of Figure 5.7-5.14, a disadvantage of using fixed soft handover thresholds is that in some locations all pilot signals may be weak, whereas in other locations, all of the pilot signals may be strong due to the localised propagation environment or terrain. Hence, using relative or normalised soft handover thresholds is expected to be advantageous in terms of overcoming this limitation. An additional benefit of using dynamic thresholds is confirmed within a fading environment, where the received pilot power may drop momentarily below a fixed threshold, thus causing unnecessary removals and additions to/from the ABS. However, these base stations may have been the only base stations in the ABS, thus ultimately resulting in a dropped call. When using dynamically controlled thresholds this scenario would not have occurred. Hence, in the next section we considered the performance of using relative received pilot power based soft handover thresholds under both non-shadowing and shadowing impaired propagation conditions.

To summarise, using fixed received pilot power thresholds in a non-shadowing environment resulted in a total network capacity of 290 users for both quality of service scenarios,

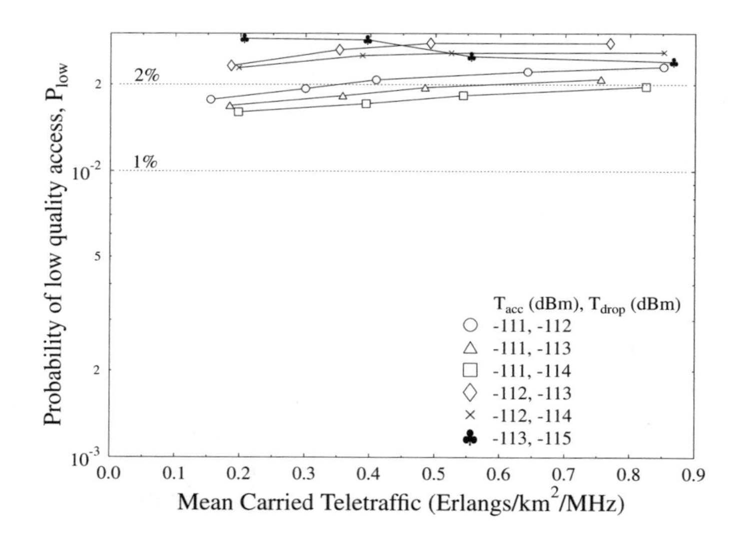

Figure 5.14: Probability of low quality access versus mean carried traffic of a CDMA based cellular network using **fixed received pilot power** based soft handover thresholds in conjunction **with 1 Hz shadowing having a standard deviation of 3 dB** for SF=16.

namely for both the conservative and lenient scenarios considered. However, this performance was severely degraded in a shadow fading impaired propagation environment, where a total network capacity of 127 users was supported in conjunction with a maximum shadow fading frequency of 0.5 Hz. Unfortunately, the network capacity could not be evaluated when using a maximum shadow fading frequency of 1.0 Hz due to the contrasting characteristics of the dropped call and low quality access probability results.

5.4.2.5 Relative Received Pilot Power Thresholds without Shadowing

Employing relative received pilot power thresholds is important in realistic propagation environments exposed to shadow fading. More explicitly, in contrast to the previously used thresholds, which were expressed in terms of dBm, i.e. with respect to 1 mW, in this section the thresholds T_{acc} and T_{drop} are expressed in terms of dB relative to the received pilot strength of the base stations in the ABS. Their employment also caters for situations, where the absolute pilot power may be too low for use in conjunction with fixed thresholds, but nonetheless sufficiently high for reliable communications. Hence, in this section we examine the performance of relative received pilot power based soft handover thresholds in a non-shadow faded environment.

The call dropping performance is depicted in Figure 5.15, which shows that reducing the soft handover thresholds, and thus increasing the time spent in soft handover, improved the call dropping performance. It was also found in the cases considered here, that simultaneously the probability of a low quality access decreased, as illustrated by Figure 5.16. However, it was also evident in both figures, that reducing the soft handover thresholds past a

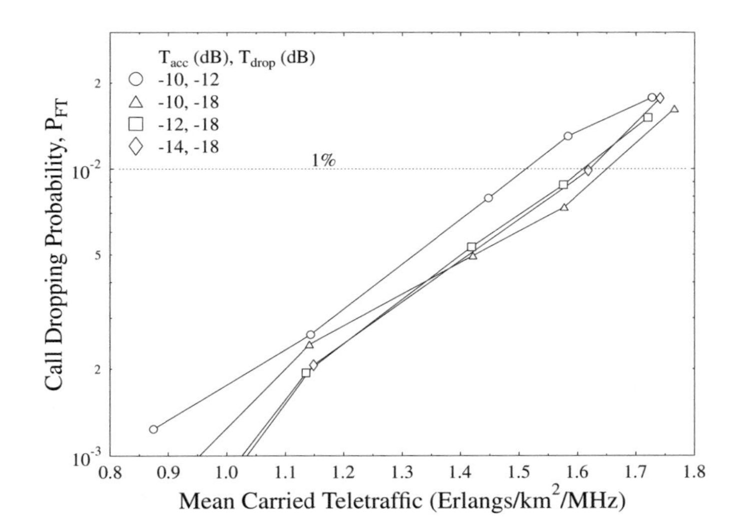

Figure 5.15: Call dropping probability versus mean carried traffic of a CDMA based cellular network using **relative received pilot power** based soft handover thresholds **without shadowing** for SF=16.

certain point resulted in degraded performance due to the extra interference incurred during the soft handover process.

Since the probability of low quality access was under the 1% threshold, the network capacity for both the lenient and conservative scenarios were the same, namely 1.65 Erlangs / km^2 / MHz or a total of 288 users over the entire simulation area of 2.86 km^2. The mean ABS size was 1.7 base stations, with a mean mobile transmission power of 4.1 dBm and an average base station transmit power of 4.7 dBm.

5.4.2.6 Relative Received Pilot Power Thresholds with 0.5 Hz Shadowing

In this section we present results obtained using relative received pilot power based soft handover thresholds in a shadowing-impaired propagation environment. The maximum shadow fading frequency was 0.5 Hz and the standard deviation of the log-normal shadowing was 3 dB.

Figure 5.17 depicts the call dropping probability for several relative thresholds and shows that by reducing both the thresholds, the call dropping performance is improved. This enables the mobile to add base stations to its ABS earlier on during the soft handover process, and to relinquish them at a much later stage than in the case of using higher handover thresholds. Therefore, using lower relative soft handover thresholds results in a longer period of time spent in soft handover, as can be seen in Figure 5.18, which shows the mean number of base stations in the ABS.

The probability of low quality access is shown in Figure 5.19, illustrating that, in general, as the relative soft handover thresholds were reduced, the probability of low quality access

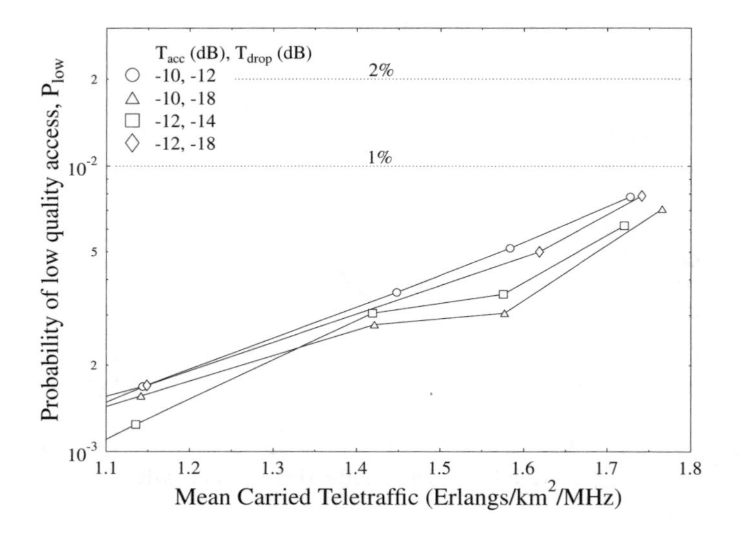

Figure 5.16: Probability of low quality access versus mean carried traffic of a CDMA based cellular network using **relative received pilot power** based soft handover thresholds **without shadowing** for SF=16.

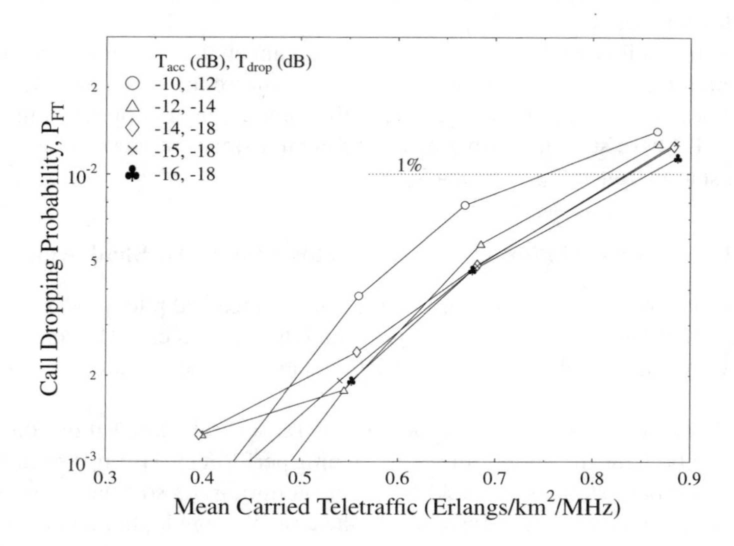

Figure 5.17: Call dropping probability versus mean carried traffic of a CDMA based cellular network using **relative received pilot power** based soft handover thresholds in conjunction **with 0.5 Hz shadowing and a standard deviation of 3 dB** for SF=16.

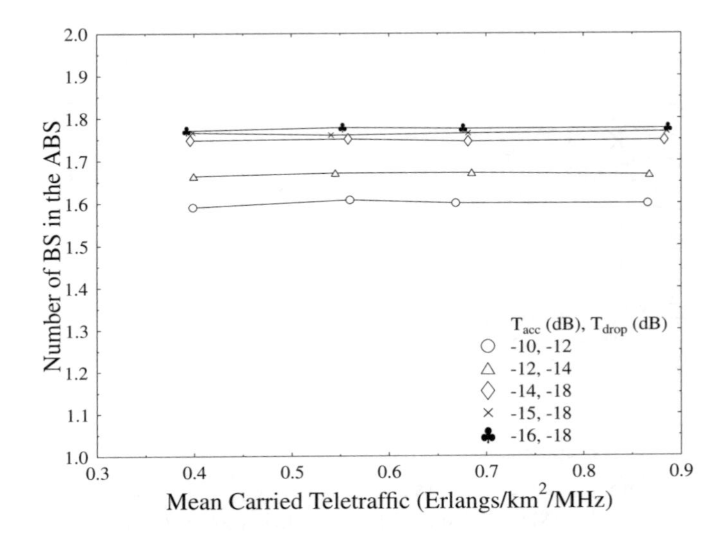

Figure 5.18: Mean number of base stations in the active base station set versus mean carried traffic of a CDMA based cellular network using **relative received pilot power** based soft handover thresholds in conjunction **with 0.5 Hz shadowing and a standard deviation of 3 dB** for SF=16.

increased. This demonstrated that spending more time in soft handover generated more co-channel interference and thus degraded the network's performance. However, the difference between the two thresholds must also be considered. For example, the probability of low quality access is higher in conjunction with $T_{acc} = -16$ dB and T_{drop}=-18 dB, than using T_{acc}=-16 dB and T_{drop}=-20 dB, since the latter scenario has a higher mean number of base stations in its ABS. Therefore, there is a point at which the soft handover gain experienced by the desired user outweighs the detrimental effects of the extra interference generated by base stations' transmissions to users engaged in the soft handover process.

Figure 5.20 shows the mean transmission powers of both the mobiles and the base stations. The mobiles are required to transmit at a lower power than the base stations, because the base stations are not subjected to downlink pilot power interference and to soft handover interference. Furthermore, the mobiles are not affected by the level of the soft handover thresholds, because only selective diversity is performed in the uplink, and hence the mobile transmits as if not in soft handover. As the soft handover thresholds were reduced, the time spent in soft handover increased and thus the mean base transmission power had to be increased in order to overcome the additional downlink interference.

The maximum network capacity of 0.835 Erlangs/km^2/MHz, or 144 users over the entire simulation area, was achieved using the soft handover thresholds of T_{acc}=-14 dB and T_{drop}=-18 dB for the conservative scenario. The mean ABS size was 1.77 base stations, while the mean mobile transmit power was -1.5 dBm and 0.6 dBm for the base stations. In the lenient scenario a maximum teletraffic load of 0.865 Erlangs / km^2 / MHz, corresponding to a total network capacity of 146 users was maintained using soft handover thresholds of T_{acc}=-16 dB

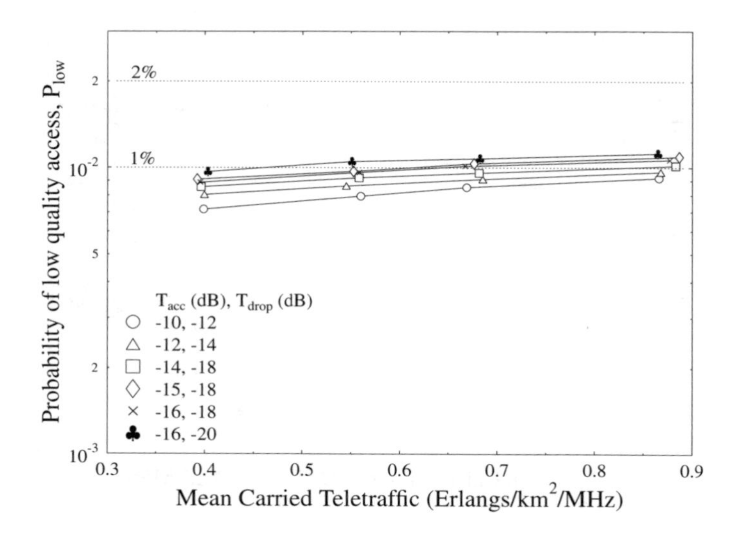

Figure 5.19: Probability of low quality access versus mean carried traffic of a CDMA based cellular network using **relative received pilot power** based soft handover thresholds in conjunction **with 0.5 Hz shadowing and a standard deviation of 3 dB** for SF=16.

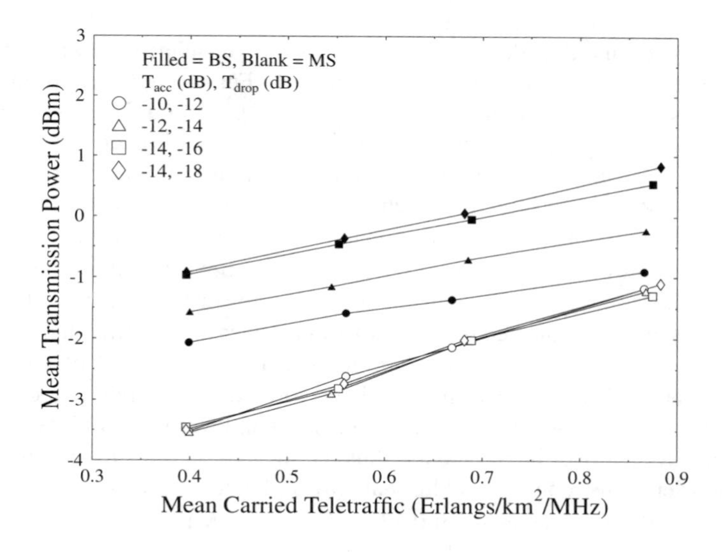

Figure 5.20: Mean transmission power versus mean carried traffic of a CDMA based cellular network using **relative received pilot power** based soft handover thresholds in conjunction **with 0.5 Hz shadowing and a standard deviation of 3 dB** for SF=16.

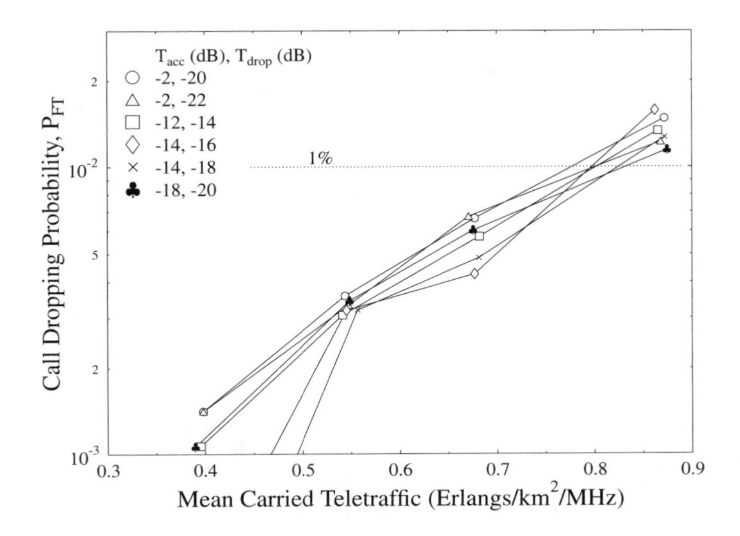

Figure 5.21: Call dropping probability versus mean carried traffic of a CDMA based cellular network using **relative received pilot power** based soft handover thresholds in conjunction **with 1 Hz shadowing and a standard deviation of 3 dB** for SF=16.

and T_{drop}=-18 dB. The mean number of base stations in the ABS was 1.78, with an average transmit power of -1.5 dBm for the mobile handset, and 1.3 dBm for the base station.

5.4.2.7 Relative Received Pilot Power Thresholds with 1.0 Hz Shadowing

In this section we present further performance results obtained using relative received pilot power based soft handover thresholds in a shadowing propagation environment. The maximum shadow fading frequency was 1.0 Hz and the standard deviation of the log-normal shadowing was 3 dB.

On comparing the call dropping probability curves seen in Figure 5.21 with the call dropping probability obtained for a maximum shadow fading frequency of 0.5 Hz in Figure 5.17 it was found that the performance of the 1.0 Hz frequency shadowing scenario was slightly worse. However, the greatest performance difference was observed in the probability of low quality access, as can be seen in Figure 5.22.

Using the soft handover thresholds which gave a good performance for a maximum shadow fading frequency of 0.5 Hz resulted in significantly poorer low quality access performance for a maximum shadowing frequency of 1.0 Hz. In order to obtain a probability of low quality access of below 1% it was necessary to use markedly different soft handover thresholds, which reduced the time spent in soft handover and hence also the size of the ABS, as illustrated in Figure 5.23.

For the conservative scenario, where the maximum probability of low quality access, P_{low}, was set to 1%, the maximum network capacity was found to be 0.69 Erlangs/km^2/MHz, equivalent to a total network capacity of 127 users, obtained using T_{drop}=-2 dB and T_{acc}=-

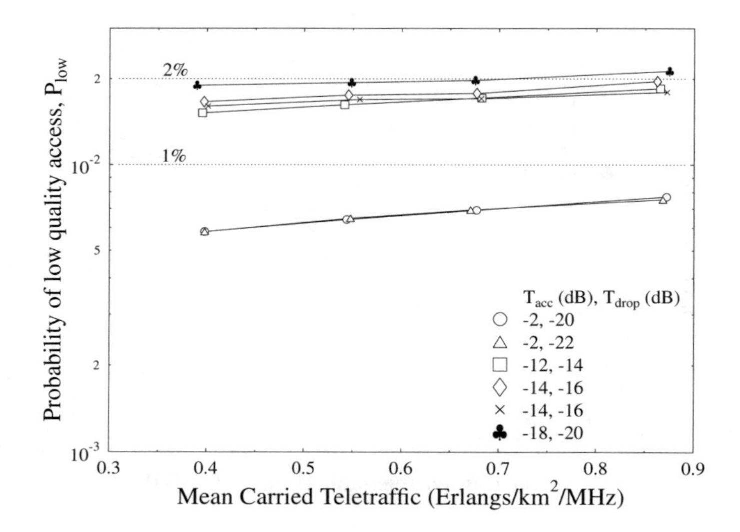

Figure 5.22: Probability of low quality access versus mean carried traffic of a CDMA based cellular network using **relative received pilot power** based soft handover thresholds in conjunction **with 1 Hz shadowing and a standard deviation of 3 dB** for SF=16.

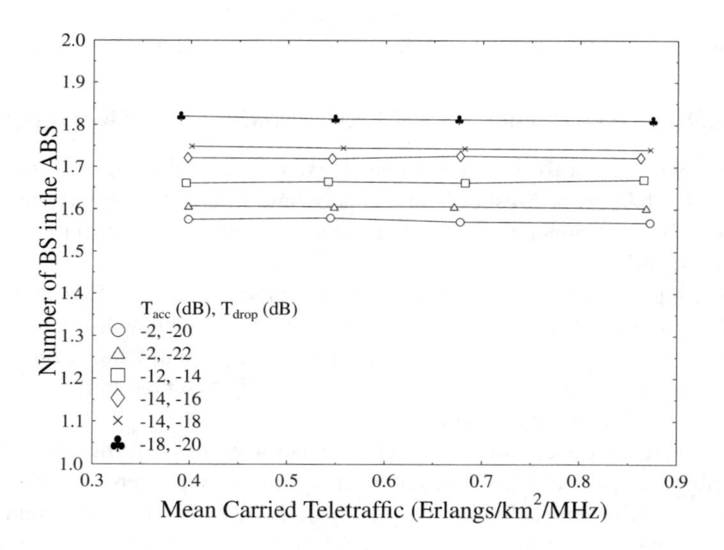

Figure 5.23: Mean number of base stations in the active base station set versus mean carried traffic of a CDMA based cellular network using **relative received pilot power** based soft handover thresholds in conjunction **with 1 Hz shadowing and a standard deviation of 3 dB** for SF=16.

16 dB. In contrast, in the lenient scenario, where the P_{low} limit was 2%, the maximum number of users supported was found to be 144, or 0.825 Erlangs/km^2/MHz, in conjunction with T_{acc}=-14 dB and T_{drop}=-18 dB.

5.4.2.8 Summary

In summary, using relative received pilot power as a soft handover metric has resulted in a significantly improved performance in comparison to that of the fixed received pilot power based results in a shadow fading environment. In the non-shadowed environment the network capacity was approximately the same as when using the fixed threshold algorithm, albeit with a slightly improved mean transmission power. Due to the time varying nature of the received signals subjected to shadow fading, using relative thresholds has been found to be more amenable to employment in a realistic propagation environment, than using fixed thresholds. In conclusion, without shadow fading the network supported a total of 288 users, whilst with a maximum shadow fading frequency of 0.5 Hz, approximately 145 users were supported by the entire network, for both the conservative and lenient scenarios. However, different soft handover thresholds were required for each situation, for achieving these capacities. At a maximum shadowing frequency of 1.0 Hz, a total of 127 users were supported in the conservative scenario, and 144 in the lenient scenario. However, again, different soft handover thresholds were required in each scenario in order to maximise the network capacity.

5.4.3 E_c/I_o Power Based Soft Handover Results

An alternative soft handover metric used to determine 'cell ownership' is the pilot to downlink interference ratio of a cell, which was proposed for employment in the 3rd generation systems [32]. The pilot to downlink interference ratio, or E_c/I_o, may be calculated thus as [374]:

$$\frac{E_c}{I_o} = \frac{P_{pilot}}{P_{pilot} + N_0 + \sum_{k=1}^{N_{cells}} P_k T_k}, \tag{5.10}$$

where P_k is the total transmit power of cell k, T_k is the transmission gain, which includes the antenna gain and pathloss as well as shadowing, N_0 is the power spectral density of the thermal noise and N_{cells} is the number of cells in the network. The advantage of using such a scheme is that it is not an absolute measurement that is used, but the ratio of the pilot power to the interference power. Thus, if fixed thresholds were used a form of admission control may be employed for new calls if the interference level became too high. A further advantage is that it takes into account the time-varying nature of the interference level in a shadowed environment.

5.4.3.1 Fixed E_c/I_o Thresholds without Shadowing

The new call blocking probability obtained when using fixed E_c/I_o soft handover thresholds without any form of shadow fading is shown in Figure 5.24, which suggests that in general, lowering the soft handover thresholds reduced the probability of a new call attempt being blocked. However, it was found that in conjunction with $T_{drop} = -40$ dB, dropping the

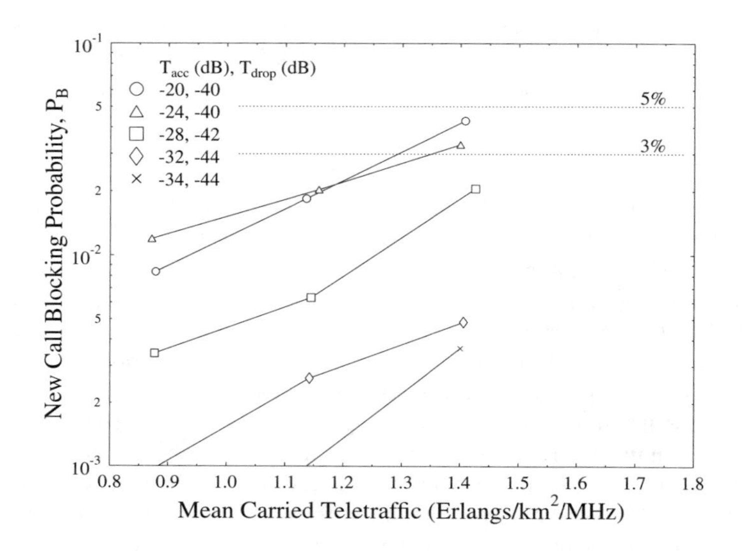

Figure 5.24: New call blocking probability versus mean carried traffic of a CDMA based cellular network using **fixed** E_c/I_o based soft handover thresholds **without shadowing** for SF=16.

threshold T_{acc} from -20 dB to -24 dB actually increased the new call blocking probability. This was attributed to the fact that the lower threshold precipitated a higher level of co-channel interference, since there was a higher mean number of base stations in the ABS, as evidenced by Figure 5.25. Therefore, since the mean level of interference present in the network is higher, when using a lower threshold, and the threshold determines the value of the pilot to downlink interference ratio at which base stations may be added to the ABS, a more frequent blocking of calls occurs. Alternatively, a lower threshold resulted in a higher level of downlink interference due to the additional interference inflicted by supporting the mobiles in soft handover, which prevented base stations from being included in the ABS due to insufficient pilot to interference 'head-room'. This then ultimately led to blocked calls due to the lack of base stations in the ABS.

Again, the mean number of base stations in the ABS is given in Figure 5.25, which illustrates that as expected, reducing the soft handover thresholds increased the proportion of time spent in soft handover, and thus reduced the mean number of base stations in the ABS. The average size of the ABS was found to decrease, as the network's traffic load increased. This was a consequence of the increased interference levels associated with the higher traffic loads, which therefore effectively reduced the pilot to interference ratio at a given point, and hence base stations were less likely to be in soft handover and in the ABS.

Figure 5.26 depicts the mean transmission powers for both the uplink and the downlink, for a range of different soft handover thresholds. These results show similar trends to the results presented in previous sections, with the required average downlink transmission power increasing, since a greater proportion of call time is spent in soft handover. Again, the mean uplink transmission power varied only slightly, since the selection diversity technique of the base stations only marginally affected the received interference power at the base stations.

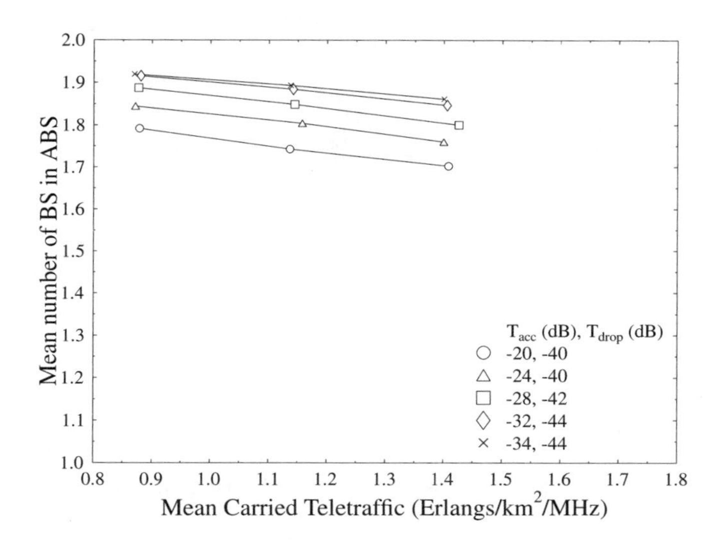

Figure 5.25: Mean number of base stations in the active base station set versus mean carried traffic of a CDMA based cellular network using **fixed** E_c/I_o based soft handover thresholds **without shadowing** for SF=16.

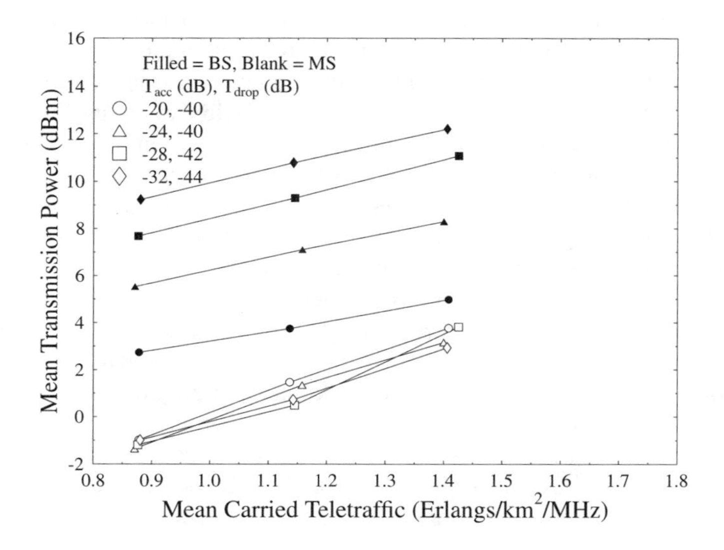

Figure 5.26: Mean transmission power versus mean carried traffic of a CDMA based cellular network using **fixed received** E_c/I_o based soft handover thresholds **without shadowing** for SF=16.

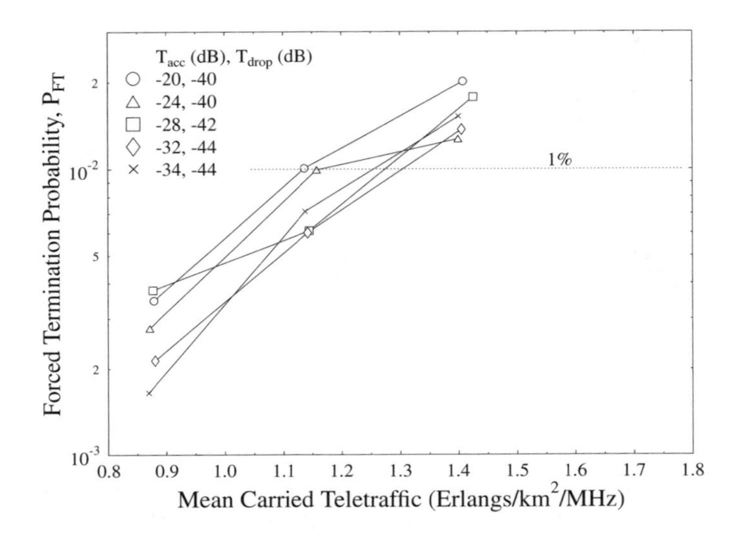

Figure 5.27: Call dropping probability versus mean carried traffic of a CDMA based cellular network using **fixed received** E_c/I_o based soft handover thresholds **without shadowing** for SF=16.

Figure 5.27 shows the call dropping performance, indicating that lowering the soft handover thresholds generally improved the call dropping performance. However, reducing the soft handover thresholds too much resulted in a degradation of the call dropping probability due to the increased levels of co-channel interference inherent when a higher proportion of the call time is spent in soft handover. This is explicitly illustrated by Figure 5.28, which indicates that reducing the soft handover thresholds caused a significant degradation in the probability of low quality access. This was a consequence of the additional co-channel interference associated with the soft handover process. The figure also shows that there is a point where the diversity gain of the mobiles obtained with the advent of the soft handover procedure outweighs the extra interference that it generates.

On the whole, the capacity of the network when using fixed E_c/I_o soft handover thresholds was lower than when using fixed received pilot power based soft handover thresholds. This can be attributed to the fact that the E_c/I_o thresholds are related to the interference level of the network, which changes with the network load and propagation conditions. Hence using a fixed threshold is sub-optimal. In the conservative scenario, the network capacity was 1.275 Erlangs/km^2/MHz, corresponding to a total network capacity of 223 users. In the lenient scenario, this increased to 1.305 Erlangs/km^2/MHz, or 231 users. In contrast, when using fixed received pilot power thresholds the entire network supported 290 users.

5.4.3.2 Fixed E_c/I_o Thresholds with 0.5 Hz Shadowing

In this section we consider fixed pilot to downlink interference ratio based soft handover thresholds in a propagation environment exhibiting shadow fading in conjunction with a max-

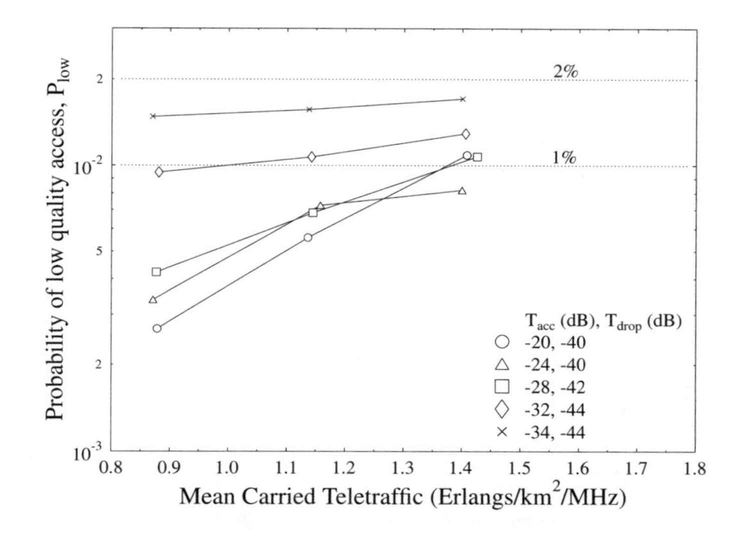

Figure 5.28: Probability of low quality access versus mean carried traffic of a CDMA based cellular network using **fixed received** E_c/I_o based soft handover thresholds **without shadowing** for SF=16.

imum fading frequency of 0.5 Hz and a standard deviation of 3 dB.

Examining Figure 5.29, which shows the call dropping probability, we see, again, that reducing the soft handover thresholds typically resulted in a lower probability of a dropped call. However, since the handover thresholds are dependent upon the interference level, there was some interaction between the handover thresholds and the call dropping rate. For example, it can be seen in the figure that when $T_{drop} = -40$ dB, the call dropping probability fell as T_{acc} was reduced from -20 dB to -24 dB. However, on lowering T_{acc} further, to -26 dB, the call dropping rate at low traffic loads became markedly higher. A similar phenomenon was observed in Figure 5.30, which shows the probability of low quality outage.

It is explicitly seen from Figures 5.29 and 5.30 that the performance of the fixed E_c/I_o soft handover threshold based scheme clearly exceeded that of the fixed received pilot power threshold based system in a shadow fading environment. The network supported a teletraffic load of 0.7 Erlangs/km^2/MHz or a total of 129 users in the conservative scenario, which rose to 0.78 Erlangs/km^2/MHz, or 140 users, in the lenient scenario. These network capacities were achieved with the aid of a mean number of active base stations in the ABS, which were 1.88 and 1.91, respectively. In order to achieve the total network capacity of 129 users in the conservative scenario, a mean mobile transmit power of -2.4 dBm was required, while the mean base station transmission power was 7 dBm. For the lenient scenario, these figures were -2.4 dBm and 8.7 dBm, respectively.

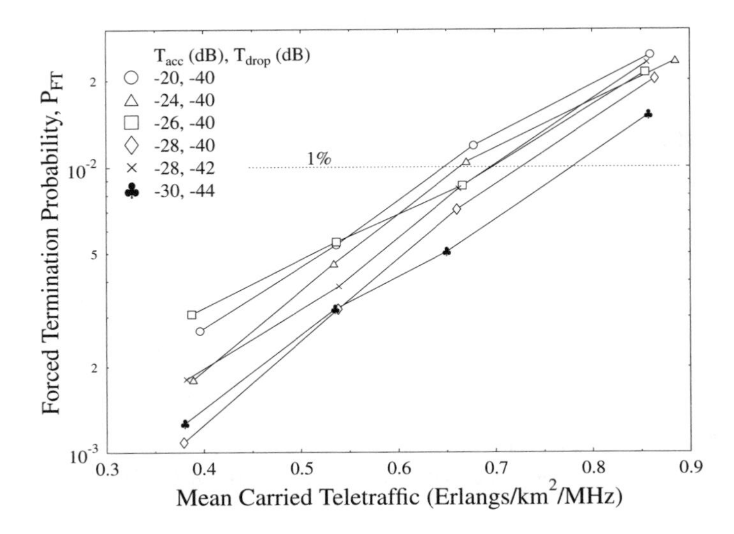

Figure 5.29: Call dropping probability versus mean carried traffic of a CDMA based cellular network using **fixed received** E_c/I_o based soft handover thresholds in conjunction **with 0.5 Hz shadowing and a standard deviation of 3 dB** for SF=16.

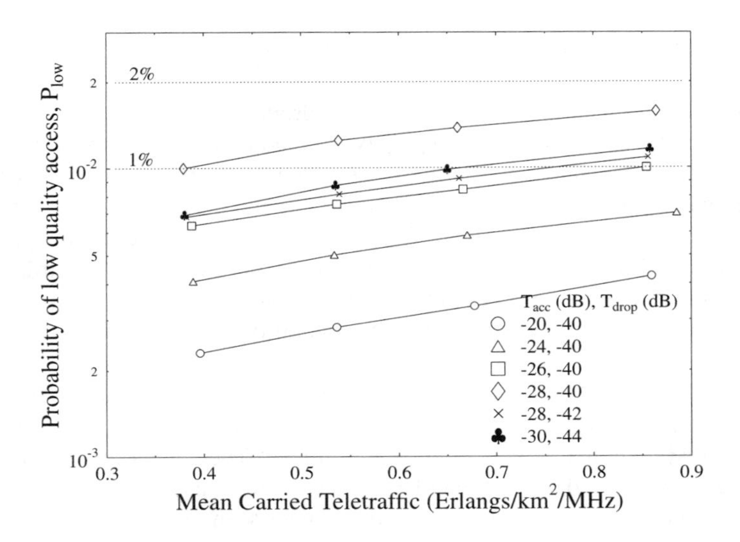

Figure 5.30: Probability of low quality access versus mean carried traffic of a CDMA based cellular network using **fixed received** E_c/I_o based soft handover thresholds in conjunction **with 0.5 Hz shadowing and a standard deviation of 3 dB** for SF=16.

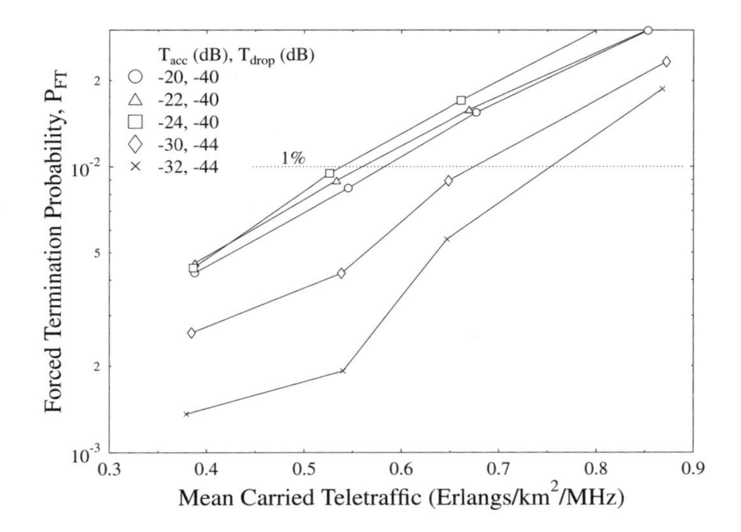

Figure 5.31: Call dropping probability versus mean carried traffic of a CDMA based cellular network using **fixed received** E_c/I_o based soft handover thresholds in conjunction **with 1.0 Hz shadowing and a standard deviation of 3 dB** for SF=16.

5.4.3.3 Fixed E_c/I_o Thresholds with 1.0 Hz Shadowing

Increasing the maximum shadow fading frequency from 0.5 Hz to 1.0 Hz resulted in an increased call dropping probability and a greater probability of low quality access, for a given level of carried teletraffic. This is clearly seen by comparing Figures 5.31 and 5.32 with Figures 5.29 and 5.30. Explicitly, Figure 5.31 and 5.32 show that reducing the soft handover threshold, T_{acc} from -20 dB to -24 dB led to both an increased call dropping probability and an increased probability of low quality access. This can be attributed to the extra co-channel interference generated by the greater proportion of call time being spent in soft handover. This is also confirmed by the increased probability of low quality access observed in Figure 5.32 for lower soft handover thresholds T_{acc} and T_{drop}.

The network capacity of the conservative scenario was 0.583 Erlangs/km^2/MHz, giving an entire network capacity of 107 users. In the lenient scenario the network supported a total of 128 users or a traffic load of 0.675 Erlangs/km^2/MHz was carried. The 107 users were serviced in conjunction with a mean ABS size of 1.86, a mean mobile transmit power of -3 dBm and a mean base station transmit power of 4.5 dBm. The 128 users supported in the lenient scenario necessitated an average mobile transmit power of -3 dBm and an average base station transmit power of 9.5 dBm. The mean number of base stations in the ABS was 1.91.

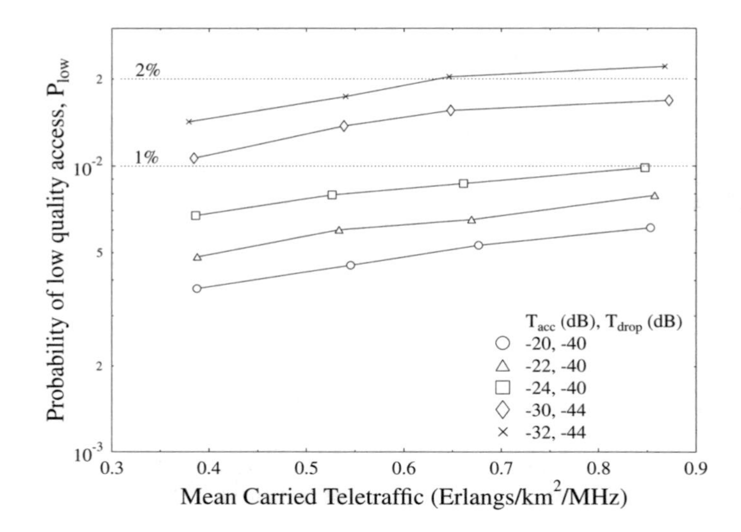

Figure 5.32: Probability of low quality access versus mean carried traffic of a CDMA based cellular network using **fixed received** E_c/I_o based soft handover thresholds in conjunction **with 1.0 Hz shadowing and a standard deviation of 3 dB** for SF=16.

5.4.3.4 Summary

In summary, a maximum network capacity of 290 users was obtained when employing the fixed E_c/I_o soft handover thresholds. This capacity was equal to that when using fixed received pilot power thresholds in the lenient scenario without shadow fading. However, in the conservative scenario the network capacity was reduced from 290 to 231 users. Nevertheless, when a realistic shadowed propagation environment was considered, using the pilot power to interference ratio based soft handover metric improved the network capacity significantly. This was particularly evident in conjunction with the maximum shadow fading frequency of 1.0 Hz, when using the fixed received pilot power thresholds no users could be supported whilst maintaining the desired call quality. In contrast, using the fixed E_c/I_o soft handover thresholds led to a total network capacity of between 107 and 128 users, for the conservative and lenient scenarios, respectively. This capacity increase was the benefit of the more efficient soft handover mechanism, which was capable of taking into account the interference level experienced, leading to a more intelligent selection of base stations supporting the call. At a maximum shadow fading frequency of 0.5 Hz the network had a maximum capacity of 129 and 140 users, for the conservative and lenient scenario, respectively, when using the fixed E_c/I_o soft handover thresholds.

5.4.3.5 Relative E_c/I_o Thresholds without Shadowing

In this section we combined the benefits of using the received E_c/I_o ratio and relative soft handover thresholds, thus ensuring that variations in both the received pilot signal strength

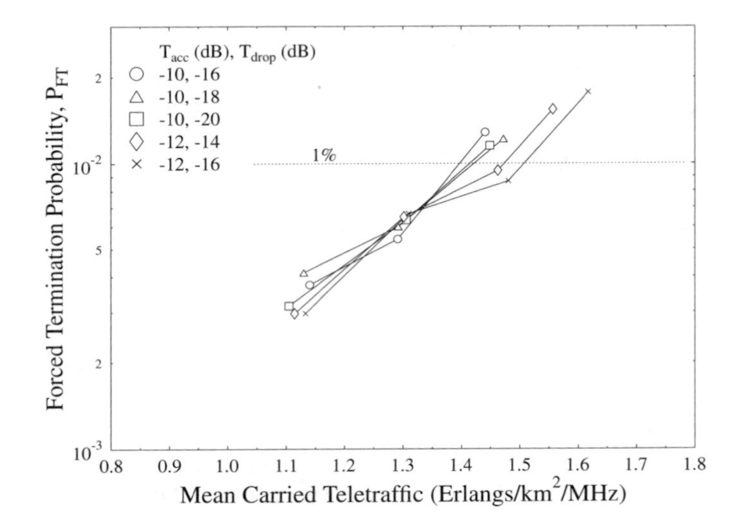

Figure 5.33: Call dropping probability versus mean carried traffic of a CDMA based cellular network using **relative received** E_c/I_o based soft handover thresholds **without shadowing** for SF=16.

and interference levels were monitored in the soft handover process.

The call dropping performance is shown in Figure 5.33, illustrating that reducing the soft handover thresholds improved the probability of dropped calls, in particular at higher traffic loads. This phenomenon is also evident in Figure 5.34, which shows the probability of a low quality outage. However, in some cases it was evident that excessive reduction of the thresholds led to increasing the co-channel interference, and hence to a greater probability of outage associated with low quality. Again, this was the consequence of supporting an excessive number of users in soft handover, which provided a beneficial diversity gain for the mobiles but also increased the amount of downlink interference inflicted by the base stations supporting the soft handovers.

The entire network supported a total of 256 users employing soft handover thresholds of T_{acc}=-12 dB and T_{drop}=-16 dB. The mean number of base stations in the active set was 1.68, and the mean mobile transmit power was 3.1 dBm. The average base station transmit power was 2.7 dBm.

5.4.3.6 Relative E_c/I_o Thresholds with 0.5 Hz Shadowing

Examining the call dropping probability graphs in Figure 5.35 shows that the probability of a dropped call was significantly lower than that of the other soft handover algorithms considered for the same propagation environment. This was because the handover algorithm was capable of taking the current interference levels into account when deciding whether to initiate a handover, additionally, the employment of the relative thresholds minimised the chances of making an inappropriate soft handover decision concerning the most suitable base

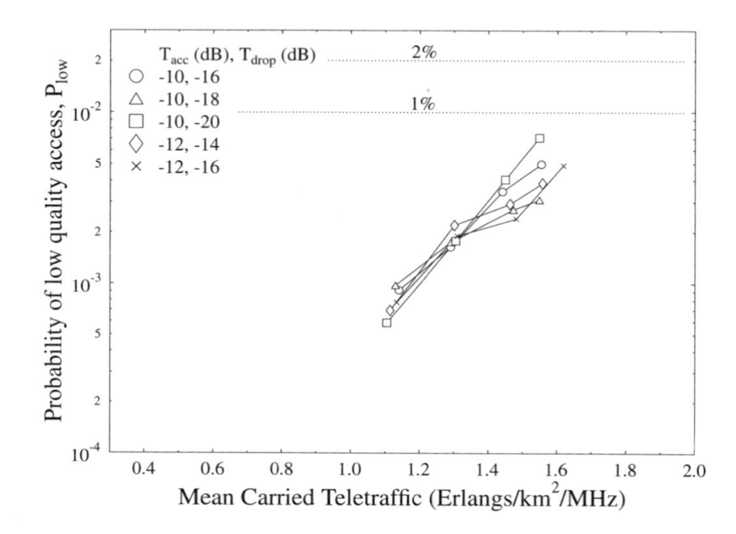

Figure 5.34: Probability of low quality access versus mean carried traffic of a CDMA based cellular network using **relative received** E_c/I_o based soft handover thresholds **without shadowing** for SF=16.

station to use. The superiority of this soft handover algorithm was further emphasised by the associated low probability of a low quality access, as illustrated in Figure 5.36, which was an order of magnitude lower than that achieved using the alternative soft handover algorithms.

When T_{acc} was set to -10 dB the ultimate capacity of the network was only marginally affected by changing T_{drop}, although some variation could be observed in the call dropping probability. Furthermore, the probability of low quality access increased for the lowest values of T_{drop}. This degradation of the probability of low quality access was due to the higher proportion of time spent in soft handover, as indicated by the correspondingly increased ABS size in Figure 5.37, which was a consequence of the associated increased co-channel interference levels.

The mean transmit power curves of Figure 5.38 exhibited a different characteristic in comparison to that observed for the other soft handover algorithms. Specifically, at low traffic loads the mean mobile transmit power was less than that of the base stations, whereas at the higher traffic loads, the mobile transmit power was greater than that of the base stations. Although, comparing this graph with Figure 5.20 revealed that the spread and the rate of change of the mobile transmit power versus the traffic load was similar in both scenarios, the mean base station transmission power was lower in Figure 5.38. This reduced base station transmission power, again demonstrated the superiority of this soft handover algorithm, which manifested itself in its more efficient use of resources.

Since the probability of low quality access fell well below the 1% threshold, both the conservative and lenient scenarios exhibited the same total network capacity, which was slightly above 150 users for the entire network. This was achieved on average with the aid of 1.65 base stations, at a mean mobile transmit power of -1.2 dBm and at a mean base station

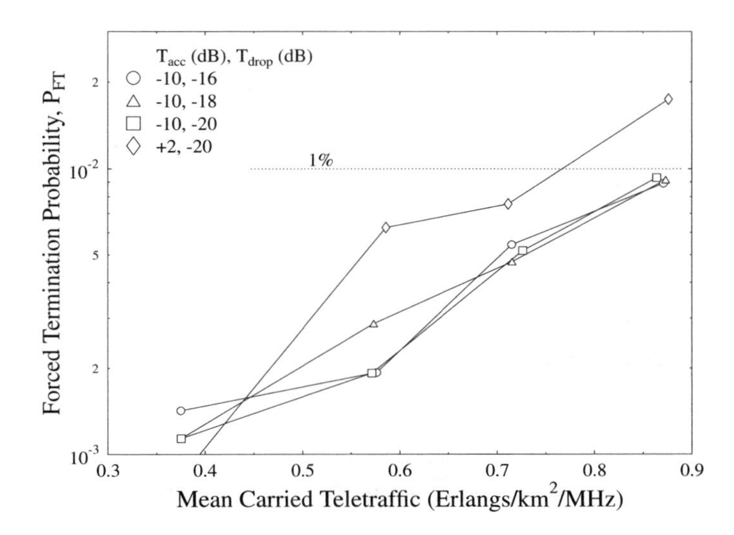

Figure 5.35: Call dropping probability versus mean carried traffic of a CDMA based cellular network using **relative received** E_c/I_o based soft handover thresholds in conjunction **with 0.5 Hz shadowing and a standard deviation of 3 dB** for SF=16.

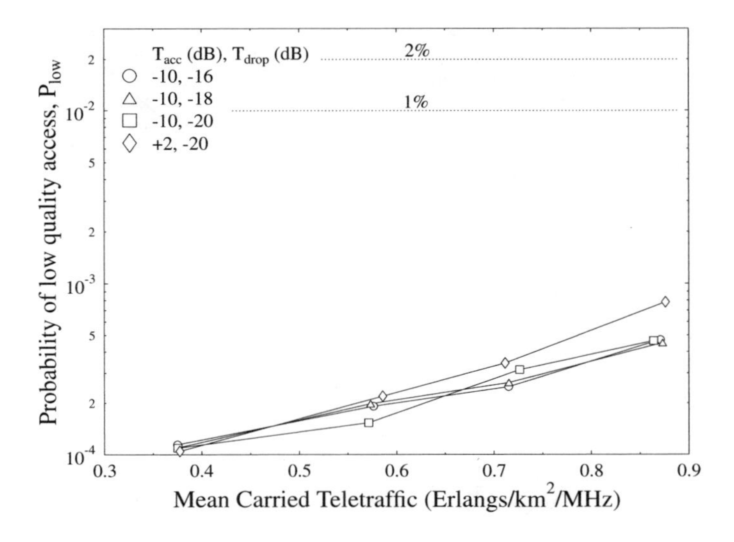

Figure 5.36: Probability of low quality access versus mean carried traffic of a CDMA based cellular network using **relative received** E_c/I_o based soft handover thresholds in conjunction **with 0.5 Hz shadowing and a standard deviation of 3 dB** for SF=16.

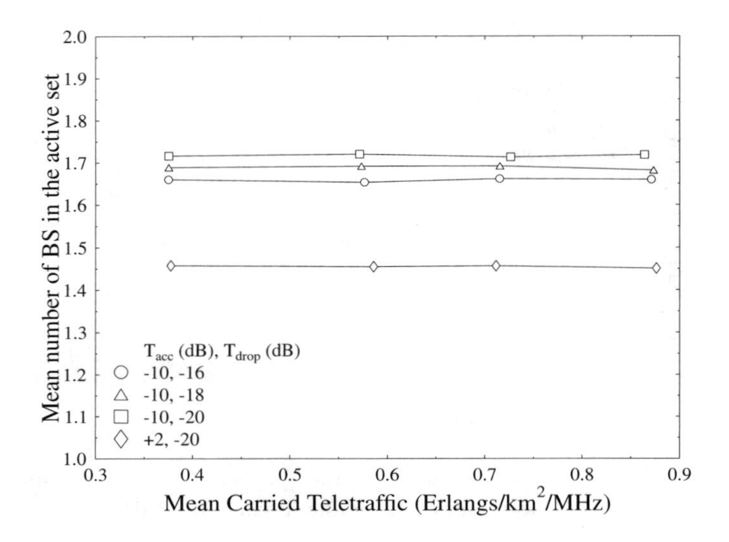

Figure 5.37: Mean number of base stations in the active base station set versus mean carried traffic of a CDMA based cellular network using **relative received** E_c/I_o based soft handover thresholds in conjunction **with 0.5 Hz shadowing and a standard deviation of 3 dB** for SF=16.

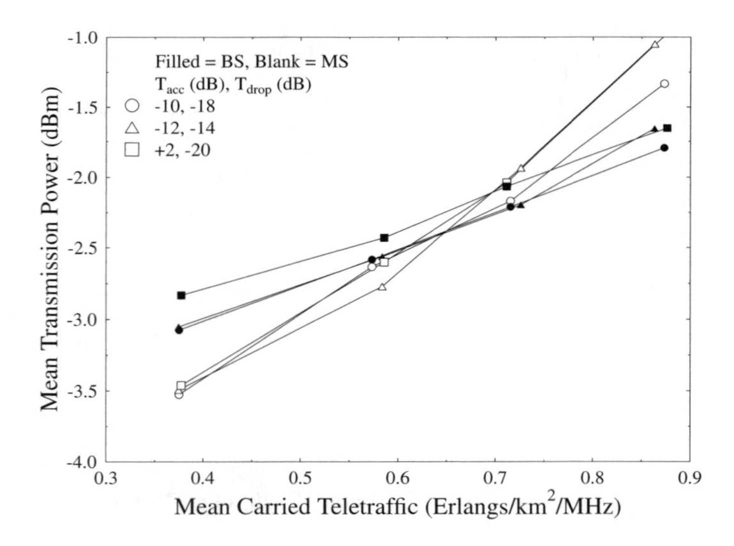

Figure 5.38: Mean transmission power versus mean carried traffic of a CDMA based cellular network using **relative received** E_c/I_o based soft handover thresholds in conjunction **with 0.5 Hz shadowing and a standard deviation of 3 dB** for SF=16.

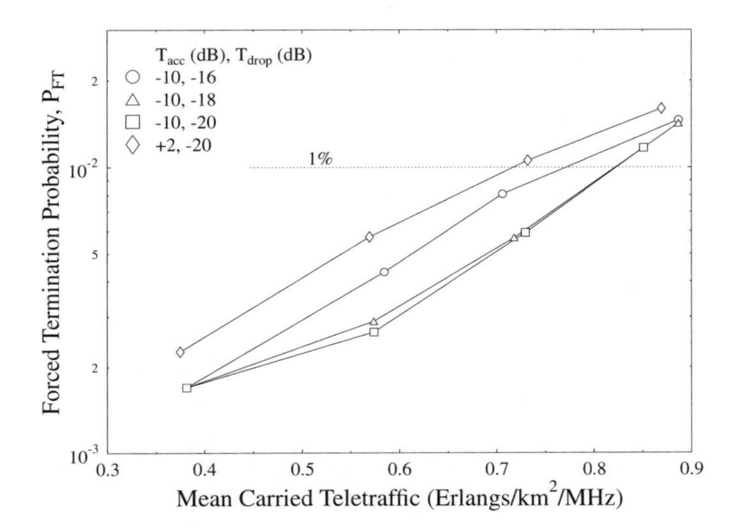

Figure 5.39: Call dropping probability versus mean carried traffic of a CDMA based cellular network using **relative received** E_c/I_o based soft handover thresholds in conjunction **with 1.0 Hz shadowing and a standard deviation of 3 dB** for SF=16.

transmit power of -1.7 dBm.

5.4.3.7 Relative E_c/I_o Thresholds with 1.0 Hz Shadowing

The call dropping probability shown in Figure 5.39 is slightly worse than that obtained in Figure 5.35 for a maximum shadow fading frequency of 0.5 Hz, with a greater performance difference achieved by altering T_{drop}. A similar performance degradation was observed for the probability of low quality access in Figure 5.40, with an associated relatively low impact due to varying the soft handover thresholds. Although not explicitly shown, we found that the mean transmission powers were similar to those required for a maximum shadow fading frequency of 0.5 Hz.

5.4.3.8 Summary

In summary, the employment of relative E_c/I_o soft handover thresholds resulted in a superior network performance and capacity under all the propagation conditions investigated. This was achieved whilst invoking the lowest average number of base stations and the minimum mean base station transmit power. A further advantage of this handover scheme is that the same soft handover thresholds excelled in all of the propagation environments studied, unlike the previously considered algorithms, which obtained their best results at different thresholds for different conditions. The entire network capacity was 256 users without shadow fading, with a mean ABS size of 1.68. At a maximum shadowing frequency of 0.5 Hz the network supported just over a total of 150 users, whilst 144 users were served by the entire network,

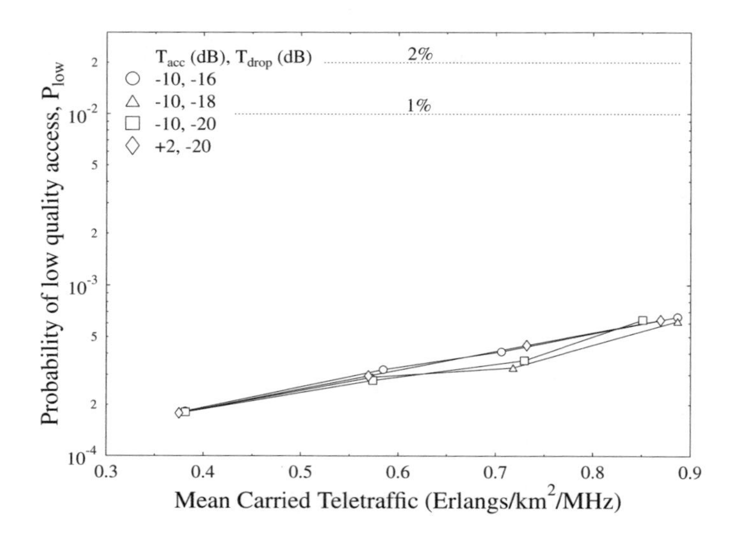

Figure 5.40: Probability of low quality access versus mean carried traffic of a CDMA based cellular network using **relative received** E_c/I_o based soft handover thresholds in conjunction **with 1.0 Hz shadowing and a standard deviation of 3 dB** for SF=16.

when a maximum shadow fading frequency of 1.0 Hz was encountered.

5.4.4 Overview of Results

Table 5.3 summarises the results obtained for the various soft handover algorithms over the three different propagation environments considered. The fixed receiver pilot power based algorithm performed the least impressively overall, as expected due to its inherent inability to cope with shadow fading. However, it did offer a high network capacity in a non-shadowed environment. Using the relative received pilot power based soft handover algorithm improved the performance under shadow fading, but different fading rates required different thresholds to meet the conservative and lenient quality criteria. The performance of the fixed E_c/I_o based soft handover algorithm also varied significantly, when using the same thresholds for the two different fading rates considered. However, the maximum network capacity achieved under the different shadow fading conditions was significantly higher, than that of the fixed received pilot power based algorithm. This benefit resulted from the inclusion of the interference levels in the handover process, which thus took into account the fading of both the signal and the co-channel interference. Combining the relative threshold based scheme with using E_c/I_o thresholds allowed us to support the highest number of users under the shadow fading conditions investigated. Whilst its performance was not the highest in the non-shadowed environment, this propagation environment is often unrealistic, and hence the relative received E_c/I_o based soft handover algorithm was chosen as the basis for our future investigations, while using the soft handover thresholds of T_{acc}=-10 dB and T_{drop}=-18 dB. The advantages of this handover algorithm were its reduced fraction of time spent in soft handover, and its

Soft handover algorithm	Shadowing	Conservative scenario P_{FT}=1%, P_{low}=1% Users	ABS	Power (dBm) MS	BS	Lenient scenario P_{FT}=1%, P_{low}=2% Users	ABS	Power (dBm) MS	BS
Fixed pilot pwr.	No	290	1.7	5.1	5.1	290	1.7	5.1	5.1
Fixed pilot pwr.	0.5 Hz, 3 dB	-	-	-	-	127	1.83	-2.0	6.5
Fixed pilot pwr.	1.0 Hz, 3 dB	-	-	-	-	-	-	-	-
Delta pilot pwr.	No	288	1.7	4.1	4.7	288	1.7	4.1	4.1
Delta pilot pwr.	0.5 Hz, 3 dB	144	1.77	-1.5	0.6	146	1.78	-1.5	1.3
Delta pilot pwr.	1.0 Hz, 3 dB	127	1.5	-2.4	-1.9	144	1.72	-1.5	0.8
Fixed E_c/I_o	No	223	1.83	2.0	10.0	231	1.86	2.0	10.3
Fixed E_c/I_o	0.5 Hz, 3 dB	129	1.88	-2.4	7.0	140	1.91	-2.4	8.7
Fixed E_c/I_o	1.0 Hz, 3 dB	107	1.86	-3.0	4.5	128	1.91	-3.0	9.5
Delta E_c/I_o	No	256	1.68	3.1	2.7	256	1.68	3.1	2.7
Delta E_c/I_o	0.5 Hz, 3 dB	$\approx$150	1.65	-1.2	-1.7	$\approx$150	1.65	-1.2	-1.7
Delta E_c/I_o	1.0 Hz, 3 dB	144	1.65	-1.1	-1.6	144	1.65	-1.1	-1.6

Table 5.3: Maximum number of mobile users that can be supported by the network, for different soft handover metrics/algorithms whilst meeting the preset quality constraints. The mean number of base stations in the Active Base station Set (ABS) is also presented, along with the mean mobile and mean base station transmit powers.

ability to perform well under both shadow fading conditions evaluated, whilst utilising the same soft handover thresholds. Since the constraining factor of these network capacity results was the probability of a dropped call, P_{FT}, which was the same for both scenarios, further network capacity results were only shown for the conservative scenario.

5.4.5 Performance of Adaptive Antenna Arrays in a High Data Rate Pedestrian Environment

In our previous investigations we endeavoured to identify the soft handover algorithm, which supports the greatest number of users, at the best call quality, regardless of the propagation conditions. In this section we study the impact of adaptive antenna arrays on the network's performance. The investigations were conducted using the relative E_c/I_o based soft handover algorithm in conjunction with T_{acc}=-10 dB and T_{drop}=-18 dB, using a spreading factor of 16. Given that the chip rate of UTRA is 3.84 Mchips/sec, this spreading factor corresponds to a channel data rate of $3.84\times10^6/16$=240 kbps. Applying $1/2$ rate error correction coding would result in an effective data throughput of 120 kbps, whereas utilising a $2/3$ rate error correction code would provide a useful throughput of 160 kps. As in the previous simulations, a cell radius of 150 m was assumed and a pedestrian walking velocity of 3 mph was used. In our previous results investigations employing adaptive antenna arrays at the base station and using a FDMA/TDMA based network, as in Chapter 4, we observed quite significant performance gains as a direct result of the interference rejection capabilities of the adaptive antenna arrays invoked. Since the CDMA based network considered here has a frequency reuse of 1, the levels of co-channel interference are significantly higher, and hence the adaptive antennas may be able to null the interference more effectively. However, the greater number of interference sources may limit the achievable interference rejection.

Network performance results were obtained using two and four element adaptive antenna

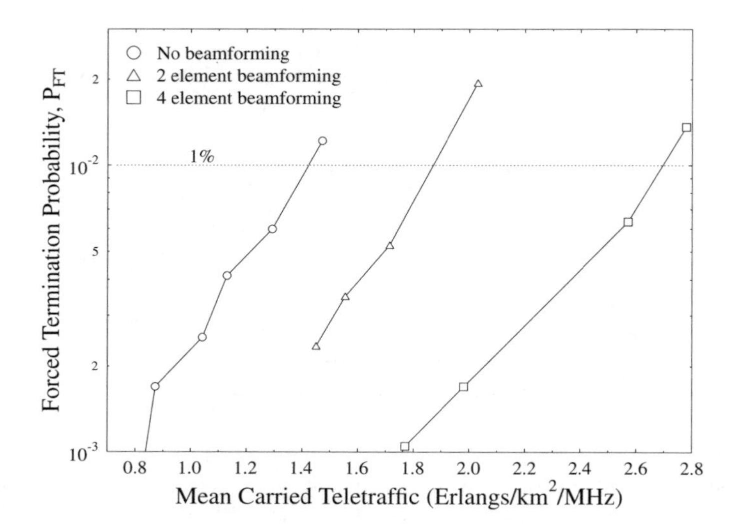

Figure 5.41: Call dropping probability versus mean carried traffic of a CDMA based cellular network using **relative received** E_c/I_o based soft handover thresholds **with and without beamforming and without shadowing** for SF=16.

arrays, both in the absence of shadow fading, and in the presence of 0.5 Hz and 1.0 Hz frequency shadow fading exhibiting a standard deviation of 3 dB. The adaptive beamforming algorithm used was the Sample Matrix Inversion (SMI) algorithm, as described in Chapter 3 and used in the FDMA/TDMA network simulations of Chapter 4. The specific adaptive beamforming implementation used in the CDMA based network was identical to that used in the FDMA/TDMA network simulations. Briefly, one of the eight possible 8-bit BPSK reference signals was used to identify the desired user, and the remaining interfering users were assigned the other seven 8-bit reference signals. The received signal's autocorrelation matrix was then calculated, and from the knowledge of the desired user's reference signal, the receiver's optimal antenna array weights were determined with the aid of the SMI algorithm. The reader is referred to Section 4.6.1 for further details. Since this implementation of the algorithm only calculated the receiver's antenna array weights, i.e. the antenna arrays weights used by the base station in the uplink, these weights may not be suitable for use in the downlink, when independent up/downlink shadow fading is experienced. Hence, further investigations were conducted, where the uplink and downlink channels were identical, in order to determine the potential performance gain that may be achieved by separately calculating the antenna array weights to be used in the downlink. The antenna array weights were re-calculated for every power control step, i.e. 15 times per UTRA data frame, due to the potential significant changes in terms of the desired signal and interference powers that may occur during one UTRA frame as a result of the possible 15 dB change in power transmitted by each user.

Figure 5.41 shows the significant reduction in the probability of a dropped call, i.e. the probability of forced termination P_{FT}, achieved by employing adaptive antenna arrays in a

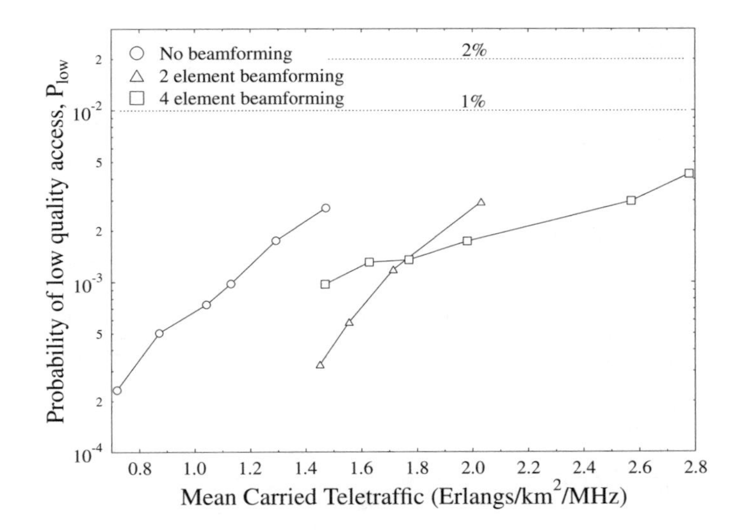

Figure 5.42: Probability of low quality access versus mean carried traffic of a CDMA based cellular network using **relative received** E_c/I_o based soft handover thresholds **with and without beamforming and without shadowing** for SF=16.

non-shadowed propagation environment. The figure has demonstrated that, even with only two antenna elements, the adaptive antenna arrays have considerably reduced the levels of co-channel interference, leading to a reduced call dropping probability. This has been achieved in spite of the numerous sources of co-channel interference resulting from the frequency reuse factor of one, which was remarkable in the light of the limited number of degrees of freedom of the two element array. Without employing antenna arrays at the base stations the network capacity was limited to 256 users, or to a teletraffic load of approximately 1.4 Erlangs/km^2/MHz. However, with the advent of two element adaptive antenna arrays at the base stations the number of users supported by the network rose by 27% to 325 users, or almost 1.9 Erlangs/km^2/MHz. Replacing the two element adaptive antenna arrays with four element arrays led to a further rise of 48%, or 88% with respect to the capacity of the network using no antenna arrays. This is associated with a network capacity of 480 users, or 2.75 Erlangs/km^2/MHz. A summary of the network capacities achieved under different conditions is given in Table 5.4.

The probability of low quality outage, presented in Figure 5.42 also exhibited a substantial improvement with the advent of two element adaptive antenna arrays. However, the performance gains obtained when invoking four element adaptive antenna arrays were more involved. It can be seen from the figure that higher traffic loads were carried with at a sufficiently low probability of a low quality occurring, and at higher traffic loads the probability of a low quality access was lower than that achieved using a two element array. However, at lower traffic loads the performance was worse than that obtained when using two element arrays, and the gradient of the performance curve was significantly lower. Further in-depth analysis of the results suggested that the vast majority of the low quality outages were oc-

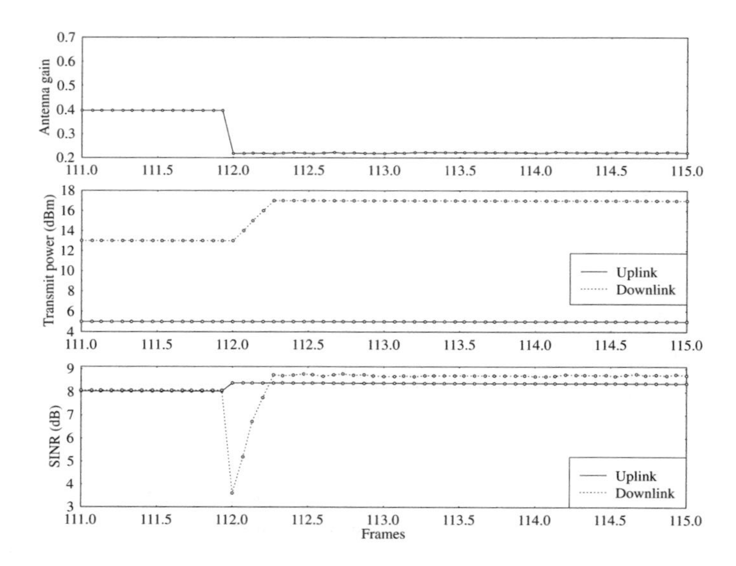

Figure 5.43: The changes in the antenna array gain, versus time, in the direction of the desired user, the up and downlink transmission powers, and the up- and downlink received SINRs, when a new call starts using four element adaptive antenna arrays without shadowing in conjunction with the **original power ramping** algorithm and SF=16.

curring when new calls started. When a user decided to commence communications with the base station, the current interference level was measured, and the target transmission power was determined in order to reach the target SINR necessary for reliable communications. However, in order to avoid disrupting existing calls the transmission power was ramped up slowly, until the target SINR was reached. A network using no adaptive antenna arrays, i.e. employing omnidirectional antennas, can be viewed as offering equal gain to all users of the network, which we assumed to be 1.0, or 0 dB. Thus, when a new call is initiated, the level of interference rises gradually, and the power control algorithm ensures that the existing users compensate for the increased level of co-channel interference by increasing their transmission power. In a network using adaptive antenna arrays, the adaptive antenna arrays are used to null the sources of interference, and in doing so the array may reduce the antenna gain in the direction of the desired user, in order to maximise the SINR. Hence a user starting a new call, even if it has low transmission power, can alter the antenna array's response, and thus the antenna gain experienced by the existing users. This phenomenon is more marked when using four element arrays since their directivity, and thus sensitivity to interfering signals, is greater.

Figure 5.43 illustrates this phenomenon, where another user starts a new call at frame 112 suddenly reducing the antenna gain in the direction of the desired user from 0.4 to just above 0.2, a drop of 3 dB. As can be seen from the figure, the downlink SINR falls sharply below the low quality outage threshold of 7.0 dB, resulting in several consecutive outages, until the downlink transmission power is increased sufficiently. The impact of reducing the initial transmission power, in order to ensure that the power ramping takes place more gently,

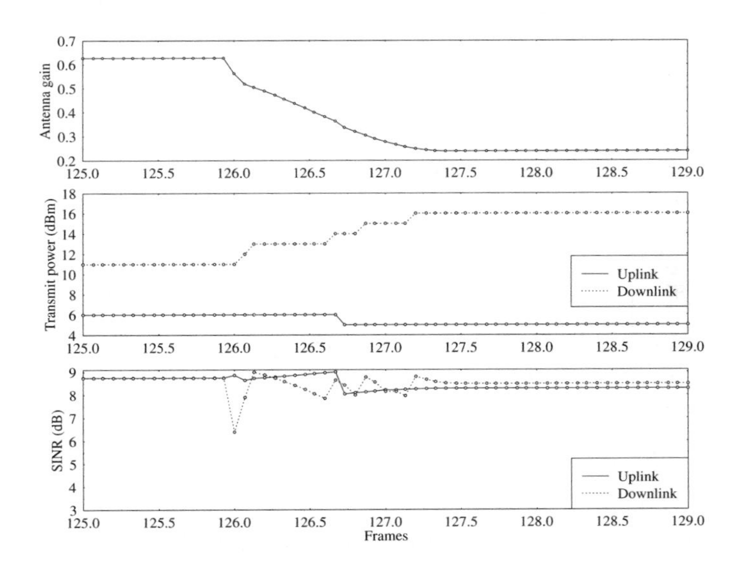

Figure 5.44: The changes in the antenna array gain, versus time, in the direction of the desired user, the up and downlink transmission powers, and the up and downlink received SINRs, when a new call starts using four element adaptive antenna arrays without shadowing in conjunction with a **slower power ramping** algorithm and SF=16.

is depicted in Figure 5.44. In this figure it can be seen that the antenna gain falls much more gently, over a prolonged period of time, thus reducing the number of low quality outages, as the downlink transmission power is increased in an effort to compensate for the lower antenna gain. It is of interest to note how the received SINR varies as the antenna gain and the power control algorithm interact, in order to maintain the target SINR.

Even though the employment of adaptive antenna arrays can result in the attenuation of the desired signal, this is performed in order to maximise the received SINR, and thus the levels of interference are attenuated more strongly, ultimately leading to the reduction of the mean transmission power, as emphasised by Figure 5.45. This figure clearly shows the lower levels of transmission power, required in order to maintain an acceptable performance, whilst using adaptive antenna arrays at the base stations. A reduction of 3 dB in the mean mobile transmission power was achieved by invoking two element antenna arrays, and a further reduction of 1.5 dB resulted from using four element arrays. These power budget savings were obtained in conjunction with reduced levels of co-channel interference, leading to superior call quality, as illustrated in Figures 5.41 and 5.42. A greater performance advantage was evident in the uplink scenario, suggesting that the selective base station diversity techniques employed in the uplink are amenable to amalgamation with adaptive antenna arrays. In contrast, the maximum ratio combining performed at the mobile inherently reduces the impact of co-channel interference, and hence benefits to a lesser extent from the employment of adaptive antenna arrays.

The impact of adaptive antenna arrays in a propagation environment subjected to shadow fading was then investigated. The associated call dropping performance is shown in Figure

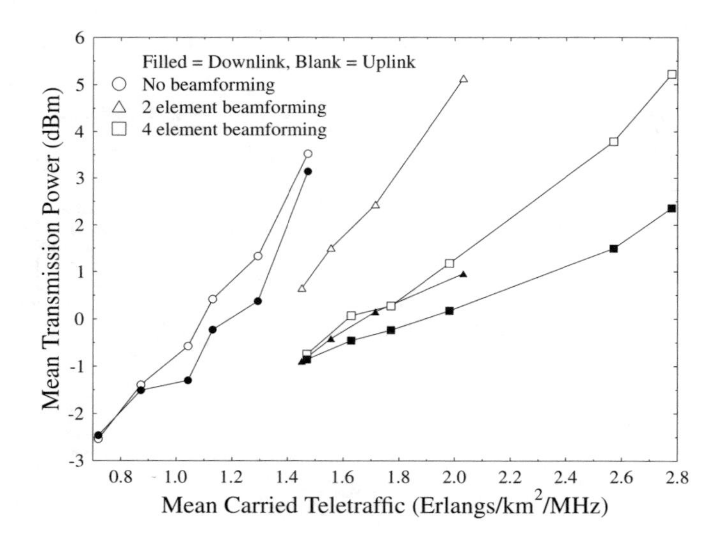

Figure 5.45: Mean transmission power versus mean carried traffic of a CDMA based cellular network using **relative received** E_c/I_o based soft handover thresholds **with and without beamforming and without shadowing** for SF=16.

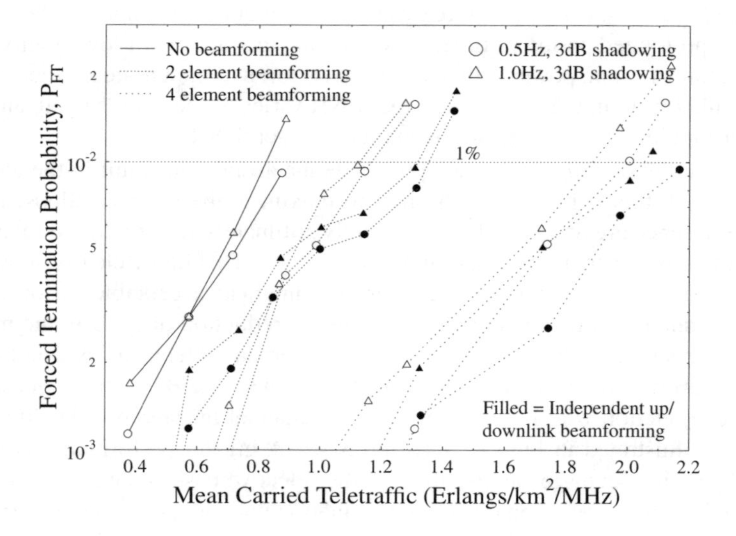

Figure 5.46: Call dropping probability versus mean carried traffic of a CDMA based cellular network using **relative received** E_c/I_o based soft handover thresholds **with and without beamforming and with shadowing having a standard deviation of 3 dB** for SF=16.

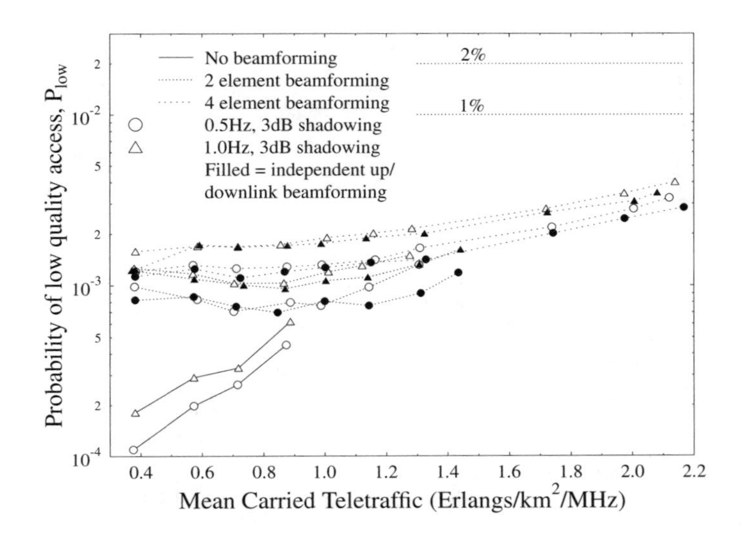

Figure 5.47: Probability of low quality access versus mean carried traffic of a CDMA based cellular network using **relative received** E_c/I_o based soft handover thresholds **with and without beamforming and with shadowing having a standard deviation of 3 dB** for SF=16.

5.46. This figure illustrates the substantial network capacity gains achieved with the aid of both two and four element adaptive antenna arrays under shadow fading propagation conditions. Simulations were conducted in conjunction with log-normal shadow fading having a standard deviation of 3 dB, and maximum shadowing frequencies of both 0.5 Hz and 1.0 Hz. As expected the network capacity was reduced at the faster fading frequency. The effect of performing independent up- and down-link beamforming, as opposed to using the base station's receive antenna array weights in the downlink was also studied, and a small, but not insignificant call dropping probability reduction can be seen in the Figure 5.46. The network supported just over 150 users, and 144 users, when subjected to 0.5 Hz and 1.0 Hz frequency shadow fading, respectively. With the application of two element adaptive antenna arrays, re-using the base station's uplink receiver weights on the downlink, these capacities increased by 35% and 40%, to 203 users and 201 users. Performing independent up- and down-link beamforming resulted in a mean further increase of 13% in the network capacity. The implementation of four element adaptive antenna arrays led to a network capacity of 349 users at a 0.5 Hz shadowing frequency, and 333 users at a 1.0 Hz shadowing frequency. This corresponded to relative gains of 133% and 131% over the capacity provided without beamforming. Invoking independent up- and down-link beamforming gave another boost of 7% and 10% to network capacity for 0.5 Hz and 1.0 Hz frequency shadowing environments, respectively, giving final network capacities of just over 375 users and 365 users.

Similar trends were observed regarding the probability of low quality outage to those found in the non-shadowing scenarios. However, the trend was much more prevalent under shadowing, due to greater variation of the received signal strengths, as a result of the shadow fading, as shown in Figure 5.47. The figure indicates that the trend is also evident, when

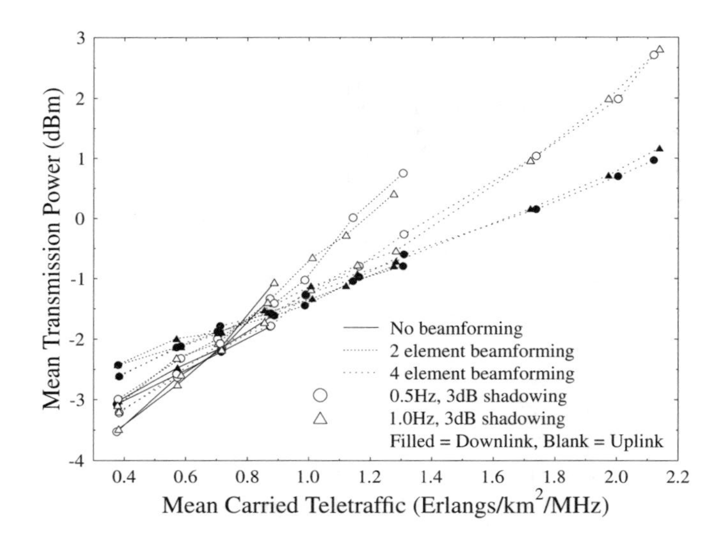

Figure 5.48: Mean transmission power versus mean carried traffic of a CDMA based cellular network using **relative received** E_c/I_o based soft handover thresholds **with and without beamforming and shadowing having a standard deviation of 3 dB** for SF=16.

using two element adaptive antenna arrays in conjunction with shadow fading. As expected, the performance deteriorated as the number of antenna elements increased, and when the maximum shadow fading frequency was increased from 0.5 Hz to 1.0 Hz. It should be noted, however that the probability of low quality access always remained below the 1% constraint of the conservative scenario, and the call dropping probability was considerably reduced by the adaptive antenna arrays.

The mean transmission power performance is depicted in Figure 5.48, suggesting that as for the non-shadowing scenario of Figure 5.45, the number of antenna elements had only a limited impact on the base stations' transmission power, although there was some reduction in the mobile stations' mean transmission power. The mean transmission powers required when using independent up- and down-link beamforming are not explicitly shown, but were slightly less than those presented here, with a mean reduction of about 0.4 dB.

A summary of the maximum network capacities of the networks considered in this section both with and without shadowing, employing beamforming using two and four element arrays is given in Table 5.4, along with the teletraffic carried and the mean mobile and base station transmission powers required.

The lower bounds of the maximum network capacities obtained under identical scenarios in conjunction with a spreading factor of 256, are also presented in Table 5.5, leading to a bit rate of 15 kbps, which is suitable for use by speech-rate users. The network capacity calculations were performed by scaling the number of users supported, as presented in Table 5.4, by the ratio of their spreading factors, i.e. 256/16=16. Further interesting user capacity figures can be inferred for a variety of target bit rates by comparing Tables 5.4, 5.5, 5.7 and 5.8 and applying the appropriate spreading factor related scaling mentioned in the context of estimating the number of 15 kbps speech users supported.

Shadowing	Beamforming:	independent up/downlink	Conservative scenario, P_{FT}=1%, P_{low}=1% Users	Traffic (Erlangs /km^2/MHz)	Power (dBm) MS	BS
No	No	-	256	1.42	3.1	2.7
No	2 elements	-	325	1.87	3.75	0.55
No	4 elements	-	480	2.75	4.55	1.85
0.5 Hz, 3 dB	No	-	$\approx$150	0.87	-1.2	-1.7
0.5 Hz, 3 dB	2 elements	No	203	1.16	0.1	-1.1
0.5 Hz, 3 dB	4 elements	No	349	2.0	2.0	0.65
0.5 Hz, 3 dB	2 elements	Yes	233	1.35	0.2	-0.8
0.5 Hz, 3 dB	4 elements	Yes	$\approx$375	2.2	2.15	0.85
1.0 Hz, 3 dB	No	-	144	0.82	-1.1	-1.6
1.0 Hz, 3 dB	2 elements	No	201	1.12	-0.3	-1.1
1.0 Hz, 3 dB	4 elements	No	333	1.88	1.6	0.5
1.0 Hz, 3 dB	2 elements	Yes	225	1.31	0.1	-0.9
1.0 Hz, 3 dB	4 elements	Yes	365	2.05	1.65	0.6

Table 5.4: Maximum mean carried traffic and maximum number of mobile users that can be supported by the network, whilst meeting the conservative quality constraints. The carried traffic is expressed in terms of normalised Erlangs (Erlang/km^2/MHz) for the network described in Table 5.2 both **with and without beamforming (as well as with and without independent up/down-link beamforming), and also with and without shadow fading having a standard deviation of 3 dB** for SF=16.

5.4.6 Performance of Adaptive Antenna Arrays and Adaptive Modulation in a High Data Rate Pedestrian Environment

In this section we build upon the results presented in the previous section by applying Adaptive Quadrature Amplitude Modulation (AQAM) techniques. The various scenarios and channel conditions investigated were identical to those of the previous section, except for the application of AQAM. Since in the previous section an increased network capacity was achieved due to using independent up- and down-link beamforming, this procedure was invoked in these simulations. AQAM involves the selection of the appropriate modulation mode in order to maximise the achievable data throughput over a channel, whilst minimising the Bit Error Ratio (BER). More explicitly, the philosophy behind adaptive modulation is the most appropriate selection of a modulation mode according to the instantaneous radio channel quality experienced [12, 13]. Therefore, if the SINR of the channel is high, then a high-order modulation mode may be employed, thus exploiting the temporal fluctuation of the radio channel's quality. Similarly, if the channel is of low quality, exhibiting a low SINR, a high-order modulation mode would result in an unacceptably high BER or FER, and hence a more robust, but lower throughput modulation mode would be employed. Therefore, adaptive modulation combats the effects of time-variant channel quality, while also attempting to maximise the achieved data throughput, and maintaining a given BER or FER. In the investigations conducted, the modulation modes of the up and downlink were determined independently, thus taking advantage of the lower levels of co-channel interference on the uplink, or of the potentially greater transmit power of the base stations.

Shadowing	Beamforming:	independent up/down-link	Users when SF=256	Traffic (Erlangs /km^2/MHz)
No	No	-	4096	22.7
No	2 elements	-	5200	29.9
No	4 elements	-	7680	44.0
0.5 Hz, 3 dB	No	-	2400	13.9
0.5 Hz, 3 dB	2 elements	No	3248	18.6
0.5 Hz, 3 dB	4 elements	No	5584	32.0
0.5 Hz, 3 dB	2 elements	Yes	3728	21.6
0.5 Hz, 3 dB	4 elements	Yes	6000	35.2
1.0 Hz, 3 dB	No	-	2304	13.1
1.0 Hz, 3 dB	2 elements	No	3216	17.9
1.0 Hz, 3 dB	4 elements	No	5328	30.1
1.0 Hz, 3 dB	2 elements	Yes	3600	21.0
1.0 Hz, 3 dB	4 elements	Yes	5840	32.8

Table 5.5: A lower bound estimate of the maximum mean traffic and the maximum number of mobile **speech-rate** users that can be supported by the network, whilst meeting the **conservative quality constraints**. The carried traffic is expressed in terms of normalised Erlangs (Erlang/km^2/MHz) for the network described in Table 5.2 both **with and without beamforming (as well as with and without independent up/down-link beamforming), and also with and without shadow fading having a standard deviation of 3 dB** for SF=256. The number of users supported in conjunction with a spreading factor of 256 was calculated by multiplying the capacities obtained in Table 5.4 by 256/16=16.

The particular implementation of AQAM used in these investigations is illustrated in Figure 5.49. This figure describes the algorithm in the context of the downlink, but the same implementation was used also in the uplink. The first step in the process was to establish the current modulation mode. If the user was invoking 16-QAM and the SINR was found to be below the Low Quality (LQ) outage SINR threshold after the completion of the power control iterations, then the modulation mode for the next data frame was 4-QAM. Alternatively, if the SINR was above the LQ outage SINR threshold, but any of the base stations in the ABS were using a transmit power within 15 dB of the maximum transmit power - which is the maximum possible power change range during a 15-slot UTRA frame - then the 4-QAM modulation mode was selected. This 'headroom' was introduced in order to provide a measure of protection, since if the interference conditions degrade, then at least 15 dB of increased transmit power would be available in order to mitigate the consequences of the SINR reduction experienced.

A similar procedure was invoked when switching to other legitimate AQAM modes from the 4-QAM mode. If the SINR was below the 4-QAM target SINR and any one of the base stations in the ABS was within 15 dB (the maximum possible power change during a 15-slot UTRA data frame) of the maximum transmit power, then the BPSK modulation mode was employed for the next data frame. However, if the SINR exceeded the 4-QAM target SINR and there would be 15 dB of headroom in the transmit power budget in excess of the extra transmit power required for switching from 4-QAM to 16-QAM, then the 16-QAM modulation mode was invoked.

And finally, when in the BPSK mode, the 4-QAM modulation mode was selected if the SINR exceeded the BPSK target SINR, and the transmit power of any of the base stations in the ABS was less than the power required to transmit reliably using 4-QAM, while being at least 15 dB below the maximum transmit power. The algorithm was activated at the end of each 15-slot UTRA data frame, after the power control algorithm had performed its 15 iterations per data frame, and thus the AQAM mode selection was performed on a UTRA transmission frame-by-frame basis. When changing from a lower-order modulation to a higher-order modulation mode, the lower-order mode was retained for an extra frame in order to ramp up the transmit power to the required level, as shown in Figure 5.50(a). Conversely, when changing from a higher-order modulation mode to a lower-order modulation mode, the lower-order modulation mode was employed whilst ramping the power down, in order to avoid excessive outages in the higher-order modulation mode due to the reduction of the transmit power, as illustrated in Figure 5.50(b).

Table 5.6 gives the BPSK, 4-QAM and 16-QAM SINR thresholds used in the simulations. The BPSK SINR thresholds were 4 dB lower than those necessary when using 4-QAM, while the 16-QAM SINR thresholds were 5.5 dB higher [367]. In other words, in moving from the BPSK modulation mode to the 4-QAM modulation mode, the target SINR, low quality outage SINR and outage SINR all increased by 4 dB. When switching to the 16-QAM mode from the 4-QAM mode, the SINR thresholds increased by 5.5 dB. However, setting the BPSK to 4-QAM and the 4-QAM to 16-QAM mode switching thresholds to a value 7 dB higher than the SINR required for maintaining the target BER/FER was necessary in order to prevent excessive outages due to sudden dramatic channel-induced variations in the SINR levels.

SINR Threshold	BPSK	4-QAM	16-QAM
Outage SINR	2.6 dB	6.6 dB	12.1 dB
Low Quality Outage SINR	3.0 dB	7.0 dB	12.5 dB
Target SINR	4.0 dB	8.0 dB	13.5 dB

Table 5.6: The target SINR, low quality outage SINR and outage SINR thresholds used for the BPSK, 4-QAM and 16-QAM modulation modes of the adaptive modem.

Performance results were obtained both with and without beamforming in a log-normal shadow fading environment, at maximum fading frequencies of 0.5 Hz and 1.0 Hz, and a standard deviation of 3 dB. A pedestrian velocity of 3 mph, a cell radius of 150 m and a spreading factor of 16 were used, as in our previous investigations.

Figure 5.51 shows the significant reduction in the probability of a dropped call, achieved by employing adaptive antenna arrays in conjunction with adaptive modulation in a log-normal shadow faded environment. The figure demonstrates that, even with the aid of a two element adaptive antenna array and its limited degrees of freedom, a substantial call dropping probability reduction was achieved. The performance benefit of increasing the array's degrees of freedom, achieved by increasing the number of antenna elements, becomes explicit from the figure, resulting in a further call dropping probability reduction. Simulations were conducted in conjunction with log-normal shadow fading having a standard deviation of 3 dB, and maximum shadowing frequencies of 0.5 Hz and 1.0 Hz. As expected, the call dropping probability was generally higher at the faster fading frequency, as demonstrated by Figure 5.51. The network was found to support 223 users, corresponding to a traffic load of 1.27

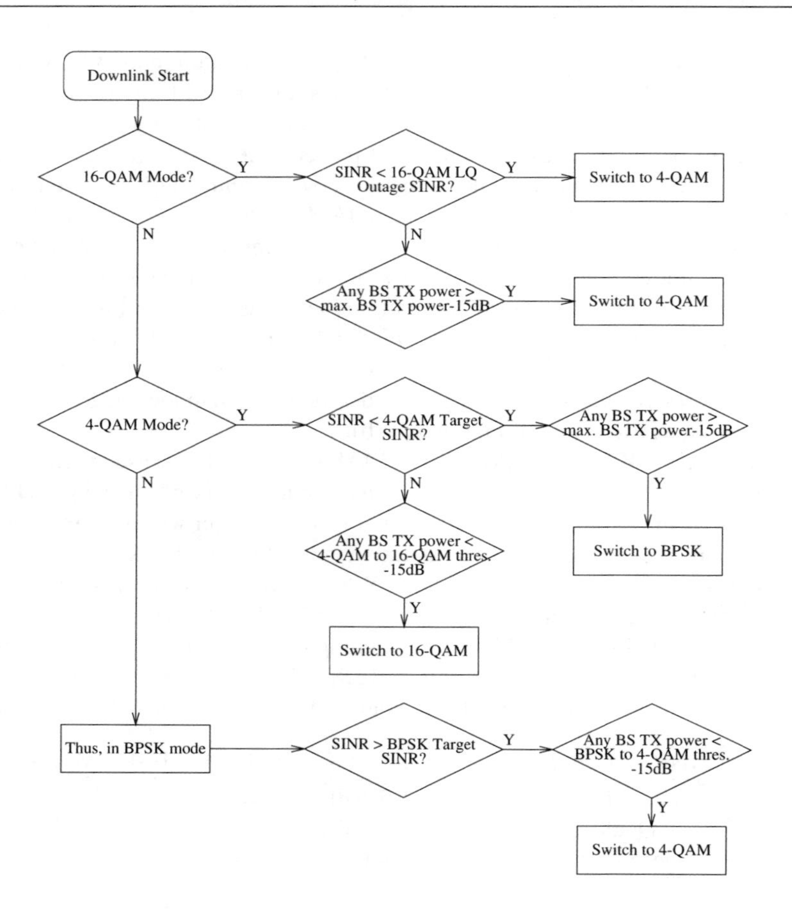

Figure 5.49: The AQAM mode switching algorithm used in the downlink of the CDMA based cellular network.

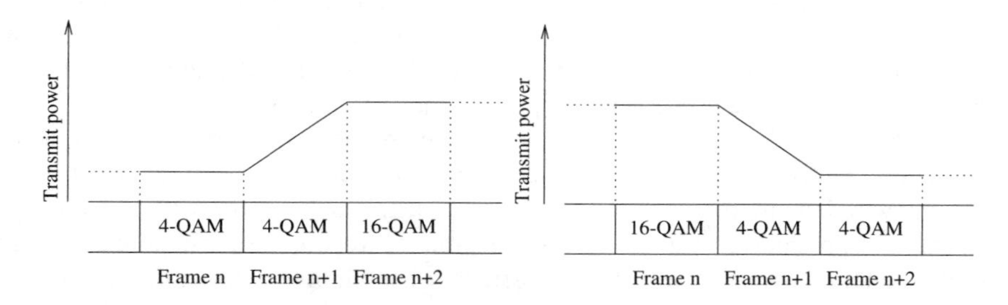

(a) Ramping up the transmit power whilst remaining in the lower order modulation mode.

(b) Ramping down the transmit power whilst switching to the lower order modulation mode.

Figure 5.50: Power ramping requirements whilst switching modulation modes.

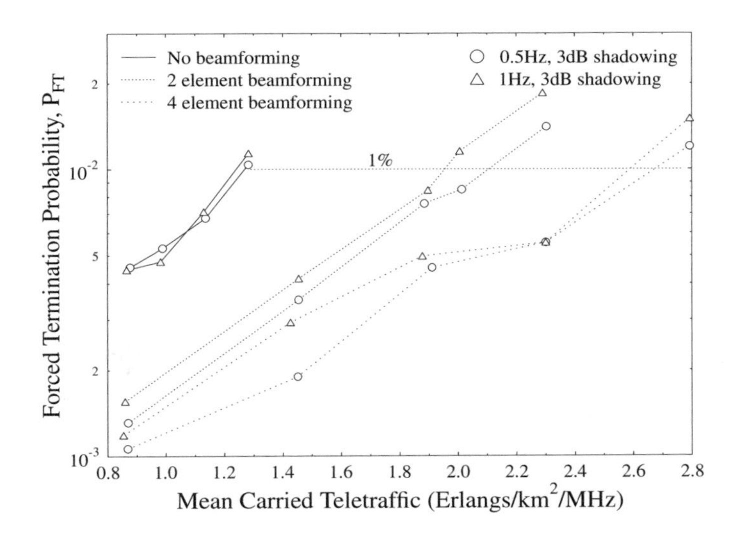

Figure 5.51: Call dropping probability versus mean carried traffic of a CDMA based cellular network using **relative received** E_c/I_o based soft handover thresholds both **with and without beamforming in conjunction with AQAM as well as with shadowing having a standard deviation of 3 dB** for SF=16. See Figure 5.46 for corresponding results without adaptive modulation.

Erlang/km^2/MHz, when subjected to 0.5 Hz frequency shadow fading. The capacity of the network was reduced to 218 users, or 1.24 Erlang/km^2/MHz, upon increasing the maximum shadow fading frequency to 1.0 Hz. On employing two element adaptive antenna arrays, the network capacity increased by 64% to 366 users, or to an equivalent traffic load of 2.11 Erlang/km^2/MHz when subjected to 0.5 Hz frequency shadow fading. When the maximum shadow fading frequency was raised to 1.0 Hz, the number of users supported by the network was 341 users, or 1.98 Erlang/km^2/MHz, representing an increase of 56% in comparison to the network without adaptive antenna arrays. Increasing the number of antenna elements to four, whilst imposing shadow fading with a maximum frequency of 0.5 Hz, resulted in a network capacity of 2.68 Erlang/km^2/MHz or 476 users, corresponding to a gain of an extra 30% with respect to the network employing two element arrays, and of 113% in comparison to the network employing no adaptive antenna arrays. In conjunction with a maximum shadow fading frequency of 1.0 Hz the network capacity was 460 users or 2.59 Erlang/km^2/MHz, which represented an increase of 35% with respect to the network invoking two element antenna arrays, or 111% relative to the identical network without adaptive antenna arrays.

The probability of low quality outage, presented in Figure 5.52, did not benefit from the application of adaptive antenna arrays, or from the employment of adaptive modulation. Figure 5.47 depicts the probability of low quality outage without adaptive modulation, and upon comparing these results to those obtained in conjunction with adaptive modulation shown in Figure 5.52, the performance degradation due to adaptive modulation can be explicitly seen. However, the increase in the probability of low quality access can be attributed to the em-

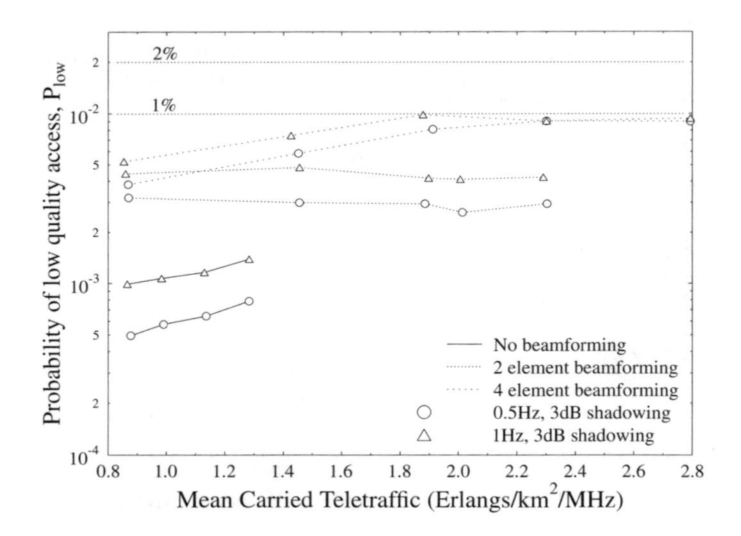

Figure 5.52: Probability of low quality access versus mean carried traffic of a CDMA based cellular network using **relative received** E_c/I_o based soft handover thresholds both **with and without beamforming in conjunction with AQAM as well as with shadowing having a standard deviation of 3 dB** for SF=16. See Figure 5.47 for corresponding results without adaptive modulation.

ployment of less robust, but higher throughput, higher-order modulation modes invoked by the adaptive modulation scheme. Hence, under given propagation conditions and using the fixed 4-QAM modulation mode a low quality outage may not occur, yet when using adaptive modulation and a higher order modulation mode, the same propagation conditions may inflict a low quality outage. This phenomenon is further exacerbated by the adaptive antenna arrays, as described in Section 5.4.5, where the addition of a new source of interference, constituted by a user initiating a new call, results in an abrupt change in the gain of the antenna in the direction of the desired user. This in turn leads to low quality outages, which are more likely to occur for prolonged periods of time when using a higher order modulation mode. Again, increasing the number of antenna elements from two to four results in an increased probability of a low quality outage due to the sharper antenna directivity. This results in a higher sensitivity to changes in the interference incident upon it.

The mean transmission power versus teletraffic performance is depicted in Figure 5.53, suggesting that the mean uplink transmission power was always significantly below the mean downlink transmission power, which can be attributed to the pilot power interference encountered by the mobiles in the downlink. This explanation can be confirmed by examining Figure 5.54, which demonstrates that the mean modem throughput in the downlink, without adaptive antenna arrays, was lower than that in the uplink even in conjunction with increased downlink transmission power. Invoking adaptive antenna arrays at the base stations reduced the mean uplink transmission power required in order to meet the service quality targets of the network. The attainable downlink power reduction increased as the number of antenna

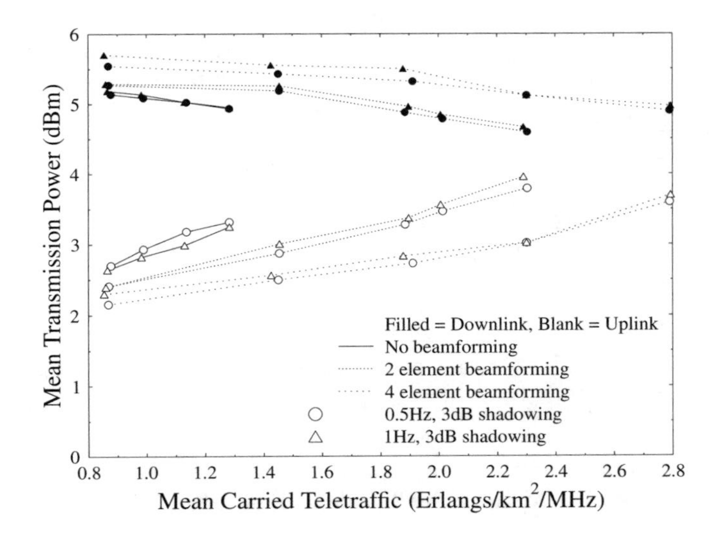

Figure 5.53: Mean transmission power versus mean carried traffic of a CDMA based cellular network using **relative received** E_c/I_o based soft handover thresholds both **with and without beamforming in conjunction with AQAM as well as with shadowing having a standard deviation of 3 dB** for SF=16. See Figure 5.48 for corresponding results without adaptive modulation.

array elements increased, as a result of the superior interference rejection achieved with the aid of a higher number of array elements. A further advantage of employing a larger number of antenna array elements was the associated increase in the mean uplink modem throughput, which became more significant at higher traffic loads. In the downlink scenario, however, increasing the number of adaptive antenna array elements led to an increased mean downlink transmission power, albeit with a substantially improved mean downlink modem throughput. This suggests that there was some interaction between the adaptive antenna arrays, the adaptive modulation mode switching algorithm and the maximal ratio combining performed at the mobiles. In contrast, simple switched diversity was performed by the base stations on the uplink, thus avoiding such a situation. However, the increase in the mean downlink transmission power resulted in a much more substantial increase in the mean downlink modem throughput, especially with the advent of the four element antenna arrays, which exhibited an approximately 0.5 BPS throughput gain over the two element arrays for identical high traffic loads which can be seen in Figure 5.54.

A summary of the maximum user capacities of the networks considered in this section in conjunction with log-normal shadowing having a standard deviation of 3 dB, with and without employing beamforming using two and four element arrays is given in Table 5.7. The teletraffic carried the mean mobile and base station transmission powers required, and the mean up- and down-link modem data throughputs achieved are also shown in Table 5.7. Similarly, the lower bounds of the maximum network capacities obtained under identical scenarios in conjunction with a spreading factor of 256, leading to a bit rate of 15 kbps,

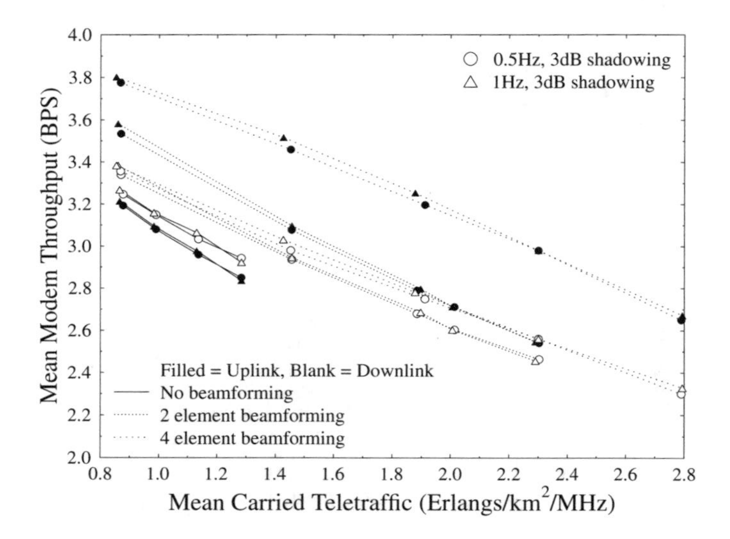

Figure 5.54: Mean modem throughput versus mean carried traffic of a CDMA based cellular network using **relative received** E_c/I_o based soft handover thresholds both **with and without beamforming in conjunction with AQAM as well as with shadowing having a standard deviation of 3 dB** for SF=16.

suitable for speech-rate users are presented in Table 5.8. The network capacity calculations were performed by scaling the number of users supported, as presented in Table 5.7, by the ratio of their spreading factors, i.e. by 256/16=16.

Shadowing	Beamforming	Users	Traffic (Erlangs /km²/MHz)	Power (dBm) MS	BS	Throughput (BPS) Uplink	Downlink
		Conservative scenario					
0.5 Hz, 3 dB	No	223	1.27	3.25	4.95	2.86	2.95
0.5 Hz, 3 dB	2 elements	366	2.11	3.55	4.7	2.56	2.66
0.5 Hz, 3 dB	4 elements	476	2.68	3.4	5.0	2.35	2.72
1.0 Hz, 3 dB	No	218	1.24	3.3	4.95	2.87	2.96
1.0 Hz, 3 dB	2 elements	341	1.98	3.5	4.9	2.62	2.73
1.0 Hz, 3 dB	4 elements	460	2.59	3.5	4.95	2.4	2.8

Table 5.7: Maximum mean carried traffic and maximum number of mobile users that can be supported by the network, whilst meeting the conservative quality constraints. The carried traffic is expressed in terms of normalised Erlangs (Erlang/km^2/MHz), for the network described in Table 5.2 both **with and without beamforming (using independent up/down-link beamforming), in conjunction with shadow fading having a standard deviation of 3 dB, whilst employing adaptive modulation techniques** for SF=16.

Shadowing	Beamforming	Conservative scenario Users	Traffic (Erlangs /km^2/MHz)
0.5 Hz, 3 dB	No	3568	20.3
0.5 Hz, 3 dB	2 elements	5856	33.8
0.5 Hz, 3 dB	4 elements	7616	42.9
1.0 Hz, 3 dB	No	3488	19.8
1.0 Hz, 3 dB	2 elements	5456	31.7
1.0 Hz, 3 dB	4 elements	7360	41.4

Table 5.8: A lower bound estimate of the maximum mean carried traffic and maximum number of mobile **speech-rate** users that can be supported by the network, whilst meeting the conservative quality constraints. The carried traffic is expressed in terms of normalised Erlangs (Erlang/km^2/MHz), for the network described in Table 5.2 both **with and without beamforming (using independent up/down-link beamforming), in conjunction with shadow fading having a standard deviation of 3 dB, whilst employing adaptive modulation techniques** for SF=256. The number of users supported in conjunction with a spreading factor of 256 was calculated by multiplying the capacities obtained in Table 5.7 by 256/16=16.

5.5 Summary and Conclusions

We commenced this chapter with a brief overview of the background behind the 3G UTRA standard. This was followed in Sections 5.2 and 5.3 by an introduction to CDMA and the techniques invoked in the UTRA standard.

Network capacity studies were then conducted in Section 5.4, which evaluated the performance of four different soft handover algorithms in the context of both non-shadowed and log-normal shadow faded propagation environments. The algorithm using relative received pilot-to-interference ratio measurements at the mobile, in order to determine the most suitable base stations for soft handover, was found to offer the highest network capacity when subjected to shadow fading propagation conditions. Hence, this algorithm and its associated parameters were selected for use in our further investigations. The impact of adaptive antenna arrays upon the network capacity was then considered in Section 5.4.5 in both non-shadowed and log-normal shadow faded propagation environments. Considerable network capacity gains were achieved, employing both two and four element adaptive antenna arrays. This work was then extended in Section 5.4.6 by the application of adaptive modulation techniques in conjunction with the previously studied adaptive antenna arrays in a log-normal shadow faded propagation environment, which elicited further significant network capacity gains.

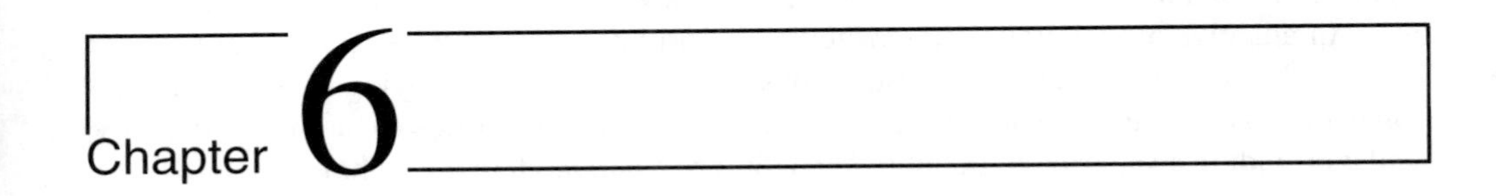

Conclusions and Further Research

6.1 Summary and Conclusions

In this book we have discussed the performance implications of adaptive antenna arrays and adaptive modulation techniques in both FDMA/TDMA and CDMA cellular mobile communications networks.

In Chapter 3 we investigated antenna arrays and adaptive beamforming algorithms. We commenced, in Section 3.2.2, by considering the possible applications of antenna arrays and their related benefits. The signal model used was then described in Section 3.2.3 and a rudimentary example of how beamforming operates was presented. Section 3.3 highlighted the process of adaptive beamforming using several different temporal reference techniques, along with the approaches used in spatial reference techniques. The challenges that must be overcome before beamforming for the downlink becomes feasible were also discussed in Section 3.3.5. Results were presented showing how the SMI, ULMS and NLMS beamforming algorithms behaved for a two element adaptive antenna in conjunction with varying eigenvalue spread and reference signal length. The SMI algorithm was shown to converge rapidly, irrespective of the eigenvalue spread. The performance of the ULMS beamformer was shown to be highly dependent upon the input signal power presented to the antenna, rendering it impractical. However, the NLMS algorithm was found to be far superior in this respect and it was later shown to approach the performance of the SMI beamformer for a three element adaptive array. A low SNR gives a poor estimate of the received signal's cross-correlation matrix, resulting in similar performance for all three algorithms. However, as the SNR improves, the SMI technique guarantees a stronger interference rejection. The SMI algorithm is more complex for a large number of antenna elements, but for a realistic number of elements, such as four, its complexity is below that of the LMS routines.

In Chapter 4 the performance gains achieved using adaptive antenna arrays at the base stations in a cellular network were investigated for both LOS and multipath environments. An exposure to modelling an adaptive array was provided in Section 4.2, before an overview of fixed and dynamic channel allocation schemes was conducted in Section 4.3. Section 4.5 then reviewed some of the different models available for simulating multipath environments, followed by a more detailed portrayal of the Geometrically Based Single-Bounce Elliptical Model (GBSBEM). The metrics used for characterising the performance of mobile cellular networks were presented under both LOS and multipath propagation conditions, with and without adaptive antenna arrays. The network capacity was found to increase when using adaptive antenna arrays, with further increases achieved due to the adoption of power control. An adaptive modulation mode switching algorithm with combined power control was developed and network capacity investigations were conducted. Employing adaptive modulation using adaptive antenna arrays was found to increase the network's capacity significantly, whilst providing a superior call quality and a higher mean modem throughput.

Our investigations in Chapter 4 initially focused on the non-wraparound or 'desert island' type networks, where the outer cells of the simulation area are subjected to lower levels of co-channel interference, a scenario that may be encountered in the suburbs of large conurbations. Simulations were carried out for the FCA algorithm, and the LOLIA using nearest base station constraints of 7 and 19, when exposed to LOS propagation conditions. The FCA algorithm offered the lowest network capacity, but benefited the most from employing adaptive antenna arrays. Specifically, the network capacity of FCA increased by 67%, when employing two element antenna arrays at the base stations, and 144%, when using four element arrays. The LOLIA using a nearest base station constraint of 7 cells supported a higher number of users, but the adaptive antenna arrays did not result in such dramatic improvements in network capacity. Explicitly, a 22% increase was observed for the two element case, and a 58% when using four elements. However, the network capacity supported by the LOLIA in conjunction with $n = 7$ always exceeded that of the FCA algorithm. When using a 19 base station constraint, the LOLIA resulted in the highest network capacity without employing adaptive antenna arrays, although the large frequency reuse distance of this algorithm resulted in a modest increase of the network capacity.

We then conducted further simulations in Section 4.6.2.2 using a more realistic 3-ray multipath propagation environment. Again, the FCA algorithm supported the lowest number of users, and gained the most from invoking adaptive antenna arrays. Using a four element array instead of a two element array led to a network capacity increase of 35%, and replacing the four element array with one employing eight elements resulted in a 24-34% increase in the number of users supported. The LOLIA employing $n = 7$ supported the greatest number of users, but did not benefit from the same capacity increases as the FCA algorithm with the advent of adaptive antenna arrays. The number of users supported increased by 18% upon upgrading the system from two to four element adaptive antenna arrays, and by between 5% and 15% upon using eight element arrays in place of the four element arrays. Using a frequency reuse constraint of 19 in conjunction with the LOLIA resulted in a network whose capacity was restricted by the high new call blocking probability associated with its large frequency reuse distance. This large frequency reuse distance led to low levels of co-channel interference, which could not be nulled effectively by the adaptive antenna arrays, and hence the network capacity did not increase by more than 5% upon doubling the number of antenna elements comprising the array. Hence, our future studies only considered the FCA algorithm and the LOLIA in conjunction with $n = 7$.

The network capacity gains accruing from the implementation of power control over the same 3-ray multipath channel, as in the previous section, were then investigated for the FCA algorithm and the LOLIA using $n = 7$. Significant network capacity increases were observed for all of the scenarios considered. Specifically, the network capacity without power control and using a given number of antenna elements, was frequently exceeded by that of an identical scenario using power control and half the number of antenna elements. On comparing otherwise identical scenarios, an increase in the network capacity of between 28% and 72% was attributed to the implementation of power control, whilst using the FCA algorithm. When employing the LOLIA and power control, the number of users supported increased by between 8.5% and 15%. The network capacity gains resulting from increasing the number of elements in the adaptive antenna arrays were reduced however, to 11% and 17% for the FCA algorithm. In contrast, the adaptive nature of the LOLIA enabled it to maintain the network capacity increases of 12-17%, achieved due to increasing the number of elements comprising the adaptive arrays.

The implementation of adaptive modulation techniques was then investigated in Section 4.6.2.4, since they allow the exploitation of good near-instantaneous channel conditions, whilst providing resilience when subjected to poor quality channels. The network capacity of the FCA algorithm was found to increase by 6-12%, when invoking adaptive modulation in conjunction with two element adaptive antenna arrays. However, when using four element adaptive antenna arrays the network capacity was reduced upon invoking adaptive modulation. This was due to the improved call dropping probability accruing from employing adaptive modulation, leading in turn to a lower number of frequency/timeslot combinations available for new calls. Since the new call blocking probability was the factor limiting the network's capacity, the capacity was reduced. This phenomenon was not observed when employing the LOLIA, which supported 43% more users on average upon invoking adaptive modulation techniques. Doubling the number of antenna elements led to an extra 20% supported users.

In summary, the network using the FCA algorithm supported 2400 users, or 14 Erlangs/km^2/MHz, in the conservative scenario, and approximately 2735 users, or 15.6 Erlangs-/km^2/MHz, in the lenient scenario. When using the LOLIA 7 channel allocation algorithm and two element adaptive antenna arrays, 3675 users (23.1 Erlangs/km^2/MHz) were carried under the conservative conditions, and 4115 users (25.4 Erlangs/km^2/MHz) under the lenient specifications. When invoking four element adaptive antenna arrays, 4460 users (27.4 Erlangs/km^2/MHz) and 4940 users (29.6 Erlangs/km^2/MHz) were supported under the conservative and lenient scenarios, respectively.

In Section 4.6.3 our investigations then led us to consider results obtained for an infinite network using the so-called 'wraparound' technique, which allows a cellular network to be simulated as if part of a much larger network, thus inflicting similar levels of co-channel interference upon all cells within the network. The FCA algorithm again supported the lowest number of users, but benefited the most from the employment of adaptive antenna arrays, resulting in network capacity increases of between 46 and 70%, when employing adaptive antenna arrays, or when using four rather than two elements. The LOLIA using a nearest base station constraint of 7, supported an extra 17-23% of users due to the application of adaptive antenna arrays at the base stations. As in the 'desert island' scenarios, the LOLIA in conjunction with a frequency reuse constraint of 19 base stations, offered the greatest network capacity without adaptive antenna arrays. However, when using two element arrays, the network capacity grew by almost 20%, since the limiting factor was the co-channel interference, not the new call blocking probability. The extra interference rejection potential offered by the four element arrays was also exploited, but was also somewhat limited, since the new call blocking probability became the capacity limiting constraint once again.

Under 3-ray multipath propagation conditions the network capacities of both the FCA algorithm and the LOLIAs were limited by the probability of low quality access, and hence invoking adaptive beamforming techniques increased the number of users supported. For an adaptive antenna array consisting of a given number of elements, the FCA algorithm supported the least number of users, and although exhibiting the greatest capacity gains due to the adaptive antenna arrays, the LOLIA 7 employing two element arrays exceeded the capacity of the FCA algorithm using eight element arrays. The LOLIA in conjunction with a frequency reuse of 19 base stations benefited from doubling the number of antenna elements from two, to four, and from four, to eight, but the network capacity was then limited by the new call blocking probability, and hence further increases in the number of antenna array elements would have had no impact on the network's capacity.

The addition of power control in the 'infinite' network was then considered under the above 3-ray multipath conditions. The capacity gains were significant for both the FCA algorithm and the LOLIA 7, when compared to our identical investigations conducted without power control. Again, the network capacity when using the FCA algorithm benefited the most, with the number of users supported increasing by between 38 and 82%, exhibiting a mean increase of 61%. However, the LOLIA 7 based network still supported the greatest number of users, although the capacity gains of the power control were limited to around 12%.

The employment of adaptive modulation techniques led to the saturation of network resources for the FCA algorithm, with the network capacity limited by the number of frequency/timeslot combinations available for new calls. Hence, increasing the number of antenna elements from two to four resulted in an increase in the mean modem throughput from 2.4 BPS to 2.7 BPS, and a small reduction in the mean transmission power. The adaptive nature of the LOLIA allowed it to fully exploit the potential of adaptive modulation and supported more than 32% extra users. The limiting factor of the LOLIA's network capacity was the requirement of a minimum mean modem throughput of 2.0 BPS.

Therefore, the FCA algorithm supported 1400 users, and carried a teletraffic load of 13.8 Erlangs/km^2/MHz in the conservative scenario and 1570 users, or 15.2 Erlangs/km^2/MHz of traffic under the lenient conditions. The LOLIA however supported an extra 35% of users, giving a network capacity of 1910 users, or 19.75 Erlangs/km^2/MHz, when using two element adaptive antenna arrays for both the conservative and lenient scenarios. Utilising four element antenna arrays at the base stations allowed 2245 users, or 23.25 Erlangs/km^2/MHz of network traffic to be supported at the required quality levels of the conservative and lenient scenarios.

Thus, the network capacity was found to substantially increase, when using adaptive antenna arrays, with further increases achieved through the adoption of power control. An adaptive modulation mode switching algorithm combined with power control was developed and network simulations were conducted. Employing adaptive modulation in conjunction with adaptive antenna arrays was found to increase the network capacity significantly, whilst providing superior call quality and a greater mean modem throughput.

Chapter 5 examined the performance of a CDMA based cellular mobile network, very similar in its nature to the FDD-mode of the proposed UTRA standard. A comparison of various soft handover algorithms was conducted in both non-shadowed and shadowed propagation environments. The algorithm that was found to offer the highest network capacity, i.e. the highest number of users supported at a given quality of service, used the relative received E_c/I_o for determining cell ownership. The impact of using adaptive antenna arrays at the base stations was then investigated, in both non-shadowed and shadowed environments for high data rate users. This work was then extended by the application of adaptive modulation techniques, in conjunction with adaptive antenna arrays.

The network capacity in terms of the number of users supported was 256 when experiencing no log-normal shadow fading and using no adaptive antenna arrays. However, with the application of two element adaptive antenna arrays the network capacity increased by 27% to 325 users, and when upgrading the system to four element arrays, the capacity of the network increased by a further 47% to 480 users. When subjected to log-normal shadow fading having a standard deviation of 3 dB in conjunction with a maximum fading frequency of 0.5 Hz, the network capacity without adaptive antennas was reduced to about 150 users. Again, invoking adaptive antenna arrays at the base stations increased the network capacity to 203 users, and 349 users, when employing two and four array elements, respectively.

We then applied independent up- and down-link beamforming. This implied determining separately the optimum weights for both the up- and the down-link, rather than re-using the antenna array weights calculated for the uplink scenario in the downlink. This measure led to further network capacity gains. Specifically, employing independent up- and down-link beamforming resulted in 15% and 7% network capacity increases, for the two and four element arrays, respectively, giving total network capacities of 349 and 375 users. Increasing the maximum shadow fading frequency from 0.5 Hz to 1.0 Hz slightly reduced the maximum number of users supported by the network, resulting in a network capacity of 144 users without beamforming, and capacities of 201 and 333 users, when invoking two and four element arrays, respectively. These absolute network capacity increases corresponded to relative network capacity gains of 40% and 131%, respectively. Again, performing independent up- and down-link beamforming increased the network capacities, with 225 and 365 users supported by the two and four element adaptive antenna arrays, respectively. Hence, these results show that applying both two and four element adaptive antenna arrays have led to significant network capacity increases both with and without log-normal shadow fading. Furthermore, the capacity of the network was found to be reduced by approximately 40%, when subjected to log-normal shadow fading having a standard deviation of 3 dB. However, increasing the maximum log-normal fading frequency from 0.5 Hz to 1.0 Hz had little impact on the total network capacity.

These results were then extended by applying adaptive modulation techniques, both with and without adaptive antenna arrays, which were performing independent up- and down-link beamforming in conjunction with log-normal shadow fading having a standard deviation of 3 dB as well as maximum fading frequencies of 0.5 Hz and 1.0 Hz. Without adaptive antenna arrays the network supported 223 users, at a mean uplink modem throughput of 2.86 BPS. The mean throughput of the downlink was 2.95 BPS. Upon increasing the maximum shadowing frequency from 0.5 Hz to 1.0 Hz the network capacity fell slightly to 218 users, whilst the mean modem throughputs remained essentially unchanged. However, invoking two element adaptive antenna arrays enhanced the network capacities by 64% upon encountering 0.5 Hz shadow fading, and by 56% when subjected to 1.0 Hz shadowing. In both cases the mean modem throughput dropped by approximately 0.3 BPS. A further 0.2 BPS reduction of the mean modem throughput occurred when applying four element adaptive antenna arrays. However, this allowed an extra 30% of users to be supported when subjected to shadow fading fluctuating at a maximum frequency of 0.5 Hz and 35% in conjunction with 1.0 Hz frequency shadowing. Therefore, these results have shown the significant network capacity increases achieved by invoking adaptive modulation techniques. These network capacity improvements have been achieved in conjunction with a higher mean modem throughput, albeit at a slightly higher mean transmission power.

The performance results obtained for the UTRA-type network of Chapter 5 were obtained for high data rate users communicating at a raw data rate of 240 kbps, using a spreading factor of 16. However, as described in Section 5.4, some exploratory investigations not presented in this book demonstrated that the increase in the number of users supported by the network, was up to a factor of two higher than expected on the basis of simple spreading factor proportionate scaling. Specifically, the expected increase in switching from a spreading factor of 16 to 256 was a factor of 256/16=16, and hence Tables 5.5 and 5.8 were presented showing the potential worst-case network capacities achieved by multiplying the high data rate results by 16. Even when considering these user capacities, the teletraffic carried by the network normalised with respect to both the occupied bandwidth and the network's area, was found to be higher than that achieved by the FDMA/TDMA based networks considered in Chapter 4.

6.2 Further Research

Future research that builds upon the investigations considered here includes applying beamforming techniques to the pilot signals, or developing a method by which the pilot signals received at the mobile may be cancelled. In future systems the carrier frequency may be sufficiently high so that two antenna elements may be incorporated into the mobile handset, thus enabling beamforming to be performed at both ends of the data link. Further research is required for optimising the AQAM mode switching criteria, which could amalgamate the power control and beamforming algorithms. This could be further developed to a joint optimisation of the adaptive modulation mode switching, power control and beamforming, and potentially could also be incorporated into multi-user detection algorithms. Additionally, the performance of multi-rate networks is worthy of investigation, especially when combined with adaptive modulation and adaptive beam-forming techniques, which are particularly suitable for mitigating the significant levels of interference inflicted by the high data rate users. Since the high rate users impose the majority of interference on the numerous low rate users, the employment of interference reduction techniques is of vital importance. This book has only considered the employment of uniform linear antenna arrays having an antenna element spacing of $\lambda/2$. However, other antenna geometries, exhibiting no symmetry and possibly relying on antenna elements, which are not omni-directional may result is higher network capacity gains. More sophisticated propagation models tailored for different environments, such as macro- and pico-cells also have to be considered. The TDD mode of UTRA offers a further rich ground for system optimisation in conjunction with various time-slot allocation techniques, whilst endeavouring to maintain the advantages of the asymmetric uplink/downlink data rate nature of the TDD mode.

As a further topic, the network performance of High Altitude Platform Stations (HAPS) [375] remains to be investigated, especially in the context of adaptive modulation, and finally, future networks may be ad-hoc [375] in nature, which currently is a promising unexplored region of research.

Glossary

AWGN	Additive White Gaussian Noise
BS	A common abbreviation for Base Station
CDMA	Code Division Multiple Access
CMA	Constant Modulus Algorithm
DCS1800	A digital mobile radio system standard, based on GSM, but operates at 1.8GHz at a lower power.
DOA	Direction Of Arrival
FDD	Frequency Division Duplex
GSM	A Pan-European digital mobile radio standard, operating at 900MHz.
HIPERLAN	High Performance Radio Local Area Network
IF	Intermediate Frequency
LMS	Least Mean Square, a stochastic gradient algorithm used in adapting coefficients of a system
MS	A common abbreviation for Mobile Station
MSE	Mean Square Error, a criterion used to optimised the coefficients of a system such that the noise contained in the received signal is minimised.
PDF	Probability Density Function
RF	Radio Frequency
RLS	Recursive Least Square

SDMA	Spatial Division Multiple Access
SINR	Signal to Interference plus Noise ratio, same as signal to noise ratio (SNR) when there is no interference.
SIR	Signal to Interference ratio
SNR	Signal to Noise Ratio, noise energy compared to the signal energy
TDD	Time Division Duplex
TDMA	Time Division Multiple Access
UMTS	Universal Mobile Telecommunication System

Bibliography

[1] M. Barrett and R. Arnott, "Adaptive antennas for mobile communications," *IEE Electronics & Communications Engineering Journal*, pp. 203–214, August 1994.

[2] S. C. Swales, M. A. Beach, D. J. Edwards, and J. P. McGeehan, "The Performance Enhancement of Multibeam Adaptive Base-Station Antennas for Cellular Land Mobile Radio Systems," *IEEE Transactions on Vehicular Technology*, vol. 39, pp. 56–67, February 1990.

[3] J. Litva and T. Lo, *Digital Beamforming in Wireless Communications*. Artech House, London, 1996.

[4] A. B. Carlson, *Communication Systems*. McGraw-Hill, 1986.

[5] J. G. Proakis, *Digital Communications*. Mc-Graw Hill International Editions, 3 ed., 1995.

[6] L. C. Godara, "Applications of Antenna Arrays to Mobile Communications, Part I: Performance Improvement, Feasibility, and System Considerations," *Proceedings of the IEEE*, vol. 85, pp. 1029–1060, July 1997.

[7] G. V. Tsoulos and M. A. Beach, "Calibration and Linearity issues for an Adaptive Antenna System," in *IEEE Proceedings of Vehicular Technology Conference*, pp. 1597–1600, 1997.

[8] B. D. V. Veen and K. M. Buckley, "Beamforming: A Versatile Approach to Spatial Filtering," *IEEE ASSP Magazine*, pp. 4–24, April 1988.

[9] A. J. Paulraj and B. C. Ng, "Space-Time Modems for Wireless Personal Communications," *IEEE Personal Communications*, pp. 36–48, February 1998.

[10] A. J. Paulraj and E. Lindskog, "Taxonomy of space-time processing for wireless networks," *IEEE Proceedings on Radar, Sonar and Navigation*, vol. 145, pp. 25–31, February 1998.

[11] R. Steele and L. Hanzo, *Mobile Radio Communications*. IEEE Press - John Wiley, 2nd ed., 1999.

[12] W. T. Webb and L. Hanzo, *Modern Quadrature Amplitude Modulation: Principles and Applications for Wireless Communications*. IEEE Press-Pentech Press, 1994. ISBN 0-7273-1701-6.

[13] L. Hanzo, W. T. Webb, and T. Keller, *Single- and Multi-Carrier Quadrature Amplitude Modulation*. John Wiley, IEEE Press, 2000.

[14] N. Anderson and P. Howard, "Technology and Transceiver Architecture Considerations for Adaptive Antenna Systems," in *Proceedings of ACTS Summit*, pp. 965–970, 1997.

[15] J. Strandell, M. Wennstrom, A. Rydberg, T. Oberg, O. Gladh, L. Rexberg, E. Sandberg, B. V. Andersson, and M. Appelgren, "Experimental Evaluation of an Adaptive Antenna for a TDMA Mobile Telephony System," in *Proceedings of PIMRC*, pp. 79–84, 1997.

[16] J. J. Monot, J. Thibault, P. Chevalier, F. Pipon, S. Mayrargue, and A. Levy, "A fully programmable prototype for the experimentation of the SDMA concept and and use of smart antennas for UMTS and GSM/DCS1800 networks," in *Proceedings of PIMRC*, (Helsinki, Finland), pp. 534–538, September 1997.

[17] M. Mizuno and T. Ohgane, "Application of Adaptive Array Antennas to Radio Communications," *Electronics and Communications in Japan, Part 1*, vol. 77, no. 2, pp. 48–56, 1994.

[18] Y. Ogawa and T. Ohgane, "Adaptive Antennas for Future Mobile Radio," *IEICE Trans. Fundamentals*, vol. E79-A, pp. 961–967, July 1996.

[19] G. V. Tsoulos, M. A. Beach, and S. C. Swales, "On the Sensitivity of the Capacity Enhancement of a TDMA system with Adaptive Multibeam Antennas," in *IEEE VTC Proceedings*, pp. 165–169, 1997.

[20] P. Leth-Espensen, P. E. Mogensen, F. Frederiksen, K. Olesen, and S. L. Larsen, "Performance of Different Combining Algorithms for a GSM System applying Antenna Arrays," in *Proceedings of ACTS Summit*, 1997.

[21] W. Jakes, ed., *Microwave Mobile Communications*. Wiley-Interscience, 1974.

[22] J. S. Blogh, P. J. Cherriman, and L. Hanzo, "Adaptive Beamforming Assisted Dynamic Channel Allocation," in *Proceedings of VTC*, (Houston, USA), pp. 199–203, May 1999.

[23] J. S. Blogh, P. J. Cherriman, and L. Hanzo, "Comparative Study of Dynamic Channel Allocation Algorithms," *IEEE Transactions on Vehicular Technology*, 2001.

[24] J. S. Blogh, P. J. Cherriman, and L. Hanzo, "Dynamic Channel Allocation Using Adaptive Antennas and Power Control," in *Proceedings of ACTS Mobile Communications Summit*, (Sorrento), pp. 943–948, June 1999.

[25] J. S. Blogh, P. J. Cherriman, and L. Hanzo, "Dynamic Channel Allocation Techniques using Adaptive Modulation and Adaptive Antennas," *Accepted for publication in IEEE Journal on Selected Areas in Communications*, 2001.

[26] J. S. Blogh, P. J. Cherriman, and L. Hanzo, "Dynamic Channel Allocation Techniques using Adaptive Modulation and Adaptive Antennas," in *Proceedings of VTC Fall*, (Amsterdam, The Netherlands), pp. 2348–2352, September 1999.

[27] J. Rapeli, "UMTS: Targets, system concept, and standardization in a global framework," *IEEE Personal Communications*, vol. 2, pp. 20–28, February 1995.

[28] L. Hanzo and J. Stefanov, "The Pan-European Digital Cellular Mobile Radio System – known as GSM," in Steele [128], ch. 8, pp. 677–765.

[29] P.-G. Andermo and L.-M. Ewerbring, "A CDMA-based radio access design for UMTS," *IEEE Personal Communications*, vol. 2, pp. 48–53, February 1995.

[30] E. Nikula, A. Toskala, E. Dahlman, L. Girard, and A. Klein, "FRAMES multiple access for UMTS and IMT-2000," *IEEE Personal Communications*, vol. 5, pp. 16–24, April 1998.

[31] T. Ojanperä and R. Prasad, ed., *Wideband CDMA for 3rd Generation Mobile Communications*. Artech House Publishers, 1998.

[32] H. Holma and A. Toskala, eds., *WCDMA for UMTS: Radio Access for Third Generation Mobile Communications*. John Wiley & Sons, Ltd., 2000.

[33] E. Berruto, M. Gudmundson, R. Menolascino, W. Mohr, and M. Pizarroso, "Research activities on UMTS radio interface, network architectures, and planning," *IEEE Communications Magazine*, vol. 36, pp. 82–95, February 1998.

[34] M. Callendar, "Future public land mobile telecommunication systems," *IEEE Personal Communications*, vol. 12, no. 4, pp. 18–22, 1994.

[35] W. Lee, "Overview of cellular CDMA," *IEEE Transactions on Vehicular Technology*, vol. 40, pp. 291–302, May 1991.

[36] K. Gilhousen, I. Jacobs, R. Padovani, A. Viterbi, L. Weaver Jr., and C. Wheatley III, "On the capacity of a cellular CDMA system," *IEEE Transactions on Vehicular Technology*, vol. 40, pp. 303–312, May 1991.

[37] R. Pickholtz, L. Milstein, and D. Schilling, "Spread spectrum for mobile communications," *IEEE Transactions on Vehicular Technology*, vol. 40, pp. 312–322, May 1991.

[38] R. Kohno, *Wireless Communications: TDMA versus CDMA*, ch. 1. Spatial and Temporal Communication Theory using Software Antennas for Wireless Communications, pp. 293–321. Kluwer Academic Publishers, 1997.

[39] A. Viterbi, *CDMA: Principles of Spread Spectrum Communication*. Addison-Wesley, June 1995. ISBN 0201633744.

[40] S. Glisic and B. Vucetic, *Spread Spectrum CDMA Systems for Wireless Communications*. Artech House, April 1997. ISBN 0890068585.

[41] R. Prasad, *CDMA for Wireless Personal Communications*. Artech House, May 1996. ISBN 0890065713.

[42] V. Garg, K. Smolik, J. Wilkes, and K. Smolik, *Applications of CDMA in Wireless/Personal Communications*. Englewood Cliffs NJ: Prentice-Hall, 1996.

[43] R. Price and E. Green Jr., "A communication technique for multipath channels," *Proceedings of the IRE*, vol. 46, pp. 555–570, March 1958.

[44] B. Sklar, "Rayleigh fading channels in mobile digital communication systems part I : Characterization," *IEEE Communications Magazine*, vol. 35, pp. 90–100, July 1997.

[45] B. Sklar, "Rayleigh fading channels in mobile digital communication systems part II: Mitigation," *IEEE Communications Magazine*, vol. 35, pp. 148–155, July 1997.

[46] F. Amoroso, "Use of DS/SS signaling to mitigate Rayleigh fading in a dense scatterer environment," *IEEE Personal Communications*, vol. 3, pp. 52–61, April 1996.

[47] W. C. Jakes, ed., *Microwave Mobile Communications.* John Wiley and Sons, 1974. ISBN 0-471-43720-4.

[48] M. Nakagami, "The m-distribution-a general formula of intensity distribution of fading," *Statistical Methods in Radio Wave Propagation*, 1960. W.C. Hoffman, ed., New York: Pergamon.

[49] H. Suzuki, "A statistical model for urban multipath propagation," *IEEE Transactions on Communications*, vol. COM-25, pp. 673–680, July 1977.

[50] "COST 207: Digital land mobile radio communications, final report." Office for Official Publications of the European Communities, 1989. Luxembourg.

[51] M. Whitmann, J. Marti, and T. Kürner, "Impact of the power delay profile shape on the bit error rate in mobile radio systems," *IEEE Transactions on Vehicular Technology*, vol. 46, pp. 329–339, May 1997.

[52] D. Greenwood and L. Hanzo, "Characterisation of mobile radio channels," in Steele [128], ch. 2, pp. 92–185.

[53] T. Eng, N. Kong, and L. Milstein, "Comparison of diversity combining techniques for Rayleigh-fading channels," *IEEE Transactions on Communications*, vol. 44, pp. 1117–1129, September 1996.

[54] M. Kavehrad and P. McLane, "Performance of low-complexity channel coding and diversity for spread spectrum in indoor, wireless communications," *AT&T Technical Journal*, vol. 64, pp. 1927–1965, October 1985.

[55] K.-T. Wu and S.-A. Tsaur, "Selection diversity for DS-SSMA communications on Nakagami fading channels," *IEEE Transactions on Vehicular Technology*, vol. 43, pp. 428–438, August 1994.

[56] L.-L. Yang and L. Hanzo, "Serial acquisition techniques for DS-CDMA signals in frequency-selective multi-user mobile channels," in *Proceedings of VTC'98 (Spring)* [376].

[57] L.-L. Yang and L. Hanzo, "Serial acquisition of DS-CDMA signals in multipath fading mobile channels." submitted to IEEE Transactions on Vehicular Technology, 1998.

[58] R. Ziemer and R. Peterson, *Digital Communications and Spread Spectrum System.* New York: Macmillan Publishing Company, 1985.

[59] R. Pickholtz, D. Schilling, and L. Milstein, "Theory of spread-spectrum communications — a tutorial," *IEEE Transactions on Communications*, vol. COM-30, pp. 855–884, May 1982.

[60] S. Rappaport and D. Grieco, "Spread-spectrum signal acquisition: Methods and technology," *IEEE Communications Magazine*, vol. 22, pp. 6–21, June 1984.

[61] E. Ström, S. Parkvall, S. Miller, and B. Ottersten, "Propagation delay estimation in asynchronous direct-sequence code division multiple access systems," *IEEE Transactions on Communications*, vol. 44, pp. 84–93, January 1996.

[62] R. Rick and L. Milstein, "Optimal decision strategies for acquisition of spread-spectrum signals in frequency-selective fading channels," *IEEE Transactions on Communications*, vol. 46, pp. 686–694, May 1998.

[63] J. Lee, *CDMA Systems Engineering Handbook.* London: Artech House Publishers, 1998.

[64] M. Varanasi and B. Aazhang, "Multistage detection in asynchronous code-division multiple-access communications," *IEEE Transactions on Communications*, vol. 38, pp. 509–519, April 1990.

[65] S. Moshavi, "Multi-user detection for DS-CDMA communications," *IEEE Communications Magazine*, vol. 34, pp. 124–136, October 1996.

[66] S. Verdú, *Multiuser Detection.* Cambridge University Press, 1998.

[67] L. Hanzo, C. H. Wong, and M. S. Yee, *Adaptive Wireless Transceivers.* John Wiley, IEEE Press, 2002. (For detailed contents please refer to http://www-mobile.ecs.soton.ac.uk.).

[68] E. Kuan and L. Hanzo, "Joint detection CDMA techniques for third-generation transceivers," in *Proceedings of ACTS Mobile Communication Summit '98*, (Rhodes, Greece), pp. 727–732, ACTS, 8–11 June 1998.

[69] E. Kuan, C. Wong, and L. Hanzo, "Burst-by-burst adaptive joint detection CDMA," in *Proceedings of VTC'98 (Spring)* [376].

[70] S. Verdú, *Multiuser Detection.* Cambridge: Cambridge University Press, 1998.

[71] F. Simpson and J. Holtzman, "Direct sequence CDMA power control, interleaving, and coding," *IEEE Journal on Selected Areas in Communications*, vol. 11, pp. 1085–1095, September 1993.

[72] M. Pursley, "Performance evaluation for phase-coded spread-spectrum multiple-access communication-part I: System analysis," *IEEE Transactions on Communications*, vol. COM-25, pp. 795–799, August 1977.

[73] R. Morrow Jr., "Bit-to-bit error dependence in slotted DS/SSMA packet systems with random signature sequences," *IEEE Transactions on Communications*, vol. 37, pp. 1052–1061, October 1989.

[74] J. Holtzman, "A simple, accurate method to calculate spread-spectrum multiple-access error probabilities," *IEEE Transactions on Communications*, vol. 40, pp. 461–464, March 1992.

[75] U.-C. Fiebig and M. Schnell, "Correlation properties of extended m-sequences," *Electronic Letters*, vol. 29, pp. 1753–1755, September 1993.

[76] J. McGeehan and A. Bateman, "Phase-locked transparent tone in band (TTIB): A new spectrum configuration particularly suited to the transmission of data over SSB mobile radio networks," *IEEE Transactions on Communications*, vol. COM-32, no. 1, pp. 81–87, 1984.

[77] A. Bateman, G. Lightfoot, A. Lymer, and J. McGeehan, "Speech and data transmissions over a 942MHz TAB and TTIB single sideband mobile radio system," *IEEE Transactions on Vehicular Technology*, vol. VT-34, pp. 13–21, February 1985.

[78] F. Davarian, "Mobile digital communications via tone calibration," *IEEE Transactions on Vehicular Technology*, vol. VT-36, pp. 55–62, May 1987.

[79] M. Moher and J. Lodge, "TCMP—a modulation and coding strategy for Rician fading channels," *IEEE Journal on Selected Areas in Communications*, vol. 7, pp. 1347–1355, December 1989.

[80] G. Irvine and P. McLane, "Symbol-aided plus decision-directed reception for PSK/TCM modulation on shadowed mobile satellite fading channels," *IEEE Journal on Selected Areas in Communications*, vol. 10, pp. 1289–1299, October 1992.

[81] A. Baier, U.-C. Fiebig, W. Granzow, W. Koch, P. Teder, and J. Thielecke, "Design study for a CDMA-based third-generation mobile system," *IEEE Journal on Selected Areas in Communications*, vol. 12, pp. 733–743, May 1994.

[82] P. Rapajic and B. Vucetic, "Adaptive receiver structures for asynchornous CDMA systems," *IEEE Journal on Selected Areas in Communications*, vol. 12, pp. 685–697, May 1994.

[83] M. Benthin and K.-D. Kammeyer, "Influence of channel estimation on the performance of a coherent DS-CDMA system," *IEEE Transactions on Vehicular Technology*, vol. 46, pp. 262–268, May 1997.

[84] M. Sawahashi, Y. Miki, H. Andoh, and K. Higuchi, "Pilot symbol-assisted coherent multistage interference canceller using recursive channel estimation for DS-CDMA mobile radio," *IEICE Transactions on Communications*, vol. E79-B, pp. 1262–1269, September 1996.

[85] J. Torrance and L. Hanzo, "Comparative study of pilot symbol assisted modem schemes," in *Proceedings of IEE Conference on Radio Receivers and Associated Systems (RRAS'95)*, (Bath), pp. 36–41, IEE, 26–28 September 1995.

[86] J. Cavers, "An analysis of pilot symbol assisted modulation for Rayleigh fading channels," *IEEE Transactions on Vehicular Technology*, vol. 40, pp. 686–693, November 1991.

[87] S. Sampei and T. Sunaga, "Rayleigh fading compensation for QAM in land mobile radio communications," *IEEE Transactions on Vehicular Technology*, vol. 42, pp. 137–147, May 1993.

[88] *The 3GPP1 website*. http://www.3gpp.org.

[89] *The 3GPP2 website*. http://www.3gpp2.org.

[90] T. Ojanperä and R. Prasad, *Wideband CDMA for Third Generation Mobile Communications*. London: Artech House, 1998.

[91] E. Dahlman, B. Gudmundson, M. Nilsson, and J. Sköld, "UMTS/IMT-2000 based on wideband CDMA," *IEEE Communications Magazine*, vol. 36, pp. 70–80, September 1998.

[92] T. Ojanperä, "Overview of research activities for third generation mobile communications," in Glisic and Leppanen [307], ch. 2 (Part 4), pp. 415–446. ISBN 0792380053.

[93] European Telecommunications Standards Institute, *The ETSI UMTS Terrestrial Radio Access (UTRA) ITU-R RTT Candidate Submission*, June 1998. ETSI/SMG/SMG2.

[94] Association of Radio Industries and Businesses, *Japan's Proposal for Candidate Radio Transmission Technology on IMT-2000: W-CDMA*, June 1998.

[95] F. Adachi, M. Sawahashi, and H. Suda, "Wideband DS-CDMA for next-generation mobile communications systems," *IEEE Communications Magazine*, vol. 36, pp. 56–69, September 1998.

[96] F. Adachi and M. Sawahashi, "Wideband wireless access based on DS-CDMA," *IEICE Transactions on Communications*, vol. E81-B, pp. 1305–1316, July 1998.

[97] A. Sasaki, "Current situation of IMT-2000 radio transmission technology study in Japan," *IEICE Transactions on Communications*, vol. E81-B, pp. 1299–1304, July 1998.

[98] P. Baier, P. Jung, and A. Klein, "Taking the challenge of multiple access for third-generation cellular mobile radio systems — a European view," *IEEE Communications Magazine*, vol. 34, pp. 82–89, February 1996.

[99] J. Schwarz da Silva, B. Barani, and B. Arroyo-Fernández, "European mobile communications on the move," *IEEE Communications Magazine*, vol. 34, pp. 60–69, February 1996.

[100] F. Ovesjö, E. Dahlman, T. Ojanperä, A. Toskala, and A. Klein, "FRAMES multiple access mode 2 — wideband CDMA," in *Proceedings of IEEE International Symposium on Personal, Indoor and Mobile Radio Communications, PIMRC'97* [377].

[101] *The UMTS Forum website*. http://www.umts-forum.org/.

[102] M. Sunay, Z.-C. Honkasalo, A. Hottinen, H. Honkasalo, and L. Ma, "A dynamic channel allocation based TDD DS CDMA residential indoor system," in *IEEE 6th International Conference on Universal Personal Communications, ICUPC'97*, (San Diego, CA), pp. 228–234, October 1997.

[103] F. Adachi, M. Sawahashi, and K. Okawa, "Tree-structured Generation of Orthogonal Spreading Codes with Different Lengths for Forward Link of DS-CDMA Mobile," *Electronics Letters*, vol. 33, no. 1, pp. 27–28, 1997.

[104] F. Adachi, K. Ohno, A. Higashi, T. Dohi, and Y. Okumura, "Coherent multicode DS-CDMA mobile Radio Access," *IEICE Transactions on Communications*, vol. E79-B, pp. 1316–1324, September 1996.

[105] L. Hanzo, C. Wong, and P. Cherriman, "Channel-adaptive wideband video telephony," *IEEE Signal Processing Magazine*, vol. 17, pp. 10–30, July 2000.

[106] P. Cherriman and L. Hanzo, "Programmable H.263-based wireless video transceivers for interference-limited environments," *IEEE Trans. on Circuits and Systems for Video Technology*, vol. 8, pp. 275–286, June 1998.

[107] C. Berrou and A. Glavieux, "Near optimum error correcting coding and decoding: turbo codes," *IEEE Transactions on Communications*, vol. 44, pp. 1261–1271, October 1996.

[108] A. Fujiwara, H. Suda, and F. Adachi, "Turbo codes application to DS-CDMA mobile radio," *IEICE Transactions on Communications*, vol. E81A, pp. 2269–2273, November 1998.

[109] M. Juntti, "System concept comparison for multirate CDMA with multiuser detection," in *Proceedings of IEEE Vehicular Technology Conference (VTC'98)* [378], pp. 18–21.

[110] S. Ramakrishna and J. Holtzman, "A comparison between single code and multiple code transmission schemes in a CDMA system," in *Proceedings of IEEE Vehicular Technology Conference (VTC'98)* [378], pp. 791–795.

[111] M. K. Simon, J. K. Omura, R. A. Scholtz, and B. K. Levitt, *Spread Spectrum Communications Handbook.* McGraw-Hill, 1994.

[112] T. Kasami, *Combinational Mathematics and its Applications.* University of North Carolina Press, 1969.

[113] A. Brand and A. Aghvami, "Multidimensional PRMA with prioritized Bayesian broadcast — a MAC strategy for multiservice traffic over UMTS," *IEEE Transactions on Vehicular Technology*, vol. 47, pp. 1148–1161, November 1998.

[114] R. Ormondroyd and J. Maxey, "Performance of low rate orthogonal convolutional codes in DS-CDMA," *IEEE Transactions on Vehicular Technology*, vol. 46, pp. 320–328, May 1997.

[115] A. Chockalingam, P. Dietrich, L. Milstein, and R. Rao, "Performance of closed-loop power control in DS-CDMA cellular systems," *IEEE Transactions on Vehicular Technology*, vol. 47, pp. 774–789, August 1998.

[116] R. Gejji, "Forward-link-power control in CDMA cellular-systems," *IEEE Transactions on Vehicular Technology*, vol. 41, pp. 532–536, November 1992.

[117] K. Higuchi, M. Sawahashi, and F. Adachi, "Fast cell search algorithm in DS-CDMA mobile radio using long spreading codes," in *Proceedings of IEEE VTC '97*, vol. 3, (Phoenix, Arizona, USA), pp. 1430–1434, IEEE, 4–7 May 1997.

[118] M. Golay, "Complementary series," *IRE Transactions on Information Theory*, vol. IT-7, pp. 82–87, 1961.

[119] V. Tarokh, H. Jafarkhani, and A. Calderbank, "Space-time block codes from orthogonal designs," *IEEE Transactions on Information Theory*, vol. 45, pp. 1456–1467, May 1999.

[120] W. Lee, *Mobile Communications Engineering.* New York: McGraw-Hill, 2nd ed., 1997.

[121] H. Wong and J. Chambers, "Two-stage interference immune blind equaliser which exploits cyclostationary statistics," *Electronics Letters*, vol. 32, pp. 1763–1764, September 1996.

[122] C. Lee and R. Steele, "Effects of Soft and Softer Handoffs on CDMA System Capacity," *IEEE Transactions on Vehicular Technology*, vol. 47, pp. 830–841, August 1998.

[123] M. Gustafsson, K. Jamal, and E. Dahlman, "Compressed mode techniques for inter-frequency measurements in a wide-band DS-CDMA system," in *Proceedings of IEEE International Symposium on Personal, Indoor and Mobile Radio Communications, PIMRC'97* [377], pp. 231–235.

[124] D. Knisely, S. Kumar, S. Laha, and S. Nanda, "Evolution of wireless data services : IS-95 to cdma2000," *IEEE Communications Magazine*, vol. 36, pp. 140–149, October 1998.

[125] Telecommunications Industry Association (TIA), *The cdma2000 ITU-R RTT Candidate Submission*, 1998.

[126] D. Knisely, Q. Li, and N. Rames, "cdma2000: A third generation radio transmission technology," *Bell Labs Technical Journal*, vol. 3, pp. 63–78, July–September 1998.

[127] Y. Okumura and F. Adachi, "Variable-rate data transmission with blind rate detection for coherent DS-CDMA mobile radio," *IEICE Transactions on Communications*, vol. E81B, pp. 1365–1373, July 1998.

[128] R. Steele, ed., *Mobile Radio Communications*. IEEE Press-Pentech Press, 1992.

[129] M. Raitola, A. Hottinen, and R. Wichman, "Transmission diversity in wideband CDMA," in *Proceedings of VTC'98 (Spring)* [376], pp. 1545–1549.

[130] J. Liberti Jr. and T. Rappaport, "Analytical results for capacity improvements in CDMA," *IEEE Transactions on Vehicular Technology*, vol. 43, pp. 680–690, August 1994.

[131] J. Winters, "Smart antennas for wireless systems," *IEEE Personal Communications*, vol. 5, pp. 23–27, February 1998.

[132] T. Lim and L. Rasmussen, "Adaptive symbol and parameter estimation in asynchronous multiuser CDMA detectors," *IEEE Transactions on Communications*, vol. 45, pp. 213–220, February 1997.

[133] T. Lim and S. Roy, "Adaptive filters in multiuser (MU) CDMA detection," *Wireless Networks*, vol. 4, pp. 307–318, June 1998.

[134] L. Wei, "Rotationally-invariant convolutional channel coding with expanded signal space, part I and II," *IEEE Transactions on Selected Areas in Comms*, vol. SAC-2, pp. 659–686, September 1984.

[135] T. Lim and M. Ho, "LMS-based simplifications to the kalman filter multiuser CDMA detector," in *Proceedings of IEEE Asia-Pacific Conference on Communications/International Conference on Communication Systems*, (Singapore), November 1998.

[136] D. You and T. Lim, "A modified blind adaptive multiuser CDMA detector," in *Proceedings of IEEE International Symposium on Spread Spectrum Techniques and Application (ISSSTA'98)* [379], pp. 878–882.

[137] S. Sun, L. Rasmussen, T. Lim, and H. Sugimoto, "Impact of estimation errors on multiuser detection in CDMA," in *Proceedings of IEEE Vehicular Technology Conference (VTC'98)* [378], pp. 1844–1848.

[138] Y. Sanada and Q. Wang, "A co-channel interference cancellation technique using orthogonal convolutional codes on multipath rayleigh fading channel," *IEEE Transactions on Vehicular Technology*, vol. 46, pp. 114–128, February 1997.

[139] P. Patel and J. Holtzman, "Analysis of a simple successive interference cancellation scheme in a DS/CDMA system," *IEEE Journal on Selected Areas in Communications*, vol. 12, pp. 796–807, June 1994.

[140] P. Tan and L. Rasmussen, "Subtractive interference cancellation for DS-CDMA systems," in *Proceedings of IEEE Asia-Pacific Conference on Communications/International Conference on Communication Systems*, (Singapore), November 1998.

[141] K. Cheah, H. Sugimoto, T. Lim, L. Rasmussen, and S. Sun, "Performance of hybrid interference canceller with zero-delay channel estimation for CDMA," in *Proceedings of Globecom'98*, (Sydney, Australia), pp. 265–270, IEEE, 8–12 Nov 1998.

[142] S. Sun, L. Rasmussen, and T. Lim, "A matrix-algebraic approach to hybrid interference cancellation in CDMA," in *Proceedings of IEEE International Conference on Universal Personal Communications '98*, (Florence, Italy), pp. 1319–1323, October 1998.

[143] A. Johansson and L. Rasmussen, "Linear group-wise successive interference cancellation in CDMA," in *Proceedings of IEEE International Symposium on Spread Spectrum Techniques and Application (ISSSTA'98)* [379], pp. 121–126.

[144] S. Sun, L. Rasmussen, H. Sugimoto, and T. Lim, "A hybrid interference canceller in CDMA," in *Proceedings of IEEE International Symposium on Spread Spectrum Techniques and Application (ISSSTA'98)* [379], pp. 150–154.

[145] D. Guo, L. Rasmussen, S. Sun, T. Lim, and C. Cheah, "MMSE-based linear parallel interference cancellation in CDMA," in *Proceedings of IEEE International Symposium on Spread Spectrum Techniques and Application (ISSSTA'98)* [379], pp. 917–921.

[146] L. Rasmussen, D. Guo, Y. Ma, and T. Lim, "Aspects on linear parallel interference cancellation in CDMA," in *Proceedings of IEEE International Symposium on Information Theory*, (Cambridge, MA), p. 37, August 1998.

[147] L. Rasmussen, T. Lim, H. Sugimoto, and T. Oyama, "Mapping functions for successive interference cancellation in CDMA," in *Proceedings of IEEE Vehicular Technology Conference (VTC'98)* [378], pp. 2301–2305.

[148] S. Sun, T. Lim, L. Rasmussen, T. Oyama, H. Sugimoto, and Y. Matsumoto, "Performance comparison of multi-stage SIC and limited tree-search detection in CDMA," in *Proceedings of IEEE Vehicular Technology Conference (VTC'98)* [378], pp. 1854–1858.

[149] H. Sim and D. Cruickshank, "Chip based multiuser detector for the downlink of a DS-CDMA system using a folded state-transition trellis," in *Proceedings of VTC'98 (Spring)* [376], pp. 846–850.

[150] J. Blogh and L. Hanzo, *3G Systems and Intelligent Networking*. John Wiley and IEEE Press, 2002. (For detailed contents please refer to http://www-mobile.ecs.soton.ac.uk.).

[151] L. Hanzo, P. Cherriman, and J. Streit, *Video Compression and Communications over Wireless Channels: From Second to Third Generation Systems and Beyond*. IEEE Press and John Wiley, 2001. (For detailed contents please refer to http://www-mobile.ecs.soton.ac.uk.).

[152] L. Hanzo, "Bandwidth-efficient wireless multimedia communications," *Proceedings of the IEEE*, vol. 86, pp. 1342–1382, July 1998.

[153] S. Nanda, K. Balachandran, and S. Kumar, "Adaptation techniques in wireless packet data services," *IEEE Communications Magazine*, vol. 38, pp. 54–64, January 2000.

[154] Research and Development Centre for Radio Systems, Japan, *Public Digital Cellular (PDC) Standard, RCR STD-27*.

[155] Telcomm. Industry Association (TIA), Washington, DC, USA, *Dual-mode subscriber equipment — Network equipment compatibility specification, Interim Standard IS-54*, 1989.

[156] Telcomm. Industry Association (TIA), Washington, DC, USA, *Mobile station — Base station compatibility standard for dual-mode wideband spread spectrum cellular system, EIA/TIA Interim Standard IS-95*, 1993.

[157] T. Ojanperä and R. Prasad, *Wideband CDMA for Third Generation Mobile Communications*. Artech House, Inc., 1998.

[158] W. Webb and R. Steele, "Variable rate QAM for mobile radio," *IEEE Transactions on Communications*, vol. 43, no. 7, pp. 2223–2230, 1995.

[159] S. Sampei, S. Komaki, and N. Morinaga, "Adaptive Modulation/TDMA scheme for large capacity personal multimedia communications systems," *IEICE Transactions on Communications*, vol. E77-B, pp. 1096–1103, September 1994.

[160] J. M. Torrance and L. Hanzo, "Upper bound performance of adaptive modulation in a slow Rayleigh fading channel," *Electronics Letters*, vol. 32, pp. 718–719, 11 April 1996.

[161] C. Wong and L. Hanzo, "Upper-bound performance of a wideband burst-by-burst adaptive modem," *IEEE Transactions on Communications*, vol. 48, pp. 367–369, March 2000.

[162] J. M. Torrance and L. Hanzo, "Optimisation of switching levels for adaptive modulation in a slow Rayleigh fading channel," *Electronics Letters*, vol. 32, pp. 1167–1169, 20 June 1996.

[163] H. Matsuoka, S. Sampei, N. Morinaga, and Y. Kamio, "Adaptive modulation system with variable coding rate concatenated code for high quality multi-media communications systems," in *Proceedings of IEEE VTC '96* [380], pp. 487–491.

[164] A. J. Goldsmith and S. G. Chua, "Variable Rate Variable Power MQAM for Fading Channels," *IEEE Transactions on Communications*, vol. 45, pp. 1218–1230, October 1997.

[165] S. Otsuki, S. Sampei, and N. Morinaga, "Square-qam adaptive modulation/TDMA/TDD systems using modulation estimation level with walsh function," *IEE Electronics Letters*, vol. 31, pp. 169–171, February 1995.

[166] J. Torrance and L. Hanzo, "Demodulation level selection in adaptive modulation," *Electronics Letters*, vol. 32, pp. 1751–1752, 12 September 1996.

[167] Y. Kamio, S. Sampei, H. Sasaoka, and N. Morinaga, "Performance of modulation-level-control adaptive-modulation under limited transmission delay time for land mobile communications," in *Proceedings of IEEE Vehicular Technology Conference (VTC'95)*, (Chicago, USA), pp. 221–225, IEEE, 15–28 July 1995.

[168] J. M. Torrance and L. Hanzo, "Latency and Networking Aspects of Adaptive Modems over Slow Indoors Rayleigh Fading Channels," *IEEE Transactions on Vehicular Technology*, vol. 48, pp. 1237–1251, July 1999.

[169] T. Ue, S. Sampei, and N. Morinaga, "Symbol rate controlled adaptive modulation/TDMA/TDD for wireless personal communication systems," *IEICE Transactions on Communications*, vol. E78-B, pp. 1117–1124, August 1995.

[170] T. Suzuki, S. Sampei, and N. Morinaga, "Space and path diversity combining technique for 10 Mbits/s adaptive modulation/TDMA in wireless communications systems," in *Proceedings of IEEE VTC '96* [380], pp. 1003–1007.

[171] K. Arimochi, S. Sampei, and N. Morinaga, "Adaptive modulation system with discrete power control and predistortion-type non-linear compensation for high spectral efficient and high power efficient wireless communication systems," in *Proceedings of IEEE International Symposium on Personal, Indoor and Mobile Radio Communications, PIMRC'97* [377], pp. 472–477.

[172] T. Ikeda, S. Sampei, and N. Morinaga, "TDMA-based adaptive modulation with dynamic channel assignment (AMDCA) for high capacity multi-media microcellular systems," in *Proceedings of IEEE Vehicular Technology Conference*, (Phoenix, USA), pp. 1479–1483, May 1997.

[173] T. Ue, S. Sampei, and N. Morinaga, "Adaptive modulation packet radio communication system using NP-CSMA/TDD scheme," in *Proceedings of IEEE VTC '96* [380], pp. 416–421.

[174] M. Naijoh, S. Sampei, N. Morinaga, and Y. Kamio, "ARQ schemes with adaptive modulation/TDMA/TDD systems for wireless multimedia communication systems," in *Proceedings of IEEE International Symposium on Personal, Indoor and Mobile Radio Communications, PIMRC'97* [377], pp. 709–713.

[175] S. Sampei, T. Ue, N. Morinaga, and K. Hamguchi, "Laboratory experimental results of an adaptive modulation TDMA/TDD for wireless multimedia communication systems," in *Proceedings of IEEE International Symposium on Personal, Indoor and Mobile Radio Communications, PIMRC'97* [377], pp. 467–471.

[176] J. Torrance and L. Hanzo, "Interference aspects of adaptive modems over slow Rayleigh fading channels," *IEEE Transactions on Vehicular Technology*, vol. 48, pp. 1527–1545, September 1999.

[177] L. Hanzo, T. H. Liew, and B. L. Yeap, *Turbo Coding, Turbo Equalisation and Space-Time Coding*. John Wiley, IEEE Press, 2002. (For detailed contents please refer to http://www-mobile.ecs.soton.ac.uk.).

[178] J. Cheung and R. Steele, "Soft-decision feedback equalizer for continuous-phase modulated signals in wide-band mobile radio channels," *IEEE Transactions on Communications*, vol. 42, pp. 1628–1638, February/March/April 1994.

[179] M. Yee, T. Liew, and L. Hanzo, "Radial basis function decision feedback equalisation assisted block turbo burst-by-burst adaptive modems," in *Proceedings of VTC '99 Fall*, (Amsterdam, Holland), pp. 1600–1604, 19-22 September 1999.

[180] M. S. Yee, B. L. Yeap, and L. Hanzo, "Radial basis function assisted turbo equalisation," in *Proceedings of IEEE Vehicular Technology Conference*, (Japan, Tokyo), pp. 640–644, IEEE, 15-18 May 2000.

[181] L. Hanzo, F. Somerville, and J. Woodard, *Voice Compression and Communications: Principles and Applications for Fixed and Wireless Channels*. IEEE Press and John Wiley, 2002. 2001 (For detailed contents, please refer to http://www-mobile.ecs.soton.ac.uk.).

[182] ITU-T, *Recommendation H.263: Video coding for low bitrate communication*, March 1996.

[183] A. Klein, R. Pirhonen, J. Skoeld, and R. Suoranta, "FRAMES multiple access mode 1 — wideband TDMA with and without spreading," in *Proceedings of IEEE International Symposium on Personal, Indoor and Mobile Radio Communications, PIMRC'97* [377], pp. 37–41.

[184] P. Cherriman, C. Wong, and L. Hanzo, "Turbo- and BCH-coded wide-band burst-by-burst adaptive H.263-assisted wireless video telephony," *IEEE Transactions on Circuits and Systems for Video Technology*, vol. 10, pp. 1355–1363, December 2000.

[185] T. Keller and L. Hanzo, "Adaptive multicarrier modulation: A convenient framework for time-frequency processing in wireless communications," *Proceedings of the IEEE*, vol. 88, pp. 611–642, May 2000.

[186] A. Klein and P. Baier, "Linear unbiased data estimation in mobile radio sytems applying CDMA," *IEEE Journal on Selected Areas in Communications*, vol. 11, pp. 1058–1066, September 1993.

[187] K. Gilhousen, I. Jacobs, R. Padovani, A. Viterbi, L. Weaver Jr., and C. Wheatley III, "On the capacity of a cellular CDMA system," *IEEE Transactions on Vehicular Technology*, vol. 40, pp. 303–312, May 1991.

[188] S. Kim, "Adaptive rate and power DS/CDMA communications in fading channels," *IEEE Communications Letters*, vol. 3, pp. 85–87, April 1999.

[189] T. Ottosson and A. Svensson, "On schemes for multirate support in DS-CDMA systems," *Wireless Personal Communications (Kluwer)*, vol. 6, pp. 265–287, March 1998.

[190] S. Ramakrishna and J. Holtzman, "A comparison between single code and multiple code transmission schemes in a CDMA system," in *Proceedings of IEEE Vehicular Technology Conference (VTC'98)* [378], pp. 791–795.

[191] M. Saquib and R. Yates, "Decorrelating detectors for a dual rate synchronous DS/CDMAchannel," *Wireless Personal Communications (Kluwer)*, vol. 9, pp. 197–216, May 1999.

[192] A.-L. Johansson and A. Svensson, "Successive interference cancellation schemes in multi-rate DS/CDMA systems," in *Wireless Information Networks (Baltzer)*, pp. 265–279, 1996.

[193] S. Abeta, S. Sampei, and N. Morinaga, "Channel activation with adaptive coding rate and processing gain control for cellular DS/CDMA systems," in *Proceedings of IEEE VTC '96* [380], pp. 1115–1119.

[194] M. Hashimoto, S. Sampei, and N. Morinaga, "Forward and reverse link capacity enhancement of DS/CDMA cellular system using channel activation and soft power control techniques," in *Proceedings of IEEE International Symposium on Personal, Indoor and Mobile Radio Communications, PIMRC'97* [377], pp. 246–250.

[195] T. Liew, C. Wong, and L. Hanzo, "Block turbo coded burst-by-burst adaptive modems," in *Proceedings of Microcoll'99, Budapest, Hungary*, pp. 59–62, 21–24 March 1999.

[196] L. Hanzo, P. Cherriman, and J. Streit, *Wireless Video Communications: From Second to Third Generation Systems, WLANs and Beyond.* IEEE Press-John Wiley, 2001. IEEE Press, 2001. (For detailed contents please refer to http://www-mobile.ecs.soton.ac.uk.).

[197] V. Tarokh, N. Seshadri, and A. Calderbank, "Space-time codes for high data rate wireless communication: Performance criterion and code construction," *IEEE Transactions on Information Theory*, vol. 44, pp. 744–765, March 1998.

[198] V. Tarokh, A. Naguib, N. Seshadri, and A. Calderbank, "Space-time codes for high data rate wireless communication: Performance criteria in the presence of channel estimation errors, mobility, and multiple paths," *IEEE Transactions on Communications*, vol. 47, pp. 199–207, February 1999.

[199] V. Tarokh, N. Seshadri, and A. Calderbank, "Space-time codes for high data rate wireless communications: Performance criterion and code construction," in *Proc IEEE International Conference on Communications '97*, (Montreal, Canada), pp. 299–303, 1997.

[200] V. Tarokh, H. Jafarkhani, and A. Calderbank, "Space-time block codes from orthogonal designs," *IEEE Transactions on Information Theory*, vol. 45, pp. 1456–1467, July 1999.

[201] T. Rappaport, ed., *Smart Antennas: Adaptive Arrays, Algorithms and Wireless Position Location*. IEEE, 1998.

[202] B. Widrow, P. E. Mantey, L. J. Griffiths and B. B. Goode, "Adaptive Antenna Systems," in *Proceedings of the IEEE*, vol. 55, pp. 2143–2159, December 1967.

[203] S. P. Applebaum, "Adaptive Arrays," *IEEE Transactions on Antennas and Propagation*, vol. AP-24, pp. 585–598, September 1976.

[204] O. L. Frost III, "An Algorithm for Linearly Constrained Adaptive Array Processing," in *Proceedings of the IEEE*, vol. 60, pp. 926–935, August 1972.

[205] I. S. Reed, J. D. Mallett and L. E. Brennan, "Rapid Convergence Rate in Adaptive Arrays," *IEEE Transactions on Aerospace Electronic System*, vol. AES-10, pp. 853–863, November 1974.

[206] J. Fernandez, I. R. Corden and M. Barrett, "Adaptive Array Algorithms for Optimal Combining in Digital Mobile Communications Systems," *IEE 8th International Conference on Antennas and Propagation*, pp. 983–986, 1993.

[207] L.C. Godara, "Applications of Antenna Arrays to Mobile Communications, Part I: Performance Improvement, Feasibility, and System Considerations," in *Proceedings of the IEEE*, vol. 85, pp. 1031–1060, July 1997.

[208] L.C. Godara, "Application of Antenna Arrays to Mobile Communications, Part II: Beam-Forming and Direction-of-Arrival Considerations," in *Proceedings of the IEEE*, vol. 85, pp. 1195–1245, August 1997.

[209] W. F. Gabriel, "Adaptive Processing Array Systems," in *Proceedings of the IEEE*, vol. 80, pp. 152–162, January 1992.

[210] A. J. Paulraj and C. B. Papadias, "Space Time Processing for Wireless Communications," *IEEE Personal Communications*, vol. 14, pp. 49–83, November 1997.

[211] J. H. Winters, "Smart Antennas for Wireless Systems," *IEEE Personal Communications*, vol. 1, pp. 23–27, February 1998.

[212] R. Kohno, "Spatial and Temporal Communication Theory Using Adaptive Antenna Array," *IEEE Personal Communications*, vol. 1, pp. 28–35, February 1998.

[213] H. Krim and M. Viberg, "Two Decades of Array Signal Processing Research," *IEEE Signal Processing Magazine*, pp. 67–94, July 1996.

[214] G. V. Tsoulos, "Smart Antennas for Mobile Communication Systems: Benefits and Challenges," *IEE Electronics and Communication Engineering Journal*, vol. 11, pp. 84–94, April 1999.

[215] Special Issue on Active and Adaptive Antennas, *IEEE Transactions on Antennas and Propagation*, vol. AP-12. March 1964.

[216] Special Issue on Adaptive Antennas, *IEEE Transaction on Antennas and Propagation*, vol. AP-24. September 1976.

[217] Special Issue on Adaptive Antennas, *IEEE Transaction on Antennas and Propagation*, vol. AP-34. March 1986.

[218] A. Paulraj, R. Roy and T. Kailath, "A Subspace Rotation Approach to Signal Parameter Estimation," in *Proceedings of the IEEE*, vol. 74, pp. 1044–1046, July 1986.

[219] J. H. Winters, "Signal Acquisition and Tracking with Adaptive Arrays in the Digital Mobile Radio System IS-54 with Flat Fading," *IEEE Transactions on Vehicular Technology*, vol. 42, pp. 373–384, November 1993.

[220] L. C. Godara and D. B. Ward, "A General Framework for Blind Beamforming," *IEEE TENCON*, pp. 1240–1243, June 1999.

[221] W. Pora, J. A. Chambers and A. G. Constantinides, "A Combined Kalman Filter and Constant Modulus Algorithm Beamformer for Fast Fading Channels," *IEEE International Conference on Acoustics, Speech and Signal Processing*, vol. 5, pp. 2925–2928, March 1999.

[222] J. H. Winters, J. Salz and R. D. Gitlin, "The Impact of Antenna Diversity on the Capacity of Wireless Communications System," *IEEE Transactions on Communications*, vol. 42, February/March/April 1994.

[223] G. V. Tsoulos, M. A. Beach and S. C. Swales, "Adaptive Antennas for Third Generation DS-CDMA Cellular Systems," *IEEE Vehicular Technology Conference*, vol. 45, pp. 45–49, 1995.

[224] L. E. Brennan and I. S. Reed, "Theory of Adaptive Radar," *IEEE Transactions on Aerospace and Electronic Systems*, vol. AES-9, pp. 237–252, March 1973.

[225] L. E. Brennan, E. L. Pugh and I. S. Reed, "Control Loop Noise in Adaptive Array Antennas," *IEEE Transactions on Aerospace and Electronic Systems*, vol. AES-7, pp. 254–262, March 1971.

[226] L. E. Brennan and I. S. Reed, "Effect of Envelope Limiting in Adaptive Array Control Loops," *IEEE Transactions on Aerospace and Electronic Systems*, vol. AES-7, pp. 698–700, July 1971.

[227] L. E. Brennan, J. Mallet and I. S. Reed, "Adaptive Arrays in Airborne MTI Radar," *IEEE Transactions on Antennas and Propagation*, vol. AP-24, pp. 607–615, September 1976.

[228] B. Widrow and J. M. McCool, "A Comparison of Adaptive Algorithms Based on the Method of Steepest Descent and Random Search," *IEEE Transactions on Antennas and Propagation*, vol. AP-24, pp. 615–637, September 1976.

[229] R. O. Schmidt, "Multiple Emitter Location and Signal Parameter Estimation," *IEEE Transactions on Antennas Propagation*, vol. AP-34, pp. 276–280, July 1986.

[230] R. Roy and T. Kailath , "ESPRIT-Estimation of Signal Parameters via Rotational Invariance Techniques," *IEEE Transactions on Acoustic, Speech and Signal Processing*, vol. ASSP-37, pp. 984–995, July 1989.

[231] L. J. Griffiths, "A Simple Adaptive Algorithm for Real-time Processing in Antenna Arrays," in *Proceedings of the IEEE*, vol. 57, pp. 1696–1704, October 1969.

[232] B. D. Van Veen and K. M. Buckley, "Beamforming: A Versatile Approach to Spatial Filtering," *IEEE Acoustic, Speech and Signal Processing Magazine*, pp. 4–24, April 1988.

[233] T. Chen, "Highlights of Statistical Signal and Array Processing," *IEEE Signal Processing Magazine*, vol. 15, pp. 21–64, September 1998.

[234] C. Q. Xu, C. L. Law, S. Yoshida, "On Nonlinear Beamforming for Interference Cancellation," in *IEEE Vehicular Technology Conference*, May 2001.

[235] A. Margarita, S. J. Flores, L. Rubio, V. Almenar and J. L. Corral, "Application of MUSIC for Spatial Reference Beamforming for SDMA in a Smart Antenna for GSM and DECT," *IEEE Vehicular Technology Conference*, May 2001.

[236] J. E. Hudson, *Adaptive Array Principles*. New York: Peter Peregrinus, Ltd., 1981.

[237] S. Haykin, *Array Signal Processing*. New Jersey: Prentice Hall, Inc., 1985.

[238] S. P. Applebaum, "Adaptive Arrays," tech. rep., Syracuse University Research Corporation, 1965. Reprinted in IEEE Transactions on Antennas and Propagation, September 1976.

[239] B. Widrow, P. E. Mantey, L. J. Griffiths, and B. B. Goode, "Adaptive Antenna Systems," *Proceedings of the IEEE*, vol. 55, pp. 2143–2159, December 1967.

[240] O. L. Frost III, "An Algorithm for Linearly Constrained Adaptive Array Processing," *Proceedings of the IEEE*, vol. 60, pp. 926–935, August 1972.

[241] L. J. Griffiths, "A Simple Adaptive Algorithm for Real-Time Processing in Antenna Arrays," *Proceedings of the IEEE*, vol. 57, pp. 1696–1704, October 1969.

[242] L. C. Godara, "Applications of Antenna Arrays to Mobile Communications, Part II: Beam-Forming and Direction-of-Arrival Considerations," *Proceedings of the IEEE*, vol. 85, pp. 1193–1245, August 1997.

[243] J. Capon, "High-resolution frequency-wavenumber spectrum analysis," *Proceedings of the IEEE*, vol. 57, pp. 1408–1418, August 1969.

[244] I. S. Reed, J. D. Mallett, and L. E. Brennan, "Rapid Convergence Rate in Adaptive Arrays," *IEEE Transactions on Aerospace and Electronic Systems*, vol. AES-10, pp. 853–863, November 1974.

[245] A. Paulraj and C. B. Papadias, "Space-Time Processing for Wireless Communications," *IEEE Signal Processing Magazine*, pp. 49–83, November 1997.

[246] J. E. Hudson, *Adaptive Array Principles*. Peregrinus, London, 1981.

[247] S. Haykin, *Adaptive Filter Theory*. Prentice-Hall International, 1991.

[248] B. Widrow and S. Steams, *Adaptive Signal Processing*. Prentice-Hall, 1985.

[249] R. A. Monzingo and T. W. Miller, *Introduction to Adaptive Arrays*. John Wiley & Sons, Inc., 1980.

[250] J. H. Winters, "Smart Antennas for Wireless Systems," *IEEE Personal Communications*, vol. 5, pp. 23–27, February 1998.

[251] U. Martin and I. Gaspard, "Capacity Enhancement of Narrowband CDMA by Intelligent Antennas," in *Proceedings of PIMRC*, pp. 90–94, 1997.

[252] A. R. Lopez, "Performance Predictions for Cellular Switched-Beam Intelligent Antenna Systems," *IEEE Communications Magazine*, pp. 152–154, October 1996.

[253] C. M. Simmonds and M. A. Beach, "Active Calibration of Adaptive Antenna Arrays for Third Generation Systems," in *Proceedings of ACTS Summit*, 1997.

[254] H. Steyskal, "Digital Beamforming Antennas: An Introduction," *Microwave Journal*, pp. 107–124, January 1987.

[255] A. Mammela, *Diversity receivers in a fast fading multipath channel*. VTT Publications, 1995.

[256] W. Hollemans, "Performance Analysis of Cellular Digital Mobile Radio Systems including Diversity Techniques," in *Proceedings of PIMRC*, pp. 266–270, 1997.

[257] W. Tuttlebee, ed., *Cordless telecommunications in Europe : the evolution of personal communications*. London: Springer-Verlag, 1990. ISBN 3540196331.

[258] H. Ochsner, "The digital european cordless telecommunications specification, DECT," in Tuttlebee [257], pp. 273–285. ISBN 3540196331.

[259] P. Petrus, J. H. Reed, and T. S. Rappaport, "Effects of Directional Antennas at the Base Station on the Doppler Spectrum," *IEEE Communications Letters*, vol. 1, pp. 40–42, March 1997.

[260] H. Krim and M. Viberg, "Two Decades of Array Signal Processing Research," *IEEE Signal Processing Magazine*, pp. 67–94, July 1996.

[261] S. Ponnekanti, A. Pollard, C. Taylor, and M. G. Kyeong, "Flexibility for the deployment of adaptive antennas in the IMT-2000 framework and enhanced interference cancellation," in *Proceedings of ACTS Summit*, 1997.

[262] J. H. Winters, J. Salz, and R. D. Gitlin, "The Impact of Antenna Diversity on the Capacity of Wireless Communication Systems," *IEEE Transactions on Communications*, vol. 42, pp. 1740–1751, February/March/April 1994.

[263] M. Barnard and S. McLaughlin, "Reconfigurable terminals for mobile communication systems," *IEE Electronics and Communication Engineering Journal*, vol. 12, pp. 281–292, December 2000.

[264] S. M. Leach, A. A. Agius, and S. R. Saunders, "The intelligent quadrifilar helix antenna," *IEE Proceedings of Microwave Antennas Propagation*, pp. 219–223, June 2000.

[265] P. Petrus, R. B. Ertel, and J. H. Reed, "Capacity Enhancement Using Adaptive Arrays in an AMPS System," *IEEE Transactions on Vehicular Technology*, vol. 47, pp. 717–727, August 1998.

[266] J. Laurila and E. Bonek, "SDMA Using Blind Adapation," in *Proceedings of ACTS Summit*, 1997.

[267] M. C. Wells, "Increasing the capacity of GSM cellular radio using adaptive antennas," *IEE Proceedings on Communications*, vol. 143, pp. 304–310, October 1996.

[268] J. H. Winters, "Signal Acquisition and Tracking with Adaptive Arrays in the Digital Mobile Radio System IS-54 with Flat Fading," *IEEE Transactions on Vehicular Technology*, vol. 42, November 1993.

[269] B. Widrow and E. Walach, "On the statistical efficiency of the LMS algorithm with nonstationary inputs," *IEEE Trans. Information Theory - Special Issue on Adaptive Filtering*, vol. 30, pp. 211–221, March 1984.

[270] Z. Raida, "Steering an Adaptive Antenna Array by the Simplified Kalman Filter," *IEEE Trans. on Antennas and Propagation*, vol. 43, pp. 627–629, June 1995.

[271] M. W. Ganz, R. L. Moses, and S. L. Wilson, "Convergence of the SMI Algorithms with Weak Interference," *IEE Trans. Antenna Propagation*, vol. 38, pp. 394–399, March 1990.

[272] H. Steyskal, "Array Error Effects in Adaptive Beamforming," *Microwave Journal*, September 1991.

[273] M. C. Vanderveen, C. B. Papadias, and A. Paulraj, "Joint Angle and Delay Estimation (JADE) for Multipath Signals Arriving at an Antenna Array," *IEEE Communications Letters*, vol. 1, pp. 12–14, January 1997.

[274] C. Passman and T. Wixforth, "A Calibrated Phased Array Antenna with Polarization Flexibility for the Tsunami (II) SDMA Field Trial," in *Proceedings of ACTS Summit*, 1997.

[275] D. N. Godard, "Self-Recovering Equalization and Carrier Tracking in Two-Dimensional Data Communication Systems," *IEEE Transactions on Communications*, vol. COM-28, pp. 1876–1875, November 1980.

[276] Z. Ding, R. A. Kennedy, B. D. O. Anderson, and C. R. Johnson Jr, "Ill-Convergence of Godard Blind Equalizers in Data Communication Systems," *IEEE Transactions on Communications*, vol. 39, pp. 1313–1327, September 1991.

[277] J. E. Mazo, "Analysis of decision-directed equalizer convergence," *Bell Systems Technical Journal*, 1980.

[278] D. Gerlach and A. Paulraj, "Adaptive Transmitting Antenna Arrays with Feedback," *IEEE Signal Processing Letters*, vol. 1, pp. 150–152, October 1994.

[279] D. Gerlach and A. Paulraj, "Base station transmitting antenna arrays for multipath environments," *Signal Processing*, pp. 59–73, 1996.

[280] T. Kanai, "Autonomous Reuse Partitioning in Cellular Systems," in *IEEE Proceedings of Vehicular Technology Conference*, vol. 2, pp. 782–785, 1992.

[281] I. Katzela and M. Naghshineh, "Channel Assignment Schemes for Cellular Mobile Telecommunication Systems: A Comprehensive Survey," *IEEE Personal Communications Magazine*, vol. 3, pp. 10–31, June 1996.

[282] M. Dell'Anna and A. H. Aghvami, "Performance of optimum and sub-optimum combining at the antenna array of a W-CDMA system," *IEEE Journal on Selected Areas in Communications*, pp. 2123–2137, December 1999.

[283] I. Howitt and Y. M. Hawwar, "Evaluation of Outage Probability Due to Cochannel Interference in Fading for a TDMA System with a Beamformer," in *Proceedings of VTC*, pp. 520–524, 1998.

[284] L. Ortigoza-Guerrero and A. H. Aghvami, "A self-adaptive prioritised hand-off DCA strategy for a microcellular environment," in *Proceedings of PIMRC*, (Helsinki, Finland), pp. 401–405, September 1997.

[285] L. Ortigoza-Guerrero and A. H. Aghvami, "A prioritised hand-off dynamic channel allocation strategy for PCS," *IEEE Transactions on Vehicular Technology*, pp. 1203–1215, July 1999.

[286] T. H. Le and H. Aghvami, "Fast channel access and DCA scheme for connection and connectionless-oriented services in UMTS," *Electronics Letters*, pp. 1048–1049, June 1999.

[287] L. Anderson, "A simulation study of some dynamic channel assignment algorithms in a high capacity mobile telecommunications system," *IEEE Trans. on Communication*, vol. 21, pp. 1294–1301, November 1973.

[288] J. I. Chuang, "Performance issues and algorithms for dynamic channel assignment," *IEEE JSAC*, vol. 11, pp. 955–963, August 1993.

[289] J. I. Chuang and N. Sollenberger, "Performance of autonomous dynamic channel assignment and power control for TDMA/FDMA wireless access," *IEEE JSAC*, vol. 12, pp. 1314–1323, October 1994.

[290] M. L. Cheng and J. I. Chuang, "Performance evaluation of distributed measurement-based dynamic channel assignment in local wireless communications," *IEEE JSAC*, vol. 14, pp. 698–710, May 1996.

[291] I. ChihLin and C. PiHui, "Local packing - distributed dynamic channel allocation at cellular base station," in *Proceedings of IEEE Globecom '93*, vol. 1, (Houston, TX, USA), pp. 293–301, Nov 29–Dec 2 1993.

[292] G. L. Stüber, *Principles of Mobile Communcation*. Kluwer Academic Publishers, 1996.

[293] A. Baiocchi, F. Delli-Priscoli, F. Grilli, and F. Sestini, "The geometric dynamic channel allocation as a practical strategy in mobile networks with bursty user mobility," *IEEE Trans. on Vech. Tech.*, vol. 44, pp. 14–23, Feb 1995.

[294] F. D. Priscoli, N. P. Magnani, V. Palestini, and F. Sestini, "Application of dynamic channel allocation strategies to the GSM cellular network," *IEEE Journal on Selected Areas in Comms.*, vol. 15, pp. 1558–1567, Oct 1997.

[295] P. J. Cherriman, F. Romiti, and L. Hanzo, "Channel Allocation for Third-generation Mobile Radio Systems," in *ACTS '98, Rhodes, Greece*, pp. 255–260, June 1998.

[296] R. B. Ertel, P. Cardieri, K. W. Sowerby, T. S. Rappaport, and J. H. Reed, "Overview of Spatial Channel Models for Antenna Array Communications Systems," *IEEE Personal Communications*, pp. 10–22, February 1998.

[297] J. C. Liberti and T. S. Rappaport, "A Geometrically Based Model for Line-Of-Sight Multipath Radio Channels," in *VTC Proceedings*, pp. 844–848, 1996.

[298] S. W. Wales, "Technique for cochannel interference suppression in TDMA mobile radio systems," *IEE Proc. Communication*, vol. 142, no. 2, pp. 106–114, 1995.

[299] J. Litva and T.Lo, *Digital Beamforming in Wireless Communications*. Artech House, London, 1996.

[300] L. Godara, "Applications of antenna arrays to mobile communications, part I: Performance improvement, feasibility, and system considerations," *Proceedings of the IEEE*, vol. 85, pp. 1029–1060, July 1997.

[301] L. Godara, "Applications of antenna arrays to mobile communications, part II: Beamforming and direction-of-arrival considerations," *Proceedings of the IEEE*, vol. 85, pp. 1193–1245, Aug 1997.

[302] E. Sourour, "Time slot assignment techniques for TDMA digital cellular systems," *IEEE Trans. Vech. Tech.*, vol. 43, pp. 121–127, Feb 1994.

[303] D. Wong and T. Lim, "Soft handoffs in CDMA mobile systems," *IEEE Personal Comms.*, pp. 6–17, December 1997.

[304] S. Tekinay and B. Jabbari, "A measurement-based prioritisation scheme for handovers in mobile cellular networks," *IEEE JSAC*, vol. 10, no. 8, pp. 1343–1350, 1992.

[305] G. P. Pollini, "Trends in handover design," *IEEE Comms. Mag.*, vol. 34, pp. 82–90, March 1996.

[306] R. C. Bernhardt, "Timeslot re-assignment in a frequency reuse TDMA portable radio system," *IEEE Tr. on Vech. Tech.*, vol. 41, pp. 296–304, August 1992.

[307] S. Glisic and P. Leppanen, eds., *Wireless Communications : TDMA versus CDMA*. Kluwer Academic Publishers, June 1997. ISBN 0792380053.

[308] A. H. M. Ross and K. S. Gilhousen, "CDMA technology and the IS-95 north american standard," in Gibson [381], ch. 27, pp. 430–448.

[309] ETSI, *Universal Mobile Telecommunications Systems (UMTS); UMTS Terrestrial Radio Access (UTRA); Concept evaluation*, Dec 1997. TR 101 146 V3.0.0.

[310] I. Katzela and M. Naghshineh, "Channel assignment schemes for cellular mobile telecommunication systems: A comprehensive survey," *IEEE Personal Comms.*, pp. 10–31, June 1996.

[311] S. Tekinay and B. Jabbari, "Handover and channel assignment in mobile cellular networks," *IEEE Comms. Mag.*, pp. 42–46, November 1991.

[312] B. Jabbari, "Fixed and dynamic channel assignment," in Gibson [381], ch. 83, pp. 1175–1181.

[313] J. Zander, "Radio resource management in future wireless networks: Requirements and limitations," *IEEE Comms. Magazine*, pp. 30–36, Aug 1997.

[314] D. Everitt, "Traffic engineering of the radio interface for cellular mobile networks," *Proc. of the IEEE*, vol. 82, pp. 1371–1382, Sept 1994.

[315] J. Dahlin, "Ericsson's multiple reuse pattern for DCS1800," *Mobile Communications International*, November 1996.

[316] M. Madfors, K. Wallstedt, S. Magnusson, H. Olofsson, P. Backman, and S. Engström, "High capacity with limited spectrum in cellular systems," *IEEE Comms. Mag.*, vol. 35, pp. 38–45, August 1997.

[317] A. Safak, "Optimal channel reuse in cellular radio systems with multiple correlated log-normal interferers," *IEEE Tr. on Vech. Tech*, vol. 43, pp. 304–312, May 1994.

[318] H. Jiang and S. S. Rappaport, "Prioritized channel borrowing without locking: a channel sharing strategy for cellular communications," *IEEE/ACM Transactions on Networking*, vol. 43, pp. 163–171, April 1996.

[319] J. Engel and M. Peritsky, "Statistically optimum dynamic server assignment in systems with interfering servers," *IEEE Trans. on Vehicular Tech.*, vol. 22, pp. 203–209, Nov 1973.

[320] M. Zhang and T. Yum, "Comparisons of channel assignment strategies in cellular mobile telephone systems," *IEEE Trans. on Vehicular Tech.*, vol. 38, pp. 211–215, Nov 1989.

[321] S. M. Elnoubi, R. Singh, and S. Gupta, "A new frequency channel assignment algorithm in high capacity mobile communications systems," *IEEE Trans. on Vehicular Tech.*, vol. 31, pp. 125–131, Aug 1982.

[322] M. Zhang and T. Yum, "The non-uniform compact pattern allocation algorithm for cellular mobile systems," *IEEE Trans. on Vehicular Tech.*, vol. 40, pp. 387–391, May 1991.

[323] S. S. Kuek and W. C. Wong, "Ordered dynamic channel assignment scheme with reassignment in highway microcell," *IEEE Trans. on Vehicular Tech.*, vol. 41, pp. 271–277, Aug 1992.

[324] T. Yum and W. Wong, "Hot spot traffic relief in cellular systems," *IEEE Journal on selected areas in Comms.*, vol. 11, pp. 934–940, Aug 1993.

[325] J. Tajima and K. Imamura, "A strategy for flexible channel assignment in mobile communication systems," *IEEE Trans. on Vehicular Tech.*, vol. 37, pp. 92–103, May 1988.

[326] ETSI, *Digital European Cordless Telecommunications (DECT)*, 1st ed., October 1992. ETS 300 175-1 – ETS 300 175-9.

[327] R. Steele, "Digital European Cordless Telecommunications (DECT) systems," in Steele [128], ch. 1.7.2, pp. 79–83.

[328] S. Asghar, "Digital European Cordless Telephone," in Gibson [381], ch. 30, pp. 478–499.

[329] A. Law and L. B. Lopes, "Performance comparison of DCA call assignment algorithms within DECT," in *Proceedings of IEEE VTC '96* [380], pp. 726–730.

[330] H. Salgado-Galicia, M. Sirbu, and J. M. Peha, "A narrowband approach to efficient PCS spectrum sharing through decentralized DCA access policies," *IEEE Personal Communications*, pp. 24–34, Feb 1997.

[331] R. Steele, J. Whitehead, and W. C. Wong, "System aspects of cellular radio," *IEEE Communications Magazine*, vol. 33, pp. 80–86, Jan 1995.

[332] D. Cox and D. Reudink, "The behavior of dynamic channel-assignment mobile communications systems as a function of number of radio channels," *IEEE Trans. on Communications*, vol. 20, pp. 471–479, June 1972.

[333] D. D. Dimitrijević and J. Vučerić, "Design and performance analysis of the algorithms for channel allocation in cellular networks," *IEEE Trans. on Vehicular Tech.*, vol. 42, pp. 526–534, Nov 1993.

[334] D. C. Cox and D. O. Reudink, "Increasing channel occupancy in large scale mobile radio systems: Dynamic channel reassignment," *IEEE Trans. on Vehicular Tech.*, vol. 22, pp. 218–222, Nov 1973.

[335] D. C. Cox and D. O. Reudink, "A comparison of some channel assignment strategies in large-scale mobile communications systems," *IEEE Trans. on Communications*, vol. 20, pp. 190–195, April 1972.

[336] S. A. Grandhi, R. D. Yates, and D. Goodman, "Resource allocation for cellular radio systems," *IEEE Trans. Vech. Tech.*, vol. 46, pp. 581–587, Aug 1997.

[337] M. Serizawa and D. Goodman, "Instability and deadlock of distributed dynamic channel allocation," in *Proceedings of IEEE VTC '93*, (Secaucus, New Jersey, USA), pp. 528–531, May 18–20 1993.

[338] Y. Akaiwa and H. Andoh, "Channel segregation - a self-organized dynamic channel allocation method: Application to tdma/fdma microcellular system," *IEEE Journal on Selected Areas in Comms.*, vol. 11, pp. 949–954, Aug 1993.

[339] E. D. Re, R. Fantacci, and G. Giambene, "Handover and dynamic channel allocation techniques in mobile cellular networks," *IEEE Trans. on Vech. Tech.*, vol. 44, pp. 229–237, May 1995.

[340] T. Kahwa and N. Georganas, "A hybrid channel assignment scheme in large-scale, cellular-structured mobile communication systems," *IEEE Trans. on Communications*, vol. 26, pp. 432–438, April 1978.

[341] J. S. Sin and N. Georganas, "A simulation study of a hybrid channel assignment scheme for cellular land-mobile radio systems with erlang-c service," *IEEE Trans. on Communications*, vol. 29, pp. 143–147, Feb 1981.

[342] S.-H. Oh and D.-W. Tcha, "Prioritized channel assignment in a cellular radio network," *IEEE Trans. on Communications*, vol. 40, pp. 1259–1269, July 1992.

[343] D. Hong and S. Rappaport, "Traffic model and performance analysis for cellular mobile radio telephone systems with prioritizes and nonprioritized handoff procedures," *IEEE Trans. on Vehicular Technology*, vol. 35, pp. 77–92, Aug 1986.

[344] R. Guérin, "Queueing-blocking system with two arrival streams and guard channels," *IEEE Trans. on Communications*, vol. 36, pp. 153–163, Feb 1988.

[345] S. Grandhi, R. Vijayan, D. Goodman, and J. Zander, "Centralized power control in cellular radio systems," *IEEE Trans. Vech. Tech.*, vol. 42, pp. 466–468, Nov 1993.

[346] J. Zander, "Performance of optimum transmitter power control in cellular radio systems," *IEEE Tr. on Vehicular Technology*, vol. 41, pp. 57–62, Feb 1992.

[347] J. Zander, "Distributed cochannel interference control in cellular radio systems," *IEEE Tr. on Vehicular Technology*, vol. 41, pp. 305–311, Aug 1992.

[348] A. Tanenbaum, "Introduction to queueing theory," in *Computer Networks*, pp. 631–641, Prentice-Hall, 2nd ed., 1989. ISBN 0131668366.

[349] D. C. Cox and D. O. Reudink, "Effects of some nonuniform spatial demand profiles on mobile radio system performance," *IEEE Trans. on Vehicular Tech.*, vol. 21, pp. 62–67, May 1972.

[350] P. J. Cherriman, *Mobile Video Communications*. PhD thesis, University of Southampton, 1998.

[351] R. C. French, "The Effect of Fading and Shadowing on Channel Reuse in Mobile Radio," *IEEE Transactions on Vehicular Technology*, vol. 28, pp. 171–181, August 1979.

[352] W. Gosling, "A simple mathematical model of co-channel and adjacent channel interference in land mobile radio," *The Radio and Electronic Engineer*, vol. 48, pp. 619–622, December 1978.

[353] R. Muammar and S. C. Gupta, "Cochannel Interference in High-Capacity Mobile Radio Systems," *IEEE Transactions on Communications*, vol. 30, pp. 1973–1978, August 1982.

[354] P. J. Cherriman and L. Hanzo, "Error-rate-based power-controlled multimode H.263-assisted video telephony," *IEEE Transactions on Vehicular Technology*, vol. 48, pp. 1726–1738, September 1999.

[355] G. Foschini and Z. Miljanic, "Distributed Autonomous Wireless Channel Assignment Algorithm with Power Control," *IEEE Transactions on Vehicular Technology*, vol. 44, pp. 420–429, August 1995.

[356] J.-I. Chuang and N. Sollenberger, "Spectrum Resource Allocation for Wireless Packet Access with Application to Advanced Cellular Internet Service," *IEEE Journal On Selected Areas in Communications*, vol. 16, pp. 820–829, August 1998.

[357] J. M. Torrance, L. Hanzo, and T. Keller, "Interference Aspects of Adaptive Modems over Slow Rayleigh Fading Channels," *IEEE Transactions on Vehicular Technology*, vol. 48, pp. 1527–1545, September 1999.

[358] C. H. Wong and L. Hanzo, "Upper-bound performance of a wideband burst-by-burst adaptive mode," *IEEE Transactions on Communications*, March 2000.

[359] "3rd Generation Partnership Project; Technical Specification Group Radio Access Network; Spreading and modulation (FDD)." 3G TS 25.213 V3.2.0 (2000-03).

[360] "3rd Generation Partnership Project; Technical Specification Group Radio Access Network; Physical layer procedures (FDD)." 3G TS 25.214 V3.2.0 (2000-03).

[361] A. D. Whalen, *Detection of signals in noise*. Academic Press, 1971.

[362] W. T. Webb and L. Hanzo, *Modern Quadrature Amplitude Modulation: Principles and Applications for Fixed and Wireless Channels*. John Wiley and IEEE Press, 1994.

[363] R. L. Pickholtz, L. B. Milstein, and D. L. Schilling, "Spread spectrum for mobile communications," *IEEE Transactions on Vehicular Technology*, vol. 40, pp. 313–322, May 1991.

[364] K. S. Gilhousen, I. M. Jacobs, R. Padovani, A. J. Viterbi, L. A. Weaver, and C. E. Wheatley, "On the capacity of a cellular CDMA system design," *IEEE Transactions on Vehicular Technology*, vol. 40, pp. 303–312, May 1991.

[365] L. Wang and A. H. Aghvami, "Optimal power allocation based on QoS balance for a multi-rate packet CDMA system with multi-media traffic," in *Proceedings of Globecom*, (Rio de Janeiro, Brazil), pp. 2778–2782, December 1999.

[366] D. Koulakiotis and A. H. Aghvami, "Data detection techniques for DS/CDMA mobile systems: A review," *IEEE Personal Communications*, pp. 24–34, June 2000.

[367] P. J. Cherriman, E. L. Kuan, and L. Hanzo, "Burst-by-burst adaptive joint-detection CDMA/H.263 based video telephony," in *Proceedings of the ACTS Mobile Communications Summit, Sorrento, Italy*, pp. 715–720, June 1999.

[368] J. Laiho-Steffens, A. Wacker, and P. Aikio, "The Impact of the Radio Network Planning and Site Configuration on the WCDMA Network Capacity and Quality of Service," in *IEEE Proceedings of Vehicular Technology Conference*, (Tokyo, Japan), pp. 1006–1010, 2000.

[369] R. D. Kimmo Hiltunen, "WCDMA Downlink Capacity Estimation," in *IEEE Proceedings of Vehicular Technology Conference*, (Tokyo, Japan), pp. 992–996, 2000.

[370] K. Sipilä, Z.-C. Honkasalo, J. Laiho-Steffens, and A. Wacker, "Estimation of Capacity and Required Transmission Power of WCDMA Downlink Based on a Downlink Pole Equation," in *IEEE Proceedings of Vehicular Technology Conference*, (Tokyo, Japan), pp. 1002–1005, 2000.

[371] "GSM 06.90: Digital cellular telecommunications system (Phase 2+)." Adaptive Multi-Rate (AMR) speech transcoding, version 7.0.0, Release 1998.

[372] S. Bruhn, E. Ekudden, and K. Hellwig, "Adaptive Multi-Rate: A new speech service for GSM and beyond," in *Proceedings of 3rd ITG Conference on Source and Channel Coding*, (Technical Univ. Munich, Germany), pp. 319–324, 17th-19th, January 2000.

[373] S. Bruhn, P. Blocher, K. Hellwig, and J. Sjoberg, "Concepts and Solutions for Link Adaptation and Inband Signalling for the GSM AMR Speech Coding Standard," in *Proceedings of VTC*, (Houston, Texas, USA), 16-20 May 1999.

[374] R. Owen, P. Jones, S. Dehgan, and D. Lister, "Uplink WCDMA capacity and range as a function of inter-to-intra cell interference: theory and practice," in *IEEE Proceedings of Vehicular Technology Conference*, vol. 1, (Tokyo, Japan), pp. 298–303, 2000.

[375] B. G. Evans and K. Baughan, "Visions of 4G," *IEE Electronics and Communication Engineering Journal*, vol. 12, pp. 293–303, December 2000.

[376] IEEE, *Proceedings of VTC'98 (Spring)*, (Houston, Texas, USA), 16–20 May 1999.

[377] IEEE, *Proceedings of IEEE International Symposium on Personal, Indoor and Mobile Radio Communications, PIMRC'97*, (Marina Congress Centre, Helsinki, Finland), 1–4 September 1997.

[378] IEEE, *Proceedings of IEEE Vehicular Technology Conference (VTC'98)*, (Ottawa, Canada), 18–21 May 1998.

[379] IEEE, *Proceedings of IEEE International Symposium on Spread Spectrum Techniques and Application (ISSSTA'98)*, (Sun City, South Africa), September 1998.

[380] IEEE, *Proceedings of IEEE VTC '96*, (Atlanta, GA, USA), 1996.

[381] J. D. Gibson, ed., *The Mobile Communications Handbook*. CRC Press and IEEE Press, 1996.

Author Index

C

D

E

F

G

K

L

M

N

O

P

R

S

T

Index

Symbols

D

E

F

G

V

W

Other Wiley and IEEE Press Books on Related Topics

- L. Hanzo, W.T. Webb, T. Keller: Single- and Multi-carrier Quadrature Amplitude Modulation: Principles and Applications for Personal Communications, WATM and Broadcasting; IEEE Press-John Wiley, 2000
- R. Steele, L. Hanzo (Ed): Mobile Radio Communications: Second and Third Generation Cellular and WATM Systems, John Wiley-IEEE Press, 2nd edition, 1999, ISBN 07 273-1406-8
- L. Hanzo, F.C.A. Somerville, J.P. Woodard: Voice Compression and Communications: Principles and Applications for Fixed and Wireless Channels; IEEE Press-John Wiley, 2001 [1]
- L. Hanzo, P. Cherriman, J. Streit: Wireless Video Communications: Second to Third Generation and Beyond, IEEE Press, 2001 [2]
- L. Hanzo, T.H. Liew, B.L. Yeap: Turbo Coding, Turbo Equalisation and Space-Time Coding, John Wiley, 2002
- L. Hanzo, C.H. Wong, M.S. Yee: Adaptive wireless transceivers: Turbo-Coded, Turbo-Equalised and Space-Time Coded TDMA, CDMA and OFDM systems, John Wiley, 2002

[1] For detailed contents please refer to http://www-mobile.ecs.soton.ac.uk
[2] For detailed contents please refer to http://www-mobile.ecs.soton.ac.uk